Photosynthesis and Production in a Changing Environment

A field and laboratory manual

Edited by D.O. Hall, J.M.O. Scurlock, H.R. Bolhàr-Nordenkampf,
R.C. Leegood and S.P. Long

CHAPMAN & HALL
London · Glasgow · New York · Tokyo · Melbourne · Madras

Published by Chapman & Hall, 2–6 Boundary Row, London SE1 8HN

Chapman & Hall, 2–6 Boundary Row, London SE1 8HN, UK

Blackie Academic & Professional, Wester Gleddens Road, Bishopbriggs, Glasgow G64 2NZ, UK

Chapman & Hall, 29 West 35th Street, New York NY10001, USA

Chapman & Hall Japan, Thomson Publishing Japan, Hirakawacho Nemoto Building, 6F, 1-7-11 Hirakawa-cho, Chiyoda-ku, Tokyo 102, Japan

Chapman & Hall Australia, Thomas Nelson Australia, 102 Dodds Street, South Melbourne, Victoria 3205, Australia

Chapman & Hall India, R. Seshadri, 32 Second Main Road, CIT East, Madras 600 035, India

---

First edition 1993

© 1993 United Nations Environment Programme

Typeset in Hong Kong by Excel Typesetters Company
Printed in England by Clays Ltd., St Ives plc

ISBN 0 412 42900 4 (HB)   0 412 42910 1 (PB)

Apart from any fair dealing for the purposes of research or private study, or criticism or review, as permitted under the UK Copyright Designs and Patents Act, 1988, this publication may not be reproduced, stored, or transmitted, in any form or by any means, without the prior permission in writing of the publishers, or in the case of reprographic reproduction only in accordance with the terms of the licences issued by the Copyright Licensing Agency in the UK, or in accordance with the terms of licences issued by the appropriate Reproduction Rights Organization outside the UK. Enquiries concerning reproduction outside the terms stated here should be sent to the publishers at the London address printed on this page.

The publisher makes no representation, express or implied, with regard to the accuracy of the information contained in this book and cannot accept any legal responsibility or liability for any errors or omissions that may be made.

A catalogue record for this book is available from the British Library

Library of Congress Cataloging-in-Publication data available

Photosynthesis and production in a changing environment: a field and laboratory manual/edited by D.O. Hall . . . [et al.].
p. cm.
Includes bibliographical references and index.
ISBN 0-412-42910-1
1. Photosynthesis – Research – Handbooks, manuals, etc. 2. Primary productivity (Biology) – Research – Handbooks, manuals, etc.
I. Hall, D.O. (David Oakley)
QK882.P5539 1993
581.1′3342 – dc20 93-27163
CIP

∞ Printed on permanent acid-free text paper, manufactured in accordance with the proposed
ANSI/NISO Z 39.48-199X and ANSI 39.48-1984

Contents

Contributors

Dr. K.A. Badcock
CSIRO Division of Forestry
Stowell Avenue
Battery Point
Tasmania 7004
AUSTRALIA

Dr. C.L. Beadle
CSIRO Division of Forestry
Stowell Avenue
Battery Point
Tasmania 7004
AUSTRALIA

Dr. R.D. Blackwell
Dept. of Biological Sciences
University of Lancaster
Bailrigg
Lancaster LA1 4YQ
UK

Prof. H.R. Bolhàr-Nordenkampf
Institute of Plant Physiology
University of Vienna
Althanstrasse 14
Postfach 291
A-1091 Vienna
AUSTRIA

Dr. A. Davie
CSIRO Division of Forestry
Stowell Avenue
Battery Point
Tasmania 7004
AUSTRALIA

Dr. G. Draxler
Institute of Plant Physiology
University of Vienna
Althanstrasse 14
Postfach 291
A-1091 Vienna
AUSTRIA

Dr. J.F. Farrar
School of Biological Sciences
University of Wales
Bangor
Gwynedd LL57 2UW
UK

Dr. I.N. Forseth
Dept. of Botany
University of Maryland
College Park
MD 20742
USA

Dr. H. Griffiths
Dept. of Agricultural and
Environmental Science
Ridley Building
University of
Newcastle-upon-Tyne
Newcastle NE1 7RU
UK

Prof. M.G. Guerrero
Institute of Plant Biochemistry
and Photosynthesis
Faculty of Biology
University of Seville
Apartado 1113
41080 Seville
SPAIN

Prof. D.O. Hall
Division of Life Sciences
King's College London
Campden Hill Road
London W8 7AH
UK

Dr. J.E. Hällgren
Dept. of Plant Physiology
Swedish University of
Agricultural Sciences
S-90183 Umea
SWEDEN

Dr. G. Hind
Dept. of Biology
Brookhaven National Laboratory
Upton, L.I.
NY 11973
USA

Dr. J.L. Honeysett
CSIRO Division of Forestry
Stowell Avenue
Battery Point
Tasmania 7004
AUSTRALIA

Dr. M.B. Jones
School of Botany
Trinity College
University of Dublin
Dublin 2
IRELAND

Dr. R.W. Langhans
Dept. of Floriculture
Cornell University
20 Plant Science Building
Ithaca
NY 14853
USA

Prof. P.J. Lea
Division of Biological Sciences
University of Lancaster
Bailrigg
Lancaster LA1 4YQ
UK

Dr. R.C. Leegood
Dept. of Animal and Plant Sciences
University of Sheffield
P.O. Box 601
Sheffield S10 2UQ
UK

Dr. P. Lindblad
Dept. of Physiological Botany
University of Uppsala
P.O. Box 540
S-75121 Uppsala
SWEDEN

Prof. S.P. Long
Dept. of Biology
University of Essex
Colchester CO4 3SQ
UK

Prof. M.M. Ludlow
CSIRO Divn. of Tropical Crops
and Pastures
200 Carmody Road
St. Lucia
Queensland 4067
AUSTRALIA

Dr. R. McMurtrie
School of Biological Science
University of New South Wales
P.O. Box 1
Kensington
NSW 2033
AUSTRALIA

Prof. P.S. Nobel
Dept. of Biology
University of California, L.A.
Los Angeles
CA 90024
USA

Prof. J.M. Norman
Dept. of Agronomy
University of Nebraska
Lincoln
NE 68583
USA

Prof. G. Öquist
Dept. of Plant Physiology
University of Umea
S-90183 Umea
SWEDEN

Dr. W. Postl
Institute of Plant Physiology
University of Vienna
Althanstrasse 14
Postfach 291
A-1091 Vienna
AUSTRIA

Dr. S.D. Prince
Dept. of Geography
Room 1113, Lefrak Hall
University of Maryland
College Park
MD 20742
USA

Dr. M.J. Roberts
Department of the Environment
Room A226, Romney House
Marsham Street
London SW1P 3PY
UK

Dr. J.M.O. Scurlock
Division of Life Sciences
King's College London
Campden Hill Road
London W8 7AH
UK

Dr. T.W. Tibbitts
Biotron
University of Wisconsin
2115 Observatory Drive
Madison
Wisconsin 53706
USA

Dr. L.L. Tieszen
Dept. of Biology
Augustana College
Sioux Falls
SD 57197
USA

Dr. A. Vonshak
Institute for Desert Research
Ben-Gurion University
Sede Boquer 84990
ISRAEL

Prof. D.A. Walker
Robert Hill Institute
University of Sheffield
26 Taptonville Road
Sheffield S10 5BR
UK

Mr. J. Woods
Division of Life Sciences
King's College London
Campden Hill Road
London W8 7AH
UK

Preface

The majority of the world's people depend on plants for their livelihood since they grow them for food, fuel, timber, fodder and many other uses. A good understanding of the practical factors which govern the productivity of plants through the process of photosynthesis is therefore of paramount importance, especially in the light of current concern about global climate change and the response of both crops and natural ecosystems.

The origins of this book lie in a series of training courses sponsored by the United Nations Environment Programme (Project No. FP/6108-88-01 (2855); 'Environment changes and the productivity of tropical grasslands'), with additional support from many international and national agencies. These intensive field and laboratory courses run for 3 weeks at a host institution in a developing country, with 25–30 trainees from the surrounding region. The invited lecturers are an international team of leading research scientists; they all give freely of their time, receiving no reward for their efforts.

At the time of writing, we have just returned from the tenth in this series of courses which began in 1978. The aim is to teach and demonstrate the latest research techniques to young scientists and technologists, who are chosen from backgrounds in agriculture, forestry, aquatic systems and ecology. Particular emphasis is given to 'hands-on' practical experience. Portable equipment is used, incorporating the latest advances in electronics. This is necessary for measuring and understanding the photosynthetic productivity of plants throughout the world, and their possible response to climate changes. We firmly believe that such monitoring and research work should be carried out at the local and regional level by locally trained people.

Following the success of our earlier book (*Techniques in Bioproductivity and Photosynthesis*; Pergamon Press, 1985), which was translated into four major languages, the editors and contributors have extensively revised the content and widened the scope of the text, so it now bears a title in line with current concern over global climate change. In particular, we have added chapters on remote sensing, controlled-environment studies, chlorophyll fluorescence, metabolite partitioning and the use of mass isotopes, all of which techniques are increasing in their application and importance to this subject area. Two chapters on modelling of plant productivity and photosynthesis are also included, since our ability to predict plant responses to climate change requires modelling of plant processes at all levels, using data gathered by the techniques described elsewhere in this book.

Each chapter takes into account the latest advances in research techniques, making allowances for readers who may have access to only limited research facilities. The relevance of the subject area and techniques to photosynthetic production and climate change are explained, practical work is described, and a list of important references provided.

The latest edition of the Equipment Review published by the Department of Biology, University of Essex, UK, is included as Appendix A. This provides invaluable information on the specifications, sources and costs of much of the equipment currently available for crop and environmental physiology. A number of useful guidelines and tables of data are also appended. In line with the increasing use of microcomputers for data analysis throughout the world, several of the contributors have provided programs suitable for IBM-compatible microcomputers (Appendices E–G). These may be purchased directly from the publishers.

D.O. Hall
J.M.O. Scurlock
King's College London

Introduction

Photosynthesis and the changing environment

D.O. Hall and S.P. Long

Photosynthesis provides the basis for all life since it drives the great cycles of CO_2 uptake and O_2 evolution. In order to benefit from these photosynthetic processes while maintaining the environment, we must understand photosynthetic mechanisms from the ecosystem level down to the primary reaction centres of membranes and enzymes. The productivity of plants determines yields of food, timber, fuelwood, fodder and numerous other plant products; it is this bioproductivity which may be affected by changing climate, with beneficial or harmful effects which we cannot yet predict. Improving plant productivity using new physiological and genetic techniques also opens up exciting new opportunities which require basic knowledge of photosynthesis if they are to be exploited to their fullest extent.

This practically oriented book provides a basis for assaying and monitoring many aspects of photosynthesis in the field and laboratory. We attempt to show how plants from trees to crops, from grasses to algae, may be analysed from various viewpoints in order to provide as complete a picture as possible of the overall functioning of plants. In this way, the productivity of the plant or the ecosystem of which it is a part may be better understood. Ultimately we may wish to model the overall productivity of plants using a mechanistic (process-based) approach, in order to better predict environmental influences on the plant and the short and long-term productivity of a plant, either alone or within its ecosystem. Thus we may be able to predict how crops, forests and other ecosystems may respond to increased CO_2 in the atmosphere, or increasing temperature, or water deficits, or shortages of nutrients, and so on.

The undoubted satisfaction of studying the various applied and basic aspects of photosynthesis is that such work can be immediately practical in the field and also immensely rewarding as an area of fundamental research. This has been the philosophy of the ten UNEP-sponsored training courses on which this book is based. Since 1978, about 240 postgraduate trainees (research workers and teachers) from 28 countries have participated in these training courses. The student–staff ratio over the three-week duration of each course has been about 2:1 in both the field and the laboratory, in order to provide intense practical transfer of information and techniques. As a result of such close interactions, the contributors to this book appreciate the necessity of clear background information on the plant processes being studied by any given technique. This is essential in order to provide a basis of understanding for research scientists using the sophisticated electronic equipment (much of it portable) which makes photosynthetic studies much faster and easier today than even a decade ago, and applicable in all parts of the world.

Over the last 2–3 years much thought has been given by plant scientists in the new International Geosphere–Biosphere Programme (IGBP) as to what basic information is needed in order to understand an ecosystem. Most of the minimum sets of measurements and monitoring necessary to allow modelling of productivity are oriented towards photosynthetic processes and are described in this book. The recommended data inputs are CO_2, temperature, precipitation, radiation, and vapour pressure (besides wind, material transport, land and water). These inputs are required to describe and model the following key properties: CO_2 flux, net primary production, H_2O flux and vapour exchanges, vegetation structure including leaf area index, and carbon distribution above and below ground (besides nutrient cycles, material and trace gas transport, plant functional types and secondary effects such as fire, pests, etc.)

Thus we see that whether we wish to study plant productivity in agriculture or forestry, or in response to changing environments, we need to have a set of basic techniques available in the field and laboratory. Only then can we better attempt to manipulate plants physiologically and/or genetically for better yields (maybe with different inputs), tolerance of stresses, ability to thrive on reduced fertiliser or water inputs, adaptability to changing climate, and so on. Each country or region should have the capability of measuring and understanding plants and their productivity. This cannot be done by others at a distance, but should be carried out locally with the most modern equipment and knowledge available.

PHOTOSYNTHESIS AND PRODUCTIVITY

Although photosynthesis is fundamental to plant productivity, many other factors modify the magnitude of productivity attained in the field. The quantitative relationship between photosynthesis and plant productivity should be considered first of all. For any crop or stand of natural vegetation, four factors determine the net biomass gain or net productivity (P_n): the quantity of incident light (**Q**), the proportion of that light intercepted by green plant organs (β), the efficiency of photosynthetic conversion of the intercepted light into biomass (ε), and respiratory losses of biomass (**R**). The relationship between plant productivity and these factors is desribed by the following equation:

$$P_n = \mathbf{Q}\beta\varepsilon - \mathbf{R}$$

For crops the *economic yield* is the amount of this productivity which is partitioned into the useful or harvested portion of the crop, e.g. the grains of cereals, the trunks of timber trees or the shoots of herbage crops. The proportion of total biomass production which is invested into the harvested parts of the plant is termed the *harvest index*. The harvest index has been increased in many crops by improved fertilisation practices and protection against pests, so ensuring that more of productivity is available for formation of the economic yield. Genetic improvement of yields has been achieved by selection of genotypes in which a larger proportion of productivity is partitioned into the harvested component, not necessarily by selection of plants with a higher total productivity. Of course, this approach is of very limited value in crops where the bulk of the plant forms the harvestable component, e.g. forage and biomass crops. In all crops, the limit to improvement of yield through increased harvest index is set by total productivity. Further increases in yield depend upon improvements in productivity itself. In natural communities, productivity is also important, both as a measure of the potential of wild species for domestication and as a measure of the total input of energy or carbon to the ecosystem.

The above equation suggests three possible means by which productivity might be increased. The amount of incident light (**Q**) is determined by the climate and is thus independent of the crop. However, the remaining three factors may be modified.

Respiratory losses of biomass (**R**) in the maintenance of existing tissues and growth of new tissue, constitute an important limitation on productivity. Recent work has shown significant differences in **R** between genotypes of herbage grasses, suggesting a promising potential for the scientific selection of genotypes with higher productivity and maximum yield potential.

The efficiency of light interception (β) is a function of the size, structure and colour of the plant canopy. Where productivity has been increased in crops, this may usually be attributed to an increase in light interception. For example, the major effect of nitrogen fertilisation in cereal crops is an increase in leaf area and duration, resulting in improved β over the growing season. Most inorganic fertilisers improve yields through their effect on leaf growth and duration, whilst many stress factors have the opposite effect. Thus, modifications to the efficiency of light interception have been achieved mostly through improved cultural practices. The I.R. varieties of rice provide an important exception, being the result of selection of genotypes with a canopy structure which gives improved light interception.

Efficiency of energy conversion (ε) is determined directly by the photosynthetic process and expresses the direct relationship between productivity and photosynthesis. ε may be measured for crops and natural communities, over periods from several days to a whole season, by combining productivity measurements with integrated measurements of the light absorbed by the canopy; or for leaves, over periods of a few minutes, by gas exchange studies. There are remarkably few instances where it has been possible to raise the maximum value of ε (ε_{max}) of a species in order to increase productivity. Furthermore, there are no proven instances of genetic improvement of ε_{max} under optimal conditions within a species. However, ε is affected by the environment: CO_2 enrichment is the one notable exception where an improvement in ε has been obtained. CO_2 enrichment greatly reduces photorespiration; this has resulted in increases in both productivity and economic yield for many glasshouse crops.

In theory, improvement of ε is the most attractive means of increasing productivity and economic yield. If this could be achieved through genetic selection, an increase in productivity would be achieved without the increased inputs of fertiliser on which many recent yield improvements have depended. Whilst improvement of ε_{max} under optimal conditions is uncertain, there is little doubt that enhancement of ε under sub-optimal conditions might be achieved. Many environmental stresses are known to lead to a decrease in the efficiency of light energy conversion, at least in the short term. In particular, photoinhibitory damage to the photosynthetic mechanism produced by combination of high light and low temperature or water stress may be significant. An important area of future crop improvement would be the identification of crop genotypes in which ε is less sensitive to such environmental stress.

Photosynthetic energy conversion describes the whole photosynthetic process from light capture on the photosynthetic membranes to CO_2 assimilation and its subsequent metabolism in the chloroplasts and elsewhere. To understand how efficiency may be improved, a fuller understanding of all levels of the photosynthetic process is required.

A RESEARCH APPROACH

When dealing with the efficency of light energy conversion into biomass in higher plants, concern often centres on such questions as why a given genotype is more productive in one environment than in another, or what the limitations to productivity are for a given genotype in a given environment. A common mistake in the scientific approach to such a problem is to look at the isolated parts of the plant first, rather than to study the whole. For example, in analysing why increased salinity decreases the productivity of a crop variety, it would be better first to study the whole plant (including below-ground parts) or the whole canopy, rather than to look at changes in single leaf rates of CO_2 assimilation, isolated ribulose bisphosphate carboxylase (Rubisco) activity, or amounts of $^{14}CO_2$ incorporated into different compounds. Even if salinity-induced changes are found, these processes may not necessarily be limiting productivity. The reduction in productivity may not have anything to do with the effect of salinity on the photosynthetic apparatus; it could equally well be an effect on leaf area or canopy structure, causing changes in the amount of light intercepted. Thus it is good practice to follow a logical sequence of steps in investigating limitations to productivity in a crop, a natural stand of plants, or a single plant. Such a logical sequence, forming a reductive analysis of limiting factors, is shown in Fig. I.1.

This book is broadly based on such a hierarchical approach. The first few chapters are concerned with the direct and indirect

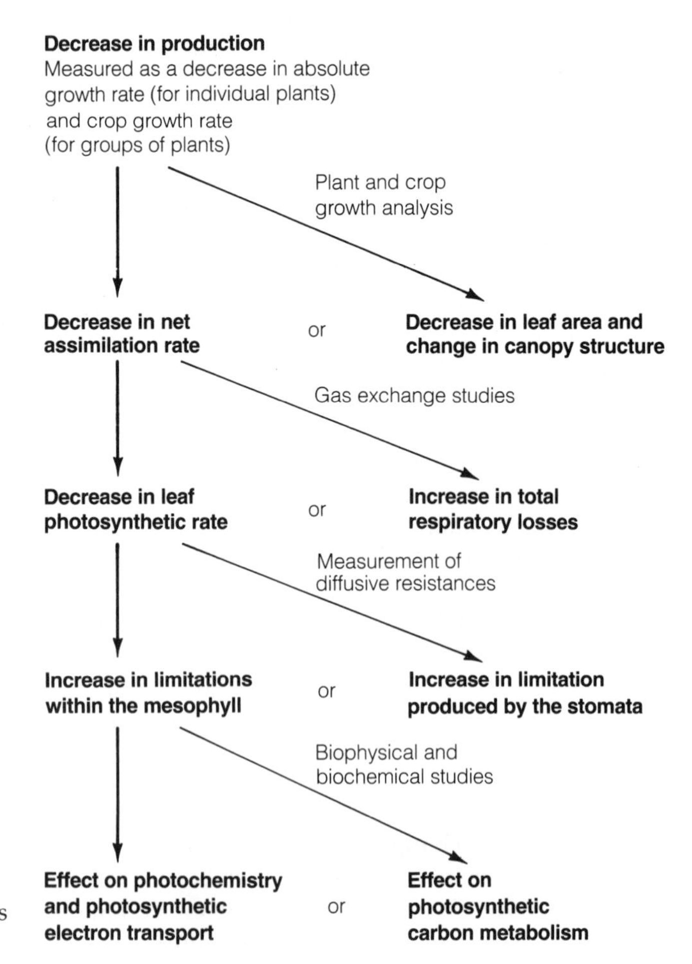

Fig. I.1. A reductive analysis of factors limiting plant/crop productivity.

measurement of whole plant productivity and its analysis, and are followed by descriptions of the determination of important factors such as light, canopy structure, water status and leaf CO_2 assimilation rate. Thus the components of the above photosynthetic equation may be evaluated. Direct measurement of plant productivity by dry matter determination may then lead to a consideration of plant metabolism and respiration, since dry matter measurements represent only the difference between what has been produced and what has been lost. A deeper insight into the underlying metabolic processes can be obtained from physiological and biochemical studies of individual aspects of the photosynthetic process. These include aspects of nitrogen assimilation (also a part of the photosynthetic process) and consideration of both nitrogen and carbon metabolism. Some of the research techniques currently used for the more laboratory-based studies of photosynthesis are described in later chapters of the book; these should lead to a better understanding of how, and indeed if, photosynthetic efficiency may be improved.

FURTHER READING

1. Beadle, C.L., S.P. Long, S.K. Imbamba, D.O. Hall and R.J. Olembo (1985) *Photosynthesis in Relation to Plant Production in Terrestrial Environments*. Natural Resources and The Environment Series, No. 18. Tycooly/Cassell, London. 156 pp.
2. Clarholm, M. and L. Bergstrom, eds. (1989) *Ecology of Arable Lands: perspectives and challenges*. Kluwer Academic Publishers, Dordrecht.
3. Hodgson, J., R.D. Baker, A. Davies, A.S. Laidlaw and J.D. Leaver (eds.) (1981) *Sward Measurement Handbook*. British Grassland Society, Maidenhead. 277 pp.
4. Houghton, J.T., G.J. Jenkins and J.J. Ephraums, eds. (1990) *Climate Change: the IPCC Scientific Assessment*. Cambridge University Press. 365 pp.
5. IGBP (1990) *Effects of Global Change on Terrestrial Ecosystems*. Global Change Report No. 11, International Geosphere-Biosphere Programme. Royal Swedish Academy of Sciences, Stockholm.
6. Jones, H.G. (1992) *Plants and Microclimate: a quantitative approach to environmental plant physiology*. 2nd edition. Cambridge University Press. 415 pp.
7. Long, S.P., M.B. Jones and M.J. Roberts, eds. (1992) *Primary Productivity of Grass Ecosystems of the Tropics and Sub-tropics*. Chapman and Hall, London. 267 pp.
8. Walker, D.A. (1992) *Energy, Plants and Man*. 2nd edition. Oxygraphics, Brighton/Packard Publishing, Chichester. 290 pp.
9. World Resources Institute (1990) *World Resources 1990/91*. Oxford University Press. 383 pp.
10. Yoshida, S., D.A. Forno, J.H. Cook and K.A. Gomez (1976) *Laboratory Manual for Physiological Studies of Rice*. 3rd edition. International Rice Research Institute, Manila. 83 pp.

1

Measurement of plant biomass and net primary production of herbaceous vegetation

M.J. Roberts, S.P. Long, L.L. Tieszen and C.L. Beadle

1.1 INTRODUCTION

The process of photosynthesis results in the conversion of light energy to chemical energy. This energy input by primary producers is used either to do work or else it is stored. The stored chemical energy is our prime concern here, since it is this biomass which can be harvested for food, fuel, fibre or other uses. Biomass is obviously of equal importance in natural ecosystems since it provides the organic molecules and energy source for all other trophic levels. This chapter is confined to the measurement of biomass and net primary production in herbaceous species. Although the theory behind the measurement of these parameters in trees and shrubs is the same, the techniques often require dimensional analysis and sampling of reference units, which are outside the scope of this discussion.

Photosynthesis and Production in a Changing Environment: a field and laboratory manual
Edited by D.O. Hall, J.M.O. Scurlock, H.R. Bolhàr-Nordenkampf, R.C. Leegood and S.P. Long
Published in 1993 by Chapman & Hall, London.
ISBN 0 412 42900 4 (HB) and 0 412 42910 1 (PB).

1.1.1 Definitions

Plant *biomass* (W) is the weight of living plant material contained above and below a unit of ground surface area at a given point in time. Production is the biomass or weight of organic matter assimilated by a community or species per unit land area per unit time. Production by photosynthetic organisms, i.e. primary production, may be expressed in two ways. *Gross primary production* (P_g) is the total amount of organic matter assimilated (including that lost in respiration). *Net primary production* (P_n) is the total amount of organic matter assimilated less that lost due to respiration (Equation 1.1), i.e. the total production which is available to other trophic levels or that which remains as stored chemical energy. Although production is expressed here in terms of dry weight of organic matter, it can be expressed as any conserved quantity, e.g. carbon or energy.

$$P_n = P_g - \mathbf{R} \qquad (1.1)$$

1.1.2 Units

The dimensions of production are mass per unit area per unit time. A wide range of units have been used to express the magnitude of dry matter production. Strictly, under SI conventions, time should be expressed as seconds (s), but days (d) or years (y) are normally more meaningful. In many ecological studies, area is now commonly expressed as m^2, although in agricultural studies the hectare (ha), equivalent to $10^4\,m^2$, is more appropriate. The relationship between some of the more common combinations of units is given below:

$$1.0\,g\,m^{-2}y^{-1} = 10\,kg\,ha^{-1}y^{-1} = 0.01\,t\,ha^{-1}y^{-1}$$

Biomass, which has dimensions of mass per unit area, is similarly expressed as $g\,m^{-2}$, $kg\,ha^{-1}$ or $t\,ha^{-1}$.

1.1.3 Principles

It is essential to distinguish P_n from standing crop or biomass (W). Unlike net primary production (which is a rate function), biomass refers to a quantity present at a fixed point in time.

From Equation 1.1, it can be seen that if P_g and **R** are measured (by determining CO_2 fluxes), an estimate of P_n may be obtained. However, such measurements are technically very difficult to make in a field situation Chapter 9). An alternative method arises from the fact that since P_n denotes a gain of material by a plant community, it may also be determined from the sum of the changes in plant biomass (ΔW) together with all the losses of plant material (e.g. death, shedding, etc.) over a given time interval. Thus:

$$P_n = \Delta W + d_l + d_g + d_e \qquad (1.2)$$

where P_n = net primary production
ΔW = change in biomass
d_l = losses by death, shedding or decomposition
d_g = loss to grazing
d_e = loss through root exudation

In most communities, decomposition and consumption by macro-invertebrates account for the only significant losses of dead material. Losses by death and shedding can therefore be summarized by:

$$d_l = \Delta D + rt\bar{D} + (\text{export} - \text{import}) \qquad (1.3)$$

where d_l = losses by death or shedding
ΔD = change in dead biomass
r = relative rate of decomposition
t = length of time interval
$\bar{D}$ = mean quantity of dead material
export = export of material (by water or wind)
import = import of material (by water or wind)

Estimation of P_n through evaluation of the terms in Equations 1.2 and 1.3 is considered later in this chapter.

1.2 SAMPLING DESIGN

Changes in biomass are usually determined by harvesting plant material at time intervals suited to the growth pattern of the species under investigation, usually about one month. Shoot (or above-ground) biomass (W_S) is commonly measured by clipping the vegetation at ground level from randomly selected quadrats, while root (or below-ground) biomass (W_R) is extracted from cores or trenches.

The area under study should be divided into a number of equal squares or rectangular plots with areas of, for example, 1.0 or $0.5\,m^2$. Each square should then be designated with a unique number. Samples can be harvested from a small area or quadrat at the centre of each square, the remaining peripheral area serving as a buffer zone.

The size of the buffer zone needed will depend on the vegetation type. Harvesting of any one quadrat should not affect adjacent quadrats. A buffer zone of width at least twice the maximum canopy height should be used, so that increased light penetration after harvesting should not affect any of the adjacent quadrats. The buffer zone must also be large enough to allow access without damaging other quadrats. In the UNEP Primary Production Studies on tropical grass-

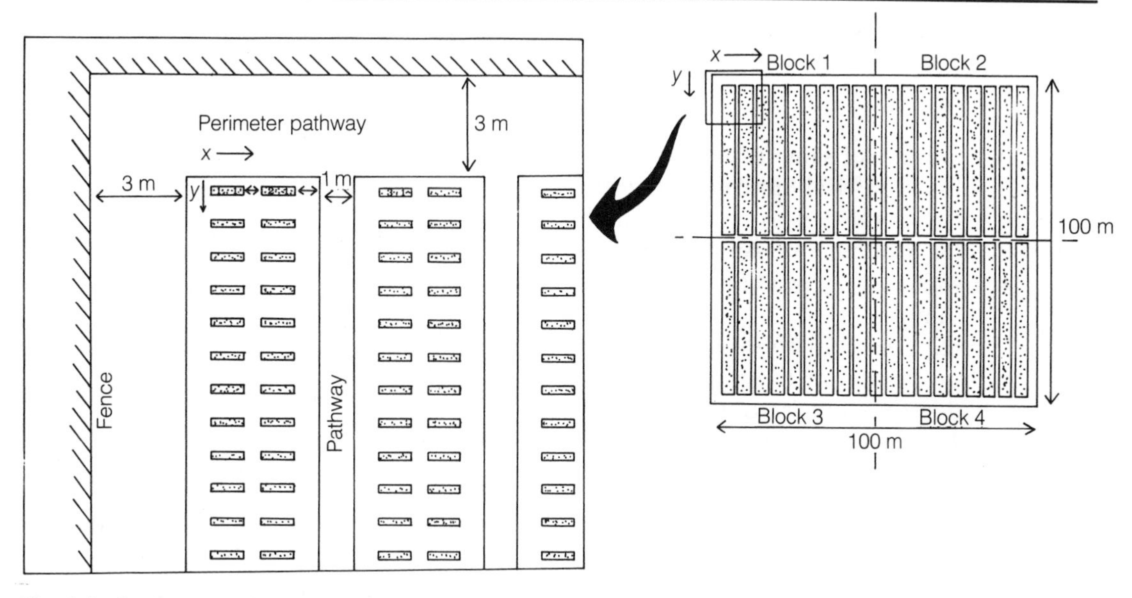

Fig. 1.1. Study area, showing a layout of paths and potential quadrats. The latter are shown stippled in the enlarged inset. The area is divided into quarters, each quarter containing 19 columns of potential quadrats running parallel to the pathway and numbered from left to right (x): 1, 2, 3, . . . etc. Each column contains 46 potential quadrats, numbered from top to bottom (y): 1, 2, 3, . . . etc. Quadrats for harvesting are selected by random x and y co-ordinates within each block.

land ecosystems, the minimum distance between any two quadrats is 0.5 m. To further reduce the risk of damage, marked pathways can be used. A map of the site should also be prepared so that selection of quadrats to be harvested may be planned before fieldwork commences.

Figure 1.1 illustrates the suggested layout of paths and quadrat locations in the one hectare study areas used in the UNEP Primary Production Studies of tropical grasslands. Vegetation should be sampled randomly using one of the following designs.

1.2.1 Fully randomized design

This design is for use at sites where there is no obvious pattern in variability of the vegetation. Quadrats (numbered as stated above) are chosen for harvest by reference to sets of random numbers such as:

i) random number tables
ii) the random number generator on a desk-top calculator
iii) computer generated pseudo-random numbers or use of the RND function in a BASIC program.

Quadrats are rejected if they fall on a pathway or if they have already been sampled.

1.2.2 Randomised block design

This design is statistically advantageous at sites where the vegetation shows an obvious gradient in form or composition across the site. Instead of selecting quadrats from throughout the area, equal numbers of quadrats are selected within each block. This procedure ensures a more even spread of samples through the study area.

1.3 MEASUREMENT OF ABOVE-GROUND BIOMASS

1.3.1 Number, area and shape of quadrats

The number of quadrats needed for a required degree of precision can be calculated by taking 10–20 trial samples and then applying the following formula:

$$n = (ts)^2/D\bar{x} \quad (1.4)$$

where n = number of samples needed
t = the statistical function *Student's* t with (N-1) degrees of freedom (N = number of samples in trial)
s = standard deviation of trial samples
D = required confidence interval as a proportion of the mean (i.e. the degree of precision required, e.g. 10% of $\bar{x}$ would be D = 0.1)
$\bar{x}$ = mean of trial samples

Optimum quadrat area is a function of the spatial pattern of variation in biomass over the study area, and of the time required for harvesting and processing of samples; the so-called 'cost'. Quadrat size may be determined by the method of Wiegert [1] who used nested quadrats of five different sizes (Fig. 1.2). Thirty sets of nested quadrats are harvested, and the time taken to sample the smallest quadrat size together with the time spent walking between quadrats, weighing material, etc. noted. For each quadrat size, the following are determined:

1. The mean biomass per m^2
2. The variance of the mean biomass (V_m) per m^2
3. The relative variance (V_r) obtained by dividing V_m for the quadrat area by V_m for the smallest quadrat area
4. The relative cost (C_r) – calculated from:

$$C_r = \frac{C_f + xC_v}{C_f + C_v} \quad (1.5)$$

where C_f = the fixed cost for each quadrat, i.e. the time spent walking between quadrats, weighing, etc. (in arbitrary units)
C_v = the time spent sampling the smallest quadrat (in the same units)
x = area of quadrat

The product V_rC_r is then plotted against quadrat area (Fig. 1.3), and the lowest value of V_rC_r used as the optimum quadrat size as shown. Random placement of quadrats should be used in conjunction with Wiegert's method.

The most efficient shape of quadrat is determined by the distribution of the foliage. The sample variance is at a minimum in non-randomly distributed foliage when rectangular quadrats are used. Circular quadrats give the smallest ratio of edge length to area enclosed, and so decrease the quantity of vegetation at the boundary of the quadrat.

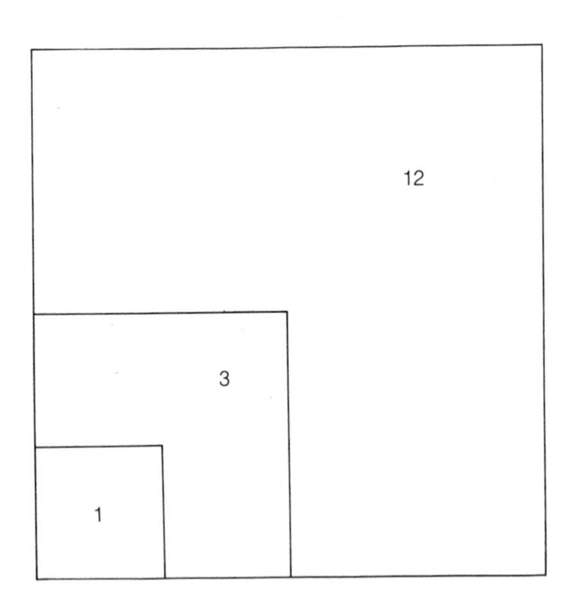

Fig. 1.2. Weigert's arrangement [1] of nested quadrats for determining optimum quadrat area for sampling vegetation in order to estimate biomass. The relative quadrat areas are 1, 3, 4, 12 and 16.

1.3.2 Harvesting

All shoot material within quadrats selected on each harvest date should be removed by clipping to ground level. Sheep shears are preferable since they may be effectively operated with one hand. Secateurs (pruning shears) will be needed to cut out any woody

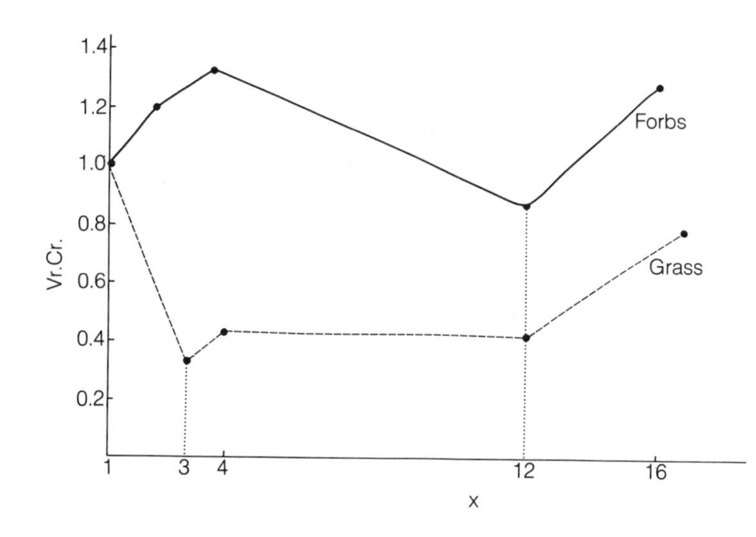

Fig. 1.3. Plot of V_rC_r against relative quadrat area (x) for Grass and Forbs in an old field in Michigan [1]. The lowest product determines the optimum relative quadrat area; 3 (0.47 m^2) and 12 (0.187 m^2) for grass and forbs respectively.

material. Harvested biomass, together with all loose dead material, is then sealed into labelled plastic bags along with a small quantity of water to prevent desiccation damage. Prior to processing, bags should be stored at 2–5°C (i.e. in a cold room or refrigerator) to minimize post-harvest weight loss through respiration.

1.3.3 Sorting

By definition biomass implies living tissue, so dead material has to be separated from the living. Leaves should be designated dead if the whole of their area has become necrotic, while necrotic tissue on otherwise green leaves should be removed. Sheaths should be removed from stems in order to clearly distinguish the state of each, since live stems can often be surrounded by dead sheaths.

1.3.4 Drying

Harvested material should be dried to constant weight at 80°C in a forced draught oven, allowed to cool in a desiccator and weighed to three significant figures, on an analytical balance if necessary. Whatever type of balance is used, regular calibration with standard weights is essential.

1.3.5 Ashing

A variable proportion of the dry weight of harvested vegetation may be inorganic material. A more meaningful expression of biomass, then, is its organic or ash-free dry weight ($W_{S,o}$ in the case of shoot material). This is obtained by burning dried samples at 500°C for 6 hours in a muffle furnace so that their organic constituents are combusted. Temperatures above 500°C may result in the volatilization of some inorganic materials. The weight of the ash remaining ($W_{S,a}$ in this case) is then used to determine the organic weight of the plant material.

1.3.6 Determination of energy content

The gross energy content of any material is defined as the number of heat units liberated when a unit weight of the material is completely burned in oxygen. Gross energy content is expressed in joules per gram dry weight of material (kJ kg^{-1}), although it is often more meaningful to use organic dry weight when making comparisons between species. Energy content per unit organic plant biomass will vary with species, organ and time of year. Most material will fall within the range 17–20 MJ kg^{-1}, but higher values are common for seeds and pollen [3].

Energy content is most commonly determined by bomb calorimetry, in which a sample of material is ignited and burned in oxygen under pressure (e.g. 3 MPa) inside a thick walled, temperature balanced and insulated stainless steel container (the 'bomb'). By comparing the rise in temperature of the container with that observed when a standard material is combusted, the calorific value of the sample can be calculated. Benzoic acid is the general standard [2], having a gross energy content of 26,447 kJ kg^{-1}.

(a) Sample preparation:
Sample dry weights must first be obtained. Care and accuracy are most important at this stage; firstly to minimise losses of volatile constituents and secondly because the variability of dry weight determination is about 20 to 100 times that of the calorific value measurement. Freeze drying and vacuum ovens have also been used [2].

The dried sample is then milled, mixed into a uniform powder and compressed into small combustion capsules using waxes of known energy content. As an aid to combustion, or where only small amounts of sample material are available, known quantities of benzoic acid may be added [2].

(b) Types of bomb calorimeter:
There are two main types of bomb calorimeter. *Adiabatic* bombs are surrounded by a water jacket whose temperature remains the same as the bomb (Fig. 1.4). Thus no heat is gained or lost by the bomb, and corrections for heat transference are unnecessary. *Non-adiabatic* bombs are able to exchange heat with their surroundings, the direction of heat flow depending on the relative temperatures of the calorimeter and its environment.

(c) Operation of the calorimeter:
The instrument is first calibrated by determining its water value (W) – the number of

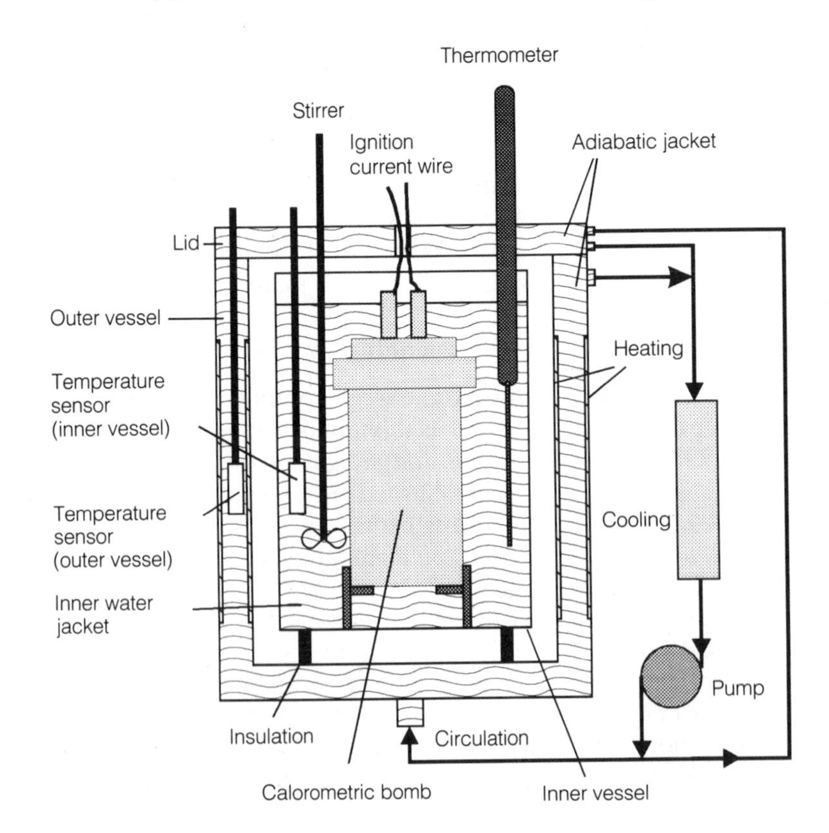

Fig. 1.4. Adiabatic bomb calorimeter (courtesy of Prof. H.R. Bolhàr-Nordenkampf).

joules required to raise the temperature of the water surrounding the bomb by 1°C. Small capsules of aceto-butyrate (0.5–1.0 g) are burned in the bomb, and the increase in water bath temperature measured. W is calculated as follows:

$$W = \frac{(VG + \Sigma c)}{\Delta T} \tag{1.6}$$

where V = gross energy content of calibration capsule
W = water value of apparatus
G = dry weight of benzoic acid
ΔT = rise in temperature of water bath
Σc = sum of additional corrections (see below)

A known weight of dried sample material is then burnt in the calorimeter, and its gross calorific value (V) obtained from:

$$V = \frac{W\Delta T - \Sigma c}{G} \tag{1.7}$$

where V = gross energy content of sample material
W = water value of apparatus
G = dry weight of sample
ΔT = rise in temperature of water bath
c = sum of additional corrections (see below)

(d) Corrections:
For both types of calorimeter, certain additional corrections may be necessary:

(1) Electrical Ignition: Electrical ignition causes an additional input of energy. This input may be measured by blank runs of the calorimeter and an appropriate subtraction made from the measurements of energy content.
(2) Burning of Firing Wire: If platinum wire is used, this correction may be considered insignificant as its gross energy content is $418\,J\,g^{-1}$. Nichrome wire may cause a somewhat larger energy input [2]; $1{,}402\,J\,g^{-1}$.
(3) Acid Formation: Nitric and sulphuric acid are formed by the oxidation of nitrogen and sulphur produced by the combustion. The acids may be obtained by washing out the calorimeter, and amounts formed estimated by titration. Errors from this source may approach 0.5%. Chiarello *et al.* [4] describe methods of correction for oxidation of N and S.

1.3.7 Carbon analysis

Two groups of techniques for determining carbon in plant material may be identified: *dry combustion*, in which the sample is completely combusted and the CO_2 evolved measured by partly or fully automatic commercial equipment (i.e. Total Organic Carbon (TOC) analysers); and *wet oxidation*, in which CO_2 is liberated from the breakdown of organic material by strong acids and oxidising agents. TOC analysers are now widely available, and are commonly coupled with simultaneous measurements of N and P (Appendix A). These techniques have largely replaced wet oxidation methods.

1.4 MEASUREMENT OF BELOW-GROUND BIOMASS

Below-ground biomass has often been neglected in biomass and primary production studies. This can be a serious omission, as below-ground biomass may account for more than 80% of total plant biomass [5]. There are many difficulties involved in sampling roots and rhizomes – not least being the separation of live and dead material, particularly in waterlogged soils where the tissue may be coated with black iron sulphide deposits. Furthermore, even with careful hand sorting, living and dead biomass of fine roots can be underestimated [6] by as much as 40%. Despite these difficulties, it would seem to be at least as important to measure below-ground biomass as it is to measure that above ground. Without this measurement, the research worker cannot determine whether an apparent increase in above-ground productivity is the result of photosynthetic gain or simply redistribution of matter from the root system.

1.4.1 Soil extraction

The most effective method is to take samples from the centre of the plots clipped for above-ground biomass measurements. There are two main ways of extracting below-ground samples:

1. Trenches, cut with spades. Here it is difficult to control depth and area required. They are impossible to use in waterlogged soils.
2. Cores, cut with corers, give better control of depth and area. Corers can be constructed from metal tubing with a hardened steel cutting edge, or (for soft soils) from plastic tubing with a sharpened end.

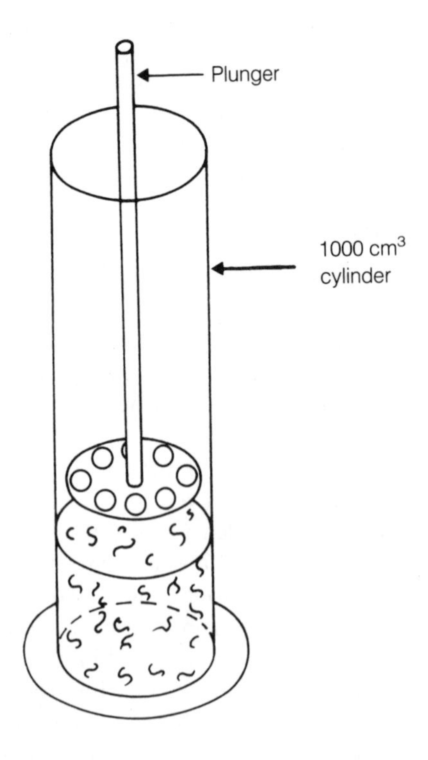

Fig. 1.5. A simple root washer. Vertical movement of the plunger disperses the soil sample: roots and organic matter are then decanted off.

Problems with corers arise from their tendency to compact the soil sample and their inability to retain the core during extraction. Compaction is reduced if the cross-sectional area of the blade is less than 10% of the total cross-sectional area. Retention is improved by decreasing the diameter of the corer, thereby increasing frictional resistance to loss of the sample.

Core samples cannot easily be obtained from soils which contain a lot of rocks or gravel. In these cases, trenches may provide the only access to below-ground plant parts. It must be admitted that under certain conditions difficulties may be so great as to make it almost impossible to obtain a statistically significant number of samples. A variety of corers for root system extraction is commercially available.

1.4.2 Separation from soil and washing

Cores should be hand washed over a sieve to remove fine particles, while large particles can be removed by flotation. Two designs of root washer are shown here (Figs. 1.5, 1.6). No attempt should be made to separate roots from soil by freezing or treatment with chemical dispersants, as these techniques can kill the roots. Iron sulphide deposits can be removed by placing the washed roots in continuously aerated water for 24 hours.

1.4.3 Separation of live roots from dead matter

In some cases, it is relatively easy to distinguish between live and dead underground biomass by differences in colour. For most species, however, this would be a highly subjective technique. Two other methods may be tried:

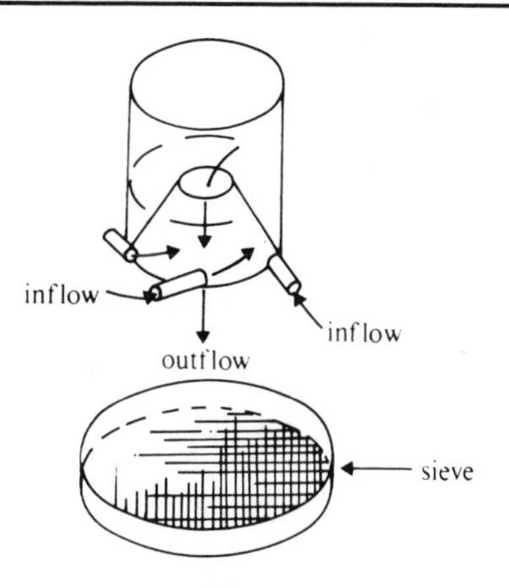

Fig. 1.6. A vortex root washer. Inflow of water swirls the roots and organic matter around so that they pass up through the outflow and fall onto the sieve for collection. Sand and larger particles fall to the base of the washer, whilst clay and silt are carried up and through the sieve.

1. *Flotation.* This depends on living root material floating and dead root material sinking when placed in solvents such as hydrogen peroxide or methanol. Both these solvents kill the roots, so the method cannot be tested by vital staining. Furthermore, the technique is species dependent, and in some species both living and dead roots sink.
2. *Vital staining.* Tetrazolium salts act as terminal electron acceptors in respiration and are reduced to their coloured form by living tissue. Roots are immersed in a 1% solution in distilled water at 30°C for 3 hours in the dark. Dead roots remain unstained.

A drawback with the vital staining method is that micro-organisms can also reduce the tetrazolium salts, so dead roots with large numbers of microbes on their surfaces can appear alive. Furthermore, old tissue may be so dark that the staining pattern may not be distinguishable. A way to overcome these problems is to make transverse cuts into the root tissue.

1.4.4 Dry weight and organic weight

After sorting, the below-ground biomass is treated in exactly the same way as the above-ground biomass. Determination of weight loss on ignition is particularly valuable for below- ground biomass, since this eliminates contamination by inorganic soil mineral particles.

1.4.5 Root 'ingrowth cores'

New root growth over a period of time may be estimated by the use of root 'ingrowth cores'. These provide a useful independent method of estimating production of below-ground organs [7], especially fine roots. 'Ingrowth cores' consist of a mesh covered cylinder of washed soil or inert material which is used to fill a circular hollow made with a soil corer. After a fixed period of time the cylinder is removed with any roots which have grown into the tube. This allows an estimate of new root production over the period for which the cylinder was in the ground, typically one growing season.

Hansson and Andren [7] describe the following method. To examine root production in a grassland over one season, 48 core holes, each 7 cm in diameter and 40 cm deep, were made in each of four plots. Mesh bags of 40 cm length, 7 cm diameter and 0.7 mm mesh, sealed at the base, were placed over a plastic tube and then inserted into each hole. The mesh material must allow root penetration, but must not decompose in the soil (nylon would be suitable). Soil which had been washed free of organic matter, dried and finally crumbled was poured into each tube, in layers. The inner plastic tube was gradually raised and a solid cylinder pushed down onto the soil to compress it against the exposed mesh. This procedure was repeated until the hole was filled. At 3-week intervals, three ingrowth cores were removed from each plot. The cores were removed by digging out the soil block surrounding the 'ingrowth core' and then it was carefully cut away from the rest of the soil, using a sharp knife to sever any roots penetrating across the mesh. The mesh can then be detached from the soil cylinder inside and the roots washed out of the soil as described previously.

1.5 NON-DESTRUCTIVE MEASUREMENT OF BIOMASS

Most estimates of P_n for herbaceous communities are based on change in biomass determined by destructive harvesting, as described in Sections 1.3 and 1.4. However, change in biomass may also be estimated non-destructively. This has the important advantage that the same quadrats may be sampled repeatedly, and thus variability between samples taken at different times is removed. The technique also allows direct determination of the death of material and amounts lost in grazing; thus P_n may be evaluated directly via Equation 1.2 if exudation losses are ignored.

In this approach, measurement of some characteristic or characteristics related to the dry weight of shoots and roots (e.g. height and length) is made at frequent intervals on material which has been mapped and marked prior to measurement. The relationship between these variables and dry weight is established from destructively harvested samples. The weight of material gained and lost can then be derived from regression equations relating weight to the characteristic measured.

Whilst this approach could also be applied to roots, it would require visual access to the root system *in situ*, e.g. by a glass wall in a trench, or clear plastic tubes. However, these will markedly alter the root environment, creating an unknown error in the estimation of root biomass. The non-destructive approach has been used for above-ground vegetation. Comparisons of destructive and non-destructive methods of measuring shoot biomass on coastal salt marshes show good agreement [8,9].

1.5.1 Estimation from stem and leaf dimensions

The procedure outlined here is based on that developed by Hussey [8] for the measurement of biomass and P_n (above-ground) in a *Puccinellia maritima* (perennial grass) salt marsh.

Quadrats are positioned by a randomized design (Section 1.2) at permanent locations within the study area. All stems are measured and marked. Numbered and colour-coded plastic bird rings provide a simple method of identifying each stem without impeding growth in stem girth. The length and width of the leaves and each stem are measured, and the location either marked with further rings or by its measured position along the stem. At suitable time intervals measurements of dimensions are repeated. New stems and leaves are noted, measured and marked. Death of leaves may also be recorded. The appearance, growth and longevity of each organ is therefore known.

To determine biomass, small destructive harvests of further quadrats within the study area are made so that regressions of stem and leaf weight against linear dimensions can be made. Jackson (pers. comm.) found by step-wise regression analyses for the grass *Spartina anglica* that the following equations adequately described the relationships of leaf (W_l) and stem (W_s) weights against leaf (l_l) and stem (l_s) lengths, respectively:

$$W_{s,i} = a + b(l_{s,i})^{3/2} \quad (1.8)$$

$$W_{l,j} = c + d(l_{l,j}), \quad (1.9)$$

where a, b, c and d are the regression constants. Biomass for any quadrat at any sampling interval will be given by:

$$B_t = \sum_{i=0}^{i=n} f_s l_{s,i} + \sum_{j=0}^{j=m} f_l l_{l,j} \quad (1.10)$$

where n and m = numbers of stems and

leaves in the quadrat, respectively

f_s and f_l = functions relating linear dimensions to weight, for stems and leaves, respectively.

The weight of material which has died may be obtained by substituting into Equation 1.10 the lengths and numbers of all leaves and stems which have died or been lost since the last recording date, using the dimensions obtained on that date. Thus P_n will be given directly for the interval between measurements.

No assumptions about the precise relationships between growth and death are made. Nor is it necessary to model decomposition of plant material since estimates of production are made on live material only; measurements on dead material never enter the calculations. However, it is assumed that the weight of a plant organ is the same just before and immediately after death, i.e. there is no translocation of material to or from the live and dead parts.

The precision with which P_n (and W) can be estimated is increased because the same samples are used throughout. This advantage of the approach increases with heterogeneity of the vegetation. However, the amount of raw data may exceed one million separate pieces of information. It is therefore essential to have a computing facility for data storage, analysis and checking. In measuring their length, stems and leaves may become physically damaged; thus the non-destructive measurement could alter longevity such that the sample is no longer representative of the whole study area. Remote sensing provides alternative non-intrusive techniques which could overcome this limitation (Chapter 2).

1.6 ESTIMATION OF LOSSES

Change in biomass (ΔW) does not in itself provide a measure of P_n. To determine P_n, measurement of ΔW must be combined with measurement of loss of biomass through death, grazing and exudation (Equation 1.2). Death of biomass will be apparent as an increase in dead vegetation (*D*) less the amount of dead vegetation which decomposes over the same time interval.

1.6.1 Decomposition

The amount of dead vegetation may be determined by the procedures outlined in Sections 1.3 and 1.4. Two main approaches have been used to assess the rate of decomposition or disappearance over a time interval.

(a) Paired plots:

This method was first described by Wiegert and Evans [10]. Several paired plots with a contiguous boundary between each pair are selected randomly. The living vegetation on each plot is removed by selective clipping. One quadrat of each pair is then selected at random, the dead material removed and its dry weight (D_i) determined. At the end of the harvest interval (t_{i+1}) any live material is carefully removed from the second quadrat and the dead material weighed (D_{i+1}). The rate of disappearance (r) can be calculated as follows:

$$r = \frac{\ln(D_i/D_{i+1})}{(t_{i+1} - t_i)} \qquad (1.11)$$

where r = instantaneous rate of disappearance of dead plant parts ($g\ g^{-1}\ day^{-1}$)

D_i = mean dry weight of dead material at t_i

D_{i+1} = mean dry weight of dead material at t_{i+1}

$(t_{i+1} - t_i)$ = harvest interval.

Equation 1.11 assumes a negative exponential model. Thus:

$$D_{i+1} = D_i \exp[-r(t_{i+1} - t_i)] \qquad (1.12)$$

so r at t_i is much higher than r at t_{i+1}. This is probably the correct model over long intervals of time but it is essential to measure r over short intervals similar to those used to measure ΔW and ΔD. Otherwise, as r decreases, D will be increasingly underestimated with time.

An alternative model for expressing the rate of disappearance makes no assumptions about the mathematical relationship between D and t. Thus:

$$r = \frac{2(D_i - D_{i+1})/(D_i + D_{i+1})}{(t_{i+1} - t_i)} \tag{1.13}$$

where r = proportion of dead material lost during time interval $(t_{i+1} - t_i)$.

This method assumes that removal of the living material at the start of the time interval does not affect the rate of decomposition. This is an important limitation, since removal of the live vegetation must alter the microclimate, the effect being greatest in the densest vegetation. Since no live vegetation is present no dead vegetation will be added during the interval.

The method also requires homogeneity within the plots. As there is inherent variability in any natural community, adjacent plots are unlikely to be identical except in crops and some monotypic communities, e.g. reeds (*Phragmites australis*) or cord-grass (*Spartina alterniflora*). The major objection to the method is therefore that D_{i+1} measured in the second plot at the end of the interval, could exceed D_i, measured in the first plot at the start of the interval. This may occur if the plots are not identical, or if regrowth and senescence occur within the second quadrat between t_i and t_{i+1}. These former problems will be alleviated by utilizing sufficient paired quadrats to show a significant change in D against the between plot variability and by reducing the interval between harvests.

In the original approach [10], the removal of live material will totally alter the microenvironment of the dead material, such that evaporative potential, temperature and intercepted radiation are altered. These will combine to create a drier environment than in adjacent undisturbed plots. The likely outcome is a reduced rate of decomposition. This underestimation is likely to be greater in tropical climates, than in the temperate climates for which the method was originally devised and tested.

Lomnicki *et al.* [11] adapted the method to overcome this limitation. Instead of removing all live material from the paired plots, all dead material (D_i) was removed from the first plot. At the end of the time interval all dead material was removed from both plots. At this time, the amount of dead material in the first plot is a measure of the quantity produced by death during the interval (D'_{i+1}) whilst that in the second plot is the total quantity remaining at the end of the interval (D_{i+1}). Equation 1.13 may be adapted for the Lomnicki modification:

$$r = \frac{2(D_i + D'_{i+1} - D_{i+1})/(D_i + D_{i+1})}{(t_{i+1} - t_i)} \tag{1.14}$$

This modification has two important advantages over the original method, the microenvironment of the second plot is unaltered and account is taken of dead material formed during the interval. These advantages will be of greater importance in tropical climates where turnover of plant biomass and alteration to evaporative demand by removal of live vegetation will be largest.

Both the original method and the modification are only applicable to above-ground vegetation. However, the litter bag technique of estimating may be applied both above-ground and below-ground.

(b) Litter bags:

For the litter bag technique, sub-samples of known weight (about 2 g dry weight) of dead roots or shoots are taken by destructive harvest. It is essential that this material is representative of the dead material present at the time of harvesting and that no material alive at the time of harvesting is included, since this will have a different rate of decomposition.

Litter bags are commonly made from nylon

mesh, although other non-biodegradable synthetic fibres could be used. The edges of the bags are either sealed by heat or sewn up with nylon thread. The mesh size is critical. An arbitrary point has to be chosen at which dead material is considered to be decomposed or lost from the category of dead vegetation. In the UNEP Primary Production Studies, passage through a mesh of 2 mm was chosen for this point. For consistency in calculation, the mesh used for the litter bags must be the same as that of the sieves over which the destructively harvested dead material is washed. It must be noted that mesh size may significantly affect decomposition rate. If large invertebrates are important decomposers in the community, then the mesh size chosen should be adequate to allow their access. Filled bags are sealed and relocated at the position in the canopy (above-ground) or in the soil horizon (below-ground) where the material occurs naturally. The bags would be best positioned according to a randomized block design, as for quadrat selection. This should be done as soon as possible after harvesting (preferably within 2 days). The bags are then removed at the time of the next harvest, the remaining oven dry-weight of material is determined, and the change in weight calculated so that r may be estimated using Equation 1.13 (or Equation 1.11 for the exponential model). The method assumes that the material placed in the bags is representative of the dead material present at t_i and that r is unaffected by placing the material in the bags. Two factors affecting the latter assumption are drying and change in microclimate.

Drying the material prior to enclosure in the bag is necessary to establish its dry weight. This could significantly alter the lability of the vegetation and the viability of the microbial population associated with it, which is in effect being partially cooked by oven drying. An alternative is to air-dry at room temperature. The dried material is then thoroughly mixed and a subsample (weighed to ±0.1 mg) oven dried (80°C) to determine the air-dry: oven-dry weight ratio. A second sub-sample of air-dried material is enclosed in an appropriate fine mesh bag and returned to the study site.

The dry weight of material placed into the litter bag (D_i) is then estimated:

$$D_i = D_{i,air}(w_{oven}/w_{air}) \tag{1.15}$$

where $D_{i,air}$ = The air-dry weight of the sub-sample of vegetation placed in the litter bags.
w_{oven}, w_{air} = the oven- and air-dry weights, respectively, of a further sub-sample of the same dead vegetation sample.

The mesh bag will alter the microclimate of the dead material, in particular by impeding air movement and evaporation. This effect may be minimized by choice of a bag of sufficient size that all vegetation fragments are in contact with the bag wall. One further problem of litter bags is that some material will be lost in handling, i.e. addition to the bag and removal from it. Control samples which are handled as described previously but retrieved immediately after placement on the plots should be used to eliminate systematic error.

The advantages of the litter bag technique over the paired plots approach is that problems arising from heterogeneity in paired plots are avoided and the method may be used both above- and below-ground.

As the bags containing below-ground material can be reinserted at an appropriate depth within the soil cores and relocated within the holes left after core extraction, the microenvironment surrounding the bags should be similar to that before sampling. With above-ground material this is more of a problem. Some vegetation may decompose while remaining in a vertical position. Anchoring litter bags at ground level may therefore not be the most appropriate way to reproduce the correct microenvironment for decomposition.

Any material which dies and decomposes within the interval will not be detected by the litter bag technique. The method may therefore underestimate r, especially when turnover is high. This error will be reduced by using shorter intervals between placement and recovery of litter bags. There is insufficient information to determine which of the two techniques (paired plots or litter bags) is the more accurate.

1.6.2 Grazing

The effects of grazing on estimates of P_n are almost impossible to account for. Study sites selected should not have been subjected to grazing in the years immediately preceding the study. It will then only be necessary to assess the effects of invertebrate herbivores, e.g. sap-sucking insects and possibly birds. If their total consumption of vegetation is small ($<1\%$) it can be assumed that the omission of d_g from Equation 1.2 will not constitute a significant source of error in the estimate of P_n.

If the study areas are sufficiently grazed to demand an accurate assessment of d_g, the methods used will be species specific. Various methods are available to exclude large vertebrates which are based on the use of exclosures. These can take the form of coarse wire mesh or electric fencing (the latter may be powered by a solar panel/battery configuration). If burrowing animals are present, the fence will require burial to an appropriate depth.

It would be impractical to exclude invertebrate herbivores or birds from a whole study area. A separate experiment would therefore be necessary to assess this form of grazing pressure. The most popular method is the use of fine mesh screens, though it is inevitable that the microclimate within the screen will be altered. Estimates of d_g are calculated as the difference between the values of W and D of screened and exposed plots using destructive or non-destructive techniques. Harvest intervals should be short to minimize the effects of microclimate changes within the screen on the estimate of d_g.

The contribution of the study sites to herbage intake by large vertebrate herbivores can also be estimated if these animals are allowed access to the study site. Experiments to measure biomass change of major invertebrate herbivores, their assimilation rates, and their contribution to rates of decomposition may assist measures of d_g and assess the effects of d_g on D [12].

1.6.3 Exudation

Remarkably little attention has been given to the final term in Equation 1.2, i.e. root exudation (d_e). No study to date has taken account of the organic material excreted into the rhizosphere in an estimate of P_n. To date, d_e has only been measured in laboratory studies by labelling techniques, or estimated from growth rates of rhizosphere microorganisms. The distribution of labelled assimilate within the plant can then be separated from that lost into the soil by exudation.

Minimum estimates suggest that 10% of the dry matter increment of roots can be exuded into the soil or assimilated by mycorrhizae [13,14]. In the absence of a suitable technique, it is not possible to overcome this problem at the present time.

1.7 ESTIMATION OF NET PRIMARY PRODUCTION

Net primary production (P_n) is given by change in biomass corrected for all losses (Equations 1.2 and 1.3). If natural death (d_l) is the major cause of loss then this will be given by the sum of change in dead vegetation and the quantity decomposed over the period. For example, where harvests of biomass and dead vegetation are made at monthly intervals together with measurements of decomposition over the same intervals, annual P_n will be given by [15]:

$$P_n = \sum_{i=1}^{i=12} (\Delta W_i + \Delta D_i + \bar{D}_i r_i). \quad (1.16)$$

In mature natural plant communities the net change in W over the year may be small; in climax communities it will be close to zero. D should also show little net change over a whole year. It is clear that P_n will depend more on estimates of d_l than W. It should therefore be realized that estimation of d_l, either through paired quadrats or decomposition, must be given equal or more emphasis than biomass estimation.

Estimation of P_n from W:

The vast majority of estimates of P_n in the literature are not based on a full analysis using equations with a theoretically sound basis (Equations 1.2 and 1.3) but are based on empirical equations valid only for a narrow range of conditions. These are summarised in Table 1.1.

Table 1.1. Methods used to estimate net primary production in salt marshes by extrapolation from biomass measurements [9]. W_{max} = maximum biomass recorded during the year; W_{min} = minimum biomass recorded during the year; D = mass of dead plant material; Δ = net change in a quantity between two sampling dates; P_n = annual net primary production; P_n^i = net primary production between two sampling dates.

1. Maximum live dry weight

$P_n = W_{max}$

Assumptions:
(i) No carry-over of biomass from one year to the next.
(ii) No death occurs before the maximum biomass is gained.

2. Maximum standing crop

$P_n = W_{max} + D$

Assumptions:
(i) No carry-over of either biomass or dead material from one year to the next.
(ii) Dead material does not decompose before the maximum biomass is obtained.

3. Maximum-minimum

$P_n = W_{max} - W_{min}$

Assumptions:
(i) As for method 1, but accounts for any carry-over of material between years.

4. International Biological Programme

$P_n = \Sigma(\Delta W)$ (negative ΔW is taken as zero)

Assumptions:
(i) Death and growth do not occur simultaneously.
(ii) P_n is never negative.

5. Smalley's method

P_n^i for any given interval is determined according to the following conditions.

IF $W > 0$ and $D > 0$	THEN $P_n^i = W + D$
IF $W < 0$ and $(W + D) > 0$	THEN $P_n^i = W + D$
IF $W > 0$ and $D < 0$	THEN $P_n^i = W$
If $(W + D) < 0$	THEN $P_n^i = 0$

$P_n = \Sigma P_n^i$

Method 1 assumes that peak biomass (W_{max}), i.e. the maximum of all values of biomass measured in the year, approximates to P_n. This assumes (1) that no carry-over occurs between years and (2) that no death occurs before W_{max} is reached, no production occurring after this. The method may be applicable to arable crops, where both assumptions may be valid, provided that there are no losses of biomass, e.g. through sequential senescence. The assumptions are clearly invalid for perennial plants and therefore most natural plant communities since there is, by definition, a carry-over of biomass from year to year. Method 3 allows for carry-over of biomass by deducting the minimum biomass (W_{min}) from the estimate. Many perennials, and communities of plants, will not show completely separate phases of growth and death. For example, in a study of shoot and leaf demography of the perennial coastal grass *Spartina anglica*, there were significant losses in every month of the year [9]. Within the tropics the assumptions underlying these methods will have even less validity. Here, there is unlikely to be one phase of growth and one separate phase of death, two phases of growth being common. Methods 4 and 5 have been designed to allow for several distinct phases of growth within a year, but they still fail to account for new shoot growth during periods of high mortality, and vice versa. Where these methods have been compared with measurements of P_n based on Equation 1.2, they may be seen to seriously underestimate P_n, by more than 50%. This underestimation results from the fact that there is a continual turnover of biomass, i.e. new material is being formed while other parts of the plant are dying. Turnover, and hence the error in P_n estimated by the methods in Table 1.1, will be greater in tropical communities. It may be concluded that P_n of herbaceous plant communities can only be determined by measurement of net gaseous exchange or by measurement of change in biomass coupled with simultaneous measurements of biomass loss. Except for arable annual crops grown in monoculture, biomass measurement alone cannot provide a valid estimate of P_n.

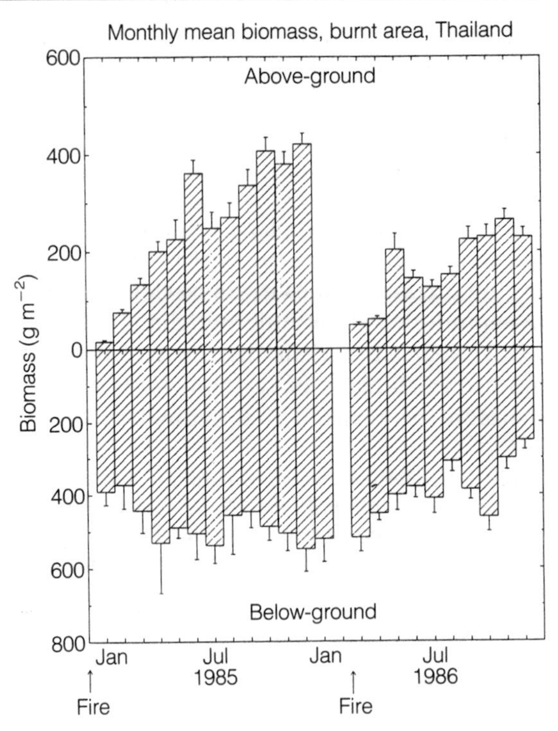

Fig. 1.7 Monthly means (±1 s.e.) from January 1985 until December 1986 of above- and below-ground biomass (dry weight) for the areas of a monsoonal grassland study site near Hat Yai in Southern Thailand [16].

Figure 1.7 illustrates the dynamics of above-ground and below-ground biomass in a humid savanna grassland in southern Thailand [16]. The monthly measurements show clearly the recovery of shoot mass following annual fires. Below-ground biomass shows an increase over the first year, but following the second year and the below-average rainfall of 1986 at this site the below-ground biomass declines by about 50% [16]. By combining these measurements with concurrent measurements of dead material and decomposition as described in the preceding sections, P_n was determined and the flow of material in the plant–soil system determined (Fig. 1.8) [17]. This shows clearly, that following the second fire production below-ground is actually negative, i.e. more mass is lost through respiration and/or retranslocation than is gained, a point which would not be apparent had biomass been measured in isolation and production extrapolated without reference to organ turnover (Table 1.1).

For large stemmed grasses, an alternative approach in determining P_n is to combine a demographic approach with destructive harvesting of individual stems. This is only efficient if the true stems are elongate, long-lived (i.e. months), and tillering occurs at discrete intervals. Piedade *et al.* [18] describe the following method used with the tropical C_4 floodplain grass *Echinochloa polystachya*, which forms a single cohort of stems each year and grows to several metres. Twenty individuals from each cohort of stems were selected by a randomised design shortly after initiation. The most recent node was marked with coloured thread, and the height and number of internodes and leaves recorded for each stem. In each subsequent month the number and length of additional internodes, the numbers of new leaves formed and old ones lost were recorded, and the most recent node marked. Concurrently, stem density was recorded monthly within 20 quadrats, 1 m × 1 m, positioned by randomised design at the start of the measurements. In parallel, a further 15 stems selected by randomised design were harvested each month, and the dry weights of internodes and leaves were determined. P_n of the cohort of shoots could then be determined for each month:

$$P_n = N_t(n_t \cdot W_{i,t} + n_t \cdot W_{l,t}), \qquad (1.17)$$

where N_t = mean stem density in month i
n_t = mean number of new internodes per stem
$W_{i,t}$ = mean mass of internodes formed in month i
$W_{l,t}$ = mean mass of leaves (including sheaths) formed in month i

1.8 EXPERIMENTAL WORK

An experiment used in the UNEP Training Courses, which can be completed by a group of students within two half-day sessions includes the determination of biomass changes, leaf area index, light interception, net primary production and the efficiency of energy storage.

The objectives are to illustrate a simple method for measuring productivity and some canopy and environmental factors which influence the magnitude of P_n. If parallel gas analysis measurements are made (Chapter 9), then P_n measured by this harvest method may be compared with P_n estimated from gas analysis data. The exercise may also illustrate simple experimental design and statistical considerations in productivity studies.

1.8.1 Requirements

Balance accurate to 1 mg (with calibration weights if possible).
Drying oven, preferably with forced circulation.
Leaf area meter or materials for measuring leaf area (e.g. graph paper or photosensitive paper)
Paper bags, protractors, metre rules, marker pens.
Line quantum sensor or tube solarimeters.
Soil corer, plastic and mesh tubes for root ingrowth cores.
Sieves to remove soil (0.5, 1.0 and 2.0 mm mesh), outside tap or buckets for washing.
Muffle furnace (550°C), forceps, tongs, crucibles.
Graph paper and calculators.
Meteorological data for the preceding 10 days

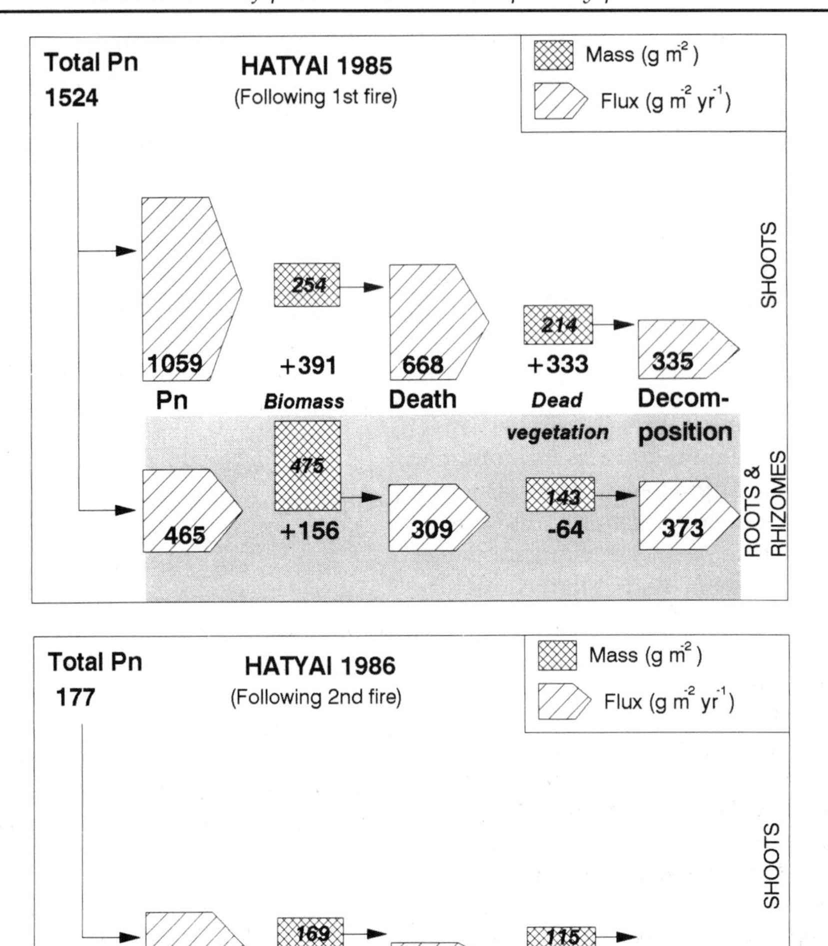

Fig. 1.8. Fluxes and mean quantities of dry mass in the grassland at Hat Yai, Thailand [17]. The arrow shaped polygons, represent the rate variables of annual fluxes (g[dry mass] $m^{-2}yr^{-1}$) and their direction, and rectangles represent the state variables of mean mass of vegetation (g[dry mass] m^{-2}). Numbers below the rectangles indicate the net change over the year in that quantity; thus +391 below the box labelled *Biomass* and *Shoots* indicates a net increase over the year of 391 g $m^{-2}yr^{-1}$ in the dry matter of shoots. Net flows are illustrated for the site for areas burnt at the beginning of 1985 and for the same areas burnt again in early 1986.

should be available, and ideally, data on productivity of the same species for other sites and conditions.

1.8.2 Plant material

Sow an excess of seeds of *Phaseolus vulgaris* or a similar species in 30 pots of 25 cm diameter (prior to sowing, insert two root ingrowth tubes in each pot. If a sterile mineral potting medium is used, it will not be necessary to wash the soil placed in the cores). Grow them under good light in a free draining soil with adequate nutrients. For the purposes of this experiment, a sandy or sandy loam soil allows relatively easy separation of roots from soil. Thin the plants to leave five healthy specimens of similar size in each pot.

1.8.3 Procedure

About 21 days after germination, make one harvest of 10 pots. Collect all plant material from each of the 10 pots in strata at 5 cm intervals (0–5 cm above soil surface, 5–10 cm, etc.). Separate leaves from stems and petioles. Keep these samples and process as detailed below.

Now establish water stress in one half of the remaining pots, by reducing or ceasing irrigation. Continue this treatment for 10 days. Position tube solarimeters or line quantum sensors above and below the canopy for each treatment, to obtain a measure of quanta or radiation intercepted. Integrate daily totals if possible.

1. At the end of the 10-day period, select five plants from each treatment by a randomised design. Determine the mean land area occupied by each pot.
2. Select one or two representative pots for detailed canopy characterisation as follows:
   (a) Count the number of leaves in each of five strata in the canopy.
   (b) Using a protractor, measure the leaf inclination of ten leaves in each stratum.
   (c) Measure the area of each leaf in each stratum, using graph paper or a leaf area meter. Number each leaf.
   (d) Measure leaf length and width and determine dry weight of laminae.
   (e) Collect stems and petioles for dry weight determination by stratum.
3. In the remaining pots, harvest all leaves (without petioles) and all stems.
4. Separate soil from all roots as follows:
   (a) Carefully remove soil which contains no roots.
   (b) Place root mass and attached soil in a large sieve (2 mm mesh). Place this sieve above another of smaller mesh; rest the sieves over a large collecting basin.
   (c) Wash soil from the roots with a jet of water. Examine both sieves and the basin to ensure that all roots are collected.
   (d) If necessary, take sub-samples for more thorough cleaning, separation, or determination of live and dead material.
5. Collect any dead leaf and stem material separately.
6. Mark each sample clearly (pot, plant, stratum number) and place in a drying oven (75°C) until a constant weight is obtained; usually about 24 hours. Determine also weight loss on combustion at 550°C for 3 hours.
7. Repeat for soil in ingrowth cores.

1.8.4 Calculation and interpretation of results

1. Calculate leaf area per pot, leaf area index and leaf inclination for each stratum. Plot the canopy profiles. Why are canopy characteristics important determinants of productivity?
2. Calculate the energy available to the community, energy intercepted by the canopy, the extinction coefficient and energy intercepted by each stratum. How do the two treatments differ?

3. Calculate P_n. If no estimate of decomposition is available, assume it to be zero.
4. Calculate the efficiency of energy incorporation for both treatments, based on
   (a) the total solar energy available and
   (b) the amount of energy intercepted.
   If it is not possible to measure the energy content of the samples, assume that 1.0 g organic matter = 18.7 kJ.
5. Do ingrowth cores underestimate root production? Compare with results for whole pots.

1.8.5 Suggested additional work

1. Use the detailed leaf area to obtain a regression equation relating leaf weight to leaf dimensions. How might dimensional analysis be used to estimate biomass?
2. Collect dead leaves, enclose in litter bags, and determine rates of decomposition in the air, on the soil surface and in the soil. Under what conditions is decomposition a significant component of total productivity? (see Equation 1.2).
3. Determine P_n as a function of leaf area index or canopy closure.

REFERENCES

1. Wiegert, R.G. (1962) The selection of an optimal quadrat size for sampling the standing crop of grasses and forbs. *Ecology* **43**, 125–129.
2. Chapman, S.B. (1976) Production ecology and nutrient budgets. In: *Methods in Plant Ecology* (ed. S.B. Chapman) pp.157–228. Blackwell, Oxford.
3. Golley, F.B. (1961) Energy values of ecological materials. *Ecology* **42**, 581–584.
4. Chiariello, N.R., H.A. Mooney and K. Williams (1989) Growth, carbon allocation and cost of plant tissues. In: *Plant Physiological Ecology: field methods and instrumentation* (Pearcy, R.W., Ehleringer, R.W., Mooney, H.A. and Rundel, P.W., eds.) pp. 327–366. Chapman and Hall, London.
5. Rodin, L.E. and N.I. Basilovic (1965) *Production and Mineral Cycling in Terrestrial Vegetation* (English translation, ed. G.E. Fogg). 253 pp. Scripta Technica Ltd., Oliver and Boyd, London.
6. Hussey, A. and S.P. Long (1982) Seasonal changes in weight of above- and below-ground vegetation and dead plant material in a salt marsh at Colne Point, Essex. *J. Ecol.* **70**, 757–772.
7. Hansson, A.-C. and O. Andren (1986) Below-ground plant production in a perennial grass ley (*Festuca pratensis* Huds.) assessed with different methods. *J. Appl. Ecol.* **23**, 657–666.
8. Hussey, A. (1980) *The Net Primary Production of an Essex Salt Marsh, with Particular Reference to Puccinellia maritima*. Ph.D. thesis, University of Essex, U.K.
9. Long, S.P. and C.F. Mason (1983) *Saltmarsh Ecology* 1st edn. Blackie, Glasgow, London.
10. Wiegert, R.G. and F.G. Evans (1964) Primary production and the disappearance of dead vegetation in an old field in South- eastern Michigan. *Ecology* **45**, 49–63.
11. Lomnicki, A., E. Bundola and K. Jankowska (1968) Modification of the Wiegert-Evans method for estimation of net primary production. *Ecology* **49**, 147–149.
12. Odum, E.P. and A.E. Smalley (1959) Comparison of population energy flow of a herbivorous and a deposit-feeding invertebrate in a salt marsh ecosystem. *Proc. Nat. Acad. Sci. Washington.* **45**, 617–622.
13. Bowen, G.D. (1980) Misconceptions, concepts and approaches to rhizosphere biology. In: *Contemporary Microbial Ecology* (eds. D.C. Ellwood, J.H. Hedger, M.J. Latham, J.M. Lynch and J.H. Slater) pp. 283–304. Academic Press, London.
14. Fogel, R. and G. Hunt (1983) Contribution of mycorrhizae and soil fungi to nutrient cycling in a Douglas Fir ecosystem. *Can. J. Forestry Research* **13**, 219–232.
15. Long, S.P. and M.B. Jones (1992) Introduction, aims, goals and general methods. In: *Primary Productivity of Grass Ecosystems of the Tropics and Sub-tropics* (S.P. Long, M.B. Jones and M.J. Roberts, eds.) pp. 1–24. Chapman and Hall, London.
16. Kamnalrut, A. and J.P. Evenson (1992) Monsoon grasslands in Thailand. In: *Primary Productivity of Grass Ecosystems of the Tropics and Sub-tropics* (S.P. Long, M.B. Jones and M.J. Roberts, eds.) pp. 100–126. Chapman and Hall, London.
17. Jones, M.B., S.P. Long and M.J. Roberts (1992) Synthesis and conclusions. In: Primary Productivity of Grass Ecosystems of the Tropics and Sub-tropics (S.P. Long, M.B. Jones and

M.J. Roberts, eds.) pp. 212–255. Chapman and Hall, London.
18. Piedade, M.T.F., W.J. Junk and S.P. Long (1991) The productivity of the C_4 grass *Echinochloa polystachya* on the Amazon Floodplain. *Ecology* **72**, 1456–1463.

2

Remote sensing of biomass and productivity

J.M.O. Scurlock and S.D. Prince

2.1 INTRODUCTION

Determination of net primary production requires frequent measurement of biomass changes by destructive harvesting (Chapter 1). Such techniques have two main drawbacks; they are time-consuming, and they limit repetitive sampling. Furthermore, they cannot usually be extended to cover large areas of vegetation because of logistics and the huge number of samples required to cope with increased sample variation. However, estimates of photosynthetic production over large areas of natural ecosystems may be obtained using *remote sensing* data combined with ground-based measurements (sometimes referred to as 'ground truth'). Such information may then be applied to estimates of carbon cycling in ecosystems and prediction of responses to climate change.

2.2 WHAT IS REMOTE SENSING?

Remote sensing is a non-destructive optical technique which relies on the fact that sunlight (direct and indirect) reflected from leaves has a characteristic spectrum, quite distinct from that reflected by wood, litter or soil. It can be carried out using hand-held, aircraft-mounted or satellite-borne instruments to give coverage at local, regional or global level. A variety of different vegetation features have been successfully estimated by remote sensing, including biomass, leaf area index, ground cover and canopy light interception. However, more remote sensing studies have been carried out on crops than on natural ecosystems.

Current research is concerned with the relationship between remote sensing measurements, light interception by a plant canopy and its net primary production (NPP). NPP can be related to the amount of light intercepted by a stand of plants. The evidence for this comes primarily from studies on relatively uniform stands of crop plants, where it has also been shown that light interception (and hence NPP) may be estimated by remote sensing [22]. If the measurement of NPP by means of remote sensing of light interception can be extended from simple crop-type canopies to the complex canopies found in many natural ecosystems, it may be possible to obtain plant bioproductivity data on a very large scale indeed. An alternative approach in natural ecosystems may be to estimate standing biomass by remote sensing at intervals throughout the year, and to incorporate these

Photosynthesis and Production in a Changing Environment: a field and laboratory manual
Edited by D.O. Hall, J.M.O. Scurlock, H.R. Bolhàr-Nordenkampf, R.C. Leegood and S.P. Long
Published in 1993 by Chapman & Hall, London.
ISBN 0 412 42900 4 (HB) and 0 412 42910 1 (PB).

data into simple models of biomass production and turnover. However, additional ground-based information on NPP and other vegetation parameters obtained by conventional field sampling is required in most cases, in order to validate the remote sensing measurements and to demonstrate their level of precision, as well as to provide any ancillary variables required by the models.

Remote sensing also offers labour-saving advantages over many alternative non-destructive techniques for estimating biomass and bioproductivity (such as the collection of many individual measurements from marked plants). Other indirect techniques which have been used for assessment of canopy structure or size include beta-particle attenuation, use of electrical capacitance meters, and distortion of the canopy with a weighted disc [39]. These techniques are generally of limited accuracy and application, although it has been reported that canopy height measurement was the best method for short temperate grassland with a high leaf area index [20].

Since its earliest applications, remote sensing has been used to estimate a number of different plant canopy variables. For example, near-infrared/red (NIR/R) reflectance ratio measured with a hand-held instrument has been correlated with leaf area index for stands of crop plants, and a similar correlation with biomass has been reported for prairie grassland [27]. NIR/R reflectance ratio has also been used to predict biomass for stands of salt-marsh vegetation [12]. Furthermore, remote sensing has been used to estimate crop grain yields and for mapping of vegetation types, land clearance and fire damage, but these applications are outside the scope of this chapter.

The application of remote sensing techniques to natural ecosystems in the tropics has been limited mainly to semi-arid regions [11,28,29,31]. Seasonal differences in canopy structure and species composition may cause changes in the relationship between remote sensing data and canopy variables such as leaf area index and biomass [13]. Thus, in the absence of ground verification, such changes may not be obvious unless the seasonal dynamics of the ecosystem are well understood.

2.2.1 Measurement of red and near-infrared reflectance

In this chapter, *irradiance* is defined as the radiant flux per unit area incident on a surface from all directions; *radiance* as the radiant flux scattered by a unit area of surface; *photon flux density* as the number of photons per unit area per unit time, incident upon or scattered by a surface; and *reflectance* as the fraction of incident flux which is reflected [4].

Radiation reflected and scattered by a plant canopy has a characteristic spectrum in the visible and short-wave infrared, distinguishable from that reflected by the canopy's surroundings. Visible radiation between 400 and 700 nm wavelength is absorbed by plant pigments, with peaks of absorbance in the blue (about 450 nm) and the red (600–680 nm), but there is a sharp increase in reflectance for wavelengths greater than about 700 nm. In contrast, the spectra of light reflected by litter, soil or water (the usual 'background' for most plant canopies) show only a gradual increase in reflectance as wavelength increases (Fig. 2.1). This contrast provides a method, both qualitative and quantitative, for assessing both the presence of a plant canopy and its physical properties.

Red reflectance (600–700 nm) tends to decrease with the amount of green vegetation present, due to absorption by chlorophyll, whereas near-infrared (NIR) reflectance (700–1000 nm) tends to increase because of light scattering by mesophyll tissue [15,21]. This spectral discontinuity may be detected when only a modest amount of plant cover exists against the background, and continues to increase with the amount of plant leaf material present. Although canopy reflectance in the red reaches an asymptote usually

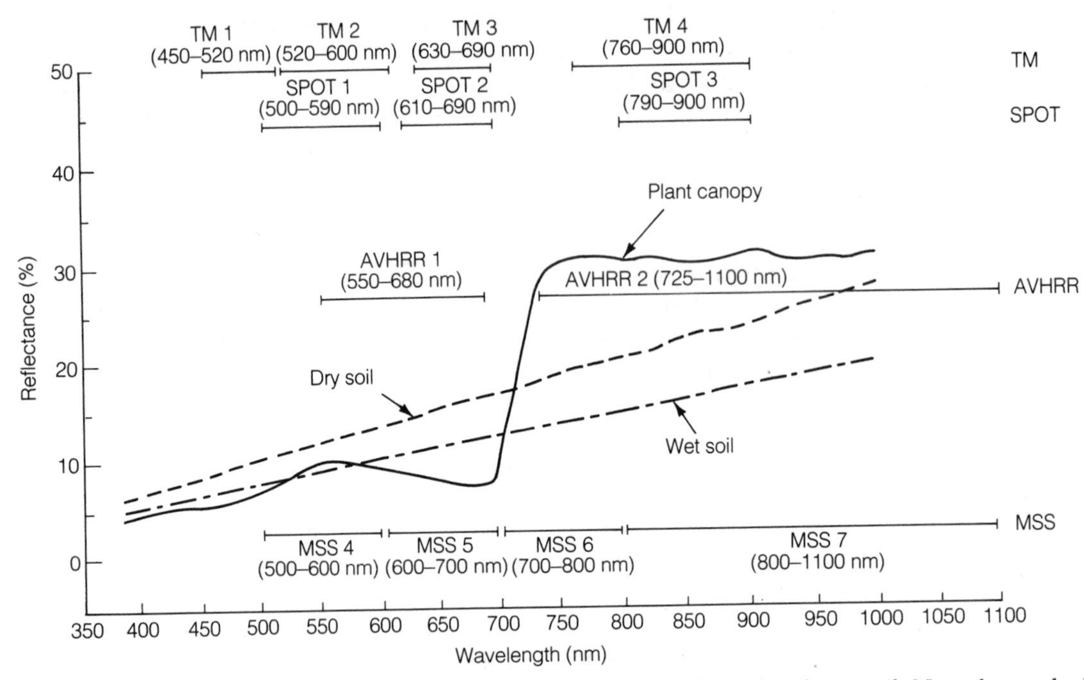

Fig. 2.1. Reflectance spectra of green vegetation (plant canopy), dry soil and wet soil. Note the marked discontinuity in vegetation reflectance around 700 nm. Soil shows no such discontinuities in reflectance; this is also true of other possible background materials such as litter, wood or water. Spectral band comparison between the Landsat multispectral scanner (MSS), Landsat Thematic Mapper (TM), French satellite SPOT, and the NOAA Advanced Very High Resolution Radiometer (AVHRR) spectrum shown overlaid on the same scale.

before 100% ground cover is attained, NIR reflectance continues to increase through this red-insensitive range, up to high values of leaf area index (LAI).

The reflected spectrum may be detected with a spectroradiometer, an instrument capable of measuring radiance in many small waveband intervals over a wide range (e.g. *LI-1800, LI-COR, Inc., Lincoln, Nebraska, USA*). However, for the purposes of measuring the extent of the spectral discontinuity, it is only necessary to measure in two specific wavebands, one in the red part of the spectrum and one in the NIR. A study of American prairie grassland used bands centred on 675 nm for red reflectance and 800 nm for NIR reflectance [27]. A review of wavebands for remote sensing concluded that 680 nm reflectance is the optimum indicator for relatively low values of canopy variables, with 775 nm optimal for medium to high values [38] (Fig. 2.2).

The use of some form of ratio of radiance in two wavebands overcomes the limited useful bandwidth of single waveband measurements, and partly compensates for the effect of variation in incident solar radiation. Thus the red/near-infrared (R/NIR) ratio gives a useful indicator which changes over a relatively wide range of values for canopy variables [38]. This ratio, which indicates the extent of the spectral discontinuity in canopy reflectance around 680–720 nm, has been correlated with ground cover and LAI. For plants where biomass is closely related to leaf area, it has also been correlated with biomass.

The most common waveband ratios in use

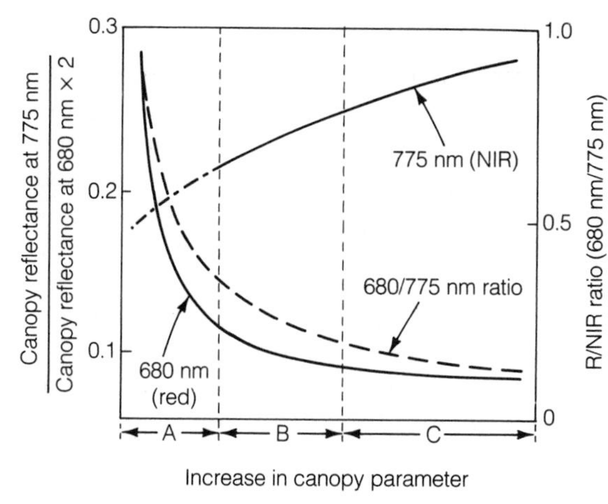

Fig. 2.2. Vegetation canopy reflectance at 680 nm (red) and 775 nm (near-infrared), and red/near-infrared ratio, plotted against increase in canopy parameter. Regions of abscissa: A – red reflectance decreasing sharply, near-infrared reflectance variable; B – red reflectance decreasing gradually, near-infrared reflectance increasing quickly; C – red reflectance asymptotic, near-infrared reflectance still increasing. Adapted from Tucker [38].

are the red/near-infrared (R/NIR) ratio (or NIR/R ratio, which is often called the Simple Ratio), and the Normalised Difference Vegetation Index (NDVI), where:

$$\text{NDVI} = \frac{\text{R} - \text{NIR}}{\text{R} + \text{NIR}}$$

Although the terms 'R' and 'NIR' usually refer to estimates of *reflectance* (reflected radiance relative to incoming solar irradiance), NDVI may be also derived from absolute values of red and NIR reflected radiance (e.g. satellite or aircraft sensors, where a measure of solar irradiance is not available).

The reflectance of individual leaves is affected by certain aspects of their physiological state, but additional factors are significant in determining canopy reflectance as a whole. The density of the canopy, and its structure (leaf angle, leaf transmittance, nature of other canopy components) are important, as well as solar angle and angle of observation [21].

Changes in leaf reflectance may occur in response to stress. Although the physiological reasons are controversial, remote sensing is nevertheless an important qualitative indicator of stress or disease because stress leads to detectable changes in canopy structure and density [18].

At low vegetation cover, background reflectance is a significant factor which can disguise or dilute the vegetation 'signal'. Different soils or other backgrounds may have different absolute values of reflectance, depending upon colour, water content and surface structure [13,24].

2.2.2 Reflectance distribution and reflectance standards

For a measured surface such as a plant canopy, the incoming solar irradiance (direct and diffuse) and its outgoing reflected radiance are most completely described as two hemispherical distributions of electromagnetic radiation, the *Bidirectional Reflectance Distribution Function* (BDRF). Although field instruments have been developed which sample reflectance for all possible source/sensor positions within these hemispherical fields [8], it is conventional to sample only a sub-set of the BDRF known as the *Bi-conical Reflectance Factor* [19,25]. Thus the sensor usually has a limited conical field of view, within which both reflected radiance and solar irradiance are measured.

Plant canopy reflected radiance is usually expressed with reference to incoming solar irradiance, determined from the radiance of a surface of known reflectance. This enables comparisons between measurements carried out in different places, or at the same place under different solar irradiation conditions (see also Section 2.3.1).

Ideally this reference surface should be a near-Lambertian reflector (one which scatters radiant energy equally in all directions) and it should show similar reflectance across the range of wavelengths of interest [19,24]. Examples of such surfaces are 'Halon' (a kind of PTFE plastic) or pressed barium sulphate powder [25]. However, Kodak Neutral Test Cards, also known as Kodak Grey Cards (*Kodak, Hemel Hempstead, U.K.*), are of value for routine field work which does not demand very strict accuracy in reflectance determination, with the low reflectance of the grey surface well-matched to typical values for vegetation and soil [24,25,26].

2.3 INSTRUMENTATION

Due to the cost and complexity of commercial spectral radiometers, a number of simple spectral ratio meters have been devised [23,24,27,32]. Unfortunately these lower-cost devices are rarely calibrated against standards, so direct comparisons between different spectral instruments are difficult. Thus there are problems of comparability between different studies, as well as a lack of standardisation in methodology [25].

With the increasing use of remote sensing for studies in the tropics, in developing countries, and in difficult terrain, it is important that ground-based instruments are rugged and truly portable. Furthermore, they should be reliable, easy to use by semi-skilled personnel and cheap to maintain or replace [24].

Remote sensing studies under the UNEP Project on Productivity of Tropical Grasslands have utilised the *Skye SKR100/110* spectral ratio meter (*Skye Instruments, Llandrindod Wells, U.K.*), which has also been used for remote sensing of temperate grasslands [20,34]. Originally designed to measure light quality in glasshouses, this instrument registers photon flux density in two 10 nm bandwidths centred on 660 nm (red) and 730 nm (NIR). These bands correspond quite well to the maxima for absorption of red radiation and reflectance of NIR radiation by green plant tissue, although a slightly longer wavelength is optimal in the NIR [38].

2.3.1 Instrument use

Plant canopy reflectance in any single waveband may be measured by comparing the amount of light reflected from the canopy under given conditions with light reflected from a reference surface (e.g. *Kodak* grey card). With the sensor mounted above the canopy on an arm attached to a tripod, the grey card may be placed at a fixed distance between the sensor and the canopy, and conveniently swung in and out of the field of view, normal to the sensor axis. Thus, for a given waveband:

$$\text{canopy reflectance} = \frac{\text{canopy radiance}}{\text{grey card radiance}} \times \text{grey card reflectance}$$

where grey card reflectance is a known quantity (about 20% in the red and NIR wavebands).

Note that many instruments used in remote sensing are *radiometers* which measure radiance, i.e. radiant flux density (e.g. *Barnes MMR, Barnes Engineering, Stamford, Connecticut, USA; Exotech-100, Exotech, Inc., Gaithersburg, Maryland, USA; Macam Photometrics Ltd., Livingston, UK*). In the case of the *Skye SKR100/110* spectral ratio meter, which measures photon flux density, the term 'radiance' in the above equation should be replaced by 'reflected photon flux density.'

The red/near-infrared reflectance ratio (R/NIR ratio), or its reciprocal the NIR/R ratio, are calculated as follows:

$$\text{R/NIR ratio} = \frac{\text{(canopy reflectance in the red, e.g. 660 nm)}}{\text{(canopy reflectance in the NIR, e.g. 730 nm)}}$$

Using the *Skye* spectral ratio meter, the ratio of red to NIR photon flux density may also be read directly. From this 'spectral ratio', R/NIR ratio can be estimated roughly by a 'short-cut' method as follows:

$$\text{R/NIR ratio} = \frac{\text{(660 nm/730 nm 'spectral ratio' of plant canopy)}}{\text{(660 nm/730 nm 'spectral ratio' of grey card)}}$$

Note that this is an approximation which assumes that the reflectance of the grey card does not change between 660 and 730 nm (in fact it increases slightly from about 20% to 21%). It may not be suitable for instruments using broader wavebands in the red and NIR [26].

The *Skye* sensor head normally has a 180° field of view, so a field stop must be fitted, such as a matt black card or plastic tube. The length of the stop tube is calculated to give a field of view of 40° (20° either side of vertical). Operating height is similarly calculated so that the vegetation canopy sample is viewed as a circle of the desired radius (Fig. 2.3). A small circular spirit level may be mounted on the sensor head to ensure that it is vertical, and a second spirit level at the edge of the grey card to confirm its correct orientation.

When using a hand-held or tripod-mounted instrument, it is obviously important that neither the operator nor the instrument cast a shadow over the viewed area. The wearing of reflective light-coloured clothing or shoes should be avoided, and highly reflective objects such as field vehicles should be kept at a distance in order not to affect the spectral quality of local irradiance.

2.4 LOW-LEVEL REMOTE SENSING IN THE FIELD

A wide variety of remote sensing applications is reported in the literature, but the consensus is what might be expected from theory – that the most precisely estimated

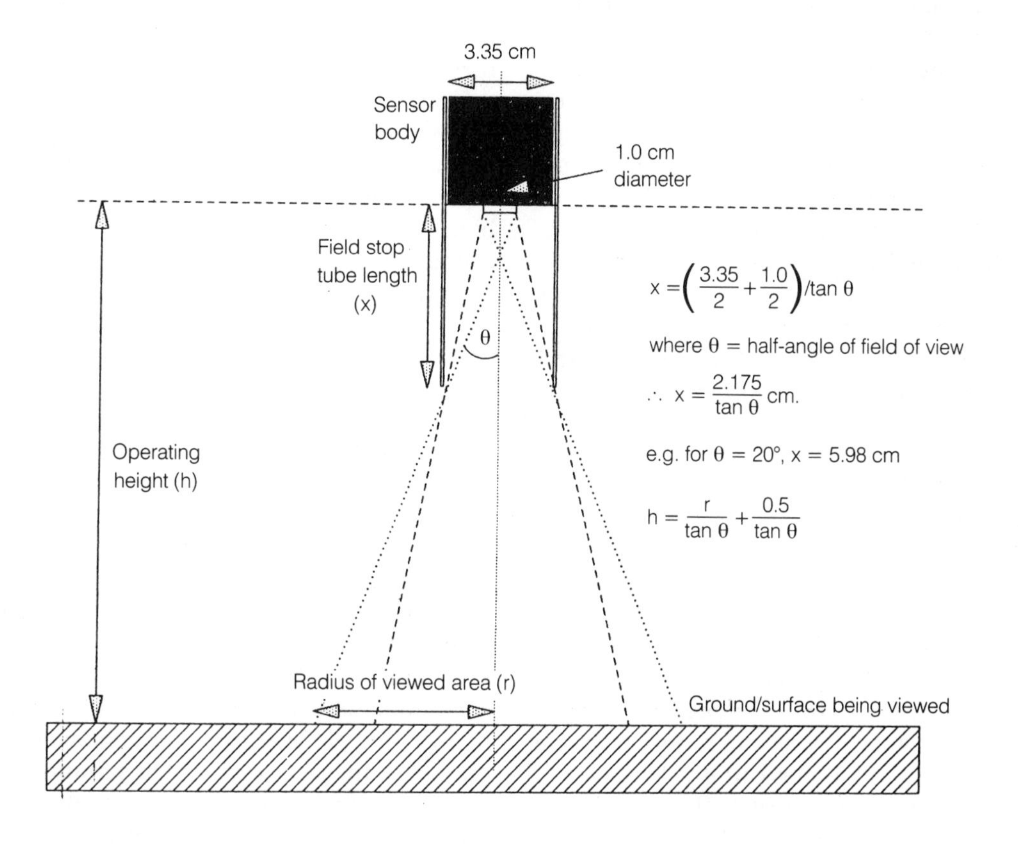

Fig. 2.3. Trigonometric calculation of field stop tube length (x), operating height (h) and radius of viewed area (r) for the *Skye Instruments* red/near-infrared sensor used in the UNEP tropical grassland study [34].

features are those related to the extent of leaves in the canopy. The best correlations with remote sensing measurements have been found using crop plants, where the relatively uniform canopies with few senescent or woody components are likely to produce little variation in reflectance data.

For a variety of different crop canopies, R/NIR ratio and Normalised Difference Vegetation Index (NDVI) have been shown to be indicators of different canopy variables, including biomass, LAI, ground cover and light interception. NDVI has also been related to light interception (estimated as a function of LAI) in stands of wheat throughout the growing season. These long-term experiments, involving a number of different planting dates, suggest that both NDVI and NIR/R ratio are relatively unaffected by changes in sun angle, extent of shadows, etc. [1]. However, the uniform structure of most crop canopies (with little or no dead matter present) facilitates the estimation of light interception, presenting a relatively homogeneous population of leaf surfaces to the remote sensing instrument.

The application of remote sensing so far to natural grass ecosystems has been limited in extent, especially in the tropics [34]. For temperate grasslands, NIR/R and R/NIR ratio have been correlated with biomass, LAI and ground cover [20,23,24,27]. Under the UNEP Project on Productivity of Tropical Grasslands, the irregular distribution of dead material in natural grassland canopies affected the precision of estimation of biomass and LAI, although the techniques had been tested previously under controlled-environment conditions [34]. Similar problems have also been reported for grasslands in Arizona, USA [16]. The presence of dead matter in plant canopies can reduce estimates of biomass and other canopy variables by increasing red reflectance [13].

The extent of such errors in remote sensing measurements can be difficult to assess, unless a great deal is known about the accumulation, decomposition and physical removal of dead matter [16]. It is also difficult to estimate accurately light interception by *live* leaves only; a way must be found to distinguish light interception by live and dead matter [7]. For complex natural ecosystem canopies, remote sensing estimates of light interception (or possibly PAR absorption) cannot be tested unless good-quality ground-based data are available on live/dead matter dynamics and the direct measurement of light interception by live photosynthetic tissues.

2.4.1 Remote sensing of net primary production

Remote sensing has been used to estimate net primary production (NPP) of a natural salt-marsh ecosystem in terms of increase in biomass [12]. Where there is likely to be little turnover of vegetation (i.e. simultaneous growth and death) during the growing season, this method may be justified – as in the case of a semi-arid environment in the African Sahel [40]. However, if turnover is high (e.g. in the humid tropics), biomass change alone is a poor indicator of both the gross and net primary production of the ecosystem.

An alternative approach is suggested by Kumar and Monteith [22], who demonstrated the relatively fixed relationship between light interception and productivity in crops of sugar beet. Productivity may therefore be estimated on the basis of remotely sensed light interception, assuming constant ratios for respiration and partitioning. Aboveground biomass accumulation in wheat has been estimated by remote sensing of light interception and measurements of incident solar radiation [2]. However, crop systems show relatively little turnover of biomass and often lack the large root systems found in the natural environment. Measurement of productivity over monthly intervals or less is therefore subject to much greater experimental error in natural ecosystems than with crops. Furthermore, accurate measurement of incident radiation is essential over short time intervals (i.e. weeks, months). Incident radiation may only be considered constant over longer periods of time, where irradiance data is available from ground measurements or via remote sensing.

In order to compare field determinations of natural ecosystem productivity with remote sensing data, relatively long-term experiments are required in order to overcome the short-term errors in determining total above-ground and below-ground NPP. One solution may be to divide the yearly cycle of plant growth and senescence into discrete periods according to environmental constraints on productivity (soil moisture, temperature, etc.); remote sensing measurements and NPP determinations would correspond to these seasonal intervals. Little work of this kind has been carried out so far; a good advance knowledge of the study areas and their plant canopy characteristics is needed, and such studies would need to be long-term in order to account for year-on-year variation.

Since plant productivity is a function of both the fraction of light actually absorbed by live matter and the amount of incident light energy (given an adequate supply of water, nutrients, etc.), claims that remote sensing can estimate global photosynthesis or global productivity should be treated with caution. Even assuming that remote sensing can estimate the *photosynthetic capacity*, i.e. the *potential productivity* of an ecosystem, there is still uncertainty about how much of this capacity is realised in practice. The UNEP Project on Productivity of Tropical Grasslands found that light energy conversion efficiencies varied significantly between different grassland ecosystems [34]. Areas of the world where productivity does approach photosynthetic capacity may be limited to those tropical regions with a relatively constant environment, where nutrient stress and other stresses are not limiting, or to those regions where a short growing season results in rapid growth and later death of the plant canopy. However, if the difficulties with estimating net light energy conversion efficiency (ε_n) can be solved for natural ecosystems, then remote sensing data could be used for modelling of NPP on a large scale, at both the regional and global level [30].

These conclusions have wide implications for the use of remote sensing for global environmental monitoring. Remote sensing by aircraft and satellite has an enormous potential for data gathering, for modelling of productivity and carbon cycling at both the local and the global level. However, its limitations must be understood; prediction of NPP may be limited to part of the year only, or to particular types of ecosystem only. The way ahead lies in increased collaboration between remote sensing workers, ecophysiologists and ecosystem modellers, through the verification of remote sensing studies with ground data collection over successive years of study.

2.5 SATELLITE REMOTE SENSING AND MODELLING

Satellites have been used for remote sensing of vegetation since the early 1970s. Resolution is limited by the size of the smallest picture element or pixel; in the case of the US NOAA satellites' Advanced Very High Resolution Radiometer (AVHRR) this is about 1100 m. Considerably higher spatial resolution, perhaps better suited to monitoring of vegetation in small sample areas, is available on the American Landsat series (80 m and 30 m) and on the French SPOT series of satellites (20 m). Details may be found in many textbooks [14,33], although these tend to concentrate on the theory and practice of satellite remote sensing, and may be rather limited in their treatment of vegetation applications. Reference books are also available, listing sources of satellite data, together with addresses of agencies, research and teaching institutions [17].

Data are usually obtained from archives stored in digital form, and manipulated using an image-processing computer system, which can sort through the data, locate the sample areas required and extract only radiance data in the wavebands of interest. Although there are many reports of the successful use of satellite remote sensing, the price and availability of data can limit its usefulness, as well as whether the user has access to the expensive equipment for data analysis.

Most remote sensing satellites follow a circular orbit which passes close to both North and South Poles and is sun-synchronous (always going over the same area at the same local time). In effect, the Earth rotates under the satellite's orbit whilst successive strips of its surface are scanned. The wavebands of greatest interest for vegetation monitoring are in the visible and near-infrared (NIR) part of the spectrum, but since these wavelengths are reflected by clouds, data cannot be collected when there is appreciable cloud cover or haze. Measurements are therefore possible only at times of clear weather, so the data record for some tropical sites is liable to be discontinuous, especially during rainy seasons.

Although the range of wavebands and the degree of resolution available from different satellite sensors may show comparative advantages and disadvantages, other factors are important if multi-temporal data are required for repeated determinations of vegetation at the same site. Although the resolution of the NOAA-AVHRR data is too coarse for detailed vegetation studies, the relatively low price and large area coverage makes this source an attractive alternative to Landsat. Furthermore, daily passes by NOAA satellites (compared with once every 18 days for Landsat) vastly improve the chances of obtaining cloud-free images, which are especially difficult to obtain for humid tropical regions. The SPOT (Systeme Probatoire d'Observation de la Terre) satellites provide a different approach to multi-temporal data collection. Although they take 26 days to cover the whole globe, their sensors can be 'steered' to view the Earth's surface obliquely, so a succession of daily images can be taken from a single site (but from different viewing angles).

Landsat or SPOT images may be preferable to NOAA-AVHRR data for monitoring of crop growth and for small-scale ecological work. However, studies such as assessment of rangeland productivity over large areas may require a combination of coarse-scale remote sensing data from the NOAA-AVHRR with more detailed data from Landsat or SPOT [9,10,28]

Thermal infrared and microwave data, which can better penetrate haze, are of increasing interest for vegetation studies. The Scanning Multifrequency Microwave Radiometer (SMMR) on the American Nimbus-7 satellite detects the difference between the vertically and horizontally polarised 'surface brightness temperature' at 37 GHz, which decreases with the amount of vegetation present. Although few ground-based comparisons have yet been carried out, this new source of data promises to complement the more common R/NIR-based vegetation indices. Such microwave polarisation data seem to be particularly sensitive to the woody component of dense canopies [6].

2.5.1 Satellite applications to vegetation monitoring

Satellite coverage of the plant canopy spectral signature is shown in Fig. 2.1. Satellite data have been used for vegetation mapping based upon spectral discrimination within multiple wavebands, and temporal discrimination using Normalised Difference Vegetation Index (NDVI) data [41]. Detection of tropical forest clearance is another application, although primary rainforest and regrowing secondary forest may be spectrally indistinguishable, and difficulties have been experienced discriminating between open forest and scrubland) [37].

Predictive equations have been developed for estimating biomass and ground cover in southern African rangelands using a Landsat-derived R/NIR ratio, but it was concluded that they applied only to sites with similar vegetation [31]. A correlation has even been demonstrated at the global level between atmospheric CO_2 variations and a satellite-derived vegetation index, on the basis that this is an indicator of intercepted radiation and therefore photosynthetic carbon flux [42]. However, as discussed in Section 2.4.1, the relationships between reflectance spectra and NPP carry a high degree of uncertainty for natural ecosystems, so it is difficult to apply satellite data in this manner without more information on how ε_n (net light energy conversion efficiency) varies on a regional scale [30].

The need for an integrated approach extending from ground-based remote sensing to satellites has already been recognised [3]. A comprehensive workplan would commence with ground data collection from (necessarily) fairly small sample sites, followed by collection and analysis of data from aircraft and then satellite-borne sensors. This way, it is possible to appreciate the gains and losses of data as the scale of the study increases from field site to continental level.

From the point of view of the plant ecophysiologist, a methodical field-based approach [3] may be preferable to immediate attempts at large-scale global and regional studies [30,42]. Although the global modelling of CO_2 flux through natural ecosystems can be greatly facilitated by satellite remote sensing, robust models will only come about if plant canopy reflectance is well understood for each category of ecosystem considered. Different canopy types within a single ecosystem have been shown to exhibit distinct relationships between biomass and NDVI; in this case a coastal wetland measured using Landsat TM bands 4 and 3 [13] (Fig. 2.4). Thus accurate models for biomass prediction should include details of canopy structure and composition; seeking a generalised relationship for an ecosystem type may not be sufficient.

Many of these points have been addressed by recent integrated studies such as the First ISLSCP Field Experiment (FIFE), which was carried out on North American tall-grass prairie grassland in 1987 and 1989 [36]. Under the aegis of the International Satellite Land Surface Climatology Project (ISLSCP), this major interdisciplinary study involved biologists, remote sensing scientists, atmospheric chemists and meteorologists in intensive monitoring of the Konza Prairie grassland site at Kansas State University, USA. Plant physiological measurements were carried out in small study sites within a larger remote sensing study area of 15 km × 15 km. Results from the FIFE, and from other regional studies planned for Canadian boreal forest and for West African dry grasslands, should indicate how the atmosphere modifies the remote sensing signals detected by satellites, and will help in developing a hierarchical approach to remote sensing of natural ecosystems.

2.5.2 Modelling of canopy reflectance

Extending from the basic principles of canopy reflectance in the red and NIR [21], complex

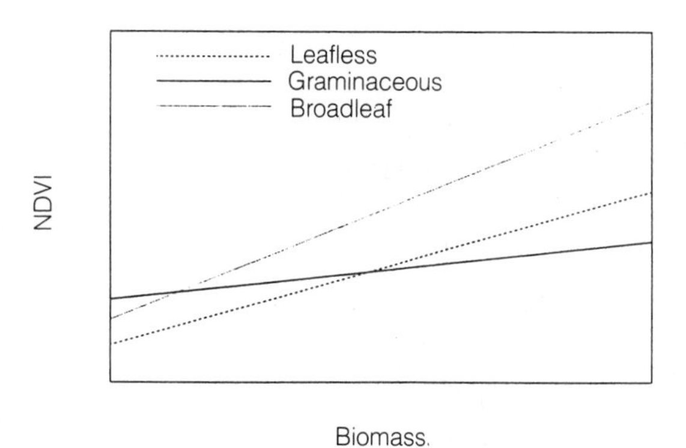

Fig. 2.4. Generalised different relationships between Normalised Difference Vegetation Index (NDVI) and biomass, exhibited by different canopy types in a coastal wetland ecosystem. After Hardisky *et al.* [13].

models have been developed which attempt to describe the effect of canopy geometry and structure on canopy reflectance, and try to link individual leaf reflectance to gross canopy reflectance [3]. On a more theoretical level, canopy reflectance in the red and NIR has been modelled and coupled to mathematical representations of photosynthetic and transpiration rates [35]. Although these studies are an active area of research, they deserve further validation against ground-based measurements; so far, model predictions have been compared only with the literature on remote sensing of crop systems [5]. A suitable modelling approach for natural ecosystems such as perennial grasslands is further complicated by a heterogeneity of canopy structure which far surpasses anything found in crops. Applications of remote sensing models at both local and global level should therefore be treated with caution, until detailed comparisons with field data are available.

2.6 EXPERIMENTAL WORK

2.6.1 Remote sensing of plant canopy variables

This simple experiment demonstrates the ability of remote sensing to detect differences in crop canopy variables such as biomass, leaf area index (LAI) and light interception. Canopies of different densities are required, preferably examples of different canopy structures.

Four to five weeks in advance, prepare a number of small field plots (each 1.0 m square) for a broad-leaved crop plant such as beans, soybeans or cowpeas. After germinating the seeds, plant them in the plots in four groups at different densities as follows:

1. 5 cm grid points (seeds 5 cm apart, in rows 5 cm apart);
2. 10 cm grid points;
3. 15 cm grid points;
4. 20 cm grid points.

Ideally, there should be at least five replicates of each 'treatment' in order to obtain statistically significant results.

A more laborious method is to establish plots of different ages, planted at intervals of one week using the same grid spacing (say 10 cm). Alternatively, plots may be subjected to different degrees of water stress during their development (using rain shelters or by withholding irrigation) in order to obtain a range of canopy densities.

Measure red/near infrared ratio (R/NIR) ratio or Normalised Difference Vegetation Index (NDVI) over each of the replicate plots for all 'treatments', using a hand-held or tripod-mounted radiometer. Take a reading from a reflectance panel (e.g. *Kodak* grey card) at the beginning and end of these measurements, or after each measurement if the spectral quality of local irradiance is changing (e.g. clouds passing across the sky). Avoid casting shadows over the viewed area, and try to keep reflective light-coloured objects (clothing, shoes, other equipment) at a distance.

After taking the remote sensing measurements, determine a number of canopy variables for each of the plots. Canopy light interception may be determined as described in Chapter 13, using miniature tube solarimeters (*TSM, Delta-T Devices, Cambridge, UK*). LAI may be estimated non-destructively using a canopy analyser (*LI-COR, Inc. USA*) or portable leaf area meter, or determined after destructive harvesting using a laboratory area meter (*Delta-T Devices, Cambridge, UK*). Above-ground biomass is measured as oven-dry weight of harvested live material, as described in Chapter 1 (keep leaves and stems separate if possible).

Plot remote sensing measurements against canopy variables, using the mean values for each treatment. Plot also against the logarithm of each canopy variable. Which gives a better fit to a straight line, the log or non-log plot? Which variable (biomass, leaf biomass, LAI, light interception) is most closely correlated with the remote sensing measurements, and why?

2.6.2 Progressive thinning method

This quick experiment demonstrates the relationship between R/NIR ratio (measured with the *Skye* spectral ratio meter) and both biomass and LAI. Progressive thinning of the plant canopy is used to generate a range of biomass and LAI values.

A fairly dense closed plant canopy is required for this experiment; ideally an area of grassland about 0.5–0.8 m in height. A broad-leaved canopy may also be suitable. A wire hoop is used to demarcate a circular sample area, 50 cm in radius (corresponding to the circular area viewed by the *Skye* sensor from 1.39 m height).

After taking an initial reading of R/NIR ratio, individual shoots or clumps of shoots are selected at random over the sample area, cut at ground level and removed, until a significant change in R/NIR ratio is noticed. Do not discriminate between live and dead matter at this stage; cut entire shoots in their natural state. After recording the new reading of R/NIR, all clipped plant material is placed in a sealed plastic bag (with a small amount of water to prevent desiccation), and the process repeated. This progressive thinning method is repeated five or six times, with a further reading of R/NIR ratio when only litter remains: thus a series of remote sensing readings are generated, representing a range of values for biomass and LAI. A reading of R/NIR ratio is also taken for bare soil after removal of the litter. Repeat the whole experiment for a further three or four circular sample areas if there is enough plant matter available.

Clipped plant material is sorted into live and dead matter on the basis of tissue necrosis, dead parts being removed from otherwise green leaves. Leaf area is determined using an area meter (e.g. *AMS, Delta-T Devices, Cambridge, UK*). Sorted plant material is washed and dried to constant weight at 80°C in a forced-draught oven. Biomass is measured as oven-dry weight (see also Chapter 1).

Plot R/NIR ratio against biomass, and R/NIR against LAI. Try also semi-log plots of R/NIR against the logarithms of biomass and LAI. Individual samples will probably be highly correlated (Fig. 2.5), but how closely correlated are the pooled data from all the samples combined? If the data are a close fit to a straight line, you can derive an equation for predicting biomass or LAI on the basis of R/NIR ratio. This equation should work well for plant canopies similar to the one you have just measured. What factors would limit its application to other types of plant canopy?

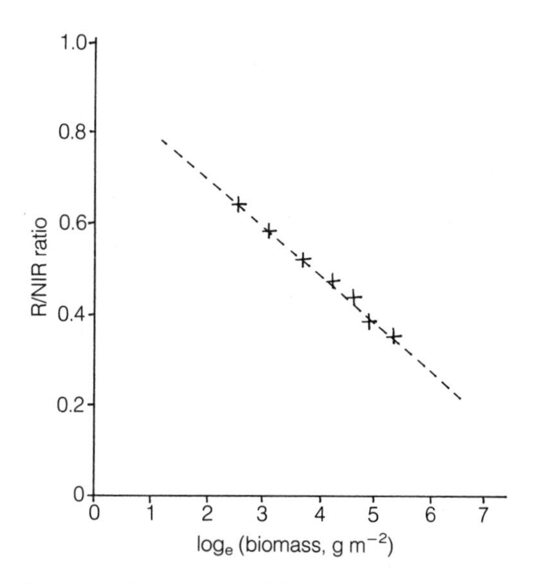

Fig. 2.5. Progressive thinning experiment: relationship between red/near-infrared reflectance ratio (R/NIR ratio) and $\log_e$(biomass) for a single sample area of grassland in Nairobi National Park, June 1986 (by kind permission of Dr. J.I. Kinyamario, University of Nairobi, Kenya).

REFERENCES

1. Asrar, G., M. Fuchs, E.T. Kanemasu and J.L Hatfield (1984) Estimating absorbed photosynthetic active radiation and Leaf Area Index from spectral reflectance in wheat. *Agronomy Journal* 76, 300–306.
2. Asrar, G., E.T. Kanemasu, R.D. Jackson and P.J. Pinter (1985) Estimation of total aboveground phytomass production using remotely sensed data. *Remote Sensing of Environment* **17**, 211–220.
3. Badhwar, G.D., R.D. MacDonald and N.C. Mehta (1986) Satellite-derived leaf area index and vegetation maps as input to global carbon cycle models – a hierarchical approach. *International Journal of Remote Sensing* **7**, 265–281.
4. Biggs, W. (1986) Radiation measurement. In: *Advanced Agricultural Instrumentation: design and use.* Martinus Nijhoff Publ., Dordrecht. pp. 3–20.
5. Choudhury, B.J. (1987) Relationships between vegetation indices, radiation absorption and net photosynthesis evaluated by a sensitivity analysis. *Remote Sensing of Environment* **22**, 209–233.
6. Choudhury, B.J. and S.D. Prince (1989) Comparison of passive microwave with near-infrared and visible data for terrestrial environmental monitoring. *International Journal of Remote Sensing* **10**, 1575–1691.
7. Daughtry, C.S.T., K.P. Gallo, S.N. Goward, S.D. Prince and W.P. Kustas (1992) Spectral estimates of absorbed radiation and phytomass production in corn and soybean canopies. *Remote Sensing of Environment* **39**, 141–152.
8. Deering, D.W. and P. Leone (1986) A sphere-scanning radiometer for rapid directional measurements of sky and ground radiance. *Remote Sensing of Environment* **19**, 1–24.
9. Goward, S.N. (1989) Satellite bioclimatology. *Journal of Climate* **2**, 710–720.
10. Graetz, R.D. (1987) Satellite remote sensing of Australian rangelands. *Remote Sensing of Environment* **23**, 313–331.
11. Grouzis, M. and M. Methy (1983) Détermination radiométrique de la phytomasse herbacée en milieu Sahélian: perspectives et limites. *Acta Oecologica: Oecolgica Plantarum* **4(18)(3)**, 241–257.
12. Hardisky, M.A., F.A. Daiber, C.T. Roman and V. Klemas (1984) Remote sensing of biomass and annual net aerial primary productivity of a salt marsh. *Remote Sensing of Environment* **16**, 91–106.
13. Hardisky, M.A., M.F. Gross and V. Klemas (1986) Remote sensing of coastal wetlands. *Bioscience* 36, 453–460.
14. Harris, R. (1987) *Satellite Remote Sensing.* Routledge and Keegan Paul, London.
15. Horler, D.N.H. and J. Barber (1981) Principles of remote sensing of plants. In: *Plants and the Daylight Spectrum.* (H. Smith, ed.). Academic Press, London. pp. 43–63.
16. Huete, A.R. and R.D. Jackson (1987) Suitability of spectral indices for evaluating vegetation characteristics on arid rangelands. *Remote Sensing of Environment* **23**, 213–232.
17. Hyatt, E. (1988) *Keyguide to Information Sources in Remote Sensing.* Mansell, London.
18. Jackson, R.D. (1986) Remote sensing of biotic and abiotic plant stress. *Annual Review of Phytopathology* **24**, 265–287.
19. Kimes, D.S. and J.A. Kirchner (1982) Irradiance measurement errors due to the assumption of a Lambertian reference panel. *Remote Sensing of Environment* **12**, 141–149.
20. King, J., E.M. Sim and G.T. Barthram (1986). A comparison of spectral reflectance and sward surface height measurements to estimate herbage mass and leaf area index in continuously stocked ryegrass pastures. *Grass and Forage Science* **41**, 251–258.
21. Knipling, E.B. (1970) Physical and physiological basis for the reflectance of visible and near-infrared radiation from vegetation. *Remote Sensing of Environment* **1**, 155–159.
22. Kumar, N. and J.L. Monteith (1981) Remote sensing of crop growth. In: *Plants and the Daylight Spectrum.* (H. Smith, ed.). Academic Press, London. pp. 133–144.
23. Mayhew, P.W., M.D. Burns and D.C. Houston (1984) An inexpensive and simple spectrophotometer for measuring grass biomass in the field. *Oikos* **43**, 62–67.
24. Milton, E.J. (1980) A portable multiband radiometer for ground data collection in remote sensing. *International Journal of Remote Sensing* **1**, 153–165.
25. Milton, E.J. (1987) Principles of field spectroscopy. *International Journal of Remote Sensing* **8**, 1807–1827.
26. Milton, E.J. (1989) On the suitability of Kodak neutral test cards as reflectance standards. *International Journal of Remote Sensing* **10**, 1041–1047.
27. Pearson, R.L., L.D. Miller and C.J. Tucker (1976) Hand-held spectral radiometer to estimate graminaceous biomass. *Applied Optics* **15**, 416–418.
28. Prince, S.D. (1986) Monitoring the vegetation of semi-arid tropical rangelands with the NOAA-7 Advanced Very High Resolution

Radiometer. *In: Remote Sensing and Tropical Land Management* (M.J. Eden and J.T. Parry, eds.). John Wiley, Chichester. pp. 307–334.

29. Prince, S.D. (1991a) Satellite remote sensing of primary production: comparison of results for Sahelian grasslands 1981–1988. *International Journal of Remote Sensing* **12**, 1301–1311.
30. Prince, S.D. (1991b) A model of regional primary production for use with coarse-resolution satellite data. *International Journal of Remote Sensing* **12**, 1313–1330.
31. Prince, S.D. and W.L. Astle (1986) Satellite remote sensing of rangelands in Botswana. I. Landsat MSS and herbaceous vegetation. *International Journal of Remote Sensing* **7**, 1533–1553.
32. Prince, S.D., P.J. Willson, D.M. Hunt and P. Halstead (1988) An integrated camera and radiometer for aerial monitoring of vegetation. *International Journal of Remote Sensing* **9**, 303–318.
33. Sabins, F.F. (1987) *Remote Sensing: Principles and Interpretation.* 2nd edn. Freeman, New York.
34. Scurlock, J.M.O. (1992) Remote sensing of primary production in natural tropical grasslands and articificial mixed-species canopies. In: *Primary Productivity of Grass Ecosystems of the Tropics and Sub-tropics* (S.P. Long, M.B. Jones and M.J. Roberts, eds.). Chapman and Hall, London. pp. 189–211.
35. Sellers, P.J. (1985) Canopy reflectance, photosynthesis and transpiration. *International Journal of Remote Sensing* **6**, 1335–1372.
36. Sellers, P.J. and F.G. Hall, eds. (1987) *Experiment Plan for the First ISLSCP Field Experiment.* NASA/Goddard Space Flight Centre, Greenbelt, MD 20771, USA.
37. Singh, A. (1987) Spectral separability of tropical forest cover classes. *International Journal of Remote Sensing* **8**, 971–979.
38. Tucker, C.J. (1979) Red and photographic infra-red linear combinations for monitoring vegetation. *Remote Sensing of Environment* **8**, 127–150.
39. Tucker, C.J. (1980) A critical review of remote sensing and other methods for non-destructive estimation of standing crop biomass. *Grass and Forage Science* **35**, 177–182.
40. Tucker, C.J., C.L. Vanpraet, M.J. Sharman and G. Van Ittersum (1985a) Satellite remote sensing of total herbaceous biomass production in the Senegalese Sahel. *Remote Sensing of Environment* **17**, 233–249.
41. Tucker, C.J., J.R.G. Townshend and T.E. Goff (1985b) African land-cover classification using satellite data. *Science* **227**, 369–375.
42. Tucker, C.J., I.Y. Fung, C.D. Keeling and R.H. Gammon (1986) Relationship between atmospheric CO*d22 variations and a satellite-derived vegetation index. *Nature* **319**, 195–199.

3

Growth analysis

C.L. Beadle

3.1 INTRODUCTION

If we wish to measure the bioproductivity of a natural ecosystem or agricultural crop, the component of ultimate interest is the net primary production or total yield of above- and below-ground parts. As pointed out in Chapter 1, it is usually necessary to restrict interest to above-ground parts. Indeed, in experiments with agricultural and forest crops it may be that the economic yield only is of importance (e.g. the grain part of cereal plants or the bole of a tree). This is commonly expressed as the Harvest index (HI):

$$\text{Harvest Index} = \frac{\text{Economic Yield}}{\text{Biological Yield}} \times 100 \tag{3.1}$$

where the biological yield has usually been considered as the total above-ground dry matter. Harvest Index is seen not only to vary between species but also within species, as a result of the growing conditions experienced between establishment and harvest (Table 3.1). As the growing environment has a considerable impact on the distribution of dry matter between the roots and the shoots (expressed as the root-to-shoot or R:S ratio), a better understanding of crop performance would be achieved if total dry matter were used as the denominator in Eqn. 3.1. Harvest Index has been manipulated in many crops, e.g. through the development of short-strawed cereals, to increase the economic yield. There is inevitably a physical and physiological limitation to this approach, however, as the plant has minimum requirements for support and photosynthetic capacity through its stem and leaves, respectively.

Harvest indices tell us little about the changes which are occurring during the growth of the crop. Photosynthetic gas-exchange analysis (Chapter 9) can be used to describe the responses of carbon fixation of single leaves and the canopy to changes in the environment. However, neither of these approaches provides information about the partitioning of dry matter into new leaf area (which can have a considerable influence on production), or about the competing sinks which determine the fate of the carbon fixed during photosynthesis, e.g. conversion into oil or carbohydrates.

In 1919, Blackman [1] defined production in terms of a compound interest law: ("If the rate of assimilation per unit area of leaf surface and the rate of respiration remain constant, and the size of the leaf system bears a relation to the dry weight of the whole plant, then the rate of production of new material, as measured by the dry weight, will follow the compound interest law.") Over the last 60 years this has been the basis for the development and refinement of the techniques of

Photosynthesis and Production in a Changing Environment: a field and laboratory manual
Edited by D.O. Hall, J.M.O. Scurlock, H.R. Bolhàr-Nordenkampf, R.C. Leegood and S.P. Long
Published in 1993 by Chapman & Hall, London.
ISBN 0 412 42900 4 (HB) and 0 412 42910 1 (PB).

Table 3.1. Yield ($t\,ha^{-1}$ dry matter) and harvest index (HI, %).

Species	Economic Yield	Biological Yield*	HI
Oryza sativa (rice)	4.4 (grain)	10.4	42
Sorghum bicolor (sorghum)			
Short season	4.8 (grain)	14.0	34
Long season	1.9 (grain)	26.6	7
Arachis hypogaea (groundnut)			
Irrigated	2.0 (pods)	4.3	47
Droughted	0.3 (pods)	1.0	17
Eucalyptus grandis (age 15 years)	131.0 (stemwood)	164.7	80

* Total above-ground dry matter at harvest.

plant-growth analysis. These techniques can answer some of the questions posed above and serve as an intermediate experimental tool between those used for gas-exchange and net primary production.

Increased concentrations of carbon dioxide in the atmosphere stimulate the growth of C_3 plants. Temperature and precipitation (through available water in soils), the weather variables most likely to be affected by climate change, also affect the storage of carbon by plants [2]. The impact of these variables and their interaction on dry matter production and its partitioning is readily studied by growth analysis. Studies of this type can help resolve the uncertainties which continue to surround the magnitude of these responses to climate change. They also provide the empirical information essential to the proper functioning of the process-based models (Chapter 14) which are being developed to predict concurrent changes to the carbon and nitrogen cycles.

3.2 BASIC PRINCIPLES

Two types of measurement only are needed for growth analysis.

(i) The plant weight. This is usually the oven-dry weight (kg) but it can be the organic matter or energy content.
(ii) The size of the assimilatory system. This is usually the leaf area (m^2) but it can be the leaf protein or chlorophyll content.

This primary data of growth analysis can be collected from individual plants or derived from whole canopies, though the destructive nature of the techniques requires the use of homogeneous sets of plants or plots. In its simplest form, plant-growth analysis requires little more than a balance, photosensitive paper and a calculator for quite detailed studies of quantitative aspects of dry-matter production. Accurate measurements are essential, however, to avoid unnecessary variation during sampling.

This quantitative description of growth is based upon several terms. As they have been developed over a number of years these terms and the symbols which represent them have evolved accordingly. The terminology used here is based upon that of Hunt [3], but modified in accordance with symbols used elsewhere in this book.

3.3 COMPONENTS OF CLASSICAL GROWTH ANALYSIS

3.3.1 Relative growth rate (R)

The basic component of growth analysis, which arose from the work of Blackman [1], is the relative growth rate (R) of the plant or crop. This is defined at any instant in time (t) as the increase of material present and is the only component of growth analysis which does not require knowledge of the size of the assimilatory system. Thus:

$$R = \frac{1}{W}\frac{dW}{dt} = \frac{d}{dt}(\ln W) \tag{3.2}$$

where W is the plant dry weight (kg DW). The relative growth rate is therefore the dry weight increase per unit of dry weight present and per unit of time, measured as the

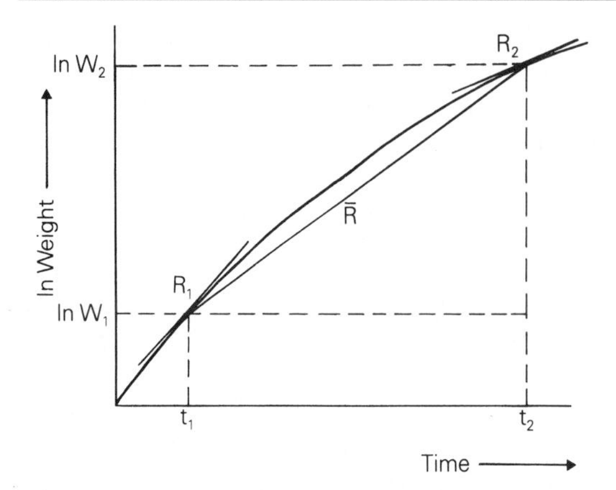

Fig. 3.1. The derivation of instantaneous and mean relative growth rates from a plot of ln W against time.

instantaneous slope of the curve of ln W against time (Fig. 3.1). In practice, the mean relative growth rate, $\bar{R}$, is measured over a discrete time interval, t_1 to t_2, which is normally no less than one day. As long as W varies continuously from t_1 to t_2, $\bar{R}$ is defined as:

$$\bar{R} = \frac{1}{t_2 - t_1} \int_{W_1}^{W_2} d(\ln W) = \frac{\ln w_2 - \ln w_1}{t_2 - t_1} \tag{3.3}$$

The relationship between R and $\bar{R}$ is shown in Fig. 3.1. It should be noted that in the special case of exponential growth, i.e. the form $W = W_0 e^{Rt}$, R is constant and equal to $\bar{R}$, where W_0 is the initial weight (when $t = 0$).

The relative growth rate serves as a fundamental measure of dry-matter production and can be used to compare the performance of species or the effects of treatments under strictly defined conditions.

Measured values of R ($kg\,kg^{-1}\,d^{-1}$) vary from 0.02 to 0.5 (from inherently slow-growing or stressed plants to fast-growing C_4 weeds), though even higher rates are obtainable in algal culture (see Chapter 21). As the measurement of $\bar{R}$ includes the initial weight of the organ, as the plant grows, the increment in dry weight becomes a decreasing proportion of the total weight. As much of this dry weight is non-productive tissue (an extreme case would be a mature tree), application of Equations 3.2 and 3.3 will simply lead to $\bar{R}$ decreasing with time and being less sensitive to applied treatments. It is then necessary to estimate $\bar{R}$ in a way which is independent of the accumulated size of the plants [4]. In experiments with crops or closely-spaced potted plants, canopy closure may also mark the useful end of the direct application of Equations 3.2 and 3.3. Thus self-shading may lead to changes in the light environment which obscure the effects of the treatments being investigated.

Relative growth rate *per se* tells us little about the causal factors which determine the performance of that plant or crop. Observed differences in growth rate could be due to assimilation or leaf area. These two potential determinants of growth are encompassed by the concepts of Unit Leaf Rate and Leaf Area Ratio.

3.3.2 Unit leaf rate (E)

The unit leaf rate, E, of a plant or crop at any instant in time t is defined as the increase of plant material (kg DW) per unit of assimilatory material s (m^2) per unit of time:

$$E = \frac{1}{s}\frac{dW}{dt} \tag{3.4}$$

The term net assimilation rate is often used interchangeably with unit leaf rate but the latter is preferred [5]. They measure the net gain in dry weight of the plant per unit leaf area per unit time, usually per day ($kg\,m^{-2}\,d^{-1}$), and differ from photosynthetic rate which measures the net gain of carbon during the light period only.

The mean unit leaf rate, $\bar{E}$, between t_1 and t_2 is given by:

$$\bar{E} = \frac{1}{t_2 - t_1} \int_{t_1}^{t_2} \frac{1}{s}\frac{dW}{dt} dt \tag{3.5}$$

Not unexpectedly, values of $\bar{E}$ are lower for C_3 than for C_4 species, which have higher

Table 3.2. Values of $\bar{E}$ ($g\,m^{-2}\,day^{-1}$) measured during periods of rapid growth.

Species	Type	$\bar{E}$
Spartina anglica (cord-grass)	C_4	18
Zea mays (maize)	C_4	16
Lolium perenne (ryegrass)	C_3	13
Phleum pratense (timothy)	C_3	9
Trifolium repens (white clover)	C_3	12
Beta vulgaris (sugar beet)	C_3	10
Pinus sylvestris (Scots pine)	C_3	5

photosynthetic rates and higher short-term growth rates than C_3 species (Table 3.2). $\bar{E}$ will also vary with the growing environment (Fig. 3.3) and age.

As Radford [6] has pointed out, this relationship cannot be integrated as above unless:

(i) the relationship between s and W is known,
(ii) the relationships between s and t, and W and t are known.

If one assumes, however, that:

(iii) over the period t_1 to t_2, s and W are linearly related, and that:
(iv) s and W are not discontinuous functions of time,

Equation 3.5 can be integrated such that:

$$\bar{E} = \frac{(W_2 - W_1)(\ln s_2 - \ln s_1)}{(s_2 - s_1)(t_2 - t_1)} \qquad (3.6)$$

Radford [6] lists further formulae for the calculation of E where the relationship between W and s is known but is not linear.

3.3.3 Leaf area ratio (F)

The leaf area ratio, F, of a plant or crop at any instant in time (t) is the ratio of the assimilatory material (m^2) per unit of plant material (kg DW) present:

$$F = \frac{s}{W} \qquad (3.7)$$

Using the same assumptions (iii) and (iv) above, the mean leaf area ratio, $\bar{F}$, is given by:

$$\bar{F} = \frac{(s_2 - s_1)(\ln W_2 - \ln W_1)}{(W_2 - W_1)(\ln s_2 - \ln s_1)} \qquad (3.8)$$

From the combination of Equations 3.4 and 3.7 it can be seen that:

$$R = E \times F = \frac{1}{W}\frac{dW}{dt} = \frac{1}{s}\frac{dW}{dt}\frac{s}{W}$$

The relative growth rate therefore consists of two components which measure the efficiency of the plant or crop as a producer of dry weight (E) and a producer of leaf area (F). Although Equation 3.9 is true at any instant, the same relationship with respect to the mean values does not usually hold (i.e. $\bar{R} \neq \bar{E} \times \bar{F}$). Thus, except under very special circumstances, the relationships between s and t, W and t, and s/W and t are not linear and rarely take the same form [6].

3.3.4 Specific leaf area and leaf weight ratio

Leaf area ratio can be redefined as:

$$F = \frac{W_1}{W} \times \frac{s}{W_1} \qquad (3.10)$$

where W_1 is the dry weight of the leaves. The two components W_1/W and s/W_1 are called the leaf weight ratio (LWR) and the specific leaf area (SLA), respectively.

LWR ($kg\,kg^{-1}$) measures the leafiness of the plant on the basis of its total dry weight. It also defines the partitioning of dry weight to leaves, a parameter which determines the potential capacity of the plant to support existing dry weight and to further increase its dry weight through photosynthesis. LWR is high during canopy development but then declines as an increasing proportion of dry matter is accumulated into storage tissues.

SLA ($m^2\,kg^{-1}$) measures the leafiness of a plant on a dry weight basis. This may decrease systematically with time as the leaves mature, but increase systematically with depth in the canopy as the light available for leaf development and light interception decreases. There is also evidence that for a given light environment, species with leaves of higher SLA (i.e. less carbon invested per unit of intercepting area) will have a higher relative growth rate (R). Values of SLA vary by an order of magnitude. The lowest may be $2\,m^2\,kg^{-1}$ (thick needles of conifers), the highest more than $50\,m^2\,kg^{-1}$. Note that leaf area should be measured as single surface or projected area (see below).

The reciprocal of SLA, the specific leaf weight (SLW, $kg\,m^{-2}$), although not used routinely in growth analysis, is a measure of the weight of leaf material per unit of leaf area. There is no clear role for SLW as a plant attribute, though it tends to be positively correlated with leaf thickness (and in some instances rates of photosynthesis per unit leaf area), often through the total number of mesophyll cells per unit leaf depth.

By way of illustration, the above concepts have been brought together in a comparison of the growth of seedlings of *Pinus sylvestris* (Scots pine), a slow-growing conifer, and *Helianthus annuus* (sunflower), a fast-growing C_3 dicotyledon (Table 3.3). An inspection of the different variables shows that the order of magnitude difference in R is partially a result of a two-fold difference in E, but mainly due to a four-fold difference in F. As there was little difference in LWR between the species (it was in fact greater in the pine than in sunflower) the most important determinant of growth was SLA. A more comprehensive example is given in section 3.5.2.

3.3.5 Leaf area index (L)

If we specifically wish to consider the productivity of crops or natural ecosystems, it is convenient to express their performance per unit land area. The leaf area ratio is therefore inappropriate and a second term is used. This is the leaf area per unit land area, or leaf area index (abbreviated to LAI or simply L). It can expressed as:

$$L = \frac{s}{G} \tag{3.11}$$

where s is the functional (green) leaf area of the crop canopy standing on ground area G. As both s and G are normally measured as areas (m^2), L is dimensionless.

s is measured as the projected area (equivalent to s_p in Chapter 6) after placing the leaf on a horizontal surface. Leaf area could also be measured as the total surface area. This will be equal to 2s for flat leaves and more than 2s for needle-shaped and succulent leaves, and photosynthetic stems. Care should be taken when making comparisons between experiments that L has been measured on the same basis (see Chapter 6 for methods of measuring s_p and L).

As explained in Chapter 6, L is the major factor determining the amount of light intercepted by the plant canopy. L varies with species and, under optimum conditions for growth, its value for a closed canopy is related to the ability of the lower leaves of the crop to intercept sufficient light to maintain a

Table 3.3. A description of seedling growth using classical growth analysis (adapted from Jarvis and Jarvis, [7])

Species	R ($kg\,kg^{-1}\,d^{-1}$)	E ($g\,m^{-2}\,d^{-1}$)	F ($m^2\,kg^{-1}$)	SLA ($m^2\,kg^{-1}$)	LWR ($kg\,kg^{-1}$)
Pinus sylvestris (Scots pine)	0.019	4.6	4.4	6.75	0.643
Helianthus annuus (sunflower)	0.136	8.5	17.7	32.75	0.542

positive carbon balance (otherwise the leaves senesce and are abscised). Some coniferous canopies are able to support $L > 15$. For deciduous forest, maxima between 6 and 8 are observed, while for annual crops maxima are between 2 and 4. All types of crops react to stress in the environment by producing canopies with lower L.

3.3.6 Crop growth rate (C)

The leaf area index can be used to calculate the instantaneous crop growth rate (at any time t) which serves as a simple index of agricultural productivity. This is defined as:

$$C = E \times L = \frac{1}{G}\frac{dW}{dt} = \frac{1}{s}\frac{dW}{dt}\frac{s}{G} \qquad (3.12)$$

and is expressed in terms of the weight per unit area and time ($kg\,m^{-2}\,s^{-1}$). C is therefore conceptually similar to R (= E × F for individual plants). As with R, C is not equal to the product of $\bar{E} \times \bar{L}$, as these components rarely take the same form. However, C does allow some approximation of the relative importance of E (the efficiency of leaves as producers of dry matter) and L (the leafiness of the crop) as determinants of the rate of growth. In general, it is concluded that C is more sensitive to L than to E [8,9].

3.3.7 Leaf area duration (*D*) and biomass duration (*Z*)

The final concepts of growth analysis which will be considered here are quantities which are normally estimated as areas beneath plots of primary data and time. The first, leaf area duration (*D*), measures the persistence of the assimilatory surface. *D* is normally calculated from the relationship between L and time (Fig. 3.2), though a similar measurement can be estimated from individual plants (*viz.* through s and t). By definition there is no instantaneous value. *D* can be determined by integral calculus in a manner similar to the determination of $\bar{E}$ and $\bar{F}$ (above). An alternative method, although not mathematically correct, is to measure the area of a trapezium under that part of the curve which is of interest. Thus:

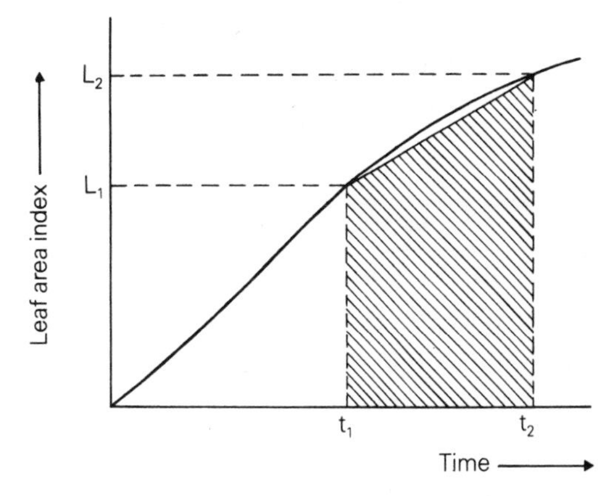

Fig. 3.2. Leaf area duration (shaded area) determined from a plot of Leaf area index against time.

$$D = \frac{(L_1 + L_2)\cdot(t_2 - t_1)}{2} \qquad (3.13)$$

As this is the product of a dimensionless unit and time, the units of *D* are time (usually expressed in days). The duration of the assimilatory area is considered to be of similar importance in the final determination of yield as L itself [10]. An approximate estimate of the total yield ($kg\,m^{-2}$) can now be defined as the product of *D* and $\bar{E}$, the mean unit leaf rate. Thus,

$$\text{Yield} = D \times \bar{E} \qquad (3.14)$$

The estimate is approximate, as in practice $\bar{E}$ cannot be determined accurately (see above). The relationship also assumes that L is functioning uniformly throughout the growing season. However, Equation 3.14 suggests that maximum rates of dry-matter production will occur when maximum L coincides with maximum $\bar{E}$, a conjunction which is not always compatible with the phenological behaviour of the crop or agronomic practice.

The second concept, biomass duration (BMD or *Z*) is analagous to *D*, but defined as the area under the weight vs. time curve (rather than the area vs. time curve) [11]. *Z* is also related to total yield in an equation similar to Eqn. 3.14, where $\bar{R}$ replaces $\bar{E}$ and Z has units of weight × time.

Direct integration of the area results in equations of the form (for Z):

$$Z = \frac{(W_2 - W_1)(t_2 - t_1)}{\ln W_2 - \ln W_1} \qquad (3.15)$$

where s replaces W for D. It should be noted that for individual plants the ratio of D to Z is the mean value of $\bar{F}$ for that period.

3.4 FUNCTIONAL VERSUS CLASSICAL GROWTH ANALYSIS

Given the support of computers and the appropriate software, classical growth analysis can be complemented by a second approach called *functional growth analysis* [3,6]. The quantitative descriptions used to define growth remain the same as those for classical growth analysis but are estimated from functions fitted to harvest data. While the two approaches share many practical considerations, the functional approach avoids the limitations of classical growth analysis discussed above. This is realized through more frequent and smaller harvests (1–3 days) which provide the larger data sets necessary to describe accurately and adequately the relationships between W and t, and s and t [12]. The grouping and pairing of plants, a necessary feature of classical growth analysis (see below) can be avoided. Experiments should therefore be designed to suit one or other of the two approaches.

The major advantage of the functional as opposed to the classical approach is that information for the whole period of interest is contained in two equations. Further functions can be developed to calculate the instantaneous values of R, E and F over the same period.

3.4.1 Fitted functions

Many functions have been used to describe the above relationships [3,13]. The simplest are polynomial expressions of the appropriate order to obtain the best fit using least-squares procedures [14]. For example, the second-order derivative would take the form:

$$\log_e W = a + b_1 t - b_2 t^2 \qquad (3.16)$$

where a, b_1 and b_2 are the coefficients. In the experiment described in Section 3.5.2 (Fig. 3.3), quadratic expressions of the above form were fitted to the primary data to describe the progression of both weight and leaf area with time. The best fit for a given data set should, of course, be determined by established statistical tests [15].

Polynomial functions are entirely empirical and the parameters which are derived are generally suitable for the observed data only. Asymptotic functions are also used in growth analysis [13], and to some extent these overcome this limitation of polynomial expressions. One of the most popular equations to be applied to the primary data is the Richards function, which takes the form:

$$W = a(1 \pm e^{(b-kt)})^{-1/n} \qquad (3.17)$$

though other forms can be derived to suit the particular requirements of the experiment [16]. The parameters of the equation, a, b, k and n, are used to provide a biologically more meaningful fit to the data.

3.5 EXPERIMENTAL INVESTIGATIONS

3.5.1 Planning and procedure

Good forward planning is essential to the success of growth analysis. As the primary objectives during the experimental phase are the measurements of growth as 'dry weight' and the size of the assimilatory system as 'leaf area' (or some surrogate) over a series of sequential harvests, their accuracy should not be obscured by unwanted variation. This is minimized, particularly in classical growth analysis, by the selection of statistically homogeneous sets of plants, pots or plots at the start of the experiment.

Plants are paired or grouped in such a way that each set to be harvested contains individuals with similar morphology and growth rate. In field crops, experimental plots of similar uniformity should be selected and similar criteria should be applied to natural vegetation if possible.

The primary difference between the classical and the functional approach is that the former requires larger harvests at less frequent intervals, with greater replication compared to the latter. Hence, with the functional approach, there is less need to group plants at the start of the experiments and to have several replicates at each harvest.

The statistical procedures applied to both approaches are those which are suitable for any experiment where statistically homogeneous sets of data are to be collected [17]. Sampling units should be randomised using a random number table or equivalent, and the experimental design will be dictated by the demands placed upon the experiment. The number of replicates at each harvest and the harvest material should be adjusted to the growth rate such that any effects of the treatments are not obscured by sampling variation. In practice, it is recommended that this interval should be timed according to changes in s, and that $s_2/s_1 = 2$ for satisfactory measurements of E [11].

For most laboratories, oven-drying may be the only option for obtaining the dry weight. As the objective is to conserve all the materials present in the fresh sample, several precautions should be taken. Slow drying at low temperature causes loss in dry weight to respiratory processes: fast drying at high temperature causes volatile compounds to be evaporated. To minimise these losses, one hour at 100°C is recommended, followed by a period at 70°C to constant weight (up to 48 h).

If further analysis of plant material is required, e.g. for mineral ions or calorific value, investment in freeze-drying equipment should be considered. Material harvested in the field should be stored and transported at low temperature to avoid any loss of material before drying. Plant material should be dried in an appropriate container, e.g. a paper bag or aluminium foil. As paper is hygroscopic, samples should be cooled under desiccation prior to weighing at room temperature.

Methods for measuring leaf area are summarised in Chapter 6.

3.5.2 Results

Temperature is a major determinant of crop growth and a significant factor in climate change models. Its impact on dry matter production and leaf area development can be studied by growth analysis. For example, the functional approach can be used to present data in the form of temperature response curves. This approach has been used to illustrate the comparative growth of C_4 and C_3 grasses (Fig. 3.3).

In this experiment, there were significant differences between R and R_1 (for total and leaf dry weight respectively) between *Lolium perenne*, the C_3 grass, and *Spartina anglica*, the C_4 grass, at all temperatures. R of both species was significantly reduced by decreasing temperatures. In contrast, E of *L. perenne* varied little with temperature whereas E of *S. anglica* was strongly dependent on temperature, such that low E could not be considered a factor contributing to the consistently lower R of the C_4 compared to the C_3 grass, except at 10°C. Further interpretation of the results in terms of F, SLA and LWR showed that the major difference between the species was the consistently lower L of *S. anglica* at all temperatures. This was primarily attributed to the lower SLA of *S. anglica*, though there was also a significant change in LWR of the C_4 grass as the partitioning of dry matter to leaves declined, while that to roots increased, as the temperature decreased.

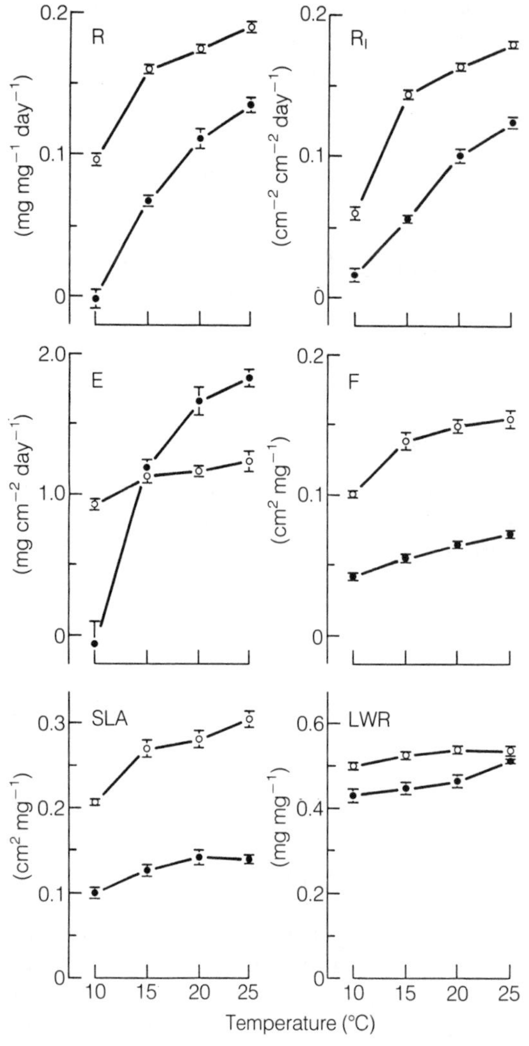

Fig. 3.3. The response of *Lolium perenne* (●) and *Spartina anglica* (○) to temperature in terms of growth analysis [18]. All points are from fitted quadratic functions describing the instantaneous relationship between the natural logarithms of dry weight (W) and leaf area (s) with time (t, Eqn. 3.16). R_l is the relative growth rate of the leaf dry weight.

This experiment was designed to study factors limiting the growth of a C_4 grass at low temperature. This approach is also suitable for the study of factors limiting growth (and species distribution) of plants in the higher temperatures being predicted for future climates, their response to other environmental variables for studies of varietal differences of crop plants or agronomic practice (e.g. mineral nutrition, spacing and irrigation), and to distinguish factors which result in plants having either fast or slow relative growth rates [19].

3.5.3 Limitations

The techniques of growth analysis are most suited to following the effects of long-lasting treatments. The limitations in accuracy which the technique offers preclude its use in the investigation of subtle changes in climate which occur in the field, as the variance attributable to sampling will often exceed that of the factor investigated. $\bar{E}$ and $\bar{R}$ may also change systematically in such a way as to obscure correlations with light and other environmental factors, except during the very early growth phase [20].

3.6 FURTHER DEVELOPMENTS

Classical and functional growth analyses are dependent on the destructive sampling of plant material. Non-destructive approaches to growth analysis can also be taken (see Chapter 2 for the use of canopy reflectance ratio 660:730 nm). For example, the linear dimensions of leaves or branches, or the diameter of stems or branches, can be measured and related to dry matter increment through regressions (allometric relationships) derived from calibration harvests. Some destructive harvesting is therefore inevitable even with this indirect approach, if the measurements are to be analysed through the quantitative descriptions of growth analysis. Measurement of root biomass and partitioning of dry matter are essentially precluded by non-destructive methods of measuring dry matter.

The growth of plants can also be described in demographic or physiological terms [21]. The former takes a modular approach to growth in terms of classes or populations of plant parts. Physiological growth analysis is the functional approach which results in the process-based models described in Chapter 14 and elsewhere [22].

The concepts of growth analysis have also been extended to encompass the nutrient productivity of plants [23]. This concept hypothesises that R is linearly related to the internal concentration of a limiting nutrient, say nitrogen or phosphorus, in the presence of non-limiting quantities of other nutrients. The control and supply of nutrients to test this hypothesis is complex but the technique has potential for the development of fertiliser regimes which add nutrients in proportion to that demanded by the current growth rate of the plant or crop [24].

REFERENCES

1. Blackman, V.H. (1919) The compound interest law and plant growth. *Annals of Botany* **33**, 353–360.
2. Goudriaan, J. and M.H. Unsworth (1990) Implications of increasing carbon dioxide and climate change for agricultural productivity and water resources. In: *Impact of Carbon Dioxide, Trace Gases and Climate Change on Agriculture* (Eds. G.H. Heichel, C.W. Stuber and D.K. Wissel. American Society of Agronomy Special Publication No. 53). pp. 111–130.
3. Hunt, R. (1978) *Plant Growth Analysis*. Studies in Biology No. 96. Edward Arnold, London. pp. 67.
4. Brand, D.G., D.F. Westman and P. Rehsher (1987) Growth analysis of perennial plants: the relative production rate and its yield components. *Annals of Botany* **59**, 45–53.
5. Evans, G.C. (1972) *The Quantitative Analysis of Plant Growth*. Blackwell Scientific, Oxford. pp. 734.
6. Radford, D.J. (1967) Growth analysis formulae – their use and abuse. *Crop Science* **7**, 171–175.
7. Jarvis, P.G. and M.S. Jarvis (1964) Growth rates of woody plants. *Physiol. Plant.* **17**, 654–666.
8. Watson, D.J. (1952) The physiological basis of variation in yield. *Advances in Agronomy* **4**, 101–145.
9. Potter, J.R. and J.W. Jones (1977) Leaf area partitioning as an important factor in growth. *Plant Physiol.* **59**, 10–14.
10. Watson, D.J. (1956) Leaf growth in relation to crop yield. In: *The Growth of Leaves* (Ed. F.L. Milthorpe) pp. 178–191. Butterworth, U.K.
11. Květ, J., J.P. Ondok, J. Nečas and P.G. Jarvis (1971) Methods of growth analysis. In: *Plant Photosynthetic Production* (Eds. Z. Šesták, J. Čatský and P.G. Jarvis) Dr. W. Junk, The Hague, Netherlands. pp. 343–39.
12. Hughes, A.P. and P.R. Freeman (1967) Growth analysis using frequent small harvests. *J. Applied Ecology* **4**, 553–560.
13. Hunt, R. (1982) *Plant Growth Curves. The Functional Approach to Growth Analysis.* Edward Arnold, London. 248 pp.
14. Nicholls, A.O. and D.M. Calder (1973) Comments on the use of regression analysis for the study of plant growth. *New Phytol.* **72**, 571–581.
15. Hunt, R., and I.T. Parsons (1974) A computer program for deriving growth functions in plant growth analysis. *J. Applied Ecology* **11**, 297–307.
16. Causton, D.R. and J.C. Venus (1981) *The Biometry of Plant Growth.* Edward Arnold, London.
17. Snedecor, G.W. and W.G. Cochrane (1972) *Statistical Methods Applied to Experiments in Agriculture and Biology.* Iowa State College, USA. pp. 593.
18. Dunn, R., S.M. Thomas, A.J. Keys and S.P. Long (1987) A comparison of the growth of the C_4 grass *Spartina anglica* and the C_3 grass *Lolium perenne* at different temperatures. *J. Exp. Bot.* **38**, 433–441.
19. Poorter, H. and Remkes, C. (1990) Leaf area ratio and net assimilation rate of 24 wild species differing in relative growth rate. *Oecologia* **83**, 553–559.
20. Monteith, J.L. (1981) Does light limit crop production? In: *Physiological processes limiting plant productivity* (Ed. C.B. Johnson) pp. 23–38. Butterworths, U.K.
21. Chiariello, N.R., H.A. Mooney and K. Williams (1989) Growth, carbon allocation and cost of plant tissues. In: *Plant Plant Physiological Ecology. Field methods and instrumentation* (Eds. R.W. Pearcy, J.R. Ehleringer, H.A. Mooney and P.W. Rundel) Chapman and Hall, London and New York. pp. 327–365.
22. Charles-Edwards, D.A. (1982) *Physiological*

Determinants of Crop Growth. Academic Press, London. pp. 161.
23. Ågren, G.I. (1985) Theory for growth of plants derived from the nitrogen productivity concept. *Physiol. Plant.* **64**, 17–28.
24. Ingestad, T. (1988) A fertilisation model based on the concepts of nutrient flux density and nutrient productivity. *Scand. J. Forest Research* **3**, 157–173.

4
Plant microclimate

M.B. Jones

4.1 GENERAL INTRODUCTION

Microclimate is the complex of environmental variables, including temperature, radiation, humidity and wind, to which vegetation is exposed. It is the climate near the surface of the Earth and it is different from the weather forecasters' macroclimate or local climate because of the influence of the Earth's surface and, most importantly, the presence of vegetation. Plants are 'coupled' to their microclimate because a change in one brings about a change in the other. This results from an exchange of force, momentum, energy or mass between the plant and its environment [1]. Two important types of coupling are (i) radiative coupling, where energy is transferred through electromagnetic vibration; and (ii) diffusive coupling, where heat, water vapour and CO_2 are exchanged across the boundary layer of the plant.

The radiant energy flux incident on a plant is coupled to the temperature of the plant by its absorptivity; the relative effectiveness of a leaf in absorbing radiant-energy. If leaf absorptivity is high, which is normally the case in the visible part of the spectrum, then the leaf temperature is tightly coupled to incident radiation, and *vice versa*. Diffusive coupling across the boundary layer can be viewed as an analogue of an electrical circuit where energy in the form of a charge moves from a high to a low 'potential' (measured as a voltage) at a rate (the current) which is inversely proportional to the resistance. Diffusive coupling at the leaf surface is illustrated in Fig. 4.1. It should be noted that both radiative and diffusive components of the microclimate are themselves coupled so that, for instance, energy from electromagnetic radiation can be consumed in evaporating water in transpiration. Consequently, radiation, air temperature, wind and humidity all interact simultaneously with the plant, so the effect of one cannot be known without specifying the state of the others. Together they determine the energy balance of the leaf.

The productivity of plants is ultimately dependent upon the influence of the microclimate on plant processes such as photosynthesis, respiration, transpiration and translocation. In order to understand how plant processes respond to the microclimate we need to be able to measure the various components of the microclimate in the natural environment. In recent years a whole range of micrometeorological instruments has been designed for this purpose [2,3,4], and in the following sections some of these will be described along with the principles upon which the measurements are based. The use of these instruments, and the analysis

Photosynthesis and Production in a Changing Environment: a field and laboratory manual
Edited by D.O. Hall, J.M.O. Scurlock, H.R. Bolhàr-Nordenkampf, R.C. Leegood and S.P. Long
Published in 1993 by Chapman & Hall, London.
ISBN 0 412 42900 4 (HB) and 0 412 42910 1 (PB).

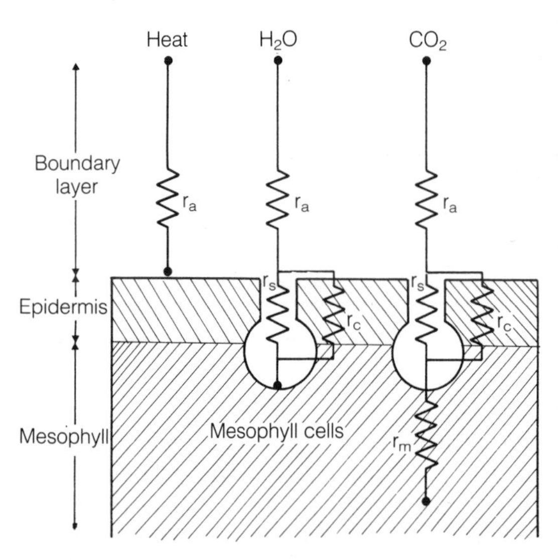

Fig. 4.1. Diagrammatic representation of diffusive coupling at the leaf surface, in the form of an electrical analogue, showing resistances to gas and heat exchange at the surface of a single leaf. r_a is the boundary layer, r_s the stomatal, r_c the cuticular and r_m the mesophyll or residual resistance. (Note that resistance is the reciprocal of conductance).

and interpretation of data collected with them, requires a basic understanding of environmental physics. Recommended textbooks on this subject include those by Monteith and Unsworth [5], Campbell [6], Woodward and Sheehy [7] and Jones [8]. Details of the characteristics of many of these instruments are listed in Appendix A of this book, and many of them are discussed more fully in Marshall and Woodward [9].

4.2 RADIATION: SOLAR AND LONG WAVE

4.2.1 Introduction

The ultimate source of energy for photosynthesis and bioproductivity is solar energy. Plants intercept solar energy for photosynthesis but normally less than 5% is used in this process; the rest of this energy heats the plant and surrounding organisms, so that solar energy also determines the temperature, through the energy balance, at which physiological processes are functioning. Apart from photosynthesis, solar radiation also influences the plant's growth and development in what are referred to as photomorphogenic, phototropic and photoperiodic responses. These normally require only very small amounts of energy to bring about a response, and different discrete parts of the radiation spectrum such as blue (450 nm), red (660 nm) and far-red (730 nm) are involved.

4.2.2 Radiation terms and units

Much confusion has existed regarding the measurement of radiation. This is partly because the units and definitions used for light measurements have been subjected to continued modifications over the years. The situation today is much more stable and Table 4.1 lists most of the widely accepted radiation terms and units.

About 98% of the radiation emitted by the sun is in the waveband from 0.3 to 3.0 μm. The energy spectrum of this radiation before it reaches the Earth's atmosphere peaks at 0.48 μm, which is consistent with a radiator or emitter with a temperature of 6000 K (Fig. 4.2). The flux of radiation (Φ) follows the Stefan–Boltzmann Law, being proportional to the fourth power of the absolute temperature (T) of the object:

$$\Phi = \varepsilon\sigma T^4$$

where ε is the emissivity (<1), σ is the Stefan–Boltzmann constant (5.6×10^{-8} W m^{-2} K^{-4}) and T is in Kelvin. The units for radiant fluxes are the units of power (W) where the term *irradiance* (I) refers to the energy flux density incident on unit surface area (W m^{-2}). Irradiance is the correct radiometric term for what is commonly and mistakenly called 'light intensity'. Strictly speaking, 'light' is that part of radiation which is visible to humans, so it is not a very appropriate term to use in plant research because

Table 4.1. Some radiation terms

Term	Symbol	Meaning	Units
Radiation or radiant energy	–	Energy transferred through space in the form of electromagnetic waves or quanta	joule (J)
Radiant flux	–	The amount of radiant energy received, emitted or transmitted per unit time	$J\,s^{-1}$ or watts (W)
Radiant flux density	Φ	The radiant flux through unit area of a plane surface	$W\,m^{-2}$
Irradiance	I	The energy flux incident on unit area of a plane surface	$W\,m^{-2}$
Photon	–	A quantum of light A mole (mol) is 6.022×10^{23} quanta or photons	–
Quantum flux density	**Q**	The number of quanta incident on unit area of a plane surface	$\mu mol\,m^{-2}\,s^{-1}$
Photosynthetically active radiation	PAR	Radiation within the band 400–700 nm	$\mu mol\,m^{-2}\,s^{-1}$ or $W\,m^{-2}$*
Short wave radiation	Φ_s	Radiation in the waveband 0.4 to 3.0 μm	$W\,m^{-2}$
Long wave radiation	Φ_l	Radiation wavelengths of 3.0 μm or greater	$W\,m^{-2}$
Net radiation	Φ_{net}	The difference between the downward and upward fluxes of total radiation	$W\,m^{-2}$

* See Appendix C for conversion factors

the action spectra of the human eye and plants are very different. The component of solar radiation used in photosynthesis falls between 400 and 700 nm and is referred to as Photosynthetically Active Radiation (PAR).

Radiant energy can be described either as waves or as discrete packets of energy called *photons*. The amount of energy contained in a photon is termed a *quantum* of energy, and is proportional to the frequency of the photon. However, in plant ecophysiology quantum and photon are often used interchangeably to mean the number of packets of light energy. When dealing with photochemical processes such as photosynthesis, the *number of photons* incident in unit time is more relevant than the energy content of the radiation. This is known as the quantum (or photon) flux area density (**Q**) and is measured in units of $\mu mol\ m^{-2}\ s^{-1}$ where a mole is Avogadro's number (6.022×10^{23}) of quanta or photons. When measuring rates of photosynthesis, it is more appropriate to express the radiation incident on the plane of the leaf in terms of photosynthetically active photon flux rather than radiant energy flux. This is because the photosynthetic rate is largely independent of the energy content of photons between 400 and 700 nm, and photons within this waveband are almost equally effective in driving the light reactions of photosynthesis [10]. However, PAR can be expressed in terms of either energy or photons.

Conversion of radiometric units to photon flux is complicated, since it will differ for each source of radiant energy. The spectral distribution of the radiant output from the source must be known. Approximate conversion factors for various light sources are given in Appendix C.

At the Earth's surface solar radiation can be divided into two components based on whether the radiation comes directly from the sun (direct) or whether it is scattered or reflected by the atmosphere and clouds (diffuse). Diffuse radiation has a different spectral composition from direct radiation because shorter wavelengths are scattered by air molecules more than long ones, giving the

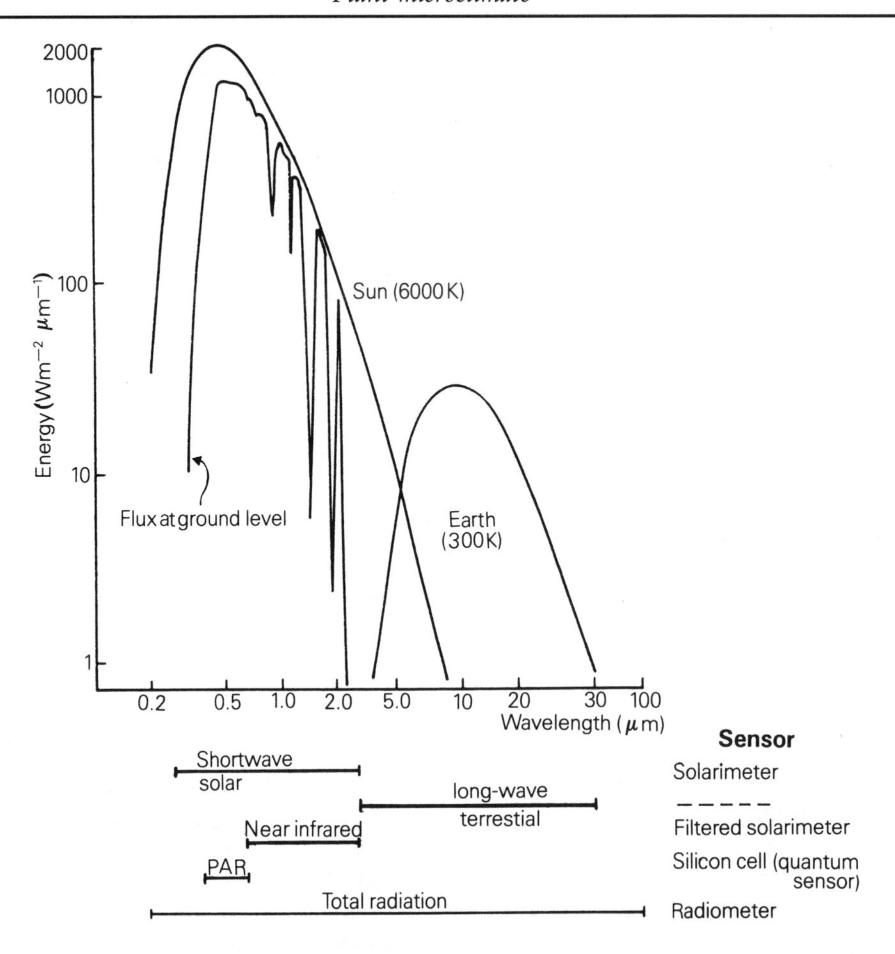

Fig. 4.2. The spectral energy distribution of the solar flux outside the Earth's atmosphere; the solar flux at ground level after attenuation by gases in the atmosphere; and the flux of radiation emitted by the Earth's surface (terrestrial flux). Depicted at the bottom of the figure are the typical ranges of instruments used to measure components of the solar and terrestrial fluxes; ----, no sensor available (adapted from Grace [12]).

blue colour to clear skies. However, larger particles such as dust and water droplets scatter all wavelengths equally, so the sky appears white when cloud-covered. The amount of diffuse radiation varies with sun angle and cloud cover, but even on clear days it contributes 10–30% of total solar irradiance. The proportion of PAR in total (direct + diffuse) radiation is about 50%; this varies little diurnally or seasonally.

Plants and any other surface on the Earth also emit radiation due to heating by the sun. According to Wien's displacement law the wavelength (μm) at which the maximum amount of radiation is emitted (λ_m) decreases as the temperature of the body increases:

$$\lambda_m = \frac{2897}{T}$$

where T is the surface temperature (K). Consequently, bodies on the Earth's surface emit long-wave radiation with a peak at approximately 9.7 μm (Fig. 4.2). There are therefore continuous fluxes of radiation from the sun during the day and between the atmosphere, plants and their surroundings at

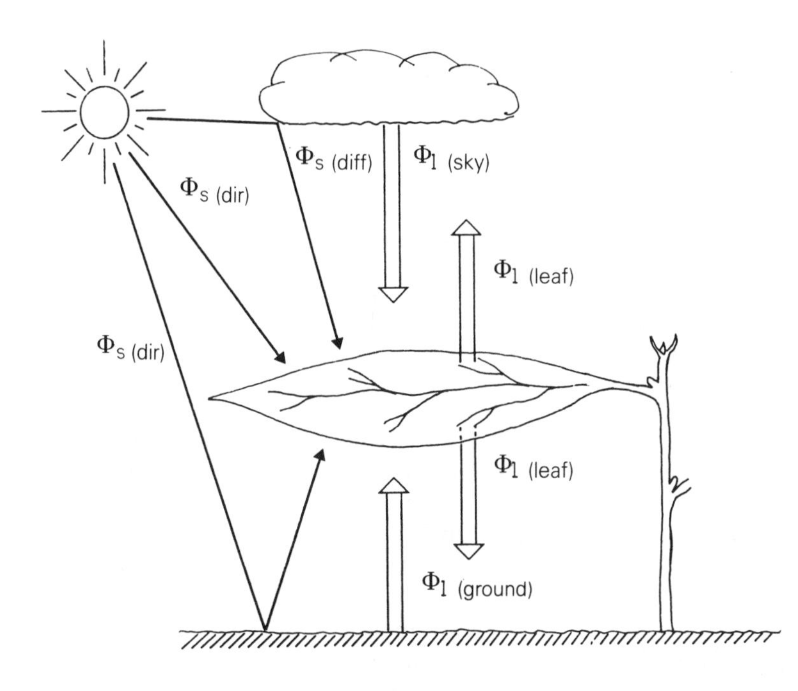

Fig. 4.3. An illustration of the short-wave (Φ_s) and long-wave (Φ_l) radiant energy fluxes between a leaf and its surroundings. (dir refers to direct and diff to diffuse radiation).

all times (Fig. 4.3). Radiation incident on a leaf or plant canopy can be absorbed, transmitted or reflected. In the PAR region of the spectrum the leaf absorbs 90% of the incident radiation whilst in the near infra-red region (0.7–3.0 μm) it transmits most of the radiation. The effect of this is to reduce the heat load from wavelengths which are not used in photosynthesis. However, in the *far* infra-red, leaves are good absorbers; thus, because good absorbers are also good emitters of radiation, they are able to dissipate excess heat very efficiently in the long-wave region of the spectrum.

4.2.3 Radiation measurements

Radiation sensors are normally either thermoelectric or photoelectric. Most instruments used for measuring solar and long-wave radiation are of the thermoelectric type and consist of different forms of thermopile arrangement. A thermopile consists of a series of alternate junctions between two dissimilar metals, e.g. copper and constantan (Fig. 4.4a). When a temperature difference exists between two sets of thermocouples a voltage is generated which is proportional to the temperature difference. When measuring solar radiation the temperature difference is created by embedding one set of junctions in a metal clamp protected from incident radiation and the other in a surface exposed to radiation, or by painting the hot and cold junctions black and white, respectively, and subjecting them both to the same radiant energy flux. The surface of the sensor is normally protected from wind and rain by glass domes whose transmittance restricts spectral sensitivity to the 0.3–3.0 μm region. An example is the *Kipp* solarimeter (or pyranometer) using a Moll thermopile, which is the standard instrument in many countries for measuring total (direct + diffuse) and diffuse (using a shade ring) solar radiation. Details of other solarimeters can be found in Monteith [3], Szeicz [2] and Fritschen and Gay [4]. When solarimeters are used with special filters (e.g. *Kodak "Wratten" 88*), they exclude visible wavelengths, so the energy in the visible region between 0.3 and 0.75 μm can be determined by difference.

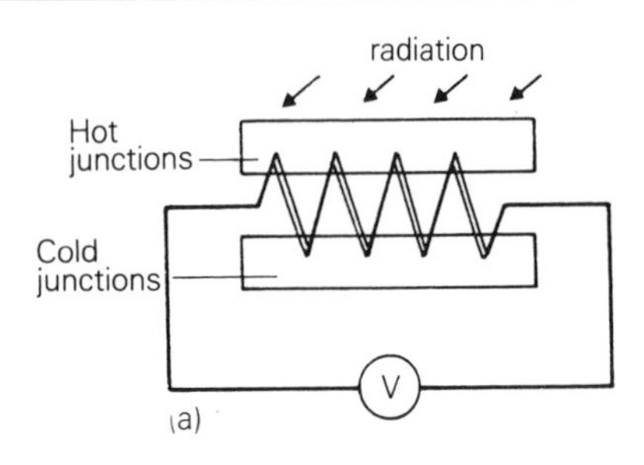

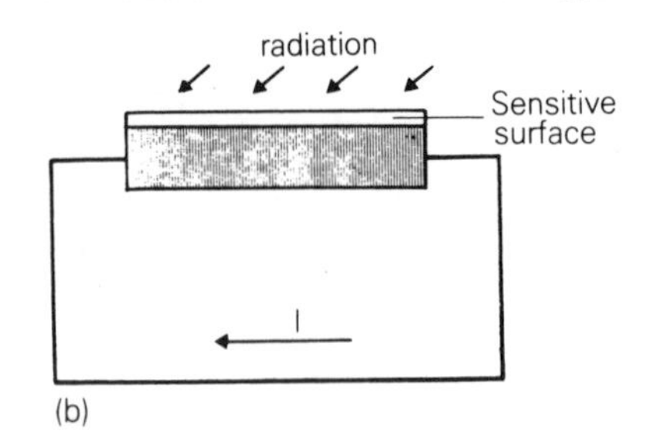

Fig. 4.4. Radiation detectors: (a) a thermoelectric thermopile, consisting of a number of thermocouples in series; (b) a photoelectric silicon semiconductor. V = voltage. I = current.

Long-wave radiation is usually measured using net radiometers which measure the difference between the total incoming and outgoing radiation fluxes at all wavelengths (Φ_{net}). When Φ_{net} is measured above a canopy, its value is the net radiation absorbed by the canopy. However, the net radiation absorbed by a layer of leaves in the canopy is the *difference* between Φ_{net} above and below this layer. The main component of a net radiometer is a flat black plate, and the temperature difference between the top and bottom surfaces, measured with a thermopile, is proportional to net irradiance. The two sensing surfaces are protected from the wind either by continuous ventilation or by inflatable domes of polyethylene, which is transparent at all wavelengths.

For measurements within plant canopies, where radiation distribution is very heterogeneous, averages can be obtained by moving a small sensor repeatedly along a track or by using a long *linear* sensor (tube solarimeters and radiometers). These linear sensors are less accurate than flat plate sensors because of greater cosine errors (see below), and should be used for relative rather than absolute measurements. Errors can be minimized by taking measurements in two directions at right angles. Measurements of canopy solar radiation interception have become increasingly important since Monteith [5] showed that the dry matter production of a plant canopy is directly proportional to the amount of PAR intercepted by the canopy. The original relationship was expressed in terms of solar energy transmission of PAR but it is preferable to use quantum units, since photosynthesis is a quantum process. It has been shown that total solar energy flux interception can be converted to quantum units for many field crops using a simple relationship [11].

In addition to the fluxes of energy through the atmosphere, there is also a vertical transfer of energy through the soil which is known as the soil-heat-flux. During daylight hours, the soil normally acts as a heat sink, so the soil-heat-flux is positive; at night, it becomes negative and of similar absolute magnitude to daylight values. It may range from 2% of Φ_{net} below dense canopies which shade the ground to more than 30% of Φ_{net} below open canopies. The vertical transfer of heat by conduction through the soil is measured with a sensor which is basically similar to a net radiometer [4]. In practice the plate consists of glass or resin, but errors can be greater than 40% because of poor conductivity matching [4].

The other type of radiation sensors which have become increasingly popular because they measure photon flux, are the photoelectric photocells. One example is the *silicon cell* which consists of a small chip of silicon that is sensitive to radiation in the visible region of the spectrum. This functions by absorbing photons, causing a release of electrons and the generation of an electric current in direct proportion to the photon

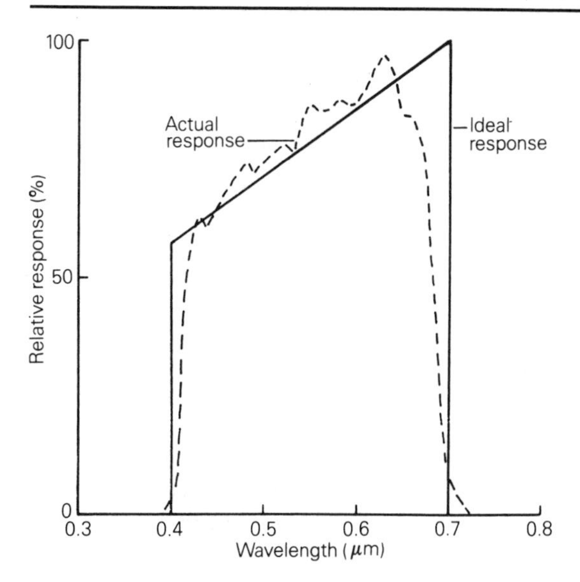

Fig. 4.5. The actual and ideal responses of a PAR quantum sensor.

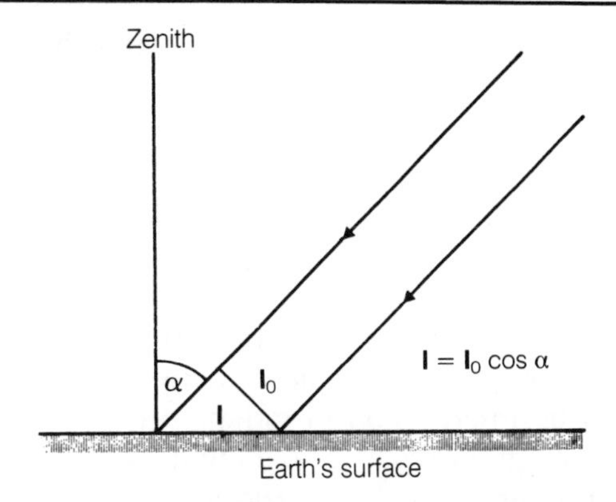

Fig. 4.6. The cosine law, showing the relationship between the angle of the sun's rays (α), the solar zenith angle and the solar irradiance on the Earth's surface (**I**), where $\mathbf{I_o}$ is the irradiance on a surface normal to the sun's rays.

flux (Fig. 4.4b). These semiconductors now have reasonable accuracy for a relatively low cost but unlike unfiltered radiometers, which have a uniform response to all wavelengths, the silicon cell typically has a peak sensitivity to photons at about 0.85–0.95 μm with a band width of 0.5 μm at 50% response. However, using suitable filters, their response can be tailored so that they measure photosynthetically active photon flux and give a good approximation to the 'ideal' quantum response (Fig. 4.5). When used with filters to exclude wavelengths outside the range 0.4–0.7 μm, the shape of the response is 'ideal' because short wavelengths contain more energy and therefore generate more current. The broken line in Fig. 4.5 shows the spectral response of the sensor adjusted for the different energy levels at the blue and red ends of the spectrum. Another type of photocell which has become commercially available more recently than the silicon cell is the gallium arsenide phosphide photodiode (GaAsp). This has a reasonably good spectral response without the need for any filters and it is cheap enough to allow the use of several sensors within or below canopies. Quantum sensors can also be arranged in a linear fashion in a similar way to the tube solarimeters so that they measure PAR within canopies (Chapter 6).

In order for radiation sensors to give accurate readings at all sun angles they should ideally obey Lambert's Cosine Law. According to this law, when radiation is incident at an angle α to the normal the irradiance (**I**) should be expressed as:

$$\mathbf{I} = \mathbf{I_o} \cos \alpha$$

where $\mathbf{I_o}$ is the irradiance on a surface normal to the sun's rays (Fig. 4.6). Different techniques are used to achieve a good cosine response, but a common method is to use a raised white perspex diffuser on top of the sensor.

Most radiation sensors give a millivolt output and they can be connected directly to recorders, data loggers or integrators. If connected to integrators they can give a total of received radiation during a given period of time. This is useful in determining the total radiation interception by a canopy for energy conversion calculations. Daily integrals of irradiance or PAR give energy or quantum flux density over a period of time, i.e. $\mathrm{J\,m^{-2}\,d^{-1}}$ or $\mathrm{mol\,m^{-2}\,d^{-1}}$. A wide range of commercially available radiation sensors is listed in Appendix A.

4.3 TEMPERATURE

4.3.1 Introduction

The temperature of the aerial parts of plants is determined by the balance between energy gain by interception of radiation (Φ_{abs}) and the energy losses by re-radiation (Φ_{rad}), convection or 'sensible' heat loss (**C**) and transpiration (**λE**, where **E** is transpiration rate and λ is the latent heat of vaporisation), so that:

$$\Phi_{abs} = \Phi_{rad} + \lambda \mathbf{E} + \mathbf{C}$$

and

$$\Phi_{net} = \Phi_{abs} - \Phi_{rad}$$

Energy storage by the plant, both physical and in the form of chemical bonds associated with photosynthesis and respiration, is generally small in quantitative terms and can usually be ignored. The temperature that a leaf will reach in a specified environment can be calculated using an iterative computing procedure [12] and the following equation, which describes **λE** and **C** in terms of an electrical circuit analogue (Fig. 4.1):

$$\Phi_{net} = \frac{\rho c_p (e_{s,T_l} - e)}{\gamma (r_s + r_{a,H_2O})} + \frac{\rho (T_l - T_a)}{r_{a,H}}$$

where ρ is the density and c_p the specific heat of air, e_{s,T_l} is the saturated vapour pressure of air at the leaf temperature (T_l), e is the water vapour pressure of free air, γ is the psychrometric constant (66 Pa °C^{-1}), r_s is the stomatal resistance, and r_{a,H_2O} and $r_{a,H}$ are the boundary layer resistances (reciprocals of conductance) to movement of water vapour and heat. The process is one of balancing the equation and finding a leaf temperature (T_l) at which $\Phi_{net} = \lambda \mathbf{E} + \mathbf{C}$. However, in practice a more convenient expression can be derived from this equation which is similar to the well-known Penman–Monteith combination equation for calculating evapotranspiration from plant canopies; this calculates leaf–air temperature difference from the sum of two terms that depend on net radiation and the vapour pressure of air respectively [8].

It is clear that plant temperature is determined by the large number of factors which influence the magnitude of Φ_{rad}, **E** and **C**. The temperature will influence most of the processes which are occurring within the plant. The units of temperature are degrees Celsius (°C) or Kelvin (K = 273 + °C). It is generally held that a measurement sensitivity of ±1°C is sufficient for analysis of plant growth and developmental processes while a sensitivity of ±0.1°C is required for precise calculations of transpiration or heat transfer determinations [13].

Physiological processes such as seed germination, photosynthesis, respiration and leaf growth all respond to temperature but it is important to be able to measure the temperature which is most relevant to the process being studied. For example, when measuring leaf expansion in grasses the temperature in the meristematic region at the base of the leaf is the most relevant measurement; this may be closer to soil temperature than air temperature because of the location of the meristems in vegetative grasses close to the soil surface [14]. The problem is made more difficult by the fact that plant temperatures can differ from air temperature by several degrees. There is also a spatial variation in temperature, commonly as a result of solar radiation interception at the top of the canopy.

In recent years there has been renewed interest in the concept of degree-days, heat units or *thermal time* in controlling plant growth and development. Plant development is assumed to show a linear response to temperature from a threshold (T_b) to an optimum (T_o), and the time taken to reach a given phenological stage is related to thermal time, defined as the integral of temperature with time. Units of thermal time (°Cd) are degree-days, calculated as the sum of the differences between daily mean temperature ($\bar{T}$) and the base temperature for each day beyond a given starting date:

$$°Cd = \sum_{0}^{n} (\bar{T} - T_b) \quad \text{for} \quad \bar{T} > T_b$$

When $\bar{T}$ is less than T_b, or if there are both upper and lower threshold temperatures, the calculation is more complicated although it is still possible to hand calculate degree-days [15].

In many natural environments, it is difficult to uncouple developmental response to temperature from other factors such as irradiance and saturation deficit. This problem can be overcome by calculating a thermal rate (p):

$$p = \zeta/(\bar{T} - T_b)$$

where ζ is the rate of response (e.g. leaf extension in $mm\,h^{-1}$). The thermal rate is expressed in units of response per unit thermal time (e.g. mm $(°C\,h)^{-1}$) and it is possible to correlate this with other environmental variables, although application of this technique to field measurement needs some caution [17]. Temperature integrations to determine degree-days can be carried out using commercially available transducers and millivolt integrators (e.g. *Type MV2, Delta-T Devices, Cambridge, UK*). The appropriate base temperature for these measurements can easily be set by moving an internal plug-on connector.

4.3.2 Temperature measurements

Temperature is measured by transducers which are based upon temperature effects on expansion, electrical or radiative responses. The two most important sources of error in temperature measurement are the effects of absorption of incoming radiation and the effect of the thermal mass of the sensor. Both these effects are more important in air than in water or when measurements are made within the plant tissue. The principal types of thermometer used in environmental physiology are listed in Table 4.2 with some of their characteristics. The term *accuracy* denotes how close the measured temperature is to the true temperature on what is known as the International Practical Temperature Scale. *Precision* indicates how closely repeated measurements at the same temperature agree with one another [16].

Liquid (normally mercury)-in-glass thermometers are the most common instruments used for measuring soil, soil surface and air temperatures. They are widely used as accurate devices in meteorological stations, but they have a large thermal mass and no facility for recording. On the other hand, the less accurate bimetallic strips used in thermographs do register on a dial or strip chart through a series of levers, and are still useful for long-term measurements of temperature.

Many temperature sensors depend on the fact that a change in temperature can alter the electrical properties of certain materials. These electrical temperature transducers are either *thermocouples* which generate a flow of electrons between two junctions of dissimilar metals if their temperatures are different (Fig. 4.4a), or *resistance thermometers* and *thermistors* in which the electrical resistance of the sensor changes with temperature.

Thermocouples are widely used for temperature measurements in biology because they are small (with low thermal mass), easy to construct and cheap. A number of types of thermocouple can be purchased or made from combinations of different metals. They have different electrical and physical properties which influence their sensitivity and suitability for different uses. Characteristics of the more common thermocouples are shown in Table 4.3.

When two thermocouple junctions are joined, the voltage (V) generated is proportional to the difference in temperature

Table 4.2. The principal types of thermometer, their accuracy and features (adapted from Bell and Rose, [16])

Type	Precision (°C)	Accuracy (°C)	Response time (s)	Calibration interval	Advantages	Disadvantages
Liquid-in-glass	1	1	30 in water	5 years	Low cost Ease of use	Fragility No electrical output
Platinum resistance	0.1	0.2	1 in water 10 in air	1 year	Stability Robustness Ease of use	Cost Size, but smaller units becoming available
Thermistor	0.1	0.3	5 in air	1 year	Low cost Ease of use Small size	Narrow temperature range of linear response
Thermocouple	0.1	0.5	1 in air	1 month	Low cost Ease of construction	Many possible sources of error Need for reference temperature
Infra-red thermometer	0.1	5	5	5 years	Non-contact measurement	Cost Poor accuracy

Table 4.3. Typical electrical properties and characteristics of thermocouples. (Adapted from Woodward and Sheehy [7]).

Thermocouple	Type	Uniformity	ϕ_{min} (mm)	Temperature °C					
				0°	10°	20°	30°	40°	50°
Copper-constantan	T	Low	0.2	0	0.39*	0.79	1.19	1.61	2.03
Chromel-alumel	K	Low	0.1	0	0.40	0.80	1.20	1.61	2.02
Chromel-constantan	E	Medium	0.05	0	0.61	1.23	1.85	2.48	3.08
Iron-constantan	J	Low	0.05	0	0.52	1.05	1.58	2.12	2.66
Plantinum-platinum/10% rhodium	S	High	0.025	0	0.06	0.11	0.17	0.24	0.30

*e.m.f. in mV with reference cold junction at 0°C.
ϕ_{min} is the smallest practicable thermocouple diameter.

between the measuring junction (sensor) and a reference junction:

$$V = k(T - T_o)$$

where T is the sensor temperature, T_o the reference temperature and k the temperature coefficient (the change in e.m.f. per unit change in temperature at the reference temperature). It is common practice to assume a linear relationship between thermocouple e.m.f. and temperature, but for more accurate work the relationship is described more precisely by a quadratic regression equation [7]. Normally thermocouples are used with the reference junction maintained at a constant temperature; this is most conveniently an ice-water mixture contained in a Dewar (vacuum) flask, which has a temperature of 0°C. Calculated values of e.m.f. for a range of temperatures with the reference junction at 0°C are given in Table 4.3. Alternatively, soil temperature at a depth of one metre is quite stable; this can be used as reference if its temperature is measured with a thermometer. More convenient than either of these references are the electronic references now available on most meters; when used in this way only one thermocouple is required to measure temperature. Unfortunately these electronic references have poor accuracy (e.g. ±1°C) although they are very convenient in portable instruments.

Thermocouples can be constructed easily, taking care to ensure a good junction between the two metals. Tin or silver can be used to solder the junctions, but silver gives the smallest junctions using borax as a flux. After soldering, the junctions can be cut with a blade under a binocular microscope to make them as small as possible. Ideally, all thermocouples should be calibrated individually because of small variations in characteristics of the wires and junctions. The accuracy of high precision thermocouples is about ±0.2°C but normal field instruments are usually specified to have an accuracy of about ±0.5°C after calibration.

Wire resistance thermometers are most often constructed from platinum, nickel or copper. Usually the commercially available platinum resistance thermometers are quite bulky, being typically 20 mm long and 3 mm in diameter. They are therefore only useful for measuring temperatures of large volumes but they are often favoured for long-term use because of high stability, relatively high accuracy (±0.2°C), resistance to weathering and an almost linear change in resistance with temperature. However, the change in resistance with temperature is relatively small, so a circuit to read voltage output must be designed with care to avoid large error resistances [2,4].

Thermistors are semiconductors, composed of sintered mixtures of metallic oxides. The resistance of thermistors decreases exponentially with temperature but with about 10 times the sensitivity of resistance thermometers. The major advantage with thermistors is that they are available in a range of sizes down to miniature bead types of 0.2 mm diameter, and the circuitry required to give a readout is relatively simple and robust. However they are generally not quite as accurate as thermocouples or platinum resistance thermometers and this is not improved much by calibration because the inaccuracy is due largely to inadequacies in the circuitry for measuring resistance.

The only non-contact method of measuring temperature is by *infra-red thermometry*, which is based upon the principle that all surfaces emit energy [13]. The flux of radiation follows the Stefan-Boltzmann law and is proportional to the fourth power of the absolute temperature of the object (Section 4.2.2). As the temperature of vegetation is about 290 K it emits long-wave radiation with a peak of emission at about 10 μm. Infra-red thermometers are typically fitted with filters which allow only radiation in the range 8–13 μm to pass to the detector. This is the atmospheric window where the transmissivity of radiation in air is high. Infra-red thermometers are expensive, difficult to calibrate and prone to errors if reflected long-wave radiation is detected. When used correctly, the accuracy at best is ±0.2°C but is more likely to be closer to ±1.0°C. They are intrinsically preferable to contact methods because the latter can alter surface temperature during measurement by simultaneous conduction between the thermometer, surface and air, possibly resulting in large errors.

4.3.3 Use of thermometers

Before use, thermometers should be calibrated over the expected range of temperature. The simplest method is to immerse the sensors in a water bath whose temperature is controlled and compare temperatures with an accurate mercury-in-glass thermometer. The infra-red thermometer cannot be calibrated in this way, but it can be set up to receive radiation from the inside of a blackened sphere immersed in a water bath whose temperature is known.

Temperatures measured are usually of air, surface, soil and tissue. The latter two are less prone to difficulties because the thermometer is immersed in the material it is sensing. For air temperatures to be measured accurately the absorption of solar and long-wave radiation should be prevented by use of a radiation shield, preferably ventilated [16]. The ideal shield should have a high reflectivity for solar radiation and a high emissivity for long-wave radiation. Aluminized 'Mylar' and clear matt white paint have been found to be the most suitable shield coverings. Surface temperature measurements such as leaf temperature are the most difficult to make accurately because they depend on a good thermal contact between the sensor and surface being measured. As the sensor normally has only a small part of its own surface in contact with the leaf it will most likely indicate a temperature that is intermediate between leaf and air temperature. Clips, springs or tapes are often used to improve contact but their presence can also lead to errors. The magnitude of errors may be as large as 2–4°C depending on the temperature difference between the leaf and the surrounding air. Further details on the use of thermometers and an analysis of the errors which might be experienced can be found in Perrier [13] and Bell and Rose [16].

4.4 HUMIDITY

4.4.1 Introduction

Humidity is a measure of the moisture content of the atmosphere. The water content of air is known as the absolute humidity (χ) and is the density of water vapour in the air in $g\,m^{-3}$. The importance of humidity to a plant's functioning is two-fold. Firstly, it determines the rate of water loss in transpiration (**E**) because:

$$\mathbf{E} = g(\chi_{air} - \chi_{leaf})$$

where g is the conductance for water vapour transfer between the evaporating surfaces within the leaf and the air. Secondly, humidity has a direct effect on the stomata of many plants, so that stomata tend to close in dry air restricting water loss but also reducing CO_2 assimilation.

4.4.2 Definitions

Because water vapour is a gas, its pressure contributes to the total measured atmospheric pressure and its partial pressure is termed water vapour pressure (e; kPa). When air above water has no extra capacity for holding water vapour the partial pressure is termed the *saturation vapour pressure* (e_s) and its density the *saturation density* (χ_s; units $g\,m^{-3}$). The saturation vapour pressure increases with temperature (Fig. 4.7). If air is cooled without change in water content, condensation occurs at its *dew-point temperature* (T_d), i.e. when $e = e_s$.

When water evaporates into less than saturated air then the temperature of the air decreases to a point which is the *wet-bulb temperature* (T′). This is the temperature to which the wet bulb falls in a psychrometer (Section 4.4.3). Its value is given by the intercept of a line of slope $-\gamma$ (where γ is the psychrometric constant), passing through the vapour pressure of the air at the dry-bulb temperature, with the curve of saturation vapour pressure against temperature (Fig. 4.7). The slope of the line differs according to whether the wet bulb is ventilated or not.

Relative humidity is the ratio of the actual vapour pressure (e) to the saturation vapour pressure (e_s) at the dry-bulb temperature (T). It is usually expressed as a percentage. However the use of this term should be discouraged as plants do not respond directly to relative humidity. *Saturation vapour pressure deficit* (D) is the difference between the saturation vapour pressure and the actual vapour pressure at the same temperature (Fig. 4.7). It is an index of the drying power of the air; the higher the deficit the greater the evaporation rate.

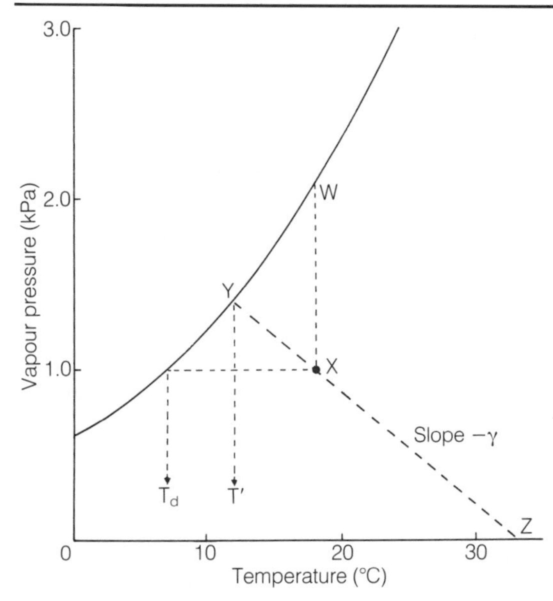

Fig. 4.7. The influence of temperature on the saturation water vapour pressure of water. The point X represents air at 18°C and 1.0 kPa vapour pressure (e). The line Y–X–Z, with a slope of $-\gamma$, gives the wet-bulb temperature (T′) where it intercepts the curve at Y (12°C). The saturation water vapour pressure deficit (D) is the difference between X and W (the saturation vapour pressure at 18°C). The dew-point (T_d) is the point at which the saturated vapour pressure is equal to X.

4.4.3 Measurements

Many different devices can be used to measure the humidity of the air. They are based on several principles including the electrical properties of sulphonated polystyrene or thin-film solid state semiconductors; wet-bulb depression; condensation of water vapour on a surface cooled to the dew-point; and infra-red absorption. Some of the instruments more widely used in the field are considered here.

Probably the most frequently used instrument for humidity measurement is the wet- and dry-bulb psychrometer. A psychrometer is a pair of identically shaped thermometers, one of which is covered with a wet sleeve. Evaporation cools the wetted sensor to the wet-bulb temperature, and the vapour pressure (e) is calculated as:

$$e = e_{s,T'} - \gamma(T - T')$$

where T′ and T are the wet- and dry-bulb temperatures, respectively, $e_{s,T'}$ is the saturation vapour pressure at the wet-bulb temperature, and γ is the psychrometric constant (equal to 66 Pa °C^{-1} at sea level in a ventilated psychrometer). Several types of psychrometer are available as commercial units, the best of which ensure efficient radiation shielding of the thermometers and minimise heat conduction along the stem of the thermometer [2]. The *Assman* psychrometer is a ventilated psychrometer containing matched thermometers; it is used for standard humidity measurements. Smaller ventilated psychrometers are commercially available for routine use above and within plant canopies (*Delta-T Devices, Cambridge, UK*). It should be noted that when using some of these instruments the constant used in the above equation may differ from the true psychrometric constant. For accurate work, the value of this constant should be determined by calibration against a standard *Assman* psychrometer. The handheld whirling or sling psychrometers are the simplest and cheapest ventilated units. In order to achieve the required aspiration rate of 3 m s^{-1} they have to be rotated at about two revolutions per second. With careful use, the best accuracy that can be achieved with a wet- and dry-bulb psychrometer is ±2% saturation deficit or ±0.15°C wet-bulb depression, whichever is the greater.

Many materials show a *change of physical dimensions* when they absorb water, and this property can be used to make instruments that measure humidity. For example, the length of animal hair increases as the air becomes wetter and decreases as the air dries; this property is used in simple mechanical hygrometers. Provided an allowance is made for the effect of temperature, hair hygrometers are usually accurate to within 5% over most of the humidity range.

The *change in electrical properties* of materials as they absorb water is used in several humidity sensors. Originally, the lithium chloride sensor was the most common type of electrical sensor. Lithium chloride is hygroscopic and the moisture content of the air determines how much water is absorbed, which in turn influences the AC resistance of the sensor. This type of sensor is susceptible to contamination by dust and other hygroscopic particles, and it suffers from a certain amount of hysteresis when wetting or drying. More recently, capacitance hygrometers, which measure the change in electrical capacitance caused by water absorption into a dielectric, have become commercially available (e.g. *Humicap*, manufactured by *Vaisala, Helsinki, Finland*) and are less temperature sensitive and show less hysteresis than other electric sensors. The main attraction of these sensors is their small size (e.g. 4 mm × 6 mm × 1 mm), as well as their linear and rapid response. Above 75% relative humidity the response is less satisfactory due to increasing non-linearity, slow response and long-term drift. When used at relative humidities below 75% their accuracy is ±2%RH.

Dew-point meters measure the temperature at which dew forms on a cooled mirror surface. Dew-point is usually determined by cooling a surface to below the point of saturation, allowing water to condense onto it, and then gradually raising the temperature until the film of condensation starts to evaporate. The temperature at which this change occurs is taken as the dew-point temperature, and the presence of the film can be detected optically or electrically. The principle source of error is contamination of the mirror surface with soluble salts, which raises the condensation point. The mirror surface must be cleaned at intervals depending on the degree of contamination of the measuring gas. Dew-point temperatures should be corrected for changes in atmospheric pressure if vapour pressure is calculated from T_d. This type of instrument is very accurate and stable but it is usually a laboratory instrument because it normally requires mains power. The accuracy achievable is about ±0.2°C.

Infra-red gas analysis can measure water vapour concentration of air as well as CO_2 (Chapter 9). The instruments are expensive but they are accurate and respond quickly. With suitable switching systems this type of instrument is almost always employed to measure concentration differences, making it very suitable for studies of vertical profiles of humidity within and above plant canopies.

4.5 WIND

Wind is the large-scale transport of air masses resulting from differences in air pressure. It is directly involved in heat and mass transfer by forced convection, so it is very important in influencing heat and gas exchange across the boundary layers of plants. Increase in wind speed decreases the boundary layer resistance over leaves (Fig. 4.1); this tends to increase evaporation and bring leaf temperature closer to air temperature. Wind is also important because it causes mechanical deformation of plants (due to the frictional drag of moving air) and because it disperses pollen, seeds and aerial pollutants. However, of all the elements of the microclimate, wind is the most spasmodic. Short-term variations in wind speed are described by the intensity of turbulence; this value represents the standard deviation of the instantaneous values, divided by the mean wind speed [18]. Turbulent air moves in packets or eddies; these are important for the movement of CO_2, H_2O and other gases in and above plant canopies.

The meteorologists' measurements of wind speed are normally made 10 m above the ground, but wind speed decreases rapidly as the plant surface or ground is approached. Wind speeds near or within vegetation are therefore likely to be much lower than at 10 m. The analysis of the profiles of mean wind speed above the canopy can be used to derive coefficients for calculating the flux of CO_2 and H_2O between the canopy and the atmosphere. However, the principles of environmental physics required to make these calculations are beyond the scope of this discussion. Furthermore, the number of expensive environmental sensors required, and to some extent the size of the data recording and processing facilities needed are beyond the budget of most research groups. Further information on crop micrometeorology may be found in Monteith and Unsworth [5].

4.5.1 Measurement of wind

A complete picture of air movements requires continuous recording of instantaneous wind speed measured in three directions; the vertical, the horizontal lateral and the horizontal perpendicular. However, these are difficult to measure, so it is generally sufficient to determine the mean value in one direction over the measurement period.

The most commonly used instrument for measuring the mean horizontal component of wind speed is the *cup anemometer*. This normally consists of three hemispherical or conical cups mounted on arms and attached to a central vertical spindle, so that they are free to rotate in the wind. The number of rotations of the cup assembly is usually measured in metres (the 'run of the wind'), and can be divided by the elapsed time to give the mean wind speed. An alternative mechanical anemometer is the vane or propellor type. The *vane anemometer* is simply a miniature windmill, consisting of a number of light vanes radially mounted on a horizontal spindle. The main sources of error with mechanical anemometers are firstly that they have a threshold below which the friction of the system prevents rotation, and secondly, that their inertia makes them overrun when the wind speed drops. Additionally, the vane anemometer is directional and consequently it must be aligned to the wind direction.

A third type of instrument is the *hot-wire anemometer*, which estimates wind speed by measuring the rate of cooling of a heated wire in moving air. These can be made very small, and are useful for measuring the rate of air flow around leaves [8], but they are often delicate and easily broken. For this reason they are not used routinely for field measurements. Further details of instruments for wind measurement can be found in Grace [18,19].

4.6 AUTOMATIC WEATHER STATIONS

In many studies, a rather broad description of the climate experienced by the plants is sufficient to begin to untangle plant/climate relationships. For these purposes, information obtained from standard meteorological sites located close to the vegetation under investigation is useful. In recent years, *automatic* weather stations have been used increasingly, especially for making measurements at remote sites. These stations can be set up on experimental sites to provide detailed measurements of the weather including solar radiation, net radiation, wind run, wind direction, air temperature, wet-bulb temperature and rainfall. An environmental data logger initiates readings, controls the sensors and stores data. Modern data loggers often incorporate more channels than are required for the basic measurements, enabling the recording of additional data such as soil temperature and outputs from micrometeorological sensors. Details of commercially available automatic weather stations may be found in Appendix A.

4.7 RECORDING

The simplest method of recording the output from micrometeorological instruments in the field is using a pencil and note-pad; and in many cases this is all that is necessary.

However, the large volume of data which often needs to be recorded necessitates some level of automation. There are, of course, many advantages in adopting automatic recording but the methods adopted must be considered in relation to the use to which the measurements will be put. For example, detailed measurements of vertical profiles of temperature, wind speed, humidity, radiation and CO_2 concentration can be used to estimate canopy evapotranspiration and CO_2 exchange [5,7], but they require complex recording facilities and are normally only possible with substantial resources of equipment and manpower. However, less intensive measurements using limited recording facilities can still give a lot of information on the relationship between the plant and its environment. Furthermore, micrometeorological measurements are not an end in themselves, and they usually need to be related to plant physiological responses such as photosynthesis, stomatal movement, water potential and leaf expansion. These responses have different time scales (e.g. seconds or minutes for photosynthesis and days for leaf expansion) and these measurements should be recorded accordingly.

The simplest automatic method of accumulating output from instruments is using analogue recorders. The most suitable of these are galvanometer recorders which can be multi-channel, and either use pen and ink as tracer or record on pressure-sensitive paper using a chopper bar. Analogue integrators can be used where detailed chart records are not necessary as they integrate small currents and voltages and can be used to determine characteristics such as degree-days and daily solar radiation integrals.

Digital data logging is perhaps the most convenient way of collecting micrometeorological data, especially where a large number of measurements are involved which can subsequently be handled by a computer. Here the analogue input from the instruments is converted into a number (by an analogue-digital converter) and stored in a memory, printed, or recorded on magnetic media such as a floppy disc or a magnetic tape. Usually a large number of inputs can be scanned in sequence to give discontinuous but frequent records of a large number of measurements (Fig. 4.8). For further discussion of this topic see Woodward and Sheehy [7] and Pearcy [20]. There are at present a number of commercially available portable data loggers which are suitable for field use (Appendix A). Most of these allow data to be transferred to portable or laboratory-based microcomputers via an RS232 serial interface. However, new models are continually being produced or updated, so it is important to obtain details from manufacturers before decisions are made about the suitability of a particular instrument for ecophysiological research. Particular attention should be paid to features such as the maximum number of available channels, scan intervals, precision (% full scale), memory storage and the maximum unattended operating period.

4.8 EXPERIMENTAL WORK

The aim of this practical work is to gain some familiarity with instruments which are used for measuring the macro- and micro-environment of plants. Much of the equipment described in this chapter can be used to build up a picture of the way in which the microclimate of a plant canopy varies through space and time.

Before the instruments can be used reliably they must be calibrated. The need for regular recalibration of most instruments cannot be overemphasized. The time intervals between calibration vary greatly from one type of instrument to another, as does the ease with which calibrations can be carried out. Recommendations from suppliers should be followed as far as possible. For some instruments, such as radiation sensors, it is likely that reliable recalibration can only be carried out by the manufacturer. However, frequent

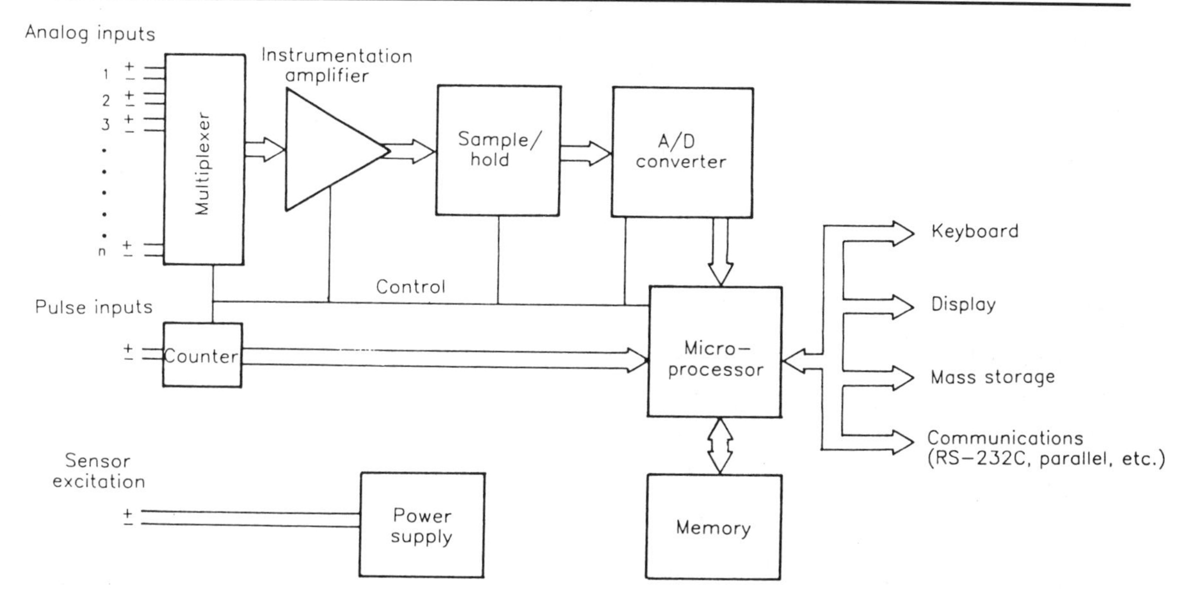

Fig. 4.8. Block diagram of a typical data logger showing the major components. For simplicity, other support components that control the flow of information to and from the microprocessor are not shown (from Pearcy [20]).

comparison with a standard sensor stored in the laboratory and only used for calibrations can ensure against overlooking significant changes in sensitivity. For other instruments, such as temperature sensors, satisfactory calibration may be obtained against a high precision (±0.05°C) mercury-in-glass thermometer (Appendix A). As an exercise make a cross comparison of the radiation sensors you intend to use and check the calibration of thermocouple, thermistor and platinum resistance thermometers.

After calibration of the instruments, the objective of this experiment is to measure the vertical profiles of different environmental factors at different times during the day in canopies of two crops of contrasting structure (e.g. maize and bean) and over bare ground. These observations are carried out in conjunction with measurements of stomatal conductance and leaf water potential (Chapter 8) with the aim of determining which factors control stomatal activity.

At different times during the day measure the following parameters at a minimum of five different levels, in and above the crop, and over bare ground.

(a) *Photosynthetically Active Radiation (PAR)* – using a point and linear quantum sensor to measure **Q**. Cross-calibrate the point and linear quantum sensors in the open. Keeping the point sensor above the canopy, place the linear sensor at progressively lower levels down to the soil surface. Measurements with the linear sensor should be made at right angles to the rows. Express the value of quantum flux at a particular height as a percentage of the incident flux.
(b) *Leaf and air temperature* – using a thermocouple thermometer to measure the temperature of the undersurface of a leaf and of the air below the leaf. Also measure soil temperature by carefully pushing the thermocouple into the soil to a depth of 1 cm. Measurements at greater depths to determine the temperature profile should be made with spear-enclosed thermistors.
(c) *Wet- and dry-bulb temperature* – using an *Assman* ventilated psychrometer or a suitable

alternative. Calculate the saturation deficit of the air (kPa) from the wet- and dry-bulb temperatures.

Plot the profiles and relate your observations to measurements of leaf stomatal conductance and water potential (Chapter 8).

REFERENCES

1. Monteith, J.L. (1981) Coupling of plants to the atmosphere. In: *Plants and their Atmospheric Environment*, 21st Symposium of the British Ecological Society (Grace, J., Ford, E.D. and Jarvis, P.G. eds.) pp. 1–29. Blackwell Scientific Publications, Oxford.
2. Szeicz, G. (1975) Instruments and their Exposure. In: *Vegetation and the Atmosphere*, Vol. 1: Principles (Monteith, J.L. ed.) pp. 229–273. Academic Press, London.
3. Monteith, J.L. (1972) *Survey of Instruments for Micrometeorology*. IBP Handbook No. 22. Blackwell Scientific Publications, Oxford.
4. Fritschen, L.J. and L.W. Gay (1979) *Environmental Instrumentation*. Springer-Verlag, New York.
5. Monteith, J.L. and M.H. Unsworth (1990) *Principles of Environmental Physics*. (2nd edn.) Edward Arnold, London.
6. Campbell, G.S. (1986) *An Introduction to Environmental Biophysics*. Springer-Verlag, New York.
7. Woodward, F.I. and J.E. Sheehy (1983) *Principles and Measurements in Environmental Biology*. Butterworths, London.
8. Jones, H.G. (1992) *Plants and Microclimate: a quantitative approach to environmental plant physiology*. 2nd edition, Cambridge University Press. 415 pp.
9. Marshall, B. and F.I. Woodward (eds.) (1985) *Instrumentation for Environmental Physiology*. Cambridge University Press.
10. McCree, K.J. (1976) A rational approach to light measurements in plant ecology. In: *Commentaries in Plant Science* (H. Smith ed.). Pergamon Press, Oxford.
11. Green, C.F. (1987) Nitrogen nutrition and wheat growth in relation to absorbed solar radiation. *Agricultural and Forest Meteorology* **41**, 207–248.
12. Grace, J. (1983) *Plant-Atmosphere Relationships*. Outline Studies in Ecology. Chapman and Hall, London.
13. Perrier, A. (1971) Leaf temperature measurement. In: *Plant Photosynthetic Production, a Manual of Methods*. (Sestak, Z., Catsky, J. and Jarvis, P.G. eds.) pp. 632–671. Dr. W. Junk, The Hague.
14. Peacock, J.M. (1975) Temperature and leaf growth in *Lolium perenne*. II. The site of temperature perception. *J. Applied Ecology* **12**, 115–123.
15. Snyder, R.L. (1985) Hand calculating degree days. *Agricultural and Forest Meteorology* **35**, 353–358.
16. Bell, C.J. and Rose, D.A. (1985) The measurement of temperature. In: *Instrumentation for Environmental Physiology* (Marshall, B. and Woodward, F.I. eds.) pp. 79–99. Cambridge University Press.
17. Ong, C.K. (1983) Response to temperature in a stand of pearl millet (*Pennisetum typhoides*) 1. Vegetative development. *J. Exp. Bot.* **34**, 322–336.
18. Grace, J. (1977) *Plant Response to Wind*. Academic Press, London.
19. Grace, J. (1985) The measurement of wind speed. In: *Instrumentation for Environmental Physiology*. (Marshall, B. and Woodward, F.I. eds.) pp. 101–121. Cambridge University Press.
20. Pearcy, R.W. (1989) Field data aquisition. In: *Plant Physiological Ecology* (Pearcy, R.W., Ehleringer, R.W., Mooney, H.A. and Rundel, P.W. eds.) pp. 15–27. Chapman and Hall, London.

5

Controlled-environment studies

T.W. Tibbitts and R.W. Langhans

5.1 INTRODUCTION

Studies of photosynthesis and plant productivity may require environmental control. This chapter outlines procedures for environmental control in structures such as growth chambers, glasshouses, and open-top chambers, for study with single plants and with canopies of plants. (See Chapter 9 for an outline of environmental control in leaf chambers.)

A number of environmental factors should be considered when developing and using controlled-environment structures. These include temperature, lighting, humidity, and carbon dioxide. The emphasis in this chapter is on growth chambers, where all factors of the environment can be controlled, although facilities with only partial environmental control, e.g. greenhouses and open-top field chambers, are also discussed. More detailed accounts of the construction and operation of controlled-environment facilities are available, with descriptions of growth chambers [9,12,14], glasshouses [10], and open-top chambers [7,11].

Photosynthesis and Production in a Changing Environment: a field and laboratory manual
Edited by D.O. Hall, J.M.O. Scurlock, H.R. Bolhàr-Nordenkampf, R.C. Leegood and S.P. Long
Published in 1993 by Chapman & Hall, London.
ISBN 0 412 42900 4 (HB) and 0 412 42910 1 (PB).

5.2 FACILITIES

5.2.1 Structures

(a) Growth chambers:
Growth chambers are generally constructed of rigid materials, including iron or aluminium framing, and metal sheeting with enamelled or painted surfaces which resist chipping or cracking. Galvanized metal should not be used, to avoid zinc contamination of the plants. The inner wall and ceiling surfaces should be painted with a high gloss white paint, or covered with a polished aluminium or mirrored surface to maximize light reflection. The walls should be insulated to reduce temperature exchange with the room. Typical designs for growth chambers are found in Carpenter and Mousley [5] and Eguchi [6].

(b) Greenhouses:
Greenhouses are constructed of glass or various types of plastic. Glass is considerably more expensive but has a life of more than 20 years, whereas plastic will require replacement at frequent intervals. Flexible plastic must be replaced every 2–3 years, whereas rigid plastic will have to be replaced every 10–15 years. Replacement of flexible plastic is necessary because it loses transmissivity and also weakens and tears with age. Use of ultraviolet (UV) absorbers in flexible plastics improves their longevity. Designs for glass and plastic houses are provided in Langhans [10].

(c) Open-top chambers:
Open-top chambers are constructed of an aluminium frame and covered with flexible plastic. The plastic is commonly a clear polyvinyl chloride (PVC) film. Significantly better long-wave transmissivity and thus better temperature maintenance is obtained with teflon film, but the cost is greatly increased. Construction details for open-top chambers are provided in Heagle [7].

5.2.2 Power

(a) Enclosed environments:
The power requirements for growth chambers are high, because of the need to provide adequate lighting for photosynthetic irradiation of the plants and the need for cooling to remove the heat generated by the lamps. Investment in good insulation against heat loss or solar heat gain may help to keep down running costs. Chambers with adequate lighting and cooling systems to grow crop plants generally require a mains voltage of 220 V, and the systems must be compatible with the existing electrical supply. The dependability of the local electricity supply is an important consideration for enclosed facilities. It may be impossible to obtain sufficient emergency generator capacity to keep the lamps and cooling systems operating. However, it is desirable to have an emergency power supply for monitoring systems, so that knowledge of the environmental conditions which exist during the power failure can be obtained (most computer data logging systems have a battery back-up).

(b) Greenhouses:
The power requirements for greenhouses are significantly less than those for enclosed environments. However, if supplementary lighting is used, power requirements will be significant – approximately 200 W m^{-2}. The requirement may be greater in plastic houses than in glasshouses because of the need for forced air ventilating fans to maintain ventilation on warm days.

(c) Open-top chambers:
The power requirement for open-top chambers is dictated by the size of the blower motor used to ventilate the structures. This is critical to the effective operation of these chambers.

5.3 CONTROL

5.3.1 Temperature

(a) Enclosed environments:
The cooling requirements of enclosed environments constitute one of the major costs of constructing and operating the facility. The large radiant loading caused by the lighting of enclosed environments necessitates large energy expenditures for cooling. In enclosed environments, 1.0–1.5 W of electricity must be expended in removing the heat from each watt of lamp power. This requirement becomes more critical during the summer or in tropical climates, because outside air cannot be used for cooling and because temperature exchange coils and cooling towers operate with significantly less efficiency. It is desirable to insulate the walls of enclosed environments, because this improves operating precision through minimizing heat exchange with the outside environment.

The cooling needs of individual growth chambers are often met by separate refrigeration units mounted on or near each chamber, usually operating with a freon (CFC) coolant. Small chambers use air-cooled refrigeration units. This adds significant heat to the room and must be accounted for in calculation of ventilation requirements. Larger chambers, and chambers operating with light banks which provide in excess of 300 μmol m^{-2} s^{-1} of photosynthetic photon flux (PPF), generally require water-cooled refrigeration units. The water is circulated through outside cooling towers with evaporative cooling. Cooling costs can be reduced if the light period, which requires the greatest amount of cooling, is scheduled during the night when outside temperatures are lower.

In facilities with multiple growth chambers, cooling is more often accomplished with a central cooling system which distributes coolant to the individual chambers. This has the advantage of increasing the operating efficiency, but if the central chilling system fails then all the growth units are affected. There are also many problems associated with piping very cold coolant around a facility, such as moisture condensing on the cold pipes if they are not carefully insulated. Moisture condensation may be partially overcome by using a coolant that is not excessively cold. However, this limits the minimum temperatures which can be attained, and reduces the effectiveness of dehumidification through condensation on the cold coils.

The temperature control range in controlled environments is commonly 5–40°C. Temperatures lower than 0°C require special construction to allow maintenance of the chamber below freezing point, and duplicate cooling coils are required to provide regular defrosting of the cooling coils. There is little need to maintain plant growth chambers above 40°C. Chambers constructed with proportional temperature controllers will provide more uniform temperature control and have smaller fluctuations in humidity level than chambers with on–off controllers. However, there is no evidence that plant growth is any different in chambers constructed with on–off controllers than in chambers with proportional controllers.

(b) Greenhouses:

Cooling in greenhouses is accomplished primarily by opening vents. These are commonly controlled by automated systems that open and close at preset temperatures. However, when outside temperatures increase during the warmer months of the year and sun intensity increases, natural ventilation cannot provide adequate cooling. Fan and pad cooling systems are therefore commonly installed. These systems involve fibrous pads on one wall over which water is run continuously, and large exhaust fans on the other side. Evaporative cooling of the air occurs as it is drawn across the moist pads. The fan should have sufficient capacity to provide one volume change per minute, and the pad surface area (m^2) is determined by dividing the value of 0.395 by the fan rate (m^3 minute^{-1}). During warm sunny periods it is usually impossible to effectively cool the greenhouse. The inside temperature can generally be held only a few degrees below outside temperatures. These systems will provide significantly better cooling in arid regions where vapour pressures of the air are low. High pressure mist or fog (2.8 Mpa; 400 psi) are also used to cool greenhouses during the warmer months. A water pump capable of producing high pressure (2.8–4.2 Mpa; 400–600 psi) should be used and one nozzle positioned every 10–15 m^2. When the system operates the whole greenhouse quickly fills with fog. The evaporation of this water cools the air.

In summer, automatic shade cloth is used alone or in conjunction with the pad and fan cooling system. The shade cloth is furled and unfurled automatically depending on the intensity of the sun. This shade cloth reduces the sunlight by at least 50% which in turn reduces heat load in the greenhouse. Mechanical air conditioning systems are so large and the electrical costs to operate the unit are so high that it makes it too expensive to install and operate. Reflective glass (as used in windows of large buildings to conserve energy) are not used in greenhouses, because of their high cost and because they cause large reductions in useful light.

The temperature control in glasshouses is maintained with either manual or automated systems that open the vents when a particular set temperature is reached during the day, and heaters come on when a set minimum temperature is reached at night. Thus temperatures often exceed the set temperature during the day when the sun is bright and outside temperatures are warm. Automated systems can be obtained to provide proportional control of vent opening to minimize temperature fluctuations in the glasshouse. The temperature control in plastic houses is similar to that in glasshouses. Fans are activated to cool, and heaters are activated to heat the structure. Computers can be incorporated to improve temperature control in both glass and plastic houses.

(c) Open-top chambers:
Cooling in open-top chambers is commonly accomplished by moving large amounts of air through the chamber. It is necessary to distribute air around all sides at the bottom of the chamber to minimize the temperature variability within the chamber. However, high air velocities increase the water deficit of plants and therefore air movement rates should be kept between 0.5 and $1.0\,m\,s^{-1}$. Care should be taken in selection of blower units so that the heat of the motor does not significantly warm the air being blown into the chamber.

No temperature controllers are generally installed in open-top chambers.

5.3.2 Lighting

(a) Enclosed environments:

Lamp Type:
The lamps used in enclosed environments should be either high pressure metal halide (MH) or high output cool white fluorescent (CWF) lamps. These lamps are used to provide an adequate balance of red and blue wavelengths for growth of a wide variety of plant species. The MH lamps provide a higher level of lighting but irradiance uniformity will not be as good as with CWF lamps. Warm-season crops which need high light, such as corn and soybeans, require the intensities produced by MH lamps for adequate plant growth. Cool season crops, such as lettuce and potatoes, will develop effectively under CWF lamps if the lamps are closely packed across the ceiling of the growing area. High pressure sodium lamps (HPS) have greater photosynthetic radiation output and better longevity, but their low level of 400–500 (blue) wavelengths makes them of questionable use in enclosed environments, unless combined with lamps with high levels of 400–500 nm wavelengths. HPS lamps have been combined with high pressure mercury lamps (HQI) and sometimes with MH lamps to provide the needed 400–500 nm wavelengths. However, the lamps must be spaced carefully, and an adequate distance provided between lamps and plants, so that a uniform overlapping of irradiation is obtained at plant level. It is desirable to include small wattage incandescent lamps to provide a low level of lighting for photoperiod control. CWF lamps have the greatest losses in intensity with ageing and HPS the least (Table 5.1). The longevity, photosynthetic efficiency, and comparative costs of different lamp types are also shown in Table 5.1.

Quantity:
The quantity of lighting required by plants is a function of the level of irradiance from the lamps and the length of time during which irradiance is provided daily. Generally a level of $400\,\mu mol\,m^{-2}\,s^{-1}$ PPF provided for 12 hour each day is sufficient for cool-season crops, and $800\,\mu mol\,m^{-2}\,s^{-1}$ PPF for 12 h each day is sufficient for warm-season crops. $1.0\,\mu mol\,m^{-2}\,s^{-1}$ of PPF is roughly equivalent to about $0.3\,W\,m^{-2}$ of photosynthetically active radiation (PAR); however, this conversion ratio varies with the type of lamp [13]. These levels can be reduced proportionately if the light period is extended. However, extension of the light period may cause undesirable growth responses in photoperiodically sensitive species.

Lighting systems should be controlled by 24-hour timers so they can be turned on and off at desired times. Settings with a precision of 10 minute intervals are generally sufficient. It is necessary to have photoperiodic control lighting (incandescent lamps) on separate

Table 5.1. Characteristics of lamps used for photosynthetic irradiation of plants.

Lamp type		Photons* (μmol s^{-1}) per watt†	Lumens† per watt	Average life (hours)‡	Reduction in output§
High pressure sodium	400W	1.47	106	24 000	10%
	1000W	1.77	128	24 000	10%
High pressure metal halide	400W	1.25	87	20 000	24%
	1000W	1.67	115	12 000	23%
Low pressure sodium	180W	1.37	144	18 000	0%¶
Cool white fluorescent	215W	0.83	67	15 000	32%
Incandescent	1000W	0.44	22	2 000	3%

* Photons of photosynthetically active radiation (400–700 nm).
† Electrical wattage for new lamps and ballast.
‡ 10 hours per start.
§ At 50% of indicated life.
¶ Generally no reduction in output during use but lamp wattage is increased significantly to avoid output reductions.

timers, to permit their use for extended photoperiods at low light levels.

Uniformity of lighting is a critical factor, particularly with incandescent (HID) lamps. HID lamps are essentially point sources of light, whereas fluorescent lamps serve as area sources of light. Reflectors on HID lamps must be carefully shaped, and located at a spacing and height which effectively disperses the light uniformly over the growing area. Care must be taken as plants grow upward in the chamber to ensure that light uniformity is maintained.

Intensity is commonly controlled by varying either the container shelf height or by varying the position of the light bank. The intensity can also be controllable by turning off individual fluorescent lamps. The intensity of HID lamps can be controlled by dimmers, although this adds significantly to the cost. Dimming of fluorescent lamps is more costly and maintenance of the system is expensive. Fluorescent lamps degrade (lose intensity) with time, so the light intensity at the plant level must be monitored at frequent intervals and corrections made to maintain the desired PPF level.

Lamp cooling:
There is no particular need for any special cooling of MH and HPS lamps provided there is provision for air movement to avoid excessive buildup of temperature in the vicinity of the lamps. However, CWF lamps should be maintained at a constant temperature, since the output is closely linked to the temperature of the lamp tube. A lamp wall temperature of 40°C (about 20°C air temperature) provides maximum output. If the temperature increases or decreases, the output is reduced. Thus control of lamp bank temperature is needed if constant output is required for research investigations.

Long-wave radiation:
MH, HPS and CWF lamps emit about twice as much long-wave radiation (>700 nm) as sunlight on the Earth's surface. Thus lamps produce significantly greater soil and leaf heating than sunlight at the same photosynthetic irradiance level, unless precautions are taken to reduce the long-wave irradiation. Heating becomes a serious problem when irradiance levels exceed about 500 μmol $m^{-2} s^{-1}$. Using a glass pan below the lamps,

it is possible to construct and maintain a water layer of about 2–4 cm in depth for absorption of this undesirable radiation [4]. This poses additional problems, since the water must be cooled to a constant temperature and maintained regularly to avoid any accumulation of particles which would reduce transmission.

(b) Greenhouses:

Lamp type:
HPS lamps are recommended for use in greenhouses because there is sufficient 400–500 nm wavelengths from the sunlight for normal growth of plants. This lamp is the most efficient and has the longest life.

Quantity:
At 40° latitude, a level of 150 μmol $m^{-2} s^{-1}$ PPF of supplementary irradiation is desirable, for cloudy days and to extend the photoperiod to 16 h. At higher latitudes, the level of supplementary light will have to be greater (about 250 μmol $m^{-2} s^{-1}$ at 50° latitude) and it should be less at lower latitudes. Greater efficiency can be obtained by monitoring the sunlight irradiance and providing supplementary irradiance to provide a reasonably constant amount of irradiance for each 24-hour period.

Supplementary lighting systems in greenhouses should be controlled by timers. Lamps should be installed in the greenhouse at a spacing and height which maintains uniform irradiance across the growing area.

(c) Open-top chambers:
No lighting control is generally provided in open-top chambers.

5.3.3 Humidity

(a) Enclosed environments:
Maintenance of a desired humidity level in enclosed environments is difficult because (i) the coils used for cooling will condense out large quantities of water and make it difficult to maintain high humidity levels, and (ii) plant transpiration water and surface water constantly adds humidity to the chamber, so it is difficult to maintain low levels of moisture. Consequently, levels of between 50 and 90% are the reasonable limits for humidity control. Special equipment is needed for effective control beyond this range.

Humidity is most commonly added with atomisers that produce very fine droplets, which vaporize to humidify the air. Various types of centrifugal atomisers are most commonly used, although ultrasonic misting is sometimes used. The use of fine nozzles and high pressure can also provide fine droplets but it is hard to maintain a consistent fine spray from nozzles. Steam is effective, but it is necessary to have a separate steam supply. Steam used for heating commonly contains additives that are toxic to plants. All humidification systems should use distilled water to avoid the accumulation of salts on chamber surfaces.

It is difficult to effectively distribute added moisture into the chamber. Nozzles are generally located in the air stream just as it enters the chamber to avoid collection and condensation of water on the interior ducts. However, plants located close to the inlet are subject to accumulation of water or heat from the steam. It is preferable to have several humidification inputs distributed around the chamber.

Humidifying chambers also make demands on the temperature control system. Addition of water droplets causes cooling of the air as water evaporates, and addition of steam adds heat to the chamber.

Dehumidification results from moisture condensation on the cooling coils. The added heat forces additional cooling and is increased by additional condensation. The extent of dehumidification can be controlled by the temperature of the coolant and the surface area of the coils. The use of proportional controls for cooling is necessary for precise humidity control. The coils must be operated

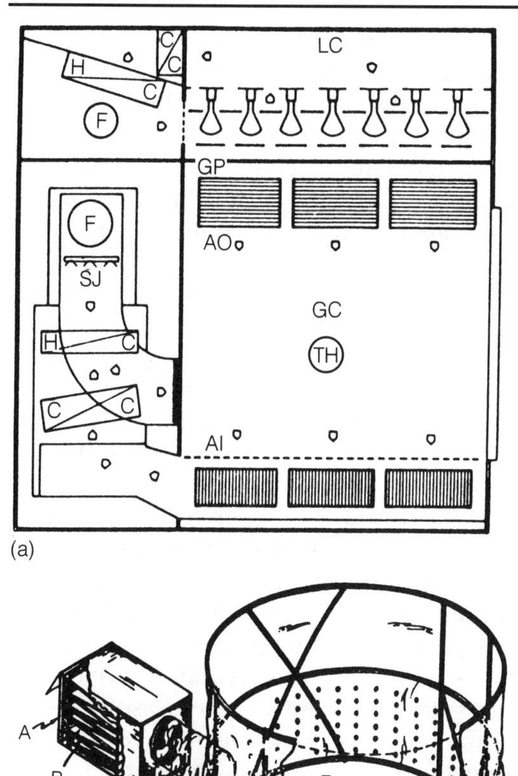

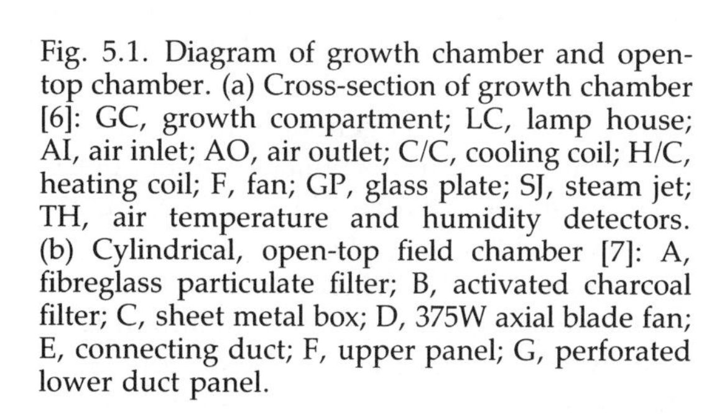

Fig. 5.1. Diagram of growth chamber and open-top chamber. (a) Cross-section of growth chamber [6]: GC, growth compartment; LC, lamp house; AI, air inlet; AO, air outlet; C/C, cooling coil; H/C, heating coil; F, fan; GP, glass plate; SJ, steam jet; TH, air temperature and humidity detectors. (b) Cylindrical, open-top field chamber [7]: A, fibreglass particulate filter; B, activated charcoal filter; C, sheet metal box; D, 375W axial blade fan; E, connecting duct; F, upper panel; G, perforated lower duct panel.

at the temperature needed to optimize both temperature and humidity control. Dehumidification beyond that possible with cooling coils is obtained with solid matrix drying units attached to the chamber. These units tend to be costly and maintenance requirements are high.

In most growth chambers, control of humidity is within a range limited by the temperature of the cooling coils. The lowest humidity will be directly related to, and can be calculated from, the amount of water in saturated air at the temperature of the cooling coils. The highest humidity can be close to saturation, but to obtain this, large amounts of water must be continuously added, because the cooling coils will constantly be condensing water from the air. High humidities can be maintained most effectively through use of cooling units with a large surface area and maintenance of relatively warm coolant with proportional controls so the temperature is no cooler than necessary. If water accumulates in the bottom of chambers, this increases the level of humidity. The dew-point temperature of the air surrounding the chambers will influence the maximum and minimum levels that can be attained within the chambers, because of the large amount of leakage in and out of the chamber. In most chambers, the control of relative humidity will have at least a 2% variation, and commonly a 5% variation, because of fluctuations in cooling coil temperatures.

(b) Greenhouses:
Humidification in greenhouses is generally accomplished with the watering system. Evaporation from the floors, bench and plants increases the humidity. The large volume of fresh air exchange in most greenhouses makes it difficult to significantly alter the humidity of greenhouses. In houses used for seed germination, or for the production of rooted cuttings, regular timed misting or fogging systems can be installed to maintain a film of water on the leaf and soil surfaces. Dehumidification in greenhouses is accomplished by exchange of the inside humid air with cooler, drier outside air. This process is very important in reducing leaf diseases such as mildew.

Accurate control of humidity is rarely attained in greenhouses. However, humidity is commonly increased by frequent hosing down of floor and bench surfaces and through use of fogging and misting.

(c) Open-top chambers:
Humidification of these facilities is not practical because of the large amount of fresh air that is constantly moved through the units.

5.3.4 Carbon dioxide

(a) Enclosed structures:
Carbon dioxide (CO_2) control is of significant importance in enclosed environments, in order to obtain precise comparisons among experimental treatments. Fluctuations in CO_2 levels occur mainly because of the respiratory activity of people and photosynthetic activity of plants, but some fluctuations result from variations in the CO_2 level of fresh air from outside.

Systems are available for increasing CO_2 levels above ambient, but systems for decreasing levels require significant maintenance, time and cost. Thus in the development of a controlled CO_2 research study, it is desirable to maintain the low level (control level) at a concentration slightly above the highest ambient level expected.

Control systems are developed around infrared gas analysers (IRGAs) that provide continuous monitoring of CO_2. The high cost of analysers generally makes it necessary to use one IRGA to monitor all rooms or chambers. The analyser is connected to all chambers through a manifold and sampling system, which samples each chamber for a short time period (1–2 minutes) in succession. During the sampling period, the control system activates the release of CO_2 into the sampled room or chamber. Many different procedures may be used to control the release of CO_2 so that uniform regulation is obtained.

The accuracy of the CO_2 control is limited by changes in atmospheric pressure. The IRGA cannot provide an accurate measurement as this occurs. Frequent calibration is necessary, at least once a day, if precise measurements are to be obtained. Atmospheric pressure can be monitored and the readings corrected for pressure differences through a computer program. Fluctuations in the water vapour present in the sampled air also reduces the accuracy of the measurements. This problem can be minimized by using vapour traps in the sampled air. However, if these are not properly maintained and/or subject to temperature fluctuations, additional errors will be introduced by sorption and desorption of CO_2 in the water trap. Some of the newer model IRGAs incorporate sensing procedures to minimize water vapour effects, thus reducing the need for a vapour trap (Chapter 9).

Carbon dioxide for supply to a chamber or room is usually obtained from cylinders of compressed CO_2. Combustion of methane or natural gas to generate CO_2 is not recommended, because this process releases varying levels of hydrocarbons that can have toxic effects upon the plants. There is also a danger to humans from accumulation of carbon monoxide if combustion is not complete. The use of frozen CO_2 is also not recommended because of the difficulties in obtaining precisely controlled release of CO_2.

Reduction in CO_2 concentration can be obtained by adding fresh ambient air, but the lowest level attainable is limited by the outside CO_2 level. To reduce CO_2 below ambient levels, it is necessary to incorporate a scrubbing system. This is generally accomplished by dry absorbent filters which contain potassium or sodium hydroxide (L. Giles, Duke University Phytotron, North Carolina, USA; pers. comm.). The filters require frequent replacement (usually daily) to maintain their effectiveness. A portion of the air stream of the room or chamber is diverted through the dry filter, and the amount diverted is controlled by the CO_2 analyser.

Accurate calibration of IRGAs is a problem because standard gases purchased from suppliers are often not determined accurately, even when special analysis has been requested. Make certain that an independent check is made on each cylinder of calibration gas, by exchange with another research worker or through use of gases traceable directly to a national standards laboratory.

(b) Greenhouses:
Supplementing of CO_2 levels in greenhouses is only practicable during cold weather, when greenhouse vents can be kept closed. During warm weather when the vents are open, continuous large dilution with ambient air makes CO_2 supplementation ineffective. Therefore the control of CO_2 levels in greenhouses cannot be maintained as precisely as in controlled environments.

Less expensive and less accurate CO_2 analysers are commonly used in commercial greenhouses, together with combustion units burning methane or natural gas to supply the CO_2. Care must be taken to ensure the natural gas does not contain significant amounts of sulphur, which will form sulphur dioxide during combustion and can cause serious injury to plants. Take care that the burners stay clean, and supply adequate fresh air to the burner in order to obtain complete combustion. Greenhouse operators generally make certain there is a small amount of fresh air venting into the greenhouse to avoid incomplete combustion and production of toxic gases.

Some large greenhouse operators use bulk CO_2 from refrigerated tanks, where returns justify the cost.

(c) Open-top chambers:
The procedures outlined above for enclosed environments are commonly used also for open-top chambers. Since the distance from the chambers to the IRGA is often quite large, it is necessary to install a separate pump to continuously draw air samples from all chambers simultaneously to reduce the time lag in sampling each chamber.

5.3.5 Air supply and movement

(a) Enclosed environments:
A significant amount of fresh air should be supplied to chambers to minimize the chance that phytotoxic contaminants will accumulate, and to maintain uniform concentrations of CO_2. This latter need arises because plants rapidly deplete CO_2 during the light period and because human breathing raises CO_2 levels. Most chambers have very significant leakage with room atmosphere and additional fresh air fans are not needed. This leakage can be as high as 5% volume exchange per minute [14]. Built-in growth rooms may not have this exchange rate and a fresh air supply should be provided. Growth chambers should be installed in an area ventilated separately from the rest of the building, to avoid contaminants present in the building from affecting plants in the chambers.

Air movement in chambers must be rapid enough to maintain uniform temperature throughout the chamber. As light levels increase, air movement should be increased to maintain temperature uniformity. Vertical air flow is most commonly used in chambers. This minimizes the horizontal temperature gradients, but increases the vertical gradients. Vertical air flow patterns are sometimes upward and other times downward. There is no evidence that direction exerts any significant effect upon plants, although vertical temperature gradients in the chamber will be greater with upward flow than with downward flow. Vertical air flow can only maintain uniform temperature control in the chamber if slotted shelves are used and if sufficient space is left between containers for good air movement. Horizontal air flow has the advantage that it produces an air flow pattern similar to that in the outside environment, but it also produces a horizontal gradient in temperature across the chamber. Horizontal air flow is desirable to maintain better temperature uniformity in tissue culture rooms that contain multiple shelves with lamps over each shelf.

(b) Greenhouses:
Fresh air is provided to greenhouses through the operation of vents in glasshouses or fans in plastic houses. Fresh air is needed for cooling and to provide adequate amounts of

CO_2 during the sunlit part of the day. In glasshouses, vents at the sides provide entry for the air and roof vents provide exhaust by convection as the air is heated within the house. In plastic houses the air is commonly distributed through large plastic ducts with holes distributed along the entire length of the duct. Air enters the duct, passes into the greenhouse and exits via the exhaust fan.

(c) Open-top chambers:
Large amounts of air are provided continuously by blower fans attached at the bottom of the chamber. Air is usually distributed through a double-layered plastic membrane which encircles the chamber to a height of one metre. Holes scattered on the inner wall surfaces provide uniform distribution of the fresh air through the chamber.

5.4 MEASUREMENT

Guidelines for measurement and reporting of the environment of controlled environments have been developed and published by growth chamber committees [2,8]. These provide specific details for recommended types of instruments and also where, when, and how measurements should be taken to describe accurately the environment under which experiments are undertaken. The recommendations developed by a growth chamber committee in the United States are shown in Table 5.2.

5.4.1 Temperature

All structures:
Many different types of temperature sensors can be used for effective temperature measurement. The principal considerations are whether they can be effectively screened from interfering radiation effects and how quickly they can respond to temperature changes. These considerations generally require that the sensors are small, protected from any hot or cold radiating surfaces, and aspirated (see also Chapter 4).

Thermocouple or resistance thermometers (thermistor, diode, or wire) are recommended because they are of small size, and have a rapid response.

The accuracy of temperature sensors (to 0.1°C) can be determined readily through use of a standard mercury thermometer and liquid calibration bath.

Temperature in the chambers should be measured at a location not affected significantly by either the growth of the plants or by different arrangements of materials in the chamber. It is advisable to have a temperature profile of each chamber, which should indicate the best location to sample the temperature of that chamber.

5.4.2 Lighting

(a) Enclosed structures:
The light level, or more accurately the photosynthetic photon flux (PPF) level, should be monitored with a quantum meter (400–700 nm waveband with a cosine corrected sensor). Use of a lux (footcandle) meter is not recommended, because these meters monitor the light sensitivity of the human eye and do not monitor accurately the photosynthetic activity of plants. It is important to establish the light gradient from end to end and from side to side within the chamber. Replicate blocks of treatments should be located so similar PPF levels are provided for each block. This may necessitate arranging plants in irregular-shaped blocks or scattering the plants of a block around in the chamber.

PPF levels should be monitored at least weekly, because the output of lamps decreases significantly with age. The average PPF level for the weekly readings should be reported as the radiation level for the experiment. Regular monitoring is also necessary to make certain that all rooms or chambers of an experiment are maintained at the same PPF level when a variable other than light is being studied.

Table 5.2. Guidelines for measurement and reporting of the environment in growth chambers*

Parameter	Typically used unit	Measurements		
		Where to take	When to take	What to report
Radiation				
PAR (photosynthetically active radiation)				
(a) Photosynthetic photon flux (PPF), 400–700 nm with cosine correction	$\mu mol\ m^{-2}\ s^{-1}$	At top of plant canopy. Obtain average over plant growing area.	At start and finish of each study and bi-weekly if studies extend beyond 14 days.	Average over containers at start of study. Decrease or fluctuation from average over course of study. Wavebands measured.
(b) Photosynthetic irradiance, 400–700 nm with cosine correction	$W\ m^{-2}$	(Same as PPF)	(Same as PPF)	(Same as PPF)
Total irradiance with cosine correction indicate bandwidth	$W\ m^{-2}$	(Same as PPF)	At start of each study.	Average over containers at start of study. Wavebands measured.
Spectral irradiance, 250–850 nm in 20 nm bandwidth with cosine correction	$W\ m^{-2}\ nm^{-1}$ or $\mu mol\ m^{-2}\ s^{-1}\ nm^{-1}$	At top of plant canopy in centre of growing area.	At start of each study.	Graph or table of irradiance for seperate wavebands.
Illuminance,† 380–780 nm with cosine correction	k lux	(Same as PPF)	At start of each study.	(Same as total irradiance)
Carbon dioxide	$mmol\ m^{-3}$	At top of plant canopy.	Hourly over the period of study.	Average of hourly average readings and range of daily average readings over the period of the study.
Watering	litre (l)	–	At times of additions.	Frequency of watering. Amount of water added per day and/or range in soil moisture content between waterings.
Substrate	–	–	–	Type of soil and amendments. Components of soil-less substrate. Container dimensions.
Nutrition	Solid media: $mol\ kg^{-1}$ or $mol\ m^{-3}$	–	At times of nutrient additions.	Nutrients added to solid media. Concentration of nutrients in liquid additions and solution culture. Amount and frequency of solution addition and renewal.

Table 5.2 Continued

Parameter	*Typically used unit*	*Measurements*		
		Where to take	*When to take*	*What to report*
pH	pH units	In saturated media, extract from media, or solution of liquid culture.	Start and end of studies in solid media. Daily in liquid culture and before each pH adjustment.	Mode and range during study.
Temperature				
Air, shielded and aspirated ($3\,m\,s^{-1}$) device	°C	At top of plant canopy. Obtain average over plant growing area.	Hourly over period of study (continuous measurement advisable).	Average of hourly average values for light and dark periods of study with range of variation over growing area.
Soil or liquid	°C	In centre of container.	Hourly during the first 24 h of the study. Start immediately after watering (monitoring over the course of the study advisable).	Average of hourly average values for the light and dark periods for the first day or over entire period of the study if taken. Location of measurement.
Atmospheric moisture				
Shielded and aspirated ($3\,m\,s^{-1}$) psychrometer, dew-point sensor or infra-red analyser	% RH, dew-point tempera-ture or $g\,m^{-3}$	At top of plant canopy in centre of growing area.	Once during each light and dark period taken at least 1 h after light changes. Monitoring over the course of study advisable.	Average of daily readings for both light and dark periods with range of diurnal variation over the period of the study (or average of hourly values if taken).
Air velocity	$m\,s^{-1}$	At top of plant canopy. Obtain maxiumum and minimum readings over plant growing area.	At start and end of studies. Take 10 successive readings at each location and average.	Average and range of readings over containers at start and end of study.
Electrical conductiviy	$dS\,m^{-1}$ (decisie-mens per metre) = $mho\,cm^{-1}$	In saturated media, extracted from media or solution of liquid culture.	Start and end of studies in solid media. Daily in liquid culture.	Average and range during study.

* Proposed by the North Central Regional Committee (NCR-101) on growth chamber use.
† Report with PAR reading **only** for historical comparison.

(b) Greenhouses:
The light in greenhouses should also be monitored with a quantum meter. It is most useful to monitor the PPF level continuously over each daily period and report the photon flux as mol $m^{-2} d^{-1}$.

(c) Open-top chambers:
Light measurement should be undertaken as in greenhouses.

5.4.3 Humidity

All structures:
Humidity is most commonly monitored and controlled with resistance or capacitance type sensors. The sensors should be maintained in aspirated housings and protected from direct lamp or sun radiation. These sensors have a nominal effective range between 50 and 90% relative humidity (RH). If they are used outside this range they should be calibrated carefully. The response of humidity sensors is subject to rapid and significant shifts in calibration and thus it is advisable to check the humidity level daily with a second sensor, preferably a wet- and dry-bulb (psychrometric) type unit (see also Chapter 4).

Accurate measurement of humidity can also be obtained with IRGAs, but these instruments are significantly more expensive than psychrometers or dew-point meters.

Accurate measurement of humidity for calibration of control sensors must be made with a psychrometer (wet- and dry-bulb instrument) or a dew-point meter. Although psychrometers can provide very accurate readings of humidity, most commercially available and moderately priced instruments require that they be read *in situ*. This makes it almost impossible to obtain accurate readings in small growth chambers because the door has to be opened for each measurement. Remotely monitored psychrometers can be obtained, but most suffer from insufficient aspiration of the wet bulb and do not provide accurate measurements. Dew-point instruments are the most useful accurate instruments for monitoring of humidity in small chambers. If the dew-point meter is positioned outside the chamber, make certain the sampling air line does not cool below the dew-point temperature of the air being sampled.

5.4.4 Carbon dioxide

All structures:
Carbon dioxide is most commonly monitored with an infrared gas analyser (IRGA) providing continuous monitoring that can be connected to a control system. A less costly method of analysis is through use of dry absorbents, although these do not provide an accurate or precise measurement. A small tube of absorbent is used which changes colour when exposed to CO_2. This is useful for obtaining an approximation of the CO_2 level, but it cannot be used for continuous monitoring or control. Carbon dioxide can be monitored accurately through gas chromatography, but this is costly and can not be readily automated for continuous measurement.

Carbon dioxide levels should be monitored at least once each hour and averaged to obtain an accurate record of the level to which plants are subjected.

REFERENCES

1. Anonomyous (1989) *Ventilation for Greenhouses*. Grow Electric Series. National Agricultural Centre, Warwickshire, U.K.
2. American Society of Agricultural Engineers (1982) Guidelines for measuring and reporting environmental parameters for plant experiments in growth chambers. ASAE Engineering Practice: EP411. pp. 406–409. In: Agricultural Engineers Yearbook 1982. American Society of Agricultural Engineers, St. Joseph, MI, USA.
3. Bishop, H.D., D. Crouch and M. Eisentraut (1981) *The Celanese Greenhouse Study*. Alberta Research Council, Alberta, Canada.
4. Bubenheim, D.L., B. Bugbee and F.B. Salisbury (1988) Radiation in controlled environ-

ments: influence of lamp type and filter material. *J. American Society of Horticultural Science* **113**, 468–474.
5. Carpenter, G.A. and L.J. Mousley (1960) The artificial illumination of environmental control chambers for plant growth. *J. Agricultural Engineering Research* **5**, 283–306.
6. Eguchi, H, ed. (1990) *Biotron Institute* (descriptive brochure). Biotron Institute, Kyushu University, Kyushu, Japan.
7. Heagle, A.S., R.B. Philbeck, H.H. Rogers and M.B. Letchworth (1979) Dispensing and monitoring ozone in open-top chambers for plant-effects studies. *Phytopathology* **69**, 15–20.
8. Krizek, D.T. (1982) Guidelines for measuring and reporting environmental conditions in controlled-environment studies. *Physiol. Plant.* **56**, 231–235.
9. Langhans, R.W., ed. (1978) *A Growth Chamber Manual*. Cornell University Press, Ithaca, NY, USA.
10. Langhans, R.W. (1990) *Greenhouse Management*, 3rd edn. Halcyon Press, Ithaca, NY, USA.
11. Olszyk, D.M., T.W. Tibbitts and W.M. Hertzberg (1980) Environment in open-top field chambers utilized for air pollution studies. *J. Environmental Quality* **9**, 610–615.
12. Payer, H.D., T. Pfirrmann and P. Mathy, eds. (1990) *Environmental Research with Plants in Closed Environments*. Air Pollution Research Report 26. Commission of the European Communities, Brussels, Belgium.
13. Thimijan, R.W. and R.D. Heins (1983) Photometric and quantum light units of measure: a review of procedures for interconversion. *HortScience* **18**, 818–822.
14. Tibbitts, T.W. and T.T. Kozlowski, eds. (1979) *Controlled Environment Guidelines for Plant Research*. Academic Press, New York, NY, USA.

6

Canopy structure and light interception

P.S. Nobel, I.N. Forseth and S.P. Long

6.1 INTRODUCTION

The rice cultivar I.R.8, so-called 'miracle rice', and its derivatives, represent one of the most significant contributions of plant physiology to the improvement of crop yields. By comparison to older varieties, the I.R. varieties have higher rates of crop photosynthesis achieved by selection of varieties with a canopy structure allowing more light to reach lower leaves. Indeed, modification of canopy structure can substantially improve crop yield by its influence on light interception by plants [2].

Canopy structure will determine the microenvironment surrounding leaves, such as radiant flux density, air temperature, soil temperature, air vapour pressure, leaf temperature, soil heat storage, wind speed, precipitation interception, and leaf wetness duration [23,25,28]. Thus, canopy structure has a major influence on the exchange of mass and energy between plants and their environment. Canopy structure refers to the amount and organization of above-ground plant material, including the size, shape and orientation of plant organs such as leaves, stems, flowers and fruits [25]. However, because of the inclusion of elements of each preceding level in the next level of organization, the complexity of canopy structural analyses increases rapidly as we proceed from individual organs to whole plants to monospecific stands to plant communities. This makes quantitative descriptions of plant canopies exceedingly difficult. In this chapter, we will address methods that are used to measure only a few important aspects of canopy structure: the leaf area index and its distribution with height, leaf inclination, and leaf orientation.

When analysing photosynthesis at the canopy level, we are confronted immediately with the great structural diversity among different types of vegetation [10,24]. Each plant community has its own unique spatial pattern for displaying photosynthetic surfaces and hence for intercepting **Q** (the photosynthetic photon flux, PPF; i.e. photons with wavelengths between 400 and 700 nm). The presence of many layers of leaves and the diurnally changing availability and direction of **Q** complicate the analysis. Thus, even approximate descriptions of **Q** interception within regular stands of arable crops grown in monoculture involve calculations based on many simplifying assumptions. Here, the

Photosynthesis and Production in a Changing Environment: a field and laboratory manual
Edited by D.O. Hall, J.M.O. Scurlock, H.R. Bolhàr-Nordenkampf, R.C. Leegood and S.P. Long
Published in 1993 by Chapman & Hall, London.
ISBN 0 412 42900 4 (HB) and 0 412 42910 1 (PB).

attenuation of **Q** down through the canopy will be treated in general outline (leaf structure is examined in Chapter 7).

6.2 RADIATION IN CANOPIES

The photosynthetic photon flux at each level in a canopy is often the major factor determining the rate of carbon dioxide (CO_2) assimilation of individual leaves. To understand the contribution made by individual leaves to canopy photosynthesis, we must first consider how **Q** varies down through the various layers of vegetation.

6.2.1 Changes in PPF within the canopy

Crop photosynthesis will not only depend on the distribution of **Q** among layers, but also on the total amount of **Q** absorbed by the canopy [10,24]. Absorption depends on leaf orientation, sun elevation in the sky, the finite width of the sun's disc, changes in spectral distribution of **Q** through the canopy, multiple reflections of **Q** within the canopy, and the arrangement of leaves in the canopy (e.g. clumped or uniform). Even in the simplest of canopies, an analysis covering all of these factors is usually too complex to be of practical use. Instead, we will assume that the decrease of **Q** down into a canopy is analogous to absorption of light by chlorophyll or other pigments in a solution, which is described by Beer's law [23]. This approximation is particularly useful when there is a random distribution of leaves horizontally, as can occur in certain plant communities of moderate density [10,22,24].

As we move down into the canopy, **Q** decreases more or less exponentially with the amount of leaf material encountered (Fig. 6.1a; Monsi and Saeki [21,22]). Thus, we must focus on the distribution of leaves in space rather than on the distance above the ground (if significant quantities of **Q** are absorbed by the stems, this must also be taken into account). For some canopies the greatest proportion of leaf area occurs near the centre (e.g. many crops and trees). Let F be the average cumulative leaf area index, i.e. total m^2 leaf area per m^2 ground area summed, as we move down through the canopy (Fig. 6.1b). As is the case for surface area used in expressing leaf photosynthetic rates (Chapter 9), the parameter F uses the area of one side of a leaf only. F is zero at the top of the canopy and takes on its maximum value at ground level, where it equals the leaf area index, L (Fig. 6.1b).

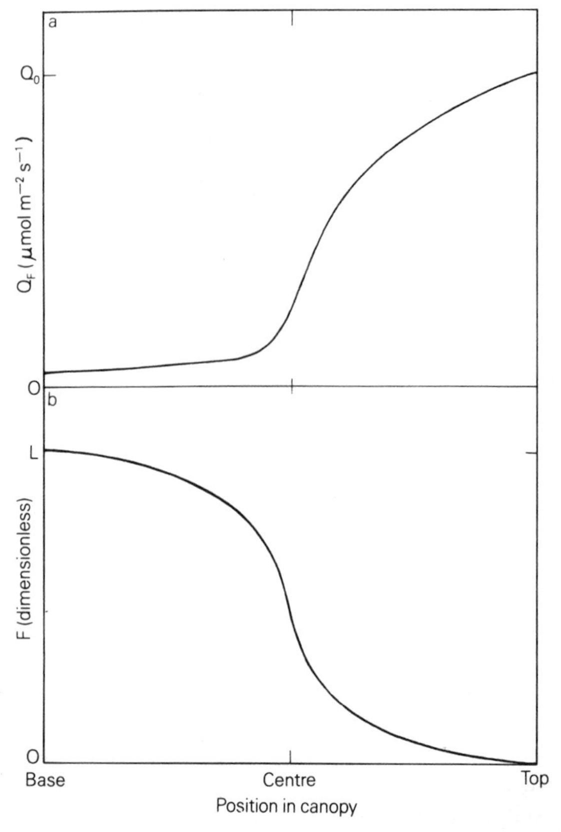

Fig. 6.1. Changes in photosynthetic photon flux density (**Q_F**, a) and cumulative leaf area index (F, b) through a canopy with most of its leaves near the centre. **Q_F** has its maximum value of **Q_0** at the top of the canopy, and F has its maximum value of L (equal to the Leaf Area Index) at the canopy base.

Q in the horizontal plane immediately above the top of the canopy is defined as **Q_0**

(Fig. 6.1a). As we move down into the canopy, **Q** decreases, primarily because of absorption by the photosynthetic pigments. At any level F within the canopy, the rate of change of **Q** is:

$$\frac{dQ}{dF} = -kQ_F \qquad (6.1a)$$

where the dimensionless parameter k represents the fraction of incident photons absorbed by unit leaf area and is referred to as the foliar absorption coefficient.

After integration, **Q** at level F (Q_F) is:

$$Q_F = Q_0 e^{-kF} \qquad (6.1b)$$

which upon taking logarithms and rearranging becomes:

$$kF = \ln(Q_0/Q_F). \qquad (6.1c)$$

If we ignore changes in spectral distributions at different levels in the vegetation, **Q** in Equation 6.1 can also be measured as the radiant energy flux.

6.2.2 Considerations for specific canopies

The foliar absorption coefficient ranges from 0.3 to 1.3 for the majority of leaf canopies. In canopies where the leaves are nearly vertical, e.g. many grasses (Fig. 6.2a), light penetrates to the lower layers readily and so k is often low, typically about 0.4. For such a canopy we may determine the cumulative leaf area index needed to absorb 95% of the PPF incident at the top of the canopy. Using Equation 6.1c, we obtain:

$$\begin{aligned} F &= \ln(Q_0/Q_F)/k \\ &= \ln[Q_0/(0.05\ Q_0)]/0.4 \\ &= 7.5 \end{aligned}$$

Thus, when the average leaf area index has the fairly high value of 7.5 for such grasses, 5% of the incident light would reach the soil surface.

We may also use Eqn. 6.1c to determine the value of k needed for absorption of 95% of Q_0 by a canopy of given L. For example, if a

Fig. 6.2. Canopies with various leaf inclinations: a) vertical leaves of grasses (erectophile); b) horizontal leaves, e.g. beans (planophile); and c) leaves varying from vertical near the top of the canopy to horizontal near the base, e.g. sugar beet.

canopy has a leaf area index of 3, the value of k required is:

$$k = \ln(Q_0/Q_L)/L$$
$$= \ln[Q_0/(0.05\ Q_0)]/3$$
$$= 1.0$$

Such a high foliar absorption coefficient applies to horizontal leaves with high chlorophyll levels (e.g. 0.5 g chlorophyll m^{-2}), which can be found in crop plants such as potato, soybean, sunflower and white clover. Canopies with most leaves in the horizontal plane are termed planophile (Fig. 6.2b), whereas canopies in which the leaves are close to vertical are termed erectophile [33] (Fig. 6.2a).

When the sun is overhead, vertical leaves absorb less **Q** per unit of leaf area than do horizontal leaves. This accounts for the low values of k for grasses, because their leaves are generally erect. Moreover, leaves tend to be vertical near the top of certain plants, e.g. sugar beet, agaves and pineapple, becoming more horizontal towards the ground (Fig. 6.2c). This architecture reduces the foliar absorption coefficient of the upper leaves, and therefore more of the **Q** incident on the plants is available for the lower leaves, analogous to the situation for the high-yielding I.R. rice cultivars. In fact, optimal **Q** utilization for photosynthesis generally occurs when the incident **Q** is distributed as uniformly as possible over the leaves, because the fraction of the leaves that are exposed to **Q** levels above photosynthetic light saturation or below light compensation is then minimized [14,20,30]. Thus, when L exceeds about 3, canopies with erect leaves near the top and horizontal leaves near the base tend to have higher productivities than canopies with random or with uniform leaf inclinations.

Our discussion presupposes that essentially all of **Q** is incident on the top of the canopy. When much **Q** comes in from the sides, as in morning or afternoon hours, or for an isolated tree or widely spaced crops, foliar absorption coefficients determined for vertically incident light should not be used in Eqn. 6.1, but rather k must be determined for other sun angles.

6.2.3 Light compensation

Let us next consider light compensation [23] for CO_2 fixation by leaves. Light compensation generally occurs at a **Q** of about 10 μmol m^{-2} s^{-1} for leaf temperatures near 20°C and an ambient CO_2 concentration of 350 μmol mol^{-1}. A moderate Q_0 of 400 μmol m^{-2} s^{-1} may be incident on a canopy for which k = 0.8 (full sunlight with the sun directly overhead on a cloudless day yields a Q_0 of about 2000 μmol m^{-2} s^{-1}). We can now calculate the cumulative leaf area index needed to reduce **Q** to the compensation point. From Eqn. 6.1c, we obtain:

$$F = \ln(Q_0/Q_F)/k$$
$$= \ln(400/10)/0.8$$
$$= 4.6$$

Thus, only the upper five 'layers' of this canopy are above light compensation for that part of the day when Q_0 is 400 μmol m^{-2} s^{-1}. Leaves below the upper five layers would be respiring more CO_2 than they are assimilating. In erectophile canopies with their lower k, a higher F is required to reach the compensation point under the same light conditions, which no doubt underlies the higher leaf area indices that are often associated with the erectophile habit. We also note that high levels of light penetrating small gaps in the canopy at specific sun angles (sunflecks) will complicate determination of the value of F at which light compensation occurs.

Leaves that are below light compensation for most of the day contribute negatively to net photosynthesis. Such leaves generally senesce, losing 30–50% of their dry weight before death and abscission [23]. Following this loss of leaves on the lower branches of trees, the branches themselves will usually die and eventually fall to the ground or be blown off by the wind, leading to a natural pruning. Many annual crop plants also show sequential senescence such that as new leaves are formed at the top of the canopy, leaves towards the base may die when the mean **Q** reaching them drops below light compensation. In this way plants can avoid the development of supra-optimal leaf area indices with its accompanying excessive self-shading.

6.2.4 Measurement of light in canopies

Chapter 4 describes the methodology for measurement of **Q** and radiant energy flux (**I**). Most instruments utilize point or circular sensors, which are appropriate where there is little horizontal variability in light quantity, as occurs in the open. However, the heterogeneity of leaf distribution in canopies leads to marked small-scale variation in **Q**, so to determine **Q** at any height in the canopy we need to make measurements at a number of locations and then obtain an average. This may be achieved by: (1) positioning an array of sensors at one height in the canopy; (2) moving a single sensor through a length of canopy; or (3) using a horizontal line or tube sensor whose output is the spatial average over the sensor length. Because of its simplicity, the latter technique is the most widely used, and newer instruments such as the *Sunfleck Ceptometer* [9] even provide measures of spatial variability of **Q** within the canopy.

Line photon sensors (e.g. the one-metre cosine and spectrally corrected *LI-COR LI-191* rod with its photoelectric detector; *LI-COR Inc., Lincoln, Nebraska, U.S.A.*) measure the average PPF falling on a line placed horizontally in the canopy. The *Sunfleck Ceptometer (Decagon Devices Inc., Pullman, Washington, U.S.A.)* consists of an array of photodiodes in a 40 or 80 cm long probe that respond only to PPF and that are scanned by a single-chip microcomputer. The readings may be used to determine average **Q**, as well as the fraction of the probe length that is in sunflecks. Tube solarimeters (e.g. the one-metre *Delta-T TSL* glass tube enclosing a strip thermopile of alternating black and white surfaces; *Delta-T Devices, Cambridge, U.K.*) measure the average solar energy flux. Instruments of the latter type are generally cheaper and relatively simple to construct [29], although filters must be used to determine the photosynthetically active part of the spectrum (the infra-red component of sunlight readily penetrates canopies and will account for most of the radiant energy near the canopy base). The appropriate sensor length or number of sampling points depends on plant spacing; for example, in a dense canopy of a fine-leaved grass, 1 m is excessive, whereas in a citrus plantation with 3-metre spacing of plants a 1 m sensor is too short (a series of readings along a line can overcome this inadequacy).

To determine k, or the absorption of **Q**, more than one sensor location must be used. Typically, a conventional photon sensor is placed facing up just above the top of the canopy to determine $\mathbf{Q}_0$, and a line or tube sensor is inserted in the canopy to determine **Q** there. To avoid systematic errors when using more than one sensor, sensors must be matched or cross-calibrated. Line and tube sensors generally are directionally sensitive, so they are not appropriate for measurement above the canopy. Directional sensitivity is less important within the canopy, because photon fluxes there tend to be more diffuse, i.e. come from all angles. When the line or tube sensor is placed at the base of the canopy, the PPF intercepted by the whole canopy ($\mathbf{Q}_a$) is:

$$\mathbf{Q}_a = \mathbf{Q}_0 - \mathbf{Q}_b \tag{6.2}$$

where $\mathbf{Q}_b$ is the PPF incident in a horizontal plane at the base of the canopy. Such a measurement tells us the absorption at one point in time. Usually, $\mathbf{Q}_a$ is integrated over a day or even a growing season to record the cumulative photosynthetic photon fluxes.

Most measurements of k and **Q** are based on the ambient PPF incident on the canopy. We could equally well base our discussion strictly on absorbed photons, where $\mathbf{Q}_0$ would represent the downward PPF minus that reflected from the canopy (the latter determined by a photon sensor facing downward placed just above the canopy). Similarly, Q_b in Eqn. 6.2 then becomes the PPF downward at the base of the canopy minus that reflected upward by the soil, i.e. the PPF absorbed by the soil. Such a change to a strictly absorbed photon basis causes the foliar absorption coefficient k to become slightly smaller compared to when incident levels of **Q** are used. The smaller k is appropriate in most studies of the quantum efficiency of CO_2 fixed or O_2 released, which are based on absorbed photons, although most studies of canopy structure and most photosynthetic response curves relating CO_2 uptake to ambient conditions are based on incident PPF.

6.3 MEASUREMENT OF CANOPY STRUCTURE

Both direct and indirect methods have been developed to quantify canopy structure [1,9,28,32]. Direct methods usually involve physical disturbance of the canopy and are very labour-intensive. However, data reduction with direct methods is usually simpler. Indirect methods exploit the fact that radiation within a canopy is determined by a combination of canopy structure and solar position [28]. Thus, when measurements of radiation within the canopy are combined with the appropriate radiative transfer theory, canopy structural properties may be inferred [9,15,17,25].

6.3.1 Direct methods

Leaf area index

At ground level, F is designated the leaf area index (L), as indicated above, which equals the projectable leaf surface area per ground area. L may be determined by harvesting all material above a quadrat marked out on the ground. (Methods of determining optimum quadrat sizes, sampling design, harvesting, as well as storing and sorting of harvested shoots are discussed in Chapter 1.) Leaf area should be measured as soon as possible after harvesting, using material kept moist to prevent shrinkage or rolling. Area is usually measured with a leaf area meter. Less expensive methods range from hand-held planimeters to area-length regressions [16] to photogravimetric methods. In the latter case, leaves are placed on photographic paper to obtain a contact print, or a photocopier is employed. The image is cut out, weighed, and area determined from the area/weight ratio for the paper used (a precision of ±2% can be achieved). Most commercial leaf area meters utilize either interruption of parallel light beams as leaves are passed between light source and sensor (e.g. *LI-COR LI-3100*, where a conveyor belt decreases processing time) or analysis of a video image of leaves against a white background (e.g. *Delta-T Area Meter*). Most instruments provide higher precision with larger leaves and may be sensitive to leaf orientation (instruments should be calibrated with shapes similar to the leaves being used).

To study the distribution of leaf area with height, sampling of stratified clips may be employed. Here, various strata are removed from the canopy separately, and the leaf area in each strata is quantified. In vegetation with a high canopy base, e.g. most orchard and forestry crops, it is simplest to cut the strata of leaves from the bottom of the canopy upward, whereas for most herbaceous communities it is easier to cut from the top downward. In the latter case, the cut herbage must be collected before it falls to lower levels in the canopy, e.g. by fitting a vacuum attachment to the cutters [27] or by unrolling cloth as a hedge trimmer is run across the vegetation. When stems or long leaves ascend and then descend, the vegetation can be 'sandwiched' in position between two boards before cutting. Where plants have rigid upright stems, the stems may be harvested and then placed on a horizontal grid to section the canopy.

Leaf inclination

Leaf inclination is the angle (α) between the leaf axis and the horizontal (Fig. 6.3). Patterns of leaf inclination within a canopy can be represented by plotting the relative frequencies of leaf inclinations, typically at 10° intervals, from 0° for a horizontal leaf to 90° for a vertical one. A planophile canopy has its greatest frequency at the lower inclination angles, e.g. α = 0°–20°, whereas an erectophile canopy would show the greatest frequency at high inclination angles, e.g. α = 70°–90° (Fig. 6.4).

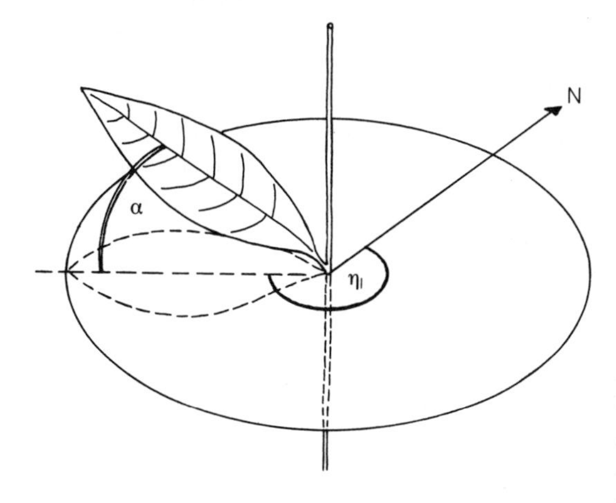

Fig. 6.3. Leaf inclination (α) is the angle formed between the long axis of the leaf (or leaf segment) and the horizontal. Leaf orientation or azimuth (η_l) is the angle formed clockwise from due north by the horizontal projection of the leaf.

Inclination may be measured directly by holding a protractor with a levelling device against the leaf, e.g. for pure stands of stiff-leaved grasses. However, for many leaves the lamina is at an angle to the midrib. Furthermore, many long leaves droop toward the tip and so display a range of inclinations. In such cases, each leaf can be divided into angle classes measured backward from the tip during cutting, and the total leaf area determined for each inclination interval (individual bags can be used for each angle class). In a crop such as maize, with high values of L and leaves that have portions of their lamina in several angle classes, the direct measurement of leaf inclination can be a tedious task. Additionally, the physical act of reaching in

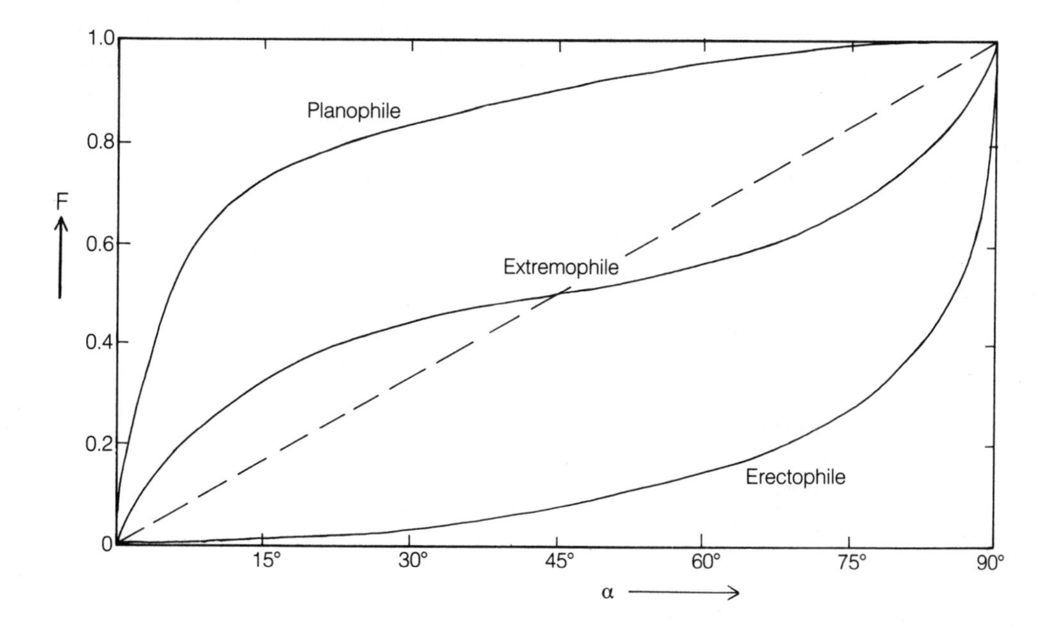

Fig. 6.4. Cumulative proportion of total leaf area (F) with leaf angle (α) from 0° to 90° for examples of three different canopy types (after de Wit [33]).

and placing a device along the leaf can disturb the parameter being measured.

Leaf orientation
In contrast to leaf inclination, leaf orientation or azimuth (the angle formed clockwise from north to the horizontal projection of the leaf axis; Fig. 6.3) is nearly random for many canopies and has received less attention [7,10]. However, orientation can be important in determining interception of PPF. Orientation may be determined directly by aligning a compass with the horizontal projection of the leaf. As with inclination, orientation may be represented by plotting the relative frequencies of leaf orientation in different intervals (typically 45°–90°) using 0° for North, 90° for East, 180° for South, and 270° for West orientations. For an initial test of non-randomness, leaf orientation can be partitioned into the four cardinal directions. Some leaves will alter their orientation through the day by solar tracking or through nyctinastic movements, most notably species of the Malvaceae and Papilionaceae [11]. Wind can also influence leaf orientation.

6.3.2 Indirect methods

Inclined point quadrats
Point quadrats through a canopy have been widely used to infer canopy structure. This involves counting the number of leaves that come into contact with a needle point as the needle passes along some line from the top of the canopy to the bottom. If 100 point quadrats through a stand have 240 vegetation contacts, the cover is 2.4. If all the leaves were horizontal, i.e. a perfect planophile canopy, then the cover would equal L. However, in an erectophile canopy, vertically determined cover is substantially lower than L, a problem that can be overcome if the point quadrats are inclined. When the angle of quadrat inclination from the horizontal is β, we obtain:

$$L = K\,N(\beta) \qquad (6.3a)$$

where K is a constant for a particular canopy and quadrat inclination angle, and N(β) is the mean contact frequency in the quadrat for inclination angle β.

Warren-Wilson [31,32] determined that K was minimally sensitive to leaf angle when quadrats were at 33° and L was given by:

$$L \equiv 1.1\,N_{33} \qquad (6.3b)$$

The error in L is 10% for mean canopy leaf angles of 0°, 36° and 90°. Errors are reduced to ±2% using point quadrats inclined at two angles, 13° and 52°:

$$L \equiv 0.23\,N_{13} + 0.78\,N_{52} \qquad (6.3c)$$

Use of three or even five inclination angles further decreases the error, but the additional work-load is rarely warranted [1,13].

Inclined point quadrats also provide a valuable non-destructive technique for studying the stratification of leaf area and even leaf angle in a canopy. For instance, distance (z_β) along a point quadrat inclined at β may be related to vertical height (z) by $z_\beta = z/\sin\beta$; to determine ΔF for a particular stratum, all point quadrat contacts made between the delimiting vertical heights are summed and then divided by the total number of point quadrats made to give the contact frequency, which is then incorporated into Eqn. 6.3. Both quadrat placement and orientation must be randomized [27]. Of particular importance is point quadrat diameter, which should be as small as practicable (achieved by mounting a fine needle at the end of a sturdy rod). Caldwell *et al.* [5,6] have developed a fibre optic point quadrat system that can rapidly quantify bunchgrass canopy architecture using an automated fibre optic probe with an effective point diameter of less than 25 μm. As the probe contacts a canopy element, the light from an LED is reflected back to a sensing photodiode; these contacts and the distance travelled by the probe between contacts are recorded.

Mean leaf inclination in the canopy, weighted on an area basis, can be determined by measurements with point quadrats inclined at two angles. Warren-Wilson [31] estimated mean leaf inclination from $\tan\alpha = \pi N_0/2N_{90}$, where N_0 and N_{90} are the leaf contact frequencies at point quadrat inclinations of 0° and 90°, respectively. Although using N_0/N_{90} is more sensitive, tables are also available [32] for estimating α from N_{13}/N_{52}.

Gap fraction analysis

At the beginning of this chapter we treated the absorption of light by a plant canopy as analogous to absorption of light by a homogeneous solution of chlorophyll. Actually, there are complicated horizontal and vertical patterns of sunlit and shaded areas within a plant canopy that change with the sun's position through the day (Fig. 6.5). The transmittance of the direct solar beam through a plant canopy may be used in an analogous fashion to inclined point quadrats to estimate L and mean leaf inclination angles [9,18].

The transmittance of direct beam radiation, $T(\theta)$, to a horizontal surface below the canopy at any given solar zenith angle, θ, may be quantified by measuring the fraction of total sunlit area (gap fraction) along a transect on the surface (Fig. 6.5). An equation similar to Eqn. 6.1b may be used to relate $T(\theta)$ to plant canopy characteristics [25]:

$$T(\theta) = e^{-k_0 L} \qquad (6.4)$$

where k_0 is the foliar absorption coefficient for direct beam radiation at solar zenith angle θ, and L is the leaf area index. Gap fraction methods are related to inclined point quadrat methods because the probability of the direct solar beam passing through the canopy is the same as that of an inclined point quadrat not contacting a leaf element as it traverses the canopy. Thus, $N(\beta) = -\ln T(90 - \beta)$, where $N(\beta)$ is the mean contact frequency for a probe inclined at β degrees from the horizontal and $T(90 - \beta)$ is the transmittance of the direct solar beam at a solar zenith angle of $\theta = (90 - \beta)$.

Gap-fraction methods assume that the leaf azimuth distribution is random, an assumption that can be applied to many crop and natural canopies [28]. From Eqn. 6.4, the transmitted fraction $T(\theta)$ of a beam of direct radiation incident on a canopy at zenith angle θ is given by [9]:

$$-\ln T(\theta) = k_0 L \qquad (6.5)$$

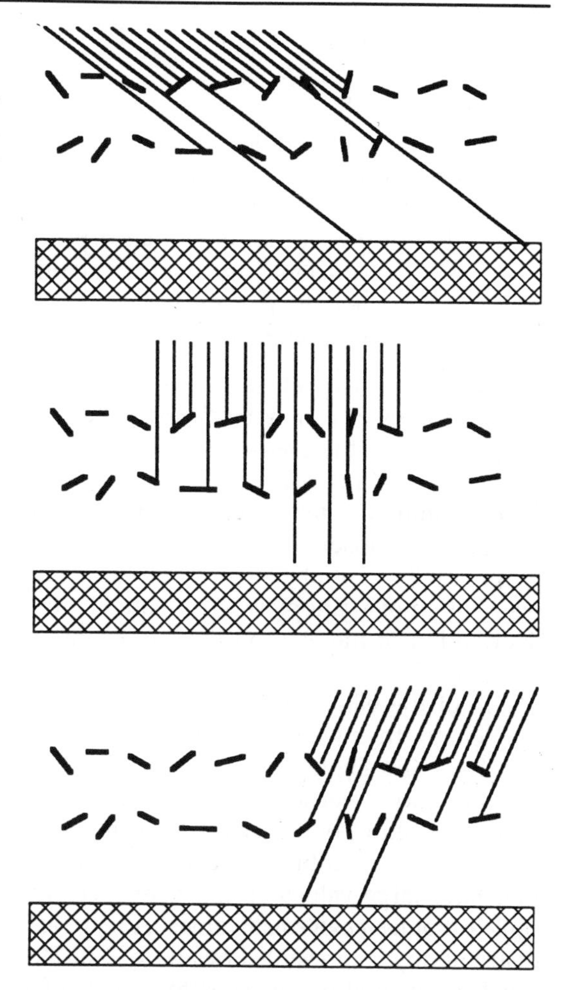

Fig. 6.5. Penetration of direct solar beam radiation into an idealized plant canopy for three different solar zenith angles. In the top panel, the gap fraction is 0.33 for the second layer of leaves and 0.13 for the soil surface. In the middle panel these values change to 0.53 and 0.20, whereas in the bottom panel the respective values are 0.40 and 0.13. These values, combined with the appropriate solar zenith angles may be used to calculate leaf area indices and mean leaf inclination [9].

where

$$L = \sum_{j=1}^{n} f_j$$

n = number of leaf angle classes
f_j = leaf area index in inclination angle class j
k_θ = direct beam foliar absorption coefficient for solar zenith angle θ and mean leaf inclination angle α.

By measuring $T(\theta)$ at several solar zenith angles, values of α may be derived that are consistent with the measured values of $T(\theta)$ by a process of inversion of Eqn. 6.5.

One commonly used inversion procedure assumes an ellipsoidal leaf angle distribution proposed by Campbell [7]. The leaf angle distribution of a canopy is represented by the distribution of the area on the surface of an ellipsoid of revolution. If a is the vertical axis length of the ellipse and b is the horizontal axis length, then we may define x as equal to b/a. Thus, the single parameter, x, representing the ratio of vertical to horizontal projections of a volume of foliage [7] determines the shape of the distribution. Planophile canopies have large values of x, whereas erectophile canopies have small values. A spherical distribution is represented by an x of 1.0. The inversion procedure searches for values of x that produce a reasonable fit to the measured gap fraction data at various solar zenith angles. The foliar absorption coefficient for particular values of and x is calculated from [7]:

$$k(\theta, x) = [(x^2 + \tan^2\theta)^{1/2}]/D \qquad (6.6a)$$

where D is the normalized ellipse area given by [8]:

$$D = x + 1.774(x + 1.182)^{-0.733} \qquad (6.6b)$$

An empirical expression relating mean leaf angle of the canopy, $\bar{\alpha}$, to x is [8]:

$$\bar{\alpha} = 9.65(3 + x)^{-1.65} \qquad (6.6c)$$

where $\bar{\alpha}$ is in radians.

A computer program (a BASIC listing is given in Campbell and Norman [9], Norman and Campbell [25], and in the manual provided with the *Decagon Devices Sunfleck Ceptometer*) may be used to insert values of x and L that minimize the expression

$$F = \Sigma[\ln T(\theta) + K(\theta, x)L]^2 \qquad (6.7)$$

subject to the constraint that $x > 0$. The minimum is found by simultaneously solving for $\delta F/\delta L = 0$ and $\delta F/\delta x = 0$, and finding x using a bisection method [9]. Once x is found, the program solves for L and mean leaf angle. Inputs into the program consist of solar zenith angles and corresponding gap fractions $[T(\theta)]$. This general approach has been used successfully to measure canopy structure in a number of different crop and community types [19,25,26].

Several instruments are now commercially available that utilize indirect methods of canopy structure analysis. The *LI-COR LAI-2000 Plant Canopy Analyser* computes L and mean leaf inclination angle from measurements of light interception made simultaneously at five angles. The radiation detector is a group of five ring-shaped sensors arranged concentrically and covered with lenses so that each sensor sees a different portion of the sky. Each sensor is filtered for light above 490 nm to minimize the contribution of reflected and transmitted light within the canopy. Light readings are made below a canopy and are divided by readings made above the canopy to compute transmittance at the five angles. The *Decagon Devices Sunfleck Ceptometer* may be used analogously to the *LI-COR Plant Canopy Analyser* if readings of gap fraction in the canopy are taken several times during the day (at different θ values). The gap fraction data are used with the computer program mentioned above [9,25] to calculate L and mean leaf inclination.

The accuracy of estimates of canopy structure made with gap fraction data depends on several conditions, including the adequacy of spatial sampling within the canopy and the appropriateness of the canopy radiation model used in inverting the data. This last consideration may be important in heliotropic canopies, which violate the assumption of random azimuthal orientation. Furthermore, these heliotropic canopies may alter their leaf inclination distribution during the day. Thus, measurements taken at different solar zenith angles will reveal only a mean daily inclination and not show changes occurring during the daytime. These changes may be quantified by measurements at different times of the day using the *LI-COR Plant Canopy Analyser*, or through the use of hemispherical fish-eye photographs taken from beneath the canopy at different times during the day [3,4]. The latter photographs may also be used to determine leaf orientation [1].

6.4 STEMS AND INFLORESCENCES

Our discussion has been limited to leaves. However, other plant organs can be green and photosynthetically active, for example, stems in grass canopies. The principles and methods outlined here for leaves can also be applied to stems and other photosynthetic organs in describing the photosynthetic aspects of canopy structure.

REFERENCES

1. Anderson, M.C. (1971) Radiation and crop structure. In: *Plant Photosynthetic Production. Manual of Methods.* (Z. Sestak, J. Catsky and P.G. Jarvis, eds.). pp. 412–466. Dr. W. Junk, The Hague, Netherlands.
2. Beadle, C.L., S.P. Long, S.K. Imbamba, D.O. Hall and R.J. Olembo (1985) *Photosynthesis in Relation to Plant Production in Terrestrial Environments.* Tycooly/Cassell, London.
3. Bonhomme, R. and P. Chartier (1972) The interpretation and automatic measurement of hemispherical photographs to obtain sunlit foliage area and gap frequency. *Israeli J. Agricultural Research* **22**, 53–61.
4. Bonhomme, R., C. Varlet Grancher, and P. Chartier (1974) The use of hemispherical photographs for determining the leaf area index of young crops. *Photosynthetica* **8**, 299–301.
5. Caldwell, M.M., T.J. Dean, R.S. Nowak, R.S. Dzurec and J.H. Richards (1983) Bunchgrass architecture, light interception, and water-use efficiency: assessment by fiber optic point quadrats and gas exchange. *Oecologia* **59**, 178–184.
6. Caldwell, M.M., G.W. Harris and R.S. Dzurec (1983) A fiber optic point quadrat system for improved accuracy in vegetation sampling. *Oecologia* **59**, 417–418.
7. Campbell, G.S. (1986) Extinction coefficients for radiation in plant canopies calculated using an ellipsoidal inclination angle distribution. *Agricultural and Forest Meteorology* **36**, 317–321.
8. Campbell, G.S. (1990) Derivation of an angle density function for canopies with ellipsoidal leaf angle distributions. *Agricultural and Forest Meteorology* **49**, 173–176.
9. Campbell, G.S., and J.M. Norman (1988) The description and measurement of plant canopy structure. In: *Plant Canopies: Their Growth, Form and Function* (G. Russell, B. Marshall, and P.G. Jarvis, eds.). pp. 1–19. Society for Experimental Biology, Seminar Series 29, Cambridge University Press, New York.
10. Duncan, W.G. (1971) Leaf angles, leaf area, and canopy photosynthesis. *Crop Sci.* **11**, 482–485.
11. Ehleringer, J. and I. Forseth (1980) Solar tracking by plants. *Science* **210**, 1094–1098.
12. Fuchs, M., G. Asrar, and E.T. Kanemasu (1984) Leaf area estimates from measurements of photosynthetically active radiation in wheat canopies. *Agricultural and Forest Meteorology* **32**, 13–22.
13. Grant, S.A. (1981) Sward components. In: *Sward Management Handbook* (J. Hodgson, R.D. Baker, A. Davies, A.S. Laidlaw and J.D. Leaver, eds.). pp. 70–91. British Grassland Society, Hurley, Berkshire, U.K.
14. Hodánová, D. (1979) Sugar beet canopy photosynthesis as limited by leaf age and irradiance. Estimation by models. *Photosynthetica* **13**, 376–385.
15. Kopec, D.M., J.M. Norman, R.C. Shearman, and M.P. Peterson (1987) An indirect method for estimating turfgrass leaf area index. *Crop Sci.* **27**, 1298–1301.
16. Kvet, J. and J.K. Marshall (1971) Assessment of leaf area and other assimilating plant surfaces. In: *Plant Photosynthetic Production. Manual of Methods.* (Z. Sestak, J Catsky and P.G. Jarvis, eds.). pp. 517–555. Dr. W. Junk, The Hague, Netherlands.
17. Lang, A.R.G., and Yueqin Xiang (1986) Estimation of leaf area index from transmission of direct sunlight in discontinuous canopies. *Agricultural and Forest Meteorology* **37**, 229–243.

18. Lang, A.R.G., Yueqin Xiang and J.M. Norman (1985) Crop structure and the penetration of direct sunlight. *Agricultural and Forest Meteorology* **35**, 83–101.
19. Lang, A.R.G. (1987) Simplified estimate of leaf area index from transmittance of the sun's beam. *Agricultural and Forest Meteorology* **41**, 179–186.
20. McMillen, C.G. and J.H. McClendon (1979) Leaf angle: an adaptive feature of sun and shade leaves. *Bot. Gazette* **140**, 437–442.
21. Monsi, M. and T. Saeki (1953) Über den Lichtfaktor in den Pflanzengesellschaften und seine Bedeutung für die Stoffproduktion. *Japanese J. Bot.* **14**, 22–52.
22. Monsi, M., Z. Uchijima and T. Oikawa (1973) Structure of foliage canopies and photosynthesis. *Ann. Rev. Ecol. System.* **4**, 301–327.
23. Nobel, P.S. (1991) Physicochemical and Environmental Plant Physiology. Academic Press, San Diego.
24. Norman, J.M. (1980) Interfacing leaf and canopy light interception models. In: *Predicting Photosynthesis for Ecosystem Models.* Vol. 2 (J.D. Hesketh and J.W. Jones, eds.). pp. 49–67. CRC Press, Boca Raton, Florida.
25. Norman, J.M. and G.S. Campbell (1989) Canopy structure. In: *Plant Physiological Ecology. Field Methods and Instrumentation.* (R.W. Pearcy, J.R. Ehleringer, H.A. Mooney and P.W. Rundel, eds.). pp. 301–325. Chapman and Hall, London, New York.
26. Perry, S.G., A.B. Fraser, D.W. Thomson and J.M. Norman (1988) Indirect sensing of plant canopy structure with simple radiation measurements. *Agricultural and Forest Meteorology* **42**, 255–278.
27. Rhodes, I. (1981) Canopy structure. In: *Sward Management Handbook* (J. Hodgson, R.D. Baker, A. Davies, A.S. Laidlaw and J.D. Leaver, eds.). pp. 141–158. British Grassland Society, Hurley, Berkshire, U.K.
28. Ross, J. (1981) *The Radiation Regime and Architecture of Plant Stands.* Dr. W. Junk, The Hague, Netherlands.
29. Szeicz, G., J.L. Monteith and J.M. dos Santos (1964) Tube solarimeters to measure radiation among plants. *J. Applied Ecology* **1**, 169–174.
30. Turitzin, S.N. and B.G. Drake (1981) The effect of a seasonal change in canopy structure on the photosynthetic efficiency of a salt marsh. *Oecologia* **48**, 79–84.
31. Warren-Wilson, J. (1959) Analysis of the distribution of foliage in grassland. In: *The Measurement of Grassland Productivity* (J.D. Irvins, ed.). pp. 51–61. Butterworths, London.
32. Warren-Wilson, J. (1963) Estimation of foliage denseness and foliage angle by inclined point quadrats. *Australian J. Bot.* **11**, 95–105.
33. Wit, C. T. de (1965) Photosynthesis of leaf canopies. *Verslagen van Landbouwkundige Onderzoekingen* No. 663. Wageningen.

7

Functional leaf anatomy

H.R. Bolhàr-Nordenkampf and G. Draxler

7.1 INTRODUCTION

The green leaves of plants are photosynthetically active organs which are able to store absorbed solar energy in reduced organic compounds. These assimilates represent the pool for both energy and compounds which have to meet the plant's requirements for growth and development. Although the dry matter production of a plant does not correlate well with net leaf photosynthesis, high-yielding crops must have high photosynthetic rates combined with energetically efficient dark respiration and assimilate partitioning which favours the harvested plant organ. Agricultural techniques therefore aim to maintain high overall photosynthetic capacity, i.e. high assimilatory rates and quantum yields, under a variety of environmental conditions. Enhancing the flexibility of photosynthesis is also the guiding principle for adaptations and modifications in leaf anatomy and morphology, generally supported by suitable leaf orientation, stand geometry and density (Chapter 6).

Several general textbooks on this subject are listed under **Further Reading** at the end of this chapter, in addition to the references cited in the text.

Photosynthesis and Production in a Changing Environment: a field and laboratory manual
Edited by D.O. Hall, J.M.O. Scurlock, H.R. Bolhàr-Nordenkampf, R.C. Leegood and S.P. Long
Published in 1993 by Chapman & Hall, London.
ISBN 0 412 42900 4 (HB) and 0 412 42910 1 (PB).

7.2 GENERAL ASPECTS OF LEAF ANATOMY

Leaf anatomy may influence net leaf photosynthesis to a large degree and thus cause great differences in light use efficiency. Like the root and stem, the leaf consists of vascular (veins), parenchymatous (mesophyll) and mechanical supporting tissues, and also dermal tissues in the form of a persistent epidermis (Fig. 7.1).

Leaves of flowering plants range in size from a few millimetres in length (e.g. *Wolffia arhiza, Lemna minor*) to two metres in some palms and consist of a blade (lamina) and a stalk (petiole) or leaf sheath. In several plants (e.g. *Fabaceae*) the petioles and the petiolules of the leaflets have pulvini which are capable of moving leaves and leaflets by potassium-flux-controlled turgor changes stimulated by internal or environmental factors.

The foliage leaves of most dicotyledonous plants have one main vein (midrib) and a reticulate venation system formed by a series of progressively smaller veins. Most monocotyledonous plants have parallel venation, in which major and minor longitudinal vascular bundles are interconnected at intervals by small commissural veins. In some species, thick-walled fibres associated with the vascular bundles are of commercial use (e.g. *Phormium tenax*, 'New Zealand flax').

The leaf is typically of determinate growth (with the exception of *Welwitschia*) with dorsiventral symmetry. Because of the usually flat leaf blade the epidermis is separated between the leaf margins into an upper (adaxial surface) and lower (abaxial surface) epidermis. Stomata can occur on both surfaces of the leaf. The chloroplast-containing mesophyll (chlorenchyma) is situated inside the leaf, between the upper and lower epidermis. In dicotyledonous plants the palisade parenchyma below the upper epidermis and the spongy parenchyma comprise the mesophyll (dorsiventral, bifacial; Fig. 7.11). In certain plants, some of them adapted to dry habitats, the palisade parenchyma is present on both sides of the leaves, thus reducing the spongy tissue down to a small strip in the centre of the mesophyll (isolateral, isobilateral; e.g. *Centaurea, Artemisia, Eucalyptus*).

Leaf structure can be influenced by light intensity resulting in so-called shade and sun leaves. Sun leaves compared to shade leaves are usually thicker, smaller, sometimes more hairy and may show up with anthocyanins in the vacuoles of the epidermal cells. Sun leaves will develop a thicker (multilayer) palisade parenchyma whereas in some thin shade leaves palisade cells are hardly differentiated.

In many monocotyledons, the distinctions between palisade and spongy tissue is not apparent. In many grasses the mesophyll cells are of uniform shape and loosely packed (C_3 grasses) but in all C_4 grasses the mesophyll cells are radially oriented and tightly packed, surrounding the chloroplast-containing bundle sheath of the vascular bundle (Kranz anatomy, e.g. *Zea, Sorghum, Saccharum*; Fig. 7.13).

The histology of gymnosperm leaves is different in some aspects. Most of the plants have evergreen xeromorphic leaves. *Cycas* has leathery dorsiventral leaves while many coniferous trees show needle-like leaves. In the centre of the uniform mesophyll the vascular bundle is situated surrounded by the transfusion tissue, which is separated by an endodermis from the chlorenchyma, comprised of a uniform matrix of mesophyll cells which can include several resin ducts. These tissues are protected by extremely thick-walled lignified hypodermis and epidermis cells (e.g. *Pinus, Picea*; Figs. 7.8 and 7.14).

7.3 EPIDERMIS

The epidermis is usually one layer of relatively unspecialised chloroplast-free cells forming a compact layer without intercellular spaces (Fig. 7.3). The outer cellulose wall rarely contains lignin (except for conifer needles) but frequently has deposits of silica salts (e.g. grasses) and is covered by pectic substances which separate the cuticle.

In the cellulose wall interfibrillar spaces occur regularly, called ectodesmata or teichodes. They are not plasmatical structures, but are supposed to serve as pathways of penetration of solutions to and from the protoplast (e.g. water, airborne nutrients and pollutants). Parts of the cell wall may become mucilaginous which enables the epidermis to act as a reversible temporary external water storage (e.g. Malvaceae, *Linum*). In some plants a multiple epidermis (multiseriate epidermis) occurs which can develop into a water storage tissue in succulent leaves (e.g. *Peperomia, Ficus*).

7.3.1 Cuticle and waxes

The epidermis has to protect the leaf against uncontrolled water loss and against damage by environmental factors such as bending of the leaf blade by wind or the attack of microorganisms. This is achieved by the compact arrangement of cells and by the incrustation of the outer wall with cutin, constituting a special layer, the cuticle.

Cutin is a fatty substance which can be stained red with Sudan IV. The thickness of the cuticle very often corresponds with the degree of xeromorphism of the leaf (e.g. *Tamerix* 7 µm, *Juncus* 2 µm). The surface of the cuticle may be smooth, rough, ridged or furrowed (Fig. 7.3). The cuticle can form species-specific ornamentation (e.g. *Eucalyptus*) which can be used to classify plants even with preserved fragments from early geological eras. Some fungal spores are able to detect suitable hosts by the species-specific structure of the cuticle.

Birefringent waxes embedded near the surface of the cuticle form a discontinuous layer interrupted by isotropic material over the anticlinal walls. The surface is depressed above the anticlinal wall and is known to lose water more easily than other regions of the leaf surface (epidermal transpiration). Waxes are synthesised by the cytoplasm and are thought to be extruded via special channels through the cell wall and the cuticle to form deposits on the surface. They may form a continuous layer or granules, rods (e.g. *Saccharum*), platelets or scales. The wax layer on leaves of the wax palm *Copernicia cerifera* can be as much as 5 mm thick and in *Agave americana* the wax deposited may weigh as much as 1.5 g per m^2 leaf surface.

The structure of the cuticle and the epicuticular wax deposit contributes to the ability of the surface to shed water and determines the boundary layer conductance. This ability is modifed by the presence of epidermal hairs.

7.3.2 Trichomes

The epidermis may develop unicellular and multicellular highly variable appendages designated as covering and glandular trichomes (hairs).

Covering trichomes:

The non-glandular trichomes form simple papillae (e.g. *Viola*), bladders (Fig. 7.1; vesicular hairs, Crassulaceae) or long (up to 5 cm) unicellular fibres which develop secondary walls. The extensive hairs of *Gossypium* seeds are used commercially (cotton). There are also more complex forms of covering trichomes known as peltate hairs (scales, e.g. *Olea*, Bromeliaceae) or as dentric, stellate or candelabrum-like hairs (e.g. Brassicaceae, *Deutzia*, *Platanus*). Sometimes the hairs are remarkably uniform and may be used to identify families, genera, species or in the analysis of interspecific hybrids.

Together with the epidermal structure the hairy cover of the epidermis cells determines the epidermal transpiration and the energy budget of the leaves. Dead dry trichomes observed with alpine plants lower water losses and wetability but can also insulate the mesophyll or may reflect excessive light, especially UV-B (e.g. alpine plants).

Together with raised stomata, live trichomes covered by a thin cuticle only are thought to enable at least some transpiration of herbs growing in the extremely humid environment of a rainforest understorey. Peltate scale-forming hairs on the leaves of epiphytic Bromeliaceae are shown to support the water and nutrient uptake by leaves. Pearl-like hairs have a similar function and are able to store water temporarily (e.g. Begoniaceae, Vitaceae, *Bauhinia*). Hooked trichomes (Fig. 7.1) facilitate attachment of climbing plants, or they may impale insects and larvae. Setaceous hairs on leaves of some carnivorous plants are sensitive to contact stimulus, causing leaf movement to catch the insect crawling on the surface (e.g. *Dionea*, *Utricularia*).

Glandular trichomes:

Secretory trichomes, often called glands, excrete a number of substances. In salt-secreting trichomes the salt is actively secreted via cuticle pores onto the surface of the cells (salt glands, e.g. *Avicennia*) or from the cytoplasm into the vacuoles of hairs, which later dry out (e.g. *Atriplex*).

Nectar-secreting trichomes show, as do many other gland cells, a dense cytoplasm rich in mitochondria and endoplasmic reticulum. Together with secretory outgrowths of the epidermis (nectaries) they occur on different parts of the floral organs (calyx, corolla, e.g. *Abutilon*, *Tropaeolum*), but also on leaf petioles (e.g. *Prunus*).

Terpene-secreting trichomes have a uniseriate stalk and a secretory head. At the stage of secretion the number of Golgi bodies increases and the oil is secreted in the subcuticular space, forming a large reservoir which may be released if the cuticle is ruptured (e.g. Labiatae, Asteraceae, Geraniaceae). The essential oils are thought to protect the leaves from herbivores. A similar function is proposed for colleteres (trichomes on bud scales), which secrete a sticky mixture of terpenes and mucilage, which may in addition reduce transpiration and wetting (e.g. *Aesculus*, *Rosa*, *Coffea*).

Stinging hairs are highly specialised trichomes. The cell wall of the needle-like hair is encrusted with silica and has a spherical tip, which breaks off when the hair is touched. The broken tip penetrates the skin easily, injecting irritant poisons (histamine, acetylcholine; e.g. *Urtica*, *Laportea*, Euphorbiaceae).

Some appendages consist of epidermal, ground and vascular tissue (emergences), such as the thorn of *Rosa*. Carnivorous plants capture and digest their prey by mucilages and enzymes from tentacles bending over the caught insect (e.g. *Biblis, Drosera*). In some carnivorous plants (e.g. *Pinguicula*) separate glandular trichomes secrete proteolytic enzymes to induce external digestion. To obtain the nutritional benefits, the digestive products are absorbed by the epidermis.

7.3.3 Stomata

The continuity of the protective epidermis is interrupted by the stomata to enable gas exchange with the atmosphere (Figs. 7.6–7.10). The exchange of water vapour, CO_2 and O_2 is largely limited to the stomatal pores, which are minute intercellular openings bounded by two kidney-shaped specialised epidermis cells, the guard cells. In many plants the stomatal apparatus includes subsidiary cells (Fig. 7.10). These adjacent cells differ morphologically from ordinary epidermis cells and are functionally connected to the guard cells (Figs. 7.6, 7.7).

The guard cells contain numerous mitochondria, Golgi bodies, vacuoles different in size, large nuclei and chloroplasts with only a few grana, which are thought to perform a CAM-like photosynthetic pathway (Chapter 16). As a result of unequal thickness and consistency of the cell walls, together with the arrangement of the wall-thickening cellulose microfibrils, turgor changes in the guard cells are responsible for the opening and closure of the stomatal pore. To initiate the opening sequence, starch is degraded in the chloroplasts of the guard cells to provide organic anions which are thought to support potassium uptake from the subsidiary cells. This enhances the osmotic value in the vacuole, giving rise to increased turgor pressure. This pressure bends the thinner and more flexible dorsal wall of the guard cells inwards towards the subsidiary cells and the central slit opens (Fig. 7.7). A loss of K^+ ions and reappearance of starch can be observed during stomatal closure.

Stomatal movements are mainly controlled by light, the internal CO_2 concentration in the sub-stomatal cavity, and by the water relations of the leaf (e.g. saturation deficit) via the formation of abscisic acid. To enable an effective control by the aperture between the guard cells above the central slit (Fig. 7.7), the front cavity and the outer cuticular ledges minimise forced transpiration, e.g. by the wind, whereas the nozzle-shaped back cavity is thought to guide the gases to and from the substomatal cavity (Fig. 7.7). The cell walls along the path are covered by cuticle down to the inner ledge.

Xeromorphic and succulent leaves show stomata in sunken positions (e.g. *Aloe, Iris*) or stomata may even be hidden in a stomatal crypt covered by dead trichomes (e.g. *Casuarina, Stipa*; Figs. 7.15, 7.16). The stomata of coniferous needles are deeply sunken and overarched by the subsidiary cells. The front cavity is filled with a three-dimensional net of wax fibres (Fig. 7.8). Raised stomata are associated with a hydrophytic habitat and are thought to be the anatomical response to moist atmosphere in general.

In leaves with parallel venation as in Gramineae and Cyperaceae the epidermis shows a ground mass of elongated narrow cells with frequently undulated anticlinal walls, silica cells and cork cells, and the stomata are arranged in rows (Fig. 7.10; e.g. Poaceae). The guard cells of Poaceae have a rather uniform specific structure. The cells are elongated and expanded on both ends while the middle portions are thick-walled with a narrow cell lumen (Fig. 7.10). The two guard cell protoplasts are interconnected via pores in the bulbous ends and therefore changes in turgor occur simultaneously to open or close a long narrow slit between the thick lignified middle part. On each side of the stoma there are triangular-shaped subsidiary cells (Fig. 7.10). This specific anatomical structure of the stomatal apparatus is thought to support the extremely fast response of grass stomata to changes in the saturation deficit of the leaves.

Stomata may occur on both sides of the leaf (amphistomatic leaf; e.g. *Sorghum, Zea, Eucalyptus*) or only on one side, usually the lower side (hypostomatic leaf; e.g. *Phaseolus, Quercus*; Table 7.1). They are less common on the upper side shown by floating leaves (epistomatic leaf; e.g. *Nymphea*). Most dicotyledonous trees have hypostomatic leaves but stomata on the upper side may emerge during adaptation to dry habitats. All grasses commonly show amphistomatic leaves, which are shown to increase their number of stomata in swampy habitats only.

Table 7.1. Stomatal frequency per square millimetre of leaf surface.

<table>
<tr><th>Species</th><th>Upper surface</th><th>Lower surface</th></tr>
<tr><td>Grasses:</td><td></td><td></td></tr>
<tr><td>Avena sativa (oats)</td><td>25–48</td><td>27–35</td></tr>
<tr><td>Phragmites communis (reed)</td><td>426–536</td><td>493–754</td></tr>
<tr><td>Saccharum officinarum (sugar cane)</td><td>59–167</td><td>176–351</td></tr>
<tr><td>Secale cereale (rye)</td><td>49</td><td>42</td></tr>
<tr><td>Triticum aestivum (wheat)</td><td>33–47</td><td>14–32</td></tr>
<tr><td>Zea mays (corn)</td><td>52–94</td><td>68–158</td></tr>
<tr><td>Herbaceous plants:</td><td></td><td></td></tr>
<tr><td>Arachis hypogaea (peanut)</td><td colspan="2">15–20</td></tr>
<tr><td>Cuscuta reflexa (devil's gut, parasite)</td><td>0</td><td>5</td></tr>
<tr><td>Datura innoxia (thorn-apple)</td><td>17–29</td><td>22–37</td></tr>
<tr><td>Helianthus annuus (sunflower)</td><td>27–326</td><td>90–408</td></tr>
<tr><td>Phaseolus vulgaris (bean)</td><td>8–16</td><td>132–184</td></tr>
<tr><td>Ranunculus glacialis (alpine)</td><td>100</td><td>33</td></tr>
<tr><td>Solanum tuberosum (potato)</td><td>10–20</td><td>230</td></tr>
<tr><td>Trifolium pratense (clover)</td><td>207</td><td>335</td></tr>
<tr><td>Succulent leaves:</td><td></td><td></td></tr>
<tr><td>Agave sp. (agave)</td><td colspan="2">8–72</td></tr>
<tr><td>Peperomia hispidula (pepper face)</td><td>0</td><td>40</td></tr>
<tr><td>Portulacca oleracea (common purslane)</td><td>0</td><td>21</td></tr>
<tr><td>Sedum acre (yellow stonecrop)</td><td>21</td><td>14</td></tr>
<tr><td>Shrubs and trees:</td><td></td><td></td></tr>
<tr><td>Artocarpus altilis (bread fruit, tropical)</td><td>(?)</td><td>500–600</td></tr>
<tr><td>Borassus flabellifer (palm tree, tropical)</td><td>22</td><td>22</td></tr>
<tr><td>Citrus sinensis (sweet orange, subtropical)</td><td>32–90</td><td>620</td></tr>
<tr><td>Coccoloba cereifera (*seaside grape, tropical)</td><td>0</td><td>160–175</td></tr>
<tr><td>Cornus suecica (dogwood, temperate)</td><td>0</td><td>70–134</td></tr>
<tr><td>Fagus silvatica (beech, temperate)</td><td></td><td></td></tr>
<tr><td>sun-exposed leaves</td><td colspan="2">130–295</td></tr>
<tr><td>shade-exposed leaves</td><td colspan="2">94–177</td></tr>
<tr><td>Hedera helix (ivy, shade, temperate)</td><td>0</td><td>105</td></tr>
<tr><td>Larrea divaricta (*creosote bush, semidesert)</td><td>104</td><td>130</td></tr>
<tr><td>Ligustrum sp. (privet, subtropical)</td><td>0</td><td>830</td></tr>
<tr><td>Myrtus communis (myrtle, subtropical)</td><td>0</td><td>158</td></tr>
<tr><td>Populus alba (poplar, temperate, riparian)</td><td>0</td><td>315</td></tr>
<tr><td>Quercus palustris (oak, swampy)</td><td>0</td><td>497</td></tr>
<tr><td>Quercus rubra (oak, temperate, dry)</td><td>0</td><td>680</td></tr>
<tr><td>Quercus triloba (oak, subtropical, dry)</td><td>0</td><td>1192</td></tr>
<tr><td>Sambucus nigra (elder, temperate)</td><td></td><td></td></tr>
<tr><td>sun-exposed leaves</td><td colspan="2">42–260</td></tr>
<tr><td>shade-exposed leaves</td><td colspan="2">12–147</td></tr>
<tr><td>Salix alba (willow, temperate, riparian)</td><td>114</td><td>(?)</td></tr>
<tr><td>Floating leaves:</td><td></td><td></td></tr>
<tr><td>Nymphea alba (water lily)</td><td>400</td><td>0</td></tr>
</table>

* the common name of a close relative.

The stomatal density per unit area is very divergent, ranging from 25 mm^{-2} in rye (*Secale*) to 1192 mm^{-2} in some oak species (*Quercus triloba*). Shade and succulent plants tend to have less stomata, while higher stomatal densities are reported for plants with xeromorphic leaf structures (Table 7.1). Small variations reported in the number of stomata per leaf for a single species suggest that higher stomatal density in sun leaves and xeromorphic leaves is partly due to the commonly smaller leaf blade.

7.4 MESOPHYLL

The photosynthetically active tissue of the leaf, which contains a very high density of chloroplasts (up to 33 million cm^{-2} on a surface area basis), is situated between the upper and the lower epidermis, consisting of the leaf parenchyma and the intercellular spaces (Fig. 7.2). While the leaflets of mosses (e.g. *Bryum*, *Hoockeria*, *Calypogeia*, analogous organs) are comprised of one cell layer only, the mesophyll of fern fronds (e.g. *Asplenium*, *Blechnum*, *Pteris*) is multicellular but undifferentiated. Although the leaves of herbaceous dicotyledonous plants often have a homogeneous chlorenchyma (e.g. *Pisum sativum*, *Lactuca sativa*, *Hosta*), in most dicots the ground tissue undergoes the usual differentiation to form palisade and spongy mesophyll (e.g. *Citrus*; Fig. 7.11).

The shape and arrangement of chlorenchyma cells can vary: soybean (*Glycine*) leaves have multi-armed palisade cells connected to the bundle sheath cells, while alfalfa (*Medicago sativa*) leaves show two compact short cell layers. Thin leaves of potato (*Solanum tuberosus*) show a single row of loosely arranged palisade cells, whereas in thin cotton leaves (*Gossypium*) the long palisade cells may comprise two-thirds of the leaf. In leathery leaves of tea (*Camellia sinensis*) or of some Myrtaceae (e.g. *Hakea*) osteosclereides extend throughout the two palisade layers to reduce the injurious effect of wilting (Fig. 7.12).

In the succulent leaves of *Peperomia*, two-thirds of the leaf's thickness is comprised of an enormous epidermal water storage tissue which is superimposed onto two narrow rows of densely packed elliptical palisade cells, followed by a well-developed spongy parenchyma with extremely lobed cells forming an impressive 'sponge'. While a comparable arrangement is seen in leaves of *Ficus* species, in some halophytic and succulent leaves an internal water storage tissue is surrounded by the chlorenchyma cells (e.g. *Agave*, *Salsola*). In submerged aquatic macrophytes the mesophyll consists of an aerenchyma with large intercellular spaces (lacunae). In some leaves of submerged plants the chloroplast-rich epidermis is the unique photosynthetically active tissue and the mesophyll may be changed to a chloroplast-free aerenchyma with storage functions (e.g. *Echinodorus*).

The mesophyll may contain laticifers, idioblasts with tannins, oil, anthocyanins or calcium oxalate crystals (e.g. subepidermal cystoliths). The vacuoles of the chlorenchyma can store a lot of secondary compounds which are valuable and may be used for pharmaceutical and cosmetic purposes.

7.4.1 Dicotyledonous leaves

In the cross-section of a dorsiventral leaf the typical palisade parenchyma cells are rod-shaped and appear to be compactly arranged in one or more rows. In the latter case the palisade cells may become shorter in the layer towards the spongy parenchyma (Fig. 7.11). In a surface parallel section the palisade cells appear rounded and only slightly attached to each other, forming intercellular spaces like spherical triangles in between. Frequently the chloroplasts are pressed against the cell walls which are exposed to intercellular channels (Fig. 7.4). The palisade tissue contains the majority of chloroplasts and has become specialised to increase photosynthetic efficiency, since chloroplasts can be oriented to form a single layer lining the cell walls of the elongated cells to enable maximal light utilisation.

Isobilateral leaves with an additional palisade tissue on the abaxial side (a xeromorphic character) lead to the anatomy of cylindrical leaves (e.g. *Hakea*) with palisade parenchyma occurring all around the periphery. Analogous arrangements can be found in needle leaves of coniferous trees ('armed palisade cells', e.g. *Pinus*) and is observed also when photosynthetically active branches replace the leaves functionally (e.g. *Casuarina*, Fig. 7.16).

The spongy parenchyma consists of chloroplast-containing cells of various shapes, sometimes irregular, forming a three-dimensional net (Fig. 7.5). The neighbouring cells are connected by lobes and branches of spongy cells mainly elongated in a direction parallel to the leaf surface.

7.4.2 Intercellular spaces

The outstanding feature of the spongy mesophyll are the well-developed intercellular spaces, which form a continuous system from the substomatal chambers to the cell walls of each cell in the chlorenchyma. Usually schizogenously developed (by cell separation along the middle lamella through the action of pectinase), the extension of the intercellular spaces from one substomatal cavity is correlated to the stomatal frequency. The volume of the intercellular system and the surface of cell walls exposed to the intercellular spaces determine the velocity of gas exchange. These demands are contradictory because a large intercellular system reduces the number of cells in the tissue and by this the internal free surface area.

Intercellular volume as a fraction of the total leaf volume has a mean value of 21% (based on a survey of 12 common agricultural and ornamental plants). A shade leaf of *Fagus silvatica* with a surface area of 36 cm^2 has a total volume of 340 mm^3, with a contribution of 80 mm^3 from the intercellular system. The leaves of the floating Araceae *Pistia stratiotes* contain 71% air (see also Fig. 7.12b), while in the succulent leaves of *Begonia hydrocotylifolia* the intercellular spaces contribute only 3.5% of the volume. The intercellular volume increases during ontogeny and shows daily changes according to the water content of the leaves (e.g. in Mediterranean evergreens the intercellular volume continues to expand until midnight).

The intercellular system contributes also to the reflection properties of the leaves. White leaves or silvery ornaments on leaves are caused by the intercellular system giving rise to total light reflection from wet cell walls facing the air spaces (e.g. *Aglaonema, Marantha*).

The ratio of the internal free surface area to the external surface area of the leaf can be as high as 17–31 in sun or xeromorphic leaves (e.g. *Citrus*). This ratio drops in mesomorphic leaves to 10–15 (e.g. *Vitis, Syringa*) and seems to reach a minimum of 7–10 in succulent and shade leaves (*Bryophyllum, Berberis*). Due to the higher cell density in the palisade, the internal free surface area is 1.6–3.5 times higher in this tissue compared to the spongy mesophyll [12].

7.4.3 Monocotyledonous leaves

Some monocotyledonous leaves are unifacial (e.g. *Iris*), whilst others may resemble the dorsiventral structure of dicotyledonous leaves (e.g. *Musa*). The basic features of hydrophytes (aerenchyma; see also Fig. 7.12b) and xerophytes (sclerenchyma; see also Fig. 7.15) are the same as those in dicotyledons. *Allium* species may develop tubular leaves with concentric layers of palisade and spongy parenchyma around the cavity in the centre.

Grass leaves:
The narrow leaf blade of grass leaves has no petiole but a sheath enclosing the stem. In the epidermis bulliform cells with thin anticlinal walls and enlarged mesophyll cells (hinge cells) participate in involution and folding of the leaf blade during excessive loss of water. As a rule grass leaves show a homogeneous mesophyll formed by parenchymatic cells which may be lobed (e.g. *Bambusa*) or even branched (e.g. *Panicum*). The variation in bundle sheaths is taxonomically significant [4] and an indicator of the type of photosynthetic pathways (Fig. 7.13).

In grasses with the C_3 pathway the vascular tissue is surrounded by two bundle sheaths. The inner or 'mestome-sheath', which lacks chloroplasts or sometimes has just a few, shows thickened lignified cell walls and replaces the missing supporting elements normally seen in the vascular bundles of dicots. The mestome sheath is encircled by a second, parenchymatic sheath of normally thin-walled cells, containing only a few chloroplasts which are sometimes smaller than those in the mesophyll. In C_3 plants, the bundle sheath cells play no significant role in photosynthetic CO_2 assimilation and metabolism. CO_2 is fixed in mesophyll cells and the mesophyll chloroplasts form starch.

In grasses with the C_4 pathway either a single or a double layered sheath may be developed. In each case the parenchymatic cells are characteristically larger than the mesophyll cells. These bundle sheath cells are rich in organelles, with large and dark green chloroplasts that often show a reduced granal size. These chloroplasts are able to synthesise large amounts of starch during the normal photoperiod.

The bundle sheath chlorenchyma is surrounded by a radially arranged mesophyll, each mesophyll cell being in direct contact with the bundle sheath or no more than one cell removed. The concentric arrangement of chlorenchyma, termed the 'Kranz syndrome', is essential for the functioning of the C_4 pathway and is thus characteristic of almost all C_4 plants. However, in some C_4 plants, such as species of the Chenopodiaceae, the Kranz syndrome is not clearly evident [6].

7.5 VASCULAR SYSTEM

The vascular bundles, branching from the vascular strands in the stem run as leaf traces before they diverge into the petioles of leaves. In the lamina these bundles may form a dichotomous (e.g. Ferns), reticulate (e.g. dicotyledons) or parallel venation (e.g. monocotyledons). In the two latter cases bundles converge and join at the leaf apex or the leaf margins. The larger veins are surrounded by collenchymatic or sclerenchymatic bundle sheath cells and act as supporting tissue for the flat expanded leaf blade.

Independent of the arrangement of xylem and phloem (collateral, bicollateral, centric) in the major veins, thinner veins have collateral bundles where xylem elements face the upper (adaxial) side and phloem elements the lower (abaxial) side of the leaf.

In a common dorsiventral leaf the vascular bundles will form a reticulate venation branching from the mid-vein. From first order lateral veins successively thinner veins diverge up to five orders. The main vein and the secondary veins develop large bundle sheath extensions which project as ribs on the lower surface of the leaf (abaxial side, see also Fig. 7.1). Fusing veins (ribs) of low order encircle intercostal areas which may enclose several small meshes formed by ultimate branching (areoles). Terminating vein endings consisting of short tracheids frequently extend into the areoles. The phloem elements may not accompany the xylem to the end.

The parenchymatic cells around the ultimate branches constitute transfer cells which support the exchange of solutes between the vascular bundles and the mesophyll cells. In a mature leaf specialised transfer cells with well-developed wall protuberances and

Fig. 7.1. *Phaseolus vulgaris* var. nanus (bush bean). Dorsiventral shade leaf. Scanning electron micrograph (SEM) of a cross-section of the blade near the mid vein. UPE, upper epidermis; PP, palisade parenchyma; SP, spongy parenchyma; LOE, lower epidermis; BSE, bundle sheath extension; V, vein; HH, hooked hairs. Bar = 100 µm.
Fig. 7.2. *Helleborus niger* (Black Hellebore, 'Christmas Rose'). Dorsiventral shade leaf. Drawing of a cross-section. UPE, upper (adaxial) epidermis; PP, palisade parenchyma; SP spongy parenchyma; LOE, lower (abaxial) epidermis; VB, vascular bundle (3rd order); i, intercellular space; ST, stoma; CH, chloroplast; T, transmitted light across the leaf (absorptivity $\alpha = 60\%$). Bar = 100 µm.
Fig. 7.3. *Helleborus niger* (Black Hellebore, 'Christmas Rose'). Dorsiventral shade leaf. Surface parallel free hand cut. Upper epidermis, ridged and furrowed cuticle. Light micrograph (LM). CW, cell wall. Bar = 50 µm.
Fig. 7.4. *Phaseolus vulgaris* var. nanus (bush bean). Dorsiventral shade leaf. High magnification of a microtome cross-section of the palisade parenchyma, redrawn in part for contrast (LM). i, intercellular spaces; CH, chloroplasts. Bar = 50 µm.
Fig. 7.5. *Helleborus niger* (Black Hellebore, 'Christmas Rose'). Dorsiventral shade leaf. Microphotograph of the living spongy parenchyma, cross-sectioned by hand, redrawn in part for contrast (LM). i, intercellular space; CH, chloroplasts; ST, stoma from inside. Bar = 50 µm.
Fig. 7.6. *Helleborus niger* (Black Hellebore, 'Christmas Rose'). Lower epidermis. Surface parallel free hand section (LM). SP, stomatal pore; CW, cell wall; GC, guard cells; BH, base of hair. Bar = 50 µm.

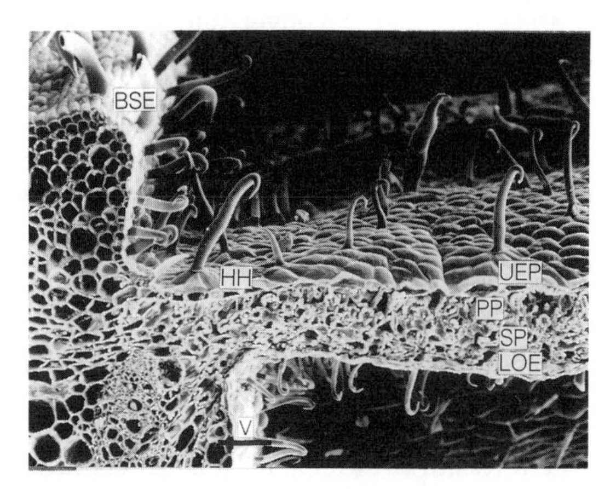

Fig. 7.1

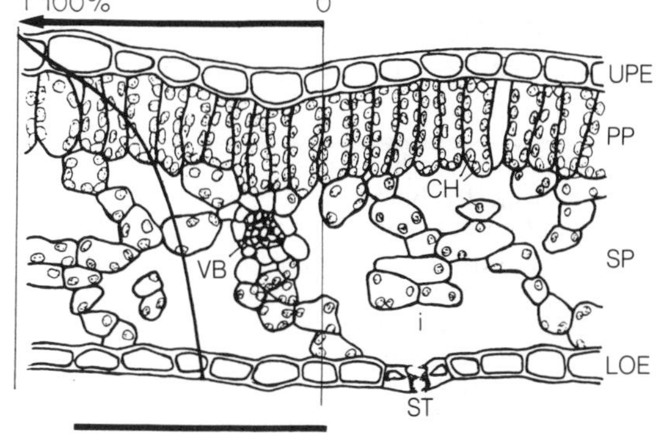

Fig. 7.2

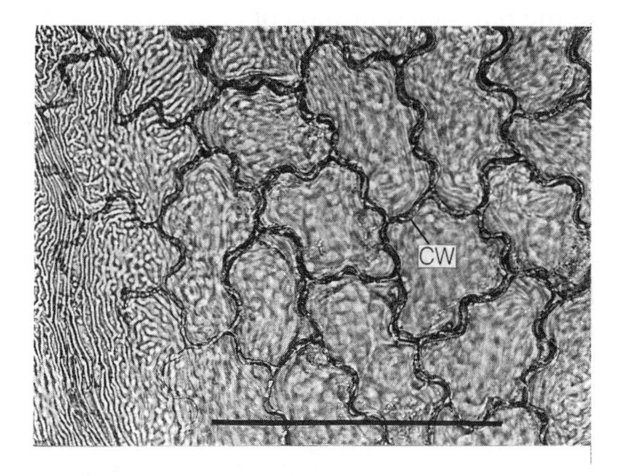

Fig. 7.3

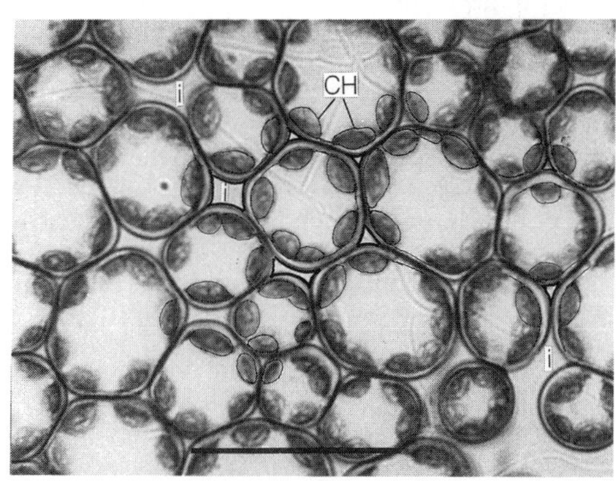

Fig. 7.4

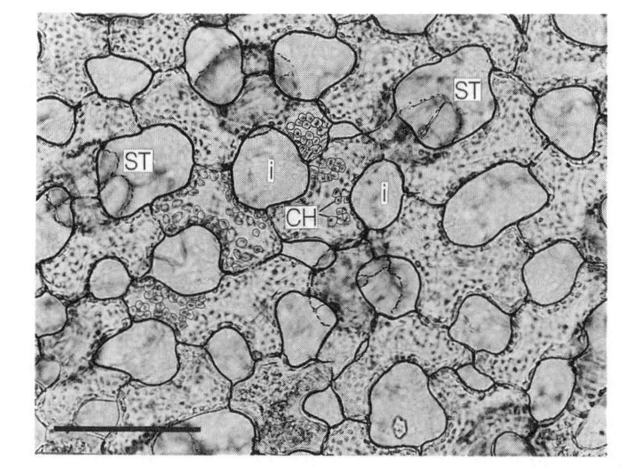

Fig. 7.5

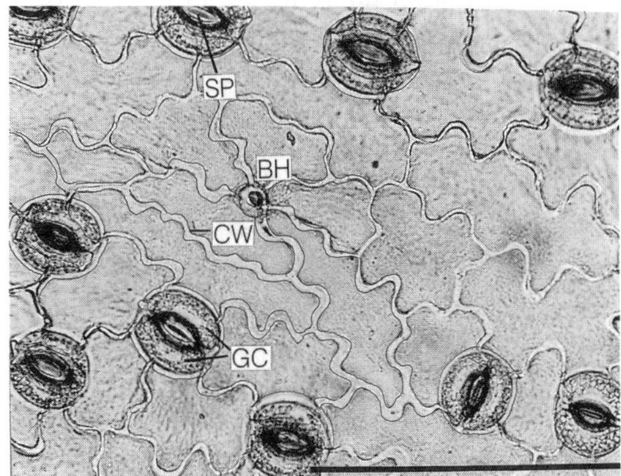

Fig. 7.6

numerous branched plasmodesmata are in contact with enlarged companion cells and together with dense phloem parenchyma cells are involved in the loading of the minor veins with assimilates.

Xylem transfer cells support the water translocation from tracheides to the cell walls of the surrounding mesophyll. In addition, parenchymatous bundle sheath extensions of minor veins conduct the water to the epidermis for lateral distribution. The smaller veins, enclosed by one or more layers of compactly arranged bundle sheath cells, are usually embedded in the upper part of the spongy parenchyma. These veins of high order (3rd and above) play the major role in the transportation of water and photosynthates, which may be controlled in some species by an endodermis (e.g. Ferns, Acanthaceae). Vein spacing (average 130 µm) is closer in leaves with a high ratio of palisade to spongy parenchyma or when bundles are missing well-developed bundle sheath extensions. Under these circumstances the total length of veins per square millimetre may reach 14 mm (e.g. *Quercus boissieri*). In C_4 plants the specialised, chloroplast containing bundle sheaths develop a suberin lamella to perform the second step of CO_2 fixation in C_4 photosynthesis (Sections 7.4.3 and 7.6.4).

7.6 FUNCTIONAL DEMANDS

The anatomical structure of a leaf must permit the absorption and conversion of a range of light quality and intensity in the visible region of solar radiation (400–700 nm). In addition to this fundamental requirement for generation of chemical energy, the substrates used in the photosynthetic process, CO_2 and H_2O, must be distributed throughout the leaf to reach each chloroplast, and the end products, assimilates and O_2, together with water vapour should find their way out of the leaf (Sections 7.3.3 and 7.4.2). To keep this functional unit in shape, supporting structures are needed (Section 7.5).

7.6.1 Adaptations to provide optimal light absorption and conversion

Light absorption by entire ecosystems, canopies, single plants and leaves is the key process for energy conversion and photoassimilation of CO_2. The regulation of light absorption and of solar energy conversion with respect to leaf anatomy takes place on three levels with time scales from seconds to months.

7.6.2 Changes in thylakoid membrane structures

If Calvin cycle activities are reduced and/or excess photons have been absorbed, Photosystem II (PSII) reaction centres will deactivate any surplus excitation energy via heat. The enforced heat deactivation leads to photoinhibition and has been shown to be related to the rapid modification of PS II centres. The thylakoid membrane is also able to rearrange its three-dimensional structural form through unstacking and restacking of grana regions whereby the distribution of light-exposed PS II centres can be changed [1].

Some of these modifications can take place in the time range of seconds and minutes and therefore they are responsible for adaptation to the daily changing light environment, including fast-moving sunflecks [9].

During plant growth leaves developed previously in full sunlight become more and more shaded by newly developed leaves and may receive less than 10% of full sunlight. As a change of arrangement in the leaf tissue is not feasible the chloroplast architecture and the pigment composition have to be changed to build a photosynthetic apparatus suited for shade conditions (Table 7.2). Apart from larger grana, more antenna chlorophyll molecules are associated with PS II reaction centres and in addition adaptations to the 'green' shade in the understorey can give rise to higher chlorophyll *b* and carotenoid content.

These modifications according to sun and shade conditions at the chloroplast level may adapt net leaf photosynthesis to irradiances equivalent to the half of full sunlight only.

As the velocity of the redox processes in the electron transport chain is limited, calculations based upon secondary electron transport ($10^{-2.5}$ s per e^-)

Table 7.2. Some typical adaptations of leaves to different light environments. Data obtained from *Fagus sylvatica*, *Sinapis alba* and *Alocasia macrorrhiza*.

	High light sun leaves	Low light shade leaves
Average leaf area (cm^2)	25	38
Number of lateral veins per leaf	16	18
Stomata frequency per mm^2 leaf area	214 ± 26	144 ± 11
Stomatal size	smaller	larger
Ratio palisade/spongy parenchymya	1.3 ± 0.1	0.8 ± 0.1
Chlorophyll content (mg 100 cm^{-2})	5.0	4.4
Relative chlorophyll content per plastid (%)	70	100
Ratio chlorophyll *a*/*b*	3.6 ± 0.2	2.7 ± 0.13
Ratio PS II versus PS I reaction centres	1.9	1.3
Chlorophyll per cytochrome *f*	320	900
Grana diameter (μm)	0.254	0.44
Variable fluorescence	1.7	1.0
Net leaf photosynthesis (mg CO_2 m^{-2} s^{-1})	0.4	0.25
Dark respiration (mg CO_2 m^{-2} s^{-1})	0.1	0.05
K_m (RuBP) (mmol)	0.234	0.36

and charge separation (10^{-4} s per e^-) at the two reaction centres indicate that about 320 excitons can be used per second to separate 320 electrons for both reaction centres. At high photon flux (2000 μmol m^{-2} s^{-1}), with an assumed absorption coefficient of α = 0.6, approximately 720 million photons μm^{-2} s^{-1} are offered and can be converted to excitons which would be able to run 22.6 million electron transport chains μm^{-2} leaf area s^{-1}. This means that for 1.0 μm^2 of granal lamellae (equivalent to 20 grana with a diameter of 0.25 μm each) there is a need for 19,000 electron transport chains, if 12 chloroplasts each with at least 100 granal lamellae are superimposed on each other in the leaf tissue. These estimates result in a leaf containing a very high number of chloroplasts: 120 million cm^{-2}. Since leaves are thought to actually contain about 33 million chloroplasts cm^{-2}, net leaf photosynthesis of C_3 plants is commonly light-saturated at less than 1000 μmol photons m^{-2} s^{-1}, and single chloroplasts may withstand only 200–300 μmol m^{-2} s^{-1}.

An outstanding feature of chloroplasts is therefore the ability to reduce the quantum yield of photochemistry (e.g. photoinhibition) via the deactivation of surplus incident (and consequently absorbed) light. If the activity of carbon metabolism is low and the regulation of light absorption at the PS II complex is not sufficient, the pool of assimilatory power will be filled up quickly. Reducing excess electrons will enhance the transfer to oxygen (Mehler reaction), forming oxygen radicals which are metabolised by superoxide dismutase and peroxidases to form water and O_2. By consuming assimilatory power (ATP and NADPH.H), photorespiration contributes to reducing the danger of photooxidative damage.

Therefore brief sunflecks or stroboscopic illumination caused by wind movement of leaves can be used quite effectively by shade-adapted leaves inside the tree crown or on the forest floor.

7.6.3 Arrangement of chloroplasts in cells and tissues

The adaptations to changes in the light environment by modifying the chloroplast's thylakoid membrane properties are accompanied by corresponding changes to the arrangement of chloroplasts in the cells. First observed with mosses, but later demonstrated to occur in higher plants as well, chloroplasts were shown to move toward the anticlinal cell walls in strong light to reduce the absorptivity of leaves. In the palisade parenchyma the chloroplasts can be clustered around the periclinal walls of the rod-shaped cells or they can be aligned along the anticlinal walls (Fig. 7.4). The latter arrangement acts as a light pipe to the spongy parenchyma situated below the palisade parenchyma in a dorsiventral leaf. The chloroplasts in the spongy parenchyma can also orient themselves according to the quantity and quality of penetrating light. Chloroplast movements can change the absorption coefficient of leaves up to 20% (e.g. *Begonia hydrocotylifolia*) and thus contribute to short-term adaptations after changes from low to high irradiance conditions, or *vice versa*.

In flatly expanded, horizontally oriented

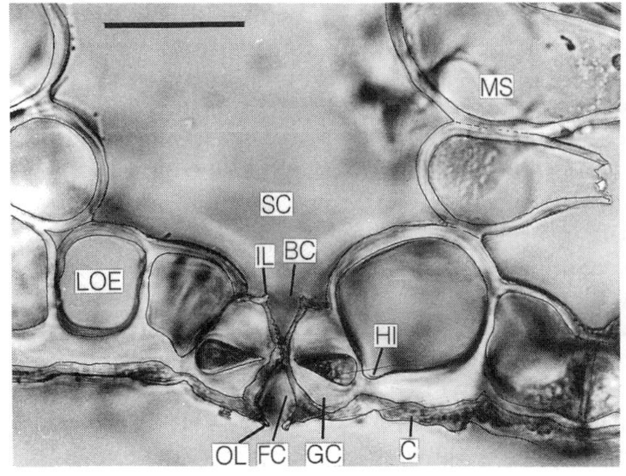

Fig. 7.7

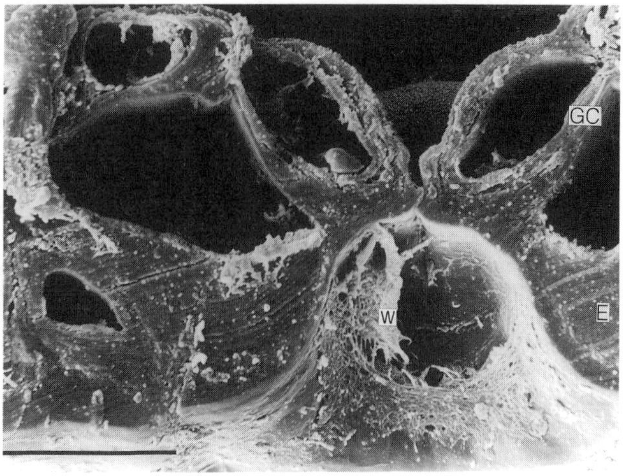

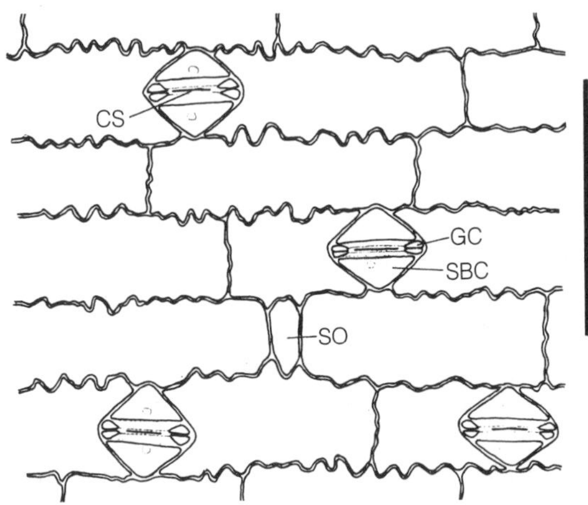

Fig. 7.8

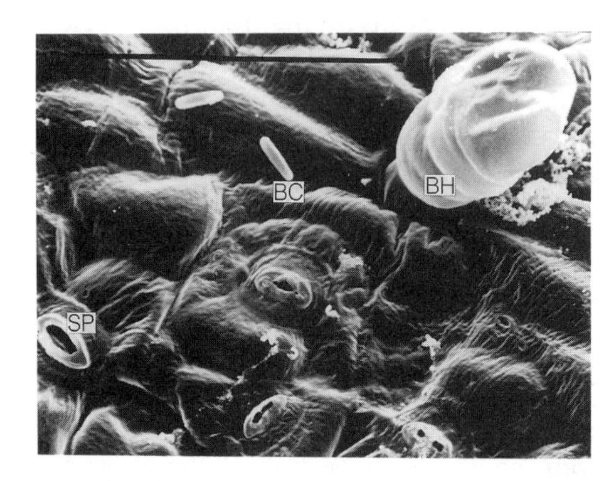

Fig. 7.9

Fig. 7.10

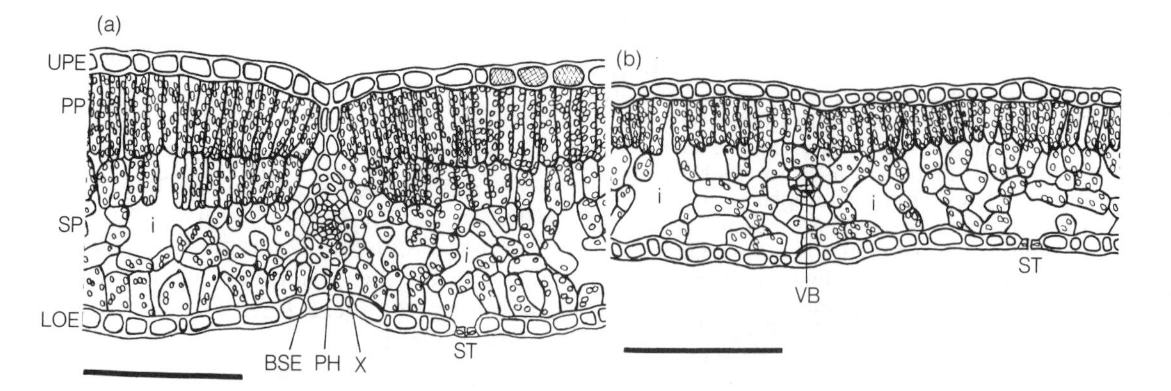

Fig. 7.11

leaf blades chloroplast movements control the light interception by changing the absorptance of cell layers, in order to vary the shading of chloroplasts in the relevant tissue situated below. Therefore chloroplasts in the spongy parenchyma are rarely submitted to photoinhibition while chloroplasts in the palisade cells above are frequently stressed by high light.

By staining cross-sections of 'sun' leaves (e.g. bush beans) with iodine (Section 7.7.5) it can be shown that in low light (e.g. early morning) starch granules are first found to occur in the palisade chloroplasts, whereas later in the day with rising light intensities starch is seen to accumulate only in the spongy parenchyma. Most plants show starch formation in leaves, generating a temporary storage for assimilates which cannot be exported. Export is controlled by the phosphate shuttle (Chapter 16) and can be slow because of inadequate phosphate supply by the cytoplasm, sometimes caused by the slow phloem loading process (Section 7.5). Large starch granules in the stroma of chloroplasts are also considered to have a shading effect on portions of grana lamellae or to be an early sign of photoinhibition.

Variation in the properties of chloroplasts in cells at different distances from the light-exposed surface can be expected, as a result of the light gradient in a leaf together with changing light quality (Fig. 7.2). This assumption is confirmed by differences in the photosynthetic rates of lower and upper leaf surfaces illuminated with equal light intensities [7,8] and the morphological and biochemical differences which have been measured [11].

The rearrangement of chloroplasts in cells of different layers in the chlorenchyma usually takes between several minutes and one hour, but if chloroplasts are submitted to photoinhibitory conditions (i.e. acting as sunscreen for the tissues below) they will undergo bleaching and the regeneration process may require an entire night or even longer.

7.6.4 Modifications of the mesophyll

The arrangement of the tissues in the mesophyll supports the adaptations seen at the chloroplast level. Dorsiventral leaves change their ratio of palisade to spongy parenchyma as a consequence of their light environment (Tables 7.3–7.5, Fig. 7.11). Compared with shade leaves, sun leaves are usually smaller and about twice as thick, with a larger contribution to palisade parenchyma (2.5 times) than spongy parenchyma (1.7 times). During the early stages of leaf ontogeny, light influences the differentiation of the leaf structure in herbaceous plants, whereas irreversibly determined buds develop in tree branches exposed to sun or shade. A loss of branches (e.g. storm damage) which opens up the crown will thus expose newly developed shade leaves to severe photoinhibition and bleaching.

In general the structure of the chlorenchyma has to support the 'light sharing sys-

Fig. 7.7. *Helleborus niger* (Black Hellebore, 'Christmas Rose'). Median cross-section of stoma (LM). LOE, lower epidermis; C, cuticle; GC, guard cell; MS, mesophyll; HI, hinge; OL, outer ledge; FC, front cavity; BC, back cavity (between FC and BC, central slit); IL, inner ledge; SC, sub-stomatal cavity. Bar = 10 μm.
Fig. 7.8. *Picea abies* (Norway spruce). Cross-section of sunken stoma (SEM). GC, guard cells; W, wax net in front cavity; E, lignified epidermis. Bar = 50 μm.
Fig. 7.9. *Phaseolus vulgaris* var. nanus (bush bean). Surface view of the lower epidermis (SEM). SP, stomatal pore; BH, bladder hair; BC, bacterium. Bar = 50 μm.
Fig. 7.10. *Zea mays* (maize). Grass leaf. Drawing of lower or upper epidermis. Arrangement of stomata and epidermis cells. GC, guard cell; SBC, subsidiary cell; CS, central slit; SO, short silicon cell. Bar = 100 μm.
Fig. 7.11. *Fagus silvatica* (beech). Drawings of cross-sections of a sun leaf (a) and a shade leaf (b). UPE, upper epidermis (in sun leaf with anthocyanin); PP, palisade parenchyma; SP, spongy parenchyma; i, intercellular space; X, xylem; PH, phloem; VB, vascular bundle; BSE, bundle sheath extension; ST, stoma; LOE, lower epidermis. Bar = 100 μm.

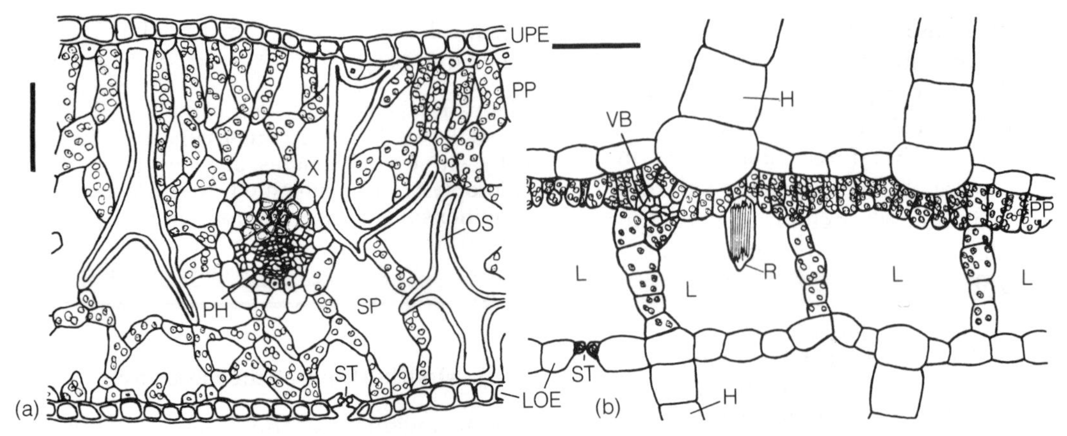

Fig. 7.12. Modifications of leaf anatomy.
(a) *Hakea salicifolia* (sweet-scented hakea). Dorsiventral xeromorphic leaf. Drawing of cross-section.
(b) *Pistia stratiotes* (water lettuce). Floating water plant (Araceae). Drawing of a leaf cross-section.
UPE, upper epidermis; ST, stoma; H, trichome; R, raphide cell; VB, vascular bundle; PP, palisade parenchyma; SP, spongy parenchyma; LOE, lower epidermis; L, lacune (air cavity); OS osteosclereides; X, xylem; PH, phloem. Bar = 100 μm.

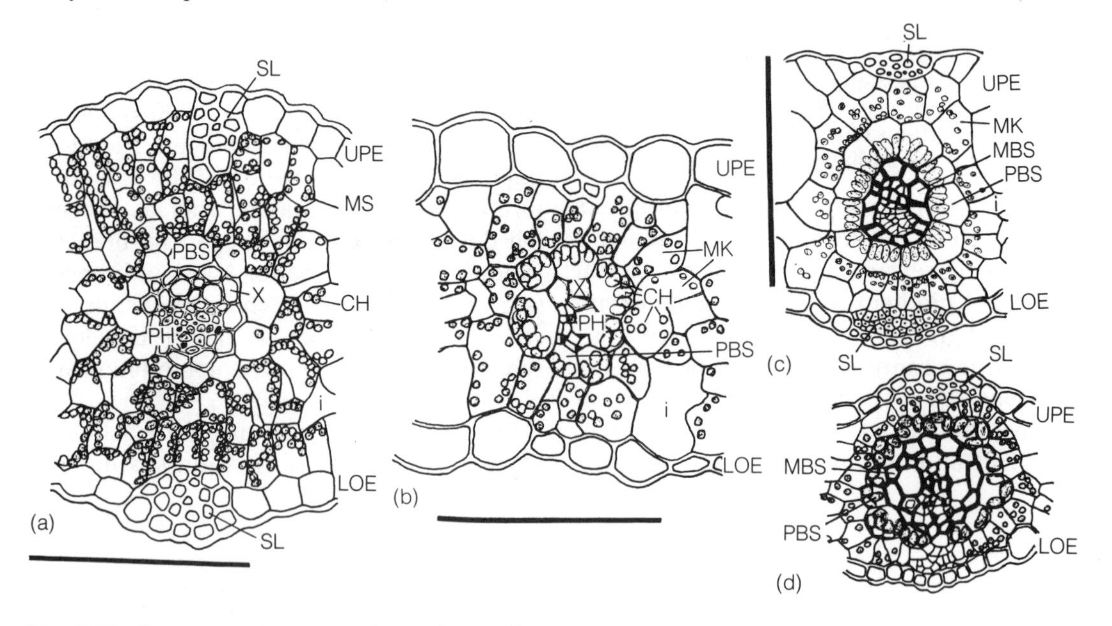

Fig. 7.13. Drawings of cross-sections of grass leaves:
(a) C_3 species (e.g. *Poa annua*);
(b) C_4 species, 'NADP-ME' type (e.g. *Zea mays*);
(c) C_4 species, 'NAD-ME' type (e.g. *Panicum capillare*);
(d) C_4 species, 'PCK' type (e.g. *Chloris guyana*).
UPE, upper epidermis; MS, mesophyll; CH, chloroplasts; i, intercellular spaces; MK, mesophyll 'Kranz' arrangement; PBS, parenchymatic bundle sheath ('Kranz' cells); MBS, mestome bundle sheath; X, xylem; PH, phloem; LOE, lower epidermis; SL, sclerenchyma. Bar = 100 μm.

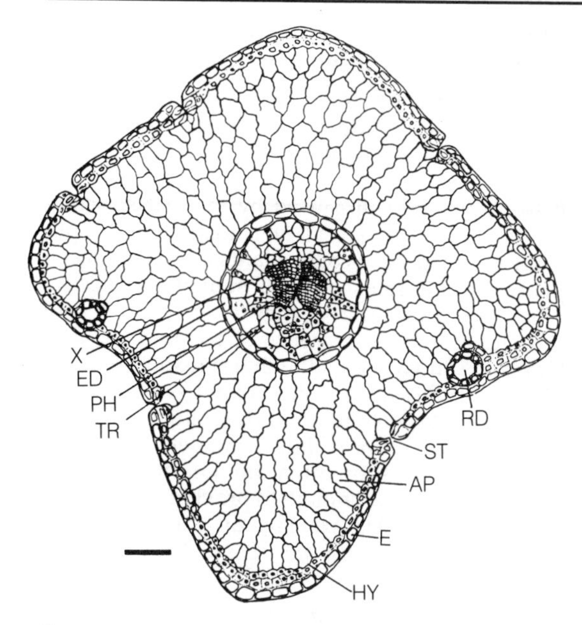

Fig. 7.14. *Picea abies* (Norway spruce). Drawing of a needle cross-section. E, sclerenchymatic epidermis; HY, hypodermis; ST, stoma (sunken); AP, armed palisade parenchyma; ED, endodermis; X, xylem; PH, phloem; TR, transfusion tissue; RD, resin duct. Bar = 100 μm.

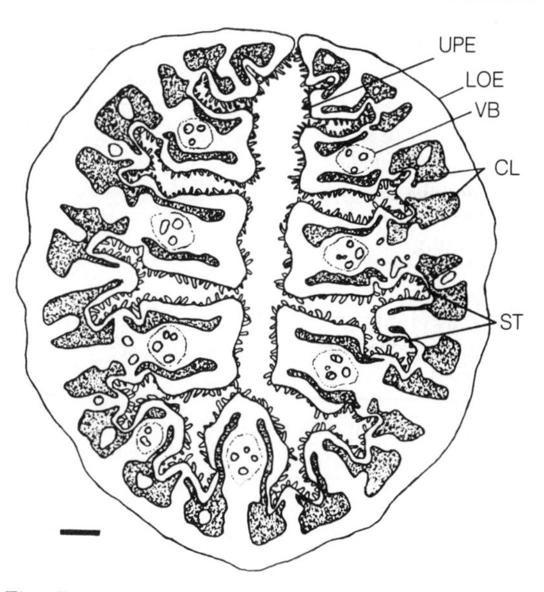

Fig. 7.15. *Stipa pennata* (feather grass). Drawing of a folded, xeromorphic grass leaf. UPE, upper epidermis with trichomes and stomata; CL, chlorenchyma; VB, vascular bundle; LOE, lower epidermis; ST, stomata. Bar = 100 μm.

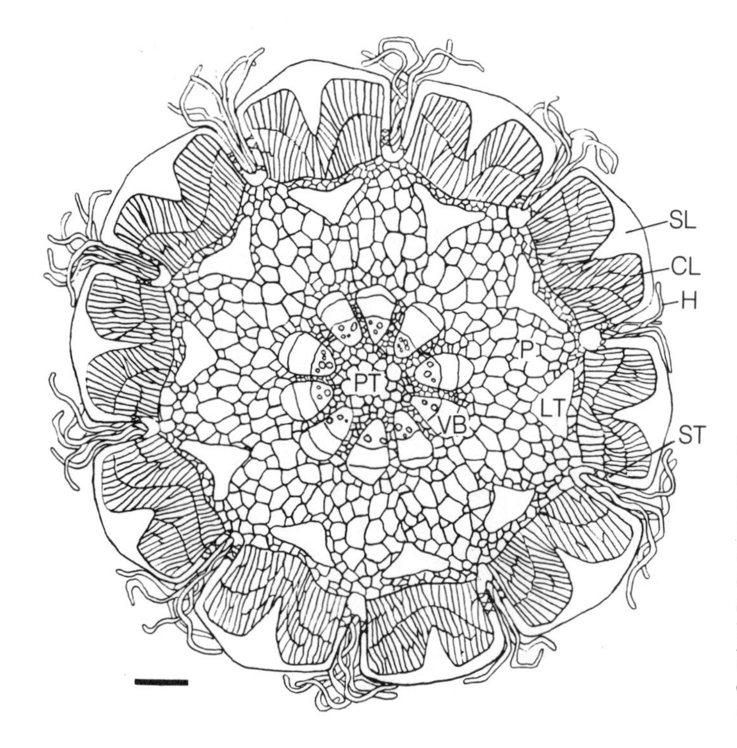

Fig. 7.16. *Casuarina equisetifolia* (Casuarina, she-oak, cassowary tree). Drawing of cross-section of photosynthetically active branch (2nd order). SL, sclerenchyma; CL, chlorenchyma (palisade parenchyma); P, storage parenchyma; LT, leaf trace bundle; VB, vascular bundle of branch; PT, pith; H, trichomes; ST, stomata in groove. Bar = 100 μm.

Table 7.3. Cell sizes (μm) in a dorsiventral leaf.

Epidermal cells	Length	Width	Height
Isodiametric (e.g. *Mitragyna stipulosa*)	20–55	12–30	20–30
Horizontally elongated, 'grass type' (e.g. *Andropogon gayanus*)	100–200	6–8	10
Vertically elongated (e.g. *Paphiopedilum niveum*)	119	82	110

Palisade parenchyma cells	Diameter	Length
Helleborus foetidus	24	123
Teucrium chamaedris	25	51
Beta vulgaris	10	44

Palisade cells per epidermal cell ('Palisade ratio')	
Average (e.g. *Coffea congensis*, *Catharanthus roseus*)	8–11
Copaifera officinalis	4
Eriodictyon californicum	6–20

Spongy parenchyma cells	Length	Height
Hoplestigma	20–70	20
Teucrium chamaedris	37	30

tem' of the cell layers at different distances from the sun-exposed leaf surface. Changes in light intensity cause the zone of maximal photosynthetic rate to move up and down within the mesophyll.

If the light gradient caused by the leaf structure corresponds with layers of chlorenchyma, optimally oriented leaves of C_3 plants may show rising photosynthetic rates up to full sunlight (e.g. *Larrea divaricata*). This reaction may be compared with the photosynthetic response of a closed stand of C_3 plants. In C_4 plants the linear response of net photosynthesis up to full sunlight is also a consequence of two superimposed light response curves of two more or less separately acting leaf surfaces. It was shown that the internal diffusive conductance is low enough to provide a separation of the intercellular systems on the upper and lower side of the leaf [7].

Internal free surface area:
All the structural adaptations in the mesophyll induced by light also influence transpiration properties and modify the diffusion pathways for CO_2 and O_2. Especially striking is the influence of variations in internal free surface area and intercellular volume. The wet cell wall surfaces exposed to intercellular spaces may absorb or evaporate water vapour and the dissolved gases CO_2 and O_2.

Table 7.4. Contribution of tissues to leaf structure (μm).

	*A	B		C	
		sun	shade	sun	shade
Cuticle		2.5	2.5		
Upper epidermis	16.8	11.5	7.5	30.0	36.0
Water storage tissue				100.0	61.0
Palisade parenchyma	52.7	70.0	42.5	196.0	110.0
Cell layers		2	1	2.5	1.5
Spongy parenchyma	48.0	60.0	56.5	159.0	135.0
Cell layers		5	4	4	4
Lower epidermis	12.0	7.5	7.5	25.0	28.0
Cuticle		2.5	2.5		
Thickness of lamina	129.5	154.0	119.0	510.0	370.0

*A, Average value of 46 *Acer* species; B, *Fagus silvatica* (Fig. 7.11); C, *Schefflera polybotrya*.

Table 7.5. Thickness of tissues (μm). Number of cell layers in parentheses.

	Leaf thickness	Palisade upper	Spongy	Palisade lower
Hosta coerulea	170.0	0	125	0
Cocoloba cereifera	580.0	155	125	145
Oenothera missouriensis	270.0	(3)	(4)	(2)
Oenothera fruticosa	260.0	(3)	(6)	(0)
Eucalyptus globulus				
Adult (sun) leaves		83 (2)	20 (1)	40 (2)
Juvenile (shade) leaves		33 (1)	66 (3)	0 (0)
Salix petandra				
Sun leaves	280.0	172 (3)	70	0
Shade leaves	148.0	68 (2)	50	0

Instead of employing the deactivation mechanisms described above, high light can be used in photosynthesis if the supply of CO_2 can keep up with the demand. Experiments with bush beans (*Phaseolus vulgaris* var. nanus) demonstrate convincingly that high light enhances the ratio of internal to external surface by 30%, to a value of 9.4:1. In the palisade parenchyma the free surface area of each cell is reduced, but it is enhanced for the tissue as a whole due to the larger number of cells, while in the spongy parenchyma a small reduction of the free surface area can be observed. Therefore it is not surprising that the maximum photosynthetic rates correlate with the ratio of internal free surface area to external leaf surface, whereas the maximum quantum yield does not change at all [12].

The internal free surface area represents the capacity of a leaf for both CO_2 absorption and transpiration. The use of these anatomical features is controlled by the movement of the stomata. Thus the ecological importance of the internal free surface area is clear; it is also demonstrated by the fact that the ratio of internal free surface area to leaf volume correlates with photosynthetic rates, with the highest ratio in C_4 plants.

(b) Sub-groupings of C_4 species

Like the variations in the internal free surface area, the arrangement of mesophyll cells around the bundle sheath cells in C_4 plants only indirectly adapts the plants to high light. It is basically a structure which enables the photosynthetic metabolism to concentrate CO_2 in the bundle sheath cells for effective secondary fixation by RubisCO (Chapter 16). In NADP-ME and PCK species, a suberin lamella in the cell wall of the bundle sheath cells acts as a very effective diffusion barrier to keep the CO_2 concentration high. It limits also the flux of various solutes through the cell wall, although numerous plasmodesmata penetrate the cell wall to establish a metabolic link to the surrounding mesophyll cells [5].

The three biochemical sub-groupings of species with the C_4 pathway of CO_2 fixation (NADP-ME, NAD-ME, PCK; Chapter 16) may be distinguished by anatomical characteristics such as the position of the chloroplasts in the bundle sheath and the number of bundle sheath layers (Fig. 7.13). These features, together with the presence of a suberin lamella in the bundle sheath cell walls, are well correlated with the photosynthetic type [3,4,10].

Species with only one bundle sheath (grasses: eu-panicoid, andropogonoid, C_4 dicots) are malate formers (Fig. 7.13b). In the mesophyll cells the first product of CO_2 fixation, oxaloacetate, is converted predominantly to malate. The malate is transported to the bundle sheath cells and is decarboxylated by NADP-dependent malic enzyme. This reaction generates CO_2, NADPH.H (reducing power) and pyruvate. Where NADPH.H is produced in the decarboxylation process no additional photoreduction in the bundle sheath chloroplasts seems to be necessary. Hence these plastids are poorly developed or have lost their grana during ontogeny. The bundle sheath chloroplasts are therefore poor in PS II. The large chloroplasts in the bundle sheath cells are centrifugal in position in grasses but centripetal in position in dicots. This category of C_4 species is termed the 'NADP-ME' type after the major decarboxylase enzyme of the sheath tissue.

All species with two bundle sheaths (grasses: chloridoid, ergastoid, C_4 dicots) are aspartate formers. If aspartate is transported to the bundle sheath cells practically no reducing power is generated in the decarboxylation process. In this case the sheath chloroplasts show well-developed grana and PS II activity. Two sub-types can be distinguished in this group.

(a) In the 'PCK' type the chloroplasts are irregular in size and centrifugal in position (Fig. 7.13d). After the transamination of aspartate to oxalacetate, CO_2 is generated by PEP carboxykinase in the sheath cell cytoplasm.

(b) In the second type the sheath cell chloroplasts are centripetal in position and no suberin lamellae can be distinguished in the cell walls (Fig. 7.13c). The aspartate from the mesophyll cells is converted first to oxalacetate and then to malate, to be decarboxylated by NAD-malic enzyme in the sheath cell mitochondria. Though reducing power is built up in the last step, these 'NAD-ME' grass species and C_4 dicots have grana and PS II activity in the chloroplasts.

The functional importance of this anatomical taxonomy in relation to the position of the chloroplasts may be questioned, but the coincidence of structure and metabolic pathways is intriguing.

7.6.5 Adaptations to dry conditions

Under conditions of water or nitrogen deficiency, plants are able to develop characteristic xeromorphic structures. In arid habitats we refer to such plants as xerophytes. The most obvious features of xeromorphic leaf structure are a low ratio of leaf surface area to leaf volume, a reduction in cell size, an increased internal free surface area of the living tissue, partly as a result of better developed palisade parenchyma, and a greater density of both the vascular system and the stomata. These characteristics facilitate high photosynthetic rates under conditions of favourable water supply.

Other xeromorphic features help to avoid uncontrolled water losses during stress periods. To reduce the transpiration rate the epidermis is covered by a thickened cuticle and waxes and bears a lot of trichomes. The stomata may be sunken (e.g. *Pinus*; Fig. 7.8) or grouped in grooves or crypts (e.g. *Nerium*). The living and transpiring tissue is often reduced in favour of a surrounding tissue of dead sclerenchymatic cells which help prevent wilting. Furthermore, the leaf blade may be involuted or folded (e.g. Gramineae; Fig. 7.15).

The xeromorphic features mentioned above are frequently accompanied by the development of special water storage tissues. These tissues are found predominantly in succulent plants. Often combined with the CAM pathway of CO_2 fixation they help the plant to survive even extreme drought and desert conditions.

7.7 EXPERIMENTS

7.7.1 Sectioning of unembedded materials by hand

The study of anatomical structures in a leaf requires cutting of sections. Despite the availability of hand microtomes and more sophisticated bench microtomes, free hand sectioning has a lot of advantages. Plant physiologists working in areas described elsewhere in this book may often use a few good hand sections made with a razor blade, in order to study the structure of the leaves on which they are working.

Cross-sections of the leaf are of most interest. To get thin sections, newly unwrapped razor blades (e.g. *Gillette Blue Blades*) must be used, and no more than about six sections should be made with each part of the edge.

Leaves are difficult to hold, even when they are folded several times. Leaf pieces should therefore be firmly held between two pieces of elder pith, fresh carrot or styrofoam. The best sections will be squashed or otherwise lost unless both the edge of the razor blade and the face of the leaf tissue are wetted before sectioning.

7.7.2 Measurements in a microscopic preparation

Transverse sections of the leaf blade, cut by hand with a razor blade, are examined in a light microscope. Look for vascular bundles and sclerenchyma cells, and examine the system of intercellular spaces in the different parts of the mesophyll. Count the layers of the palisade parenchyma cells, compare the thickness of the palisade parenchyma with the spongy parenchyma and calculate the ratio.

Measurements of actual length in a microscopic preparation are performed using an eyepiece ruler (glass disc with scale) or an ocular micrometer. The arbitrarily divided ruler ('micrometer') on the aperture stop in the eyepiece must be calibrated by means of a stage micrometer for each objective (lens) in use. The micrometer slide is marked with a scale commonly 1 mm long, divided into 100 parts each of 10 μm. By rotating the eyepiece and moving the micrometer slide, superimpose the starting points of the two scales and calibrate the micrometer. Count the number of ocular divisions which coincide best with a division on the stage micrometer. Thus, using a ×10 magnification lens, if 5 points (equal to 50 μm) on the micrometer slide equal 14 arbitary divisions on the eyepiece scale, then one division on the eye piece ruler equals 3.57 μm (50/14).

Make a drawing of a part of the cross-section and label it with the measured thicknesses of the tissues.

7.7.3 Stomatal position, frequency and width

Using a hand-cut cross-section of the leaf blade, find the stomata and describe their position and arrangement (raised, sunken, in crypts, protected by trichomes or wax; hypostomatic, hyperstomatic or amphistomatic).

In order to count the number of stomata per unit leaf area, a good incident light microscope is needed. If the leaves have a blade without many trichomes, and if the stomata are not too deeply sunken, you can count the stomata on artificial replicas made with *Celloidin* (cellulose nitrate in ether), silicon rubber monomer, *Rhoplex AC-33* (an acrylic polymer emulsion) or any transparent paint or varnish. Put a drop of the water-soluble varnish on the surface of the leaf, and pull off the dry transparent film with forceps after a few minutes. Now count the stomatal impressions on this replica with the help of a calibrated grid in the eyepiece of the microscope. Calculate the leaf surface area covered by the grid on the replica from a comparative measurement with the objective slide ruler. Take replicas of different parts of the leaf, because stomatal frequency may vary greatly across its width. Less than 60 stomata per mm^2 and more than 600 stomata per mm^2 indicate that the plant grows in an extreme habitat (xeromorphic or hygromorphic, respectively). However, do not measure stomatal widths by the replica technique or from epidermal strips. It is better to use an incident light microscope with a mirror (reflecting) objective for such measurements. Alternatively stomatal aperture may be determined by indirect methods as detailed in Chapter 5.

An estimate of the relative stomatal aperture can be obtained by infiltration using xylol, alcohol (ethanol) and paraffin oil. Put a drop of each solution on the upper and lower surface of the leaf and judge the degree of infiltration by the size of the area darkened by the liquid entering the internal air spaces.

7.7.4 Staining of sections

To improve the information gained from hand sections, a simple and rapid staining process can be used with polychromatic dyes like Toluidine Blue O or Gentian Violet (comparable to Methyl Violet and Crystal Violet):

Toluidine Blue O: 0.05% in benzoate buffer at pH 4.4 (0.125 g benzoic acid, 0.145 g sodium benzoate, in 100 ml distilled water).

Gentian Violet: 1% in distilled water. (As the dye is a component of copying-ink pencils, 5 mm of the pencil lead can be dissolved to give an appropriate solution.)

Stain the section for 1–5 minutes in a drop of one of the solutions and rinse in running water. Toluidine Blue O has a surface-staining effect and therefore gives good results even with thick sections, which will retain the colours of chlorophylls and anthocyanins. Gentian Violet will stain the cytoplasm (leucoplasts and mitochondria) and the cell walls. Lignin-free cell walls can be destained in 50% ethanol containing a few drops of hydrochloric acid (2 ml HCl in 100 ml ethanol). Sections may be dehydrated and preserved as described below.

Dehydrate in each of the given ethanol concentrations for 2 minutes; 30%, 60%, 96%.

Transfer the section into terpineol or isopropanol for 3 minutes. Wash thoroughly in xylol for 2 minutes and mount in Canada Balsam, Malinol or Eukit.

The best hand sections may be used for succedan staining with Safranin and 'Astra Blau' as follows:

Stain for 5–15 minutes in an aqueous solution of 1% safranin. Rinse thoroughly in distilled water.

Stain in an aqueous solution of 0.05% 'Astra Blau' (containing 2 g tartaric acid in 100 ml).

Proceed with dehydration and mounting as described above. The tissue will show lignified cell walls in bright red, nuclei in pink and the cellulose cell walls in blue.

There is no need for thin sections if a fluorescence microscope can be used. Many plant tissues contain molecules that can be made to fluoresce when irradiated with blue-violet light. Cellulose fibrils can be stained by Calcofluor (the whitener in washing powders) to emit blue fluorescence. Lignin and cutin are able to bind Auramine O, a fluorochrome that fluoresces yellow in blue light.

Leaf material can be fixed to obtain better sections and/or to preserve it for later use. In general, it is assumed that glutaraldehyde fixation gives the best results. Signs of damage are rarely observed, even in the electron microscope. Cell walls fix well in 60% ethanol, and fixation in a mixture of ethanol – acetic acid (1:2) preserves tissue in a suitable state for anatomical studies using the light microscope. Leaves with a well-developed aerenchyma should be vacuum infiltrated. The fixative is coagulant and cannot be used to study cytoplasmic structures. To preserve leaf material for several years a mixture of equal parts ethanol, glycerin and water may be used.

7.7.5 Differentiation between C_3 and C_4 plants by staining starch granules with iodine *in situ*

Cross-sections of the blades of leaves of various types are stained by immersion in iodine solution. Many plants will form starch in the chloroplasts if the leaf has been illuminated for a long time, if net leaf photosynthesis has been proportionally high and if phosphate supply has become limiting because of slow translocation of photosynthate by the phloem loading process. In such leaves chloroplasts will be stained dark blue by iodine. This occurs in chloroplasts of the mesophyll cells of all C_3 plants. In C_4 plants such starch accumulation occurs predominantly in the 'Kranz' bundle-sheath cells. The detection of starch in this way is more distinct if chlorophyll is extracted from the leaves, using hot alcohol, before staining. In this case cellulose and proteins may appear brown whereas pectin, cutin and cork show a yellowish colour. Sections can be stained subsequently with Gentian Violet.

Solution for iodine stain (Lugol): First dissolve 2.0 g of potassium iodide (KI) in 5 ml & water, then add 1.0 g of iodine and make up to 300 ml with water.

7.7.6 Vital staining for Photosystem II activity

The presence of PS II activity can be demonstrated by using the reduction of tetranitro-blue-tetrazolium chloride (TNBT). Cross-sections of the leaf blade, which need not be very thin, are cut by hand and infiltrated under vacuum with the staining solution (see below). After 5–20 minutes of illumination, under a microscope for instance, the areas in the leaf transverse section containing chloroplasts with high PS II activity will be a dark-blue colour. This occurs first in all wounded cells, then in the mesophyll cells of all plants and finally in the granal bundle sheath plastids of the C_4 grasses of the NAD-ME and PCK groups. The agranal bundle sheath plastids of the NADP-ME species turn blue only after several hours of illumination. Misleading results may occur if infiltration is poor because of residual air in the intercellular spaces. Note that the bundle-sheath cells which are longitudinally elongated are more likely to be injured during sectioning than the more rounded mesophyll cells. Thus C_4 plants, especially dicots, may show some coloured bundle-sheath chloroplasts even after a short period of illumination [2].

Stock solutions:
(A) 0.1% TNBT ($1.0 \, mg \, ml^{-1}$). (Do not make up more than 10 ml at one time: store the dye powder and solution in refrigerator.)
(B) 0.1 M phosphate buffer, pH 6.0.
(C) 0.3 M sucrose.
Staining solution:
1 part A + 3 parts B + 1 part C.

7.7.7 Starch prints

After 2 days in the dark, the chloroplasts in the leaf are generally destarched. Such a leaf can be used to produce high resolution starch prints or pictures. A photographic negative with high contrast is mounted on the upper surface of the leaf and exposed to light. Starch will be formed only in those areas which are actually reached by light. The starch formation will be proportional to incident light intensity. Thus if the leaf is cut off, killed and extracted in hot water followed by hot alcohol (in a water bath), and stained with iodine, the negative will be reproduced as a positive image. The high resolution image obtained indicates that starch is generally only formed in those cells which are actually illuminated, and that translocation of photosynthate from an illuminated chloroplast to a non-illuminated chloroplast does not readily occur, even when they both lie in the same cell.

7.7.8 Demonstration of oxygen evolution from whole plants

Fill a large glass container with a 0.01% indigo carmine (indigo disulphonate) solution. While stirring constantly, carefully add single drops of a 10% sodium dithionite solution until the blue indigo carmine is just reduced to the yellow form. Place a whole plant or a branch into this solution and seal the container, excluding air. After several minutes of illumination, if there is no excess sodium dithionite, blue areas will appear around all green parts of the plant. The reason for this is that the small amounts of O_2 produced by photosynthesis inside the leaf are enough to reoxidize the yellow indigo carmine back to the blue form.

ACKNOWLEDGEMENTS

We are very grateful to Dr. C. Critchley (University of Queensland) for encouraging comments when she was reading and correcting the manuscript. We also wish to thank Dr. J. Wilson (CSIRO, Brisbane) for his comments and Dr. U. Rohrhofer (University of Vienna) for the light microscope and SEM-micrographs of *Phaseolus vulgaris*. Figures were drawn by Mag. W. Ruppert (University of Vienna).

REFERENCES

1. Anderson, J.M. (1986) Photoregulation of the composition, function and structure of thylakoid membranes. *Ann. Rev. Plant Physiol.* **37**, 93–136.
2. Downton, W.J.S., J.A. Berry and E.B. Tregunna (1970) C_4 photosynthesis: non-cyclic electron flow and grana development in bundle sheath chloroplasts. *J. Plant Physiol.* **63**, 194–198.
3. Eastman, P.A.K., N.G. Dengler and C.A. Peterson (1988) Suberized bundle sheaths in grasses (*Poaceae*) on different photosynthetic types. I. Anatomy, ultrastructure and histochemistry. II. Apoplastic permeability. *Protoplasma* **142**, 92–126.
4. Ellis, R.P. (1977) Distribution of the Kranz syndrome in the southern African Ergrostoideae and Panicoideae according to bundle sheath cytology. *Agroplantae* **9**, 73–110.
5. Hatch, M.D. (1988) C_4 photosynthesis: a unique blend of modified biochemistry, anatomy and ultrastructure. *Biochim. Biophys. Acta* **895**, 81–106.
6. Hattersley, P.W. (1987) Variations in the photosynthetic pathway. In: *Grass Systematics and Evolution*. (Soderstrom, T.R., Hilu, K.W., Campbell. C.S. and Barkworth, H.E., eds.) Smithsonian Inst. Press, Washington DC. pp. 49–64.

7. Long, S.P., P.K. Farage, H.R.Bolhàr-Nordenkampf and U. Rohrhofer (1989) Separating the contribution of the upper and lower mesophyll to photosynthesis in *Zea mays* leaves. *Planta* **177**, 207–216.
8. Postl, W. and H.R. Bolhàr-Nordenkampf (1990) The light response of CO_2 fixation separated for the upper and lower side of a maize leaf. In: *Current Research in Photosynthesis*. Vol.IV. (M. Baltscheffsky, ed.) Kluwer Academic Publ., Dordrecht. pp. 31–34.
9. Pearcy, R.W. (1990) Sunflecks and photosynthesis in plant canopies. *Ann. Rev. Plant. Physiol. Plant. Mol. Biol.* **41**, 421–453.
10. Pendergast, H.D.V. and P.W. Hattersley (1987) Australian C_4 grasses (Poaceae): Leaf blade anatomical features in relation to C_4 acid decarboxylation types. *Australian J. Bot.* **35**, 355–382.
11. Terashima, I., S. Sakaguchi and N. Hara (1986) Intra-leaf and intracellular gradients in chloroplast ultrastructure of dorsiventral leaves illuminated from the adaxial or abaxial side during their development. *Plant Cell Physiol.* **27**, 1023–1031.
12. Turrell, F.M. (1936) The area of the internal exposed surface of dicotyledon leaves. *American J. Bot.* **23**, 255–264.

FURTHER READING

Cutter, E.G. (1973) *Plant Anatomy. I. Cells and Tissues; II. Organs*. Edward Arnold, London.

Evans, J.R., S. von Caemerer and W.W. Adams III (1988) Ecology of photosynthesis in sun and shade. *Australian J. Plant. Physiol.* **15**, 1–2.

Edwards, G. and D.A. Walker (1983) *C3, C4: Mechanisms, Cellular and Environmental Regulation of Photosynthesis*. Blackwell Scientific, Oxford.

Esau, K. (1977) *Anatomy of Seed Plants*, 2nd edn. John Wiley and Sons, New York.

Fahn, A. (1982) *Plant Anatomy*, 3rd edn. Pergamon Press, Oxford.

Johansen, D.A. (1940) *Plant Microtechnique*. McGraw-Hill, New York.

Larcher, W. (1975) *Physiological Plant Ecology*. Springer-Verlag, Berlin.

Napp-Zinn, K. (1973/74) Anatomie des Blattes. Blattanatomie der Angiospermen. In: *Handbuch der Pflanzenanatomie*. Vol. VIII 2A/1+2 (Linsbauer, ed.) Gebr. Bornträger, Berlin.

O'Brien, T.P. and M.E. McCully (1981) *The Study of Plant Structure, Principles and Selected Methods*. Termarcarphi (Pty) Ltd., Melbourne.

Purvis, M.J., D.C. Collier and D. Walls (1966) *Laboratory Techniques in Botany*. Butterworth, London.

8

Water relations

C.L. Beadle, M.M. Ludlow and J.L. Honeysett

8.1 STOMATAL CONDUCTANCE

8.1.1 Introduction

In order to absorb carbon dioxide (CO_2) for photosynthesis, plants expose wet surfaces to a dry atmosphere and in consequence suffer evaporative water loss. The resultant cooling, nevertheless, often accounts for a considerable proportion of heat dissipation by leaves and is probably essential for maintaining equable temperatures for photosynthesis. Too much water loss would result in dehydration. Plants have therefore evolved leaves with an epidermis composed of a relatively impermeable cuticle and turgor-operated valves, called stomata. The epidermis not only reduces rates of CO_2 and water-vapour exchange, but it also provides a means of controlling assimilation and transpiration through the size of the stomatal pores. Thus stomata play a pivotal role in controlling the balance between water loss and carbon gain, i.e. biomass production. Measurements of the size of the stomatal opening (stomatal aperture) or of the resistance to CO_2 and water-vapour (H_2O) transfer between the atmosphere and the internal tissue of the leaf imposed by the stomata (stomatal resistance) are important in many studies of biomass production. This is particularly the case in cropping situations where it is important to maximize water-use efficiency which we define as the mass of CO_2 assimilated (or dry weight gained) per unit mass of water transpired.

Carbon dioxide, which is currently rising in concentration at a rate of over 1 μmol $mol^{-1} a^{-1}$, has a direct effect on photosynthesis through its role as a substrate, and indirect effects on both photosynthesis and transpiration through its effect on stomatal conductance. The net effect is that photosynthesis will tend to increase while stomatal conductance will tend to decrease with an increase in CO_2 concentration. The water-use efficiency, particularly of C_3 plants, will tend to increase as a result. The mechanism which results in closure of stomata with increasing CO_2 concentration has been the subject of much investigation but still remains speculative [1]. There is also some evidence that stomatal density (numbers per unit area) is reduced with increasing CO_2 concentration [2].

Stomatal conductance responds to many weather variables, either alone or more usually in combination. As these include light, temperature and atmospheric deficit, there is some inevitability that climate change will have an impact on stomatal conductance through these variables as well. It is not possible at present to say in which direction conductance will change, as the complexities of the feedback and feed-forward phenomena of climate change have yet to be resolved

Photosynthesis and Production in a Changing Environment: a field and laboratory manual
Edited by D.O. Hall, J.M.O. Scurlock, H.R. Bolhàr-Nordenkampf, R.C. Leegood and S.P. Long
Published in 1993 by Chapman & Hall, London.
ISBN 0 412 42900 4 (HB) and 0 412 42910 1 (PB).

with respect to the weather. The changes realized may also vary with latitude and longitude.

8.1.2 Resistance or conductance?

The Ohm's law analogy of CO_2 and H_2O diffusion is described in Chapters 4 and 9. The restriction to the movement of CO_2 and H_2O offered by the stomata is defined there as a resistance (conceptually similar to an electrical resistance). The size of the stomatal resistance (r_s) is often compared with that of the boundary layer (r_b) and the intracellular processes for CO_2 transfer. Comparison of resistances is both theoretically correct and biologically meaningful [3]. However, if the limitation offered by stomata is being compared with the flux of CO_2 and H_2O, or being correlated with some biological or environmental variable such as leaf water status or deficit, it is more meaningful and less prone to misinterpretation to express it as a conductance (= 1/resistance) rather than as resistance [3,4,5]. Fluxes are proportional to conductances (g) but inversely proportional to resistances (r). However, many instruments are calibrated against physical resistances. Thus r is usually measured and then g is calculated. Unfortunately this can lead to large errors when low r is converted to give high g because small systematic or random errors are relatively large when r is small. Stomatal conductance can be obtained by determining the size of the stomatal aperture or by measuring the rate of gaseous loss of water vapour.

8.1.3 Methods for stomatal aperture

Stomatal aperture is usually measured by direct microscopic observation or by the extent or rate of infiltration of organic solvents [6].

Direct observation:
For a given leaf, plant or variety, the length and depth of stomata do not vary among stomata in mature tissues. Instead, most of the changes in aperture are associated with changes in width. In practice it is not possible to make a direct microscopic observation of stomata and at the same time preserve natural conditions. The change in conditions could alter the stomatal aperture. However, it is possible to make a stomatal impression before stomata have time to react by applying a quick-drying substance to the leaf surface [6,7,8]. The size of the stomatal aperture can be measured under a microscope either from the impression (where the stomatal pore is represented as a raised area in what is equivalent to a photographic negative) or a positive which is made by painting the negative with a substance such as cosmetic nail varnish (stomatal pores appear as holes in the positive). Stomatal aperture can be converted into an equivalent diffusive resistance (or conductance) using the following equation [9,10]:

$$r\ (\mathrm{s\,m^{-1}}) = \frac{s(L + 0.25\,\pi D)}{dna}$$

where s = leaf area (m^2);
L = length of stomatal tube (m);
d = diameter of stomatal pore (m);
n = stomatal density (m^{-2});
a = area of stomatal pore (m^2);
d = diffusivity of water vapour in air ($m^2\,s^{-1}$).

Infiltration by liquids:
A series of mixtures of two liquids (0–100%) is made, one with a high, the other low, viscosity. The mixtures are then applied to leaf surfaces in sequence from the most to the least viscous. The first mixture to infiltrate the leaf surfaces is an index of the degree of stomatal opening. This index can be correlated for each species with aperture obtained by direct observation or with diffusive resistance [6]. The infiltration method is simple and cheap, but of limited accuracy.

8.1.4 Methods for rate of water vapour loss

Stomatal conductance can be calculated from rates of water-vapour loss. The most accurate way is by measuring water-vapour loss from leaves enclosed in leaf chambers using gas-exchange techniques [6] (see also Chapter 9). However, gas exchange, which is mainly a laboratory technique, is expensive and requires good technical support, especially if used in the field. Simpler, though less accurate techniques are required for field measurements where many determinations are necessary if statistically acceptable estimates of conductance are the goal of the study [11]. In addition, instruments should be rugged, portable, battery-operated and relatively inexpensive [6]. Three such techniques will be described here.

(a) Cobalt chloride paper

Paper impregnated with cobalt chloride is blue when dry and pink when moist. The time taken for colour change when the paper is held against a leaf surface is an index of the rate of water loss, and hence stomatal conductance. The technique is quick, cheap, but only semi-quantitative. The time for the change in colour is linearly correlated with diffusive resistance measured with a diffusion porometer for sorghum ($r = 0.91$) and soybean ($r = 0.92$). The relationship may, however, vary with species and environmental conditions.

Cobalt chloride paper is prepared by adding a few drops of dilute HCl to 100 ml of 3% $CoCl_2$ in distilled water. Whatman No. 2 filter paper is placed in the solution for one minute and then dried between new filter papers at 45°C until blue. The paper should be stored in a sealed container with silica gel [6,9].

(b) Mass-flow porometers

There are basically two types of porometers, *viz.* mass-flow and diffusion porometers. The mass-flow porometer measures the rate at which air is forced through (i.e. across) leaves under pressure [6]. Mass-flow porometers are simple, cheap, and usually do not involve electronic circuitry. Therefore they are useful in teaching, and for work at remote field sites. However, they have the following disadvantages; (i) they are best used for comparative rather than absolute measurement because of the errors and limitations; (ii) use is mainly restricted to amphistomatous leaves (with stomata on both surfaces) although they can be used on hypostomatous leaves with special precautions; (iii) leakage of air from the apparatus, or more importantly, at the point of leaf attachment, can cause serious errors.

The resistance to mass flow can be correlated with stomatal aperture [6] or with diffusive resistance (see below) for each species, depending upon stomatal distribution and the gas being considered. The viscous (or mass) flow resistance (Ω; $kg\,m^{-2}\,s^{-1}$) is given by:

$$\Omega = \Delta p/f$$

where Δp and f are, respectively, the pressure gradient ($kg\,m\,s^{-2}$) and the flow of air across the leaf ($m^3\,s^{-1}$). A simple and rough generalization of the relationship between diffusive and viscous flow resistance to water vapour transfer is:

$$r_s \propto n\Omega$$

where n is estimated empirically to be 0.4.

(c) Diffusion porometers

Diffusion porometry is based on measurement of the rate of water-vapour loss from a leaf or portion of a leaf enclosed in a porometer chamber [6] and the resistance measured is the diffusive resistance. The rate of loss is determined from the rate of increase in humidity measured as a transit time (transit-time porometer), or from the rate at which dry air is added to offset the increase in humidity due to transpiration (null-balance porometer). In both approaches, water loss occurs from both the stomata and the cuticle. It is generally assumed that most of the loss occurs from the stomata, but the cuticular

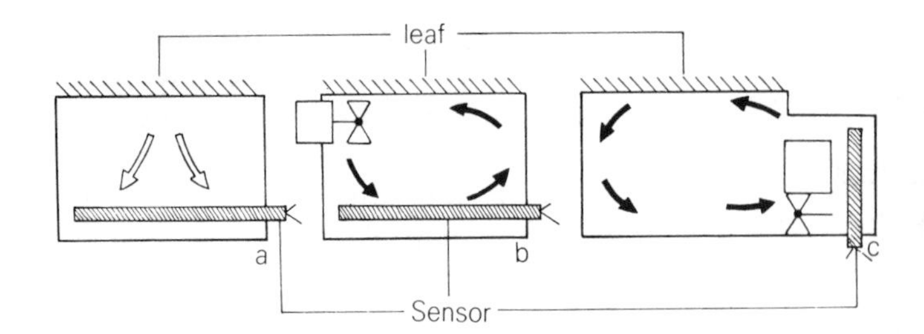

Fig. 8.1 Principles of operation of three types of diffusion porometer: (a) non-aspirated or non-ventilated, (b) aspirated by a fan within the chamber and (c) aspirated by pushing air through a by-pass containing the sensor and back into the chamber.

component becomes increasingly important as stomata close [12]. A further assumption is that the internal air spaces are saturated with water vapour, and thus if the leaf temperature is known, the concentration of water vapour on the inside of the stomata is equal to the saturation vapour pressure at that temperature.

8.1.5 Principles and calibration of diffusion porometers

Transit-time porometers:
These porometers have a humidity sensor in the chamber enclosing the leaf. Sensors may be lithium chloride (no longer used commercially), sulphonated polystyrene, or solid-state thin-film semiconductors. The chamber can be non-aspirated (non-ventilated; Fig. 8.1a) or it can have a small fan which either stirs the air within the chamber (Fig. 8.1b) or forces air though a bypass containing the sensor and back into the leaf chamber (Fig. 8.1c).

Aspirated porometers have a much lower r_b, and this increases the sensitivity for measuring r_s. Moreover, the air in the chamber is well mixed. Both these characteristics are necessary to measure needle-like leaves or small branches which cannot be measured with the non-aspirated instruments.

Non-aspirated porometers are simple, relatively cheap, and are quite suitable for broad-leaved plants. The major commercial instrument of this type is the *Delta-T Automatic Porometer AP4* with optional microprocessor (Delta-T Devices, Cambridge, U.K.).

Both aspirated and non-aspirated porometers are markedly affected by temperature. During measurements, the difference between leaf and sensor temperature should be less than 1°C because isothermal conditions are usually assumed. In practice, this is achieved by shading the leaf for 1–2 minutes before measurement, and keeping the cup shaded during measurement. Moreover, these porometers have to be calibrated at accurately controlled temperatures (better than 0.1°C) and over the range of temperatures to be encountered in the field. They are calibrated against plastic or metal plates drilled with holes of precise diameter and therefore of known resistance. These calibration plates should be kept clean at all times. The porometer is clamped on one side of the plate and wet blotting paper against the other side. The rate of increase of humidity at each temperature is measured for a series of combinations of size and number of holes. The time for humidity to increase over a particular range (e.g. 40–45% relative humidity), hence the transit time, is plotted against the calculated physical resistance (Fig. 8.2). A series of straight lines with a common intercept on the x-axis (which is the cup resistance) is obtained. The slopes of the lines are then plotted against temperature (Fig. 8.3). A polynomial can be fitted to these relationships and used to calculate the slope for each measurement temperature. Alternatively, the slope can be obtained graphically and substituted in the

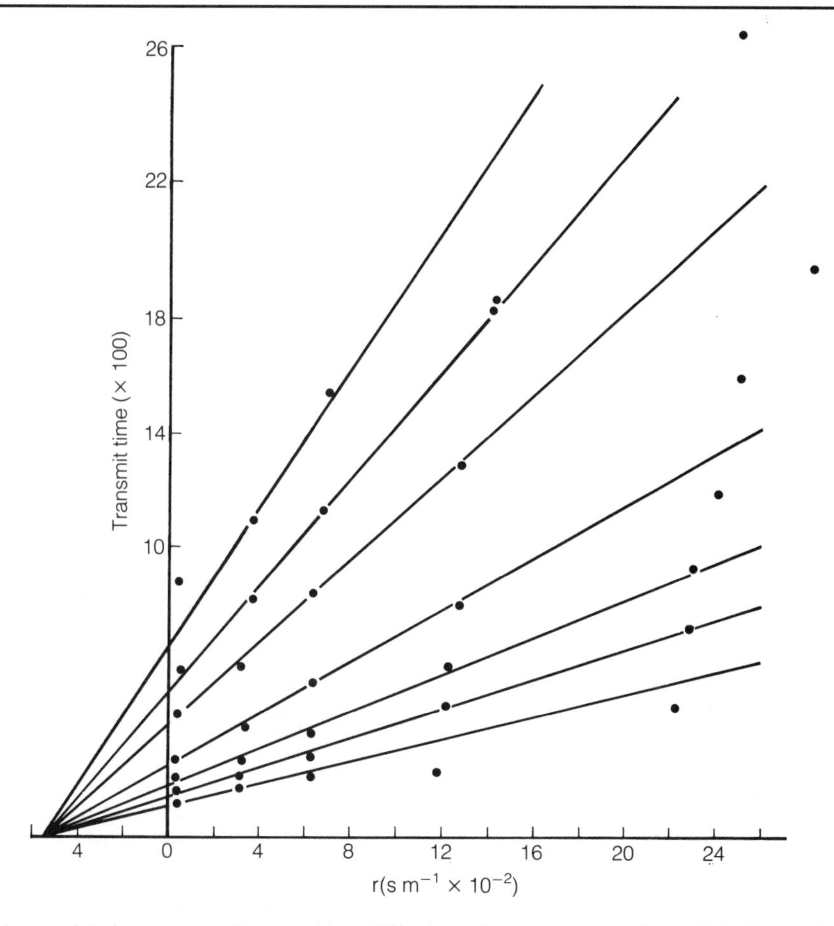

Fig. 8.2. Relationship between resistance (r) to diffusion of water vapour through holes on the calibration plate and time for relative humidity to increase over a predetermined range (transit time), determined at a range of temperatures. The point where the lines intersect is the cup resistance which is independent of temperature.

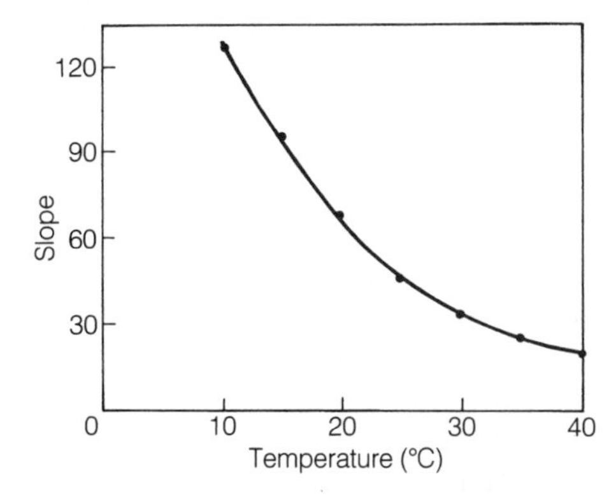

Fig. 8.3. Relationship between temperature and the slope of the line which describes the relation between resistance and transit time in Fig. 8.2.

following equation to calculate stomatal resistance, r_s (s m^{-1}; or conductance, g_s, in m s^{-1}):

$$r_s = \frac{\Delta t}{slope_t} - r_{cup}$$

where Δt is the transit time (in counts or seconds), $slope_t$ is the slope for each measurement temperature and r_{cup} is the cup resistance determined from the relationship between transit time and calculated physical resistance (Fig. 8.2). If the leaf temperature is higher than the chamber temperature, the resistance will be underestimated and *vice versa*; hence the need to keep temperature differences to a minimum.

Calibration is the most difficult part of diffusion porometry based on transit-time porometers, and frequent recalibration may be necessary. The method based on physical resistance described above or alternatively one based on the injection of known amounts of water vapour into the porometer cup can be used. Commercial manufacturers of these porometers provide both equipment for and details of calibration techniques.

Null-balance porometers:
Null-balance (or continuous flow) porometers are aspirated to ensure a rapid approach and accurate measurement of the null or balance point. In the more common type, the rate at which dry air (0% relative humidity) is added to the porometer cup or chamber to balance the increase in humidity due to transpiration and keep the relative humidity at the balance point constant, is used to calculate stomatal resistance [6,13]. (Alternatively the relative humidity at the balance point using a constant flow rate of dry air can be used [14]).

If a flow rate of dry air, f ($m^3 s^{-1}$) is required to maintain a relative humidity h in a chamber enclosing a leaf of area s (m^2), the rate of transpiration (**E**; $kg\,m^{-2}\,s^{-1}$) would be:

$$\mathbf{E} = f\,C_a/s$$

where C_a is the water vapour density ($kg\,m^{-3}$) of the air in the chamber at the balance point. At steady state, **E** can be written as the Ohm's Law analogy equation:

$$\mathbf{E} = (C_l - C_a)/r_l$$

where C_l is the water vapour density corresponding to the saturation vapour pressure at the leaf temperature, and r_l (= $r_s + r_b$) is the resistance to water vapour flow from the leaf. Combining these two equations gives:

$$\mathbf{E} = ((C_l - C_a) - 1)s/f$$

If the cup is properly aspirated, the boundary layer resistance (r_b) is small. It can either be neglected, or determined by replacing the leaf with wet blotting paper or similar material [15]. Moreover, if the leaf and air temperatures are similar, C_l/C_a becomes 1/h, the reciprocal of relative humidity. Thus stomatal resistance (r_s) is:

$$r_s = ([(1/h) - 1]s/f) - r_b$$

Thus the units of r_s are $s\,m^{-1}$ (and of its reciprocal g_s, $m\,s^{-1}$). These were the units used originally in the development of the diffusion theory of water loss from leaves. Alternatively, g_s can be expressed in terms of $mol\,m^{-2}\,s^{-1}$. This unit and its derivation is explained in Chapter 9. Conversion between $mol\,m^{-2}\,s^{-1}$ and $m\,s^{-1}$ is given by:

$$g_s\,(m\,s^{-1}) = g_s\,(mol\,m^{-2}\,s^{-1}) \times \frac{8.314(T + 273)}{p \times 1000}$$

where T and p are the temperature (°C) and pressure (kPa) recorded during measurement.

Calibration of the porometer involves calibration of flow rate rather than resistance. This simply requires a bubble flow meter and stop watch. It may however be necessary to calibrate the humidity sensor. The simplest method is to place the cup over saturated salt solutions at constant temperature.

The advantages of null-balance compared to transit-time porometers are as follows:

(i) Humidity is constant during measurement and it is a steady state rather than dynamic measurement. This could be important if stomata respond to humidity. Commercial instruments (e.g. *LICOR LI-1600*, LICOR Inc., Lincoln, Nebraska, U.S.A.) are now available which allow the balance point to be set at ambient humidity. (Note that transit-time porometers now offer several ranges of humidity over which transit time can be measured.)
(ii) Measurements can be done at different constant humidities.
(iii) Calibration is more direct and less dependent on assumptions and uncertainties.
(iv) They can be used with needle-like leaves or small branches.

The disadvantages of null-balance over other types of porometer are that they are more complex, they require more power and better technical support, and they are more expensive.

8.1.6 Measurement

Diffusion porometers are robust and well adapted for both field and laboratory measurement. Modern instruments can be supplied with data storage and transfer facilities to allow one-person operation (the *AP4* and *LI-1600* described above feature an RS-232 interface as optional and standard fitment, respectively). Manual recording of data will need the support of a second person to minimize the time required to complete a set of measurements. As both the transit-time and null-balance porometers rely on humidity change in the chamber, the period of time required to take a reading on a given leaf tends to increase as conductance decreases.

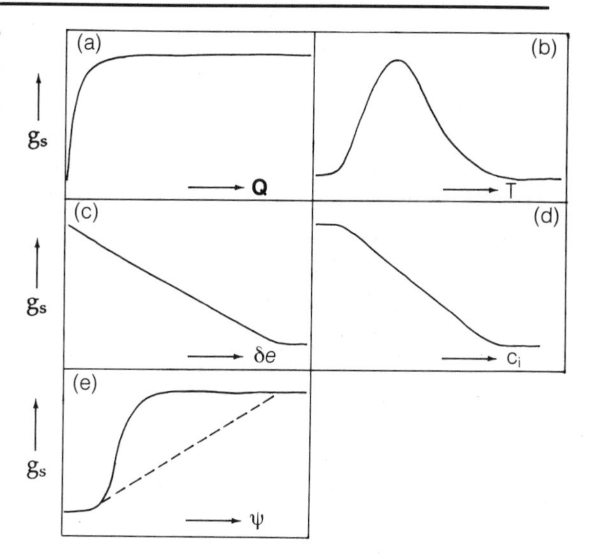

Fig. 8.4. Representative responses of stomatal conductance (g_s) to a range of environmental variables: (a) light, **Q**; (b) temperature, T; (c) vapour pressure deficit, δe; (d) intercellular CO_2 concentration, c_i; and (e) leaf water potential, ψ. The reponse may vary between species (see broken lines) and may be modified by stress.

Stomatal conductance is a function of many weather variables (Fig. 8.4) [16]. Stomata are sensitive to light quality and quantity (flux density), to temperature and atmospheric deficit and to plant water status. The extent of any response to these variables is dependent on species, age, and any adaptive response that the plant or crop may have undergone prior to measurement. The shape of the relationship between conductance and a given variable therefore requires experiments in controlled environments at given values of other variables. Field measurements with porometers can give a description of basic trends with extensive sampling over a long period of time [17].

Porometers are not designed for seeking an insight into the relationship between stomatal conductance (g_s) and ambient CO_2 concentration (c). It should also be noted that stomata sense the CO_2 concentration of the intercellular air spaces (c_i), which requires a measurement of photosynthetic rate (**A**) as well as of g_s. Gas-exchange systems and CO_2 porometers (portable field instruments which simultaneously measure **A** and g_s) are more suited to this type of study (Chapter 9). Experiments suggest that the response of g_s to c and c_i is once again species-dependent and is modified by other variables [1]. The way plants adapt with respect to g_s during long-term increases in CO_2 concentration has yet to be resolved.

8.2 PLANT WATER STATUS

8.2.1 Introduction

Although the amount of water used directly in the biochemical reactions of photosynthesis is small compared with that transpired or stored by plants at any one time, plant water status strongly influences plant growth and biomass production particularly through its effect on leaf and root extension. The rate of photosynthesis of a crop canopy (which can be used to predict biomass production) will also decline under water stress because of stomatal closure and the effects of water deficits on chloroplast processes [18].

In general, biomass production is directly proportional to the supply and use of water. Therefore measurement of plant water status is an important part of understanding biomass production and, in conjunction with a consideration of soil water status, for maximising yield under irrigation.

Whereas it is generally accepted that water moves through the soil-plant-atmosphere system along gradients of water potential, there is still argument as to whether the water content or the water potential has the greater effect on physiological activity and on survival. Ideally, both content and potential should be measured in experiments, though the nature of the experiment may dictate that only one of the two is measured.

8.2.2 Water content

The amount of water in plant material can be expressed in a number of ways [19]. All are based on the measurement of fresh weight (W_f) at time of sampling, dry weight (W_d, usually oven-dry weight at 70°C) and turgid weight (W_t). Turgid weight is obtained by floating leaves or leaf disks on water at the light compensation point until constant weight is reached. Thus:

$$\text{water content} = \frac{(W_f - W_d)}{W_d} \times 100\%;$$

$$\text{relative water content } (R^*) = \frac{W_f - W_d}{W_t - W_d} \times 100\%;$$

$$\text{water saturation deficit} = (100 - R^*)\%;$$

$$\text{turgid weight/dry weight ratio} = W_t/W_d.$$

Determining water content requires relatively simple, cheap equipment and many samples can be taken as replicates or across a large number of treatments. However, the technique for determining relative water content requires considerable skill and precision to obtain accurate results and preliminary trials should be made to determine the best method of arriving at W_t as this may vary with species [20]. In addition, there are sometimes difficulties in interpreting results, because there is not much information on the relationship between relative water content and the rates of physiological processes.

There are also methods for non-destructive measurement of water content, such as the β-ray absorption technique. Besides good technical support, this approach requires calibration against direct measurements of water content, such as those above [21].

8.2.3 Water potential

Water potential, ψ, is defined as the potential energy (joules) per unit mass of water (m^3) with reference to pure water at zero potential. Thus:

$$\psi = \frac{\mu_w - \mu_w^o}{\bar{V}_w}$$

where μ_w is the chemical potential or amount of free energy contributed by one mole of water in the system at constant temperature and pressure. As this cannot be measured directly, it is compared to the chemical potential of pure water, μ_w^o, at atmospheric pressure and the same temperature as the system being studied. It is considered most useful to express this difference, $\mu_w - \mu_w^o$, as the work required to move one mole of water from some point in the system to a pool of pure water at atmospheric pressure and the same temperature, on a per unit volume basis. Hence $\bar{V}_w$ is the partial molal volume of water (volume of one mole of water). As 1.0 Newton = 1.0 J m^{-2}, water potential is expressed in units of pressure ($N\,m^{-2}$ = Pascals, symbol Pa or more commonly MPa). Water in most biological systems has less potential energy than pure water. This results in negative values for water potential.

The two main methods of measuring water potential are the pressure chamber and psychrometric techniques (see tables in Appendix A). Simpler but less accurate methods such as liquid exchange (Shardakov's method) are also available [19].

Thermocouple psychrometry is based on placing plant tissue (or soil) into a small chamber, and allowing the water potential to come to equilibrium with the air in the chamber. The vapour pressure of this air is then measured by wet-bulb psychrometry or dew-point hygrometry [21,22]. These instruments are calibrated using blotting paper soaked in solutions of known osmotic potential. They are potentially accurate and can be used in the field but they are expensive, complex and require good technical support. Also, plant material may take many hours to reach equilibrium. This greatly reduces the number of measurements which can be made. These techniques are best used in the laboratory.

The pressure chamber is simple, cheap, rugged and ideally suited for field studies [23]. A leaf cut from the plant is placed in the chamber with the cut end projecting through the hole in a rubber bung. The pressure applied to the leaf or branch to return the water interface to where it was before detachment, is equal and opposite to the tension in the xylem of the intact plant. Because the osmotic potential of xylem sap is usually greater than $-0.02\,\text{MPa}$, the hydrostatic pressure in the xylem is equal to the water potential. The pressure-chamber technique has been reviewed extensively [23,24] and is now the most widely used method for characterizing plant water status.

8.2.4 Components of water potential

The water potential, ψ, of a plant (or leaf) consists of two major components; osmotic potential (π), arising from the presence of dissolved solutes in the cell and turgor potential (P), arising from pressure exerted on the cells by their walls. A third component, the matric potential (τ), can also drive osmosis, but compared to values of π and P, its contribution in the symplast is negligible (τ mainly contributes to ψ in the apoplasm as a function of activity close to the charged surfaces of the cell wall). Each is expressed in pressure units (MPa). Thus:

$$\psi = \pi + P\ (+\ \tau)$$

π of the vacuolar sap is always negative whereas P is always positive or zero. The relationship between them is described by a Höfler diagram (Fig. 8.5). Note that when $P = 0$, the cell is at incipient plasmolysis and ψ is equal to π. The most efficient method of evaluating π in the field is by means of sap extraction. π can be measured by thermocouple psychrometry or hygrometry (see above) after the sap has been expressed or after the material has been frozen and crushed to a fine powder. A laborious but more accurate method is the experimental determination of the pressure-volume relationships of individual shoots or leaves. As this technique can be used to derive several water relations parameters it is considered in detail in Section 8.2.5.

A technique is also available for measuring the turgor potential of cells directly through use of the pressure probe [25]. However, this is not an extensively used field technique and so P is generally inferred as the difference between the more common measures of ψ and π. There are limitations to this approach. Sap-extraction techniques used above to estimate π result in the dilution of symplastic water by apoplastic water which is low in solutes. As a result π is overestimated and the inferred value of P is underestimated and often negative. (Pressure-volume relationships for individual shoots and leaves suggest that negative turgors do not occur at water potentials below incipient plasmolysis.) Values of P calculated by difference between ψ and π are therefore only useful for making comparisons between species or treatments on a relative basis.

8.2.5 Pressure–volume curves

Pressure–volume curves and their analyses to determine the components of water relations are an extension of the simple use of the pressure chamber to measure water

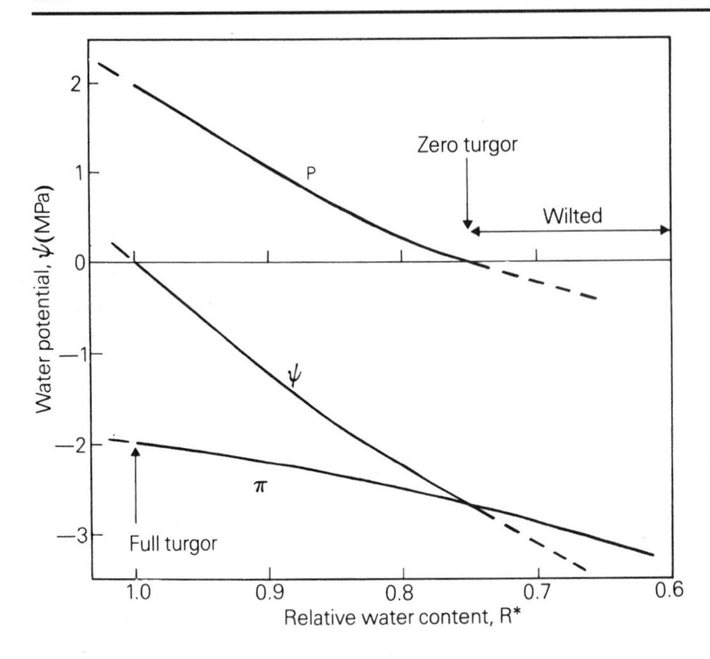

Fig. 8.5 Höfler diagram illustrating the relationships between total water potential (ψ), turgor potential (P), osmotic potential (π) and relative water content (R*) as a cell or tissue loses water from a fully turgid state. The dotted line below zero turgor represents possible negative turgor in rigid cells.

potential. In this technique a shoot or leaf is rehydrated. The tissue is then allowed to dehydrate to a point well beyond incipient plasmolysis. At periodic intervals of declining ψ either the volume of expressed sap is weighed, the shoot or leaf remaining in the pressure chamber [26] or the weight of the shoot or leaf measured directly [27]. The inverse of water potential is then plotted against the volume expressed or the relative water content (Fig. 8.6).

From this relationship it can be predicted that the slope $d(1/\psi)/dV$ or $d(1/\psi)/dR^*$ will be linear when $P = 0$. The fitted line is thus an expression of the relationship between $1/\pi$ and R* and the y-intercept when $R^* = 1.0$, $1/\pi$ at full turgor (Fig. 8.6). This calculated value of π can then be subtracted from the measured values of ψ to obtain P. Thus relationships between P and R* (Fig. 8.7), and P and ψ, can also be obtained.

The advantage of this method is that it yields a value for undiluted cell sap [28]. Its disadvantage is that it is only possible for one person working with one pressure chamber to determine these relationships for six to eight shoots or leaves per day.

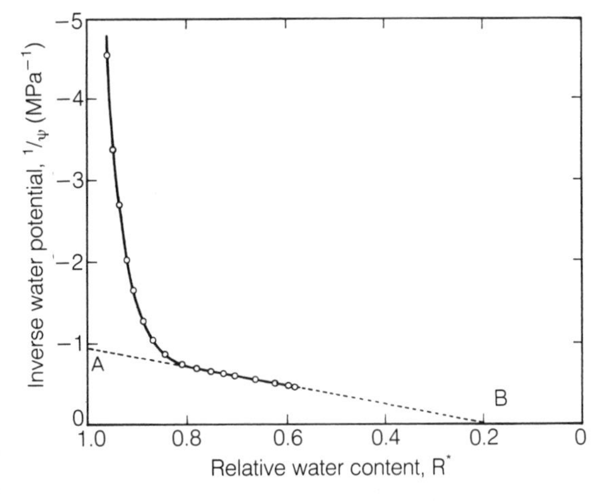

Fig. 8.6. A typical relationship between the inverse of water potential ($1/\psi$) as a function of relative water content (R*). The fitted line represents the relationship between $1/\pi$ and R* when $P = 0$, the intercept A on the y-axis, $1/\pi$ at full turgor ($R^* = 1.0$), and the intercept B on the x-axis, the fraction of water in the apoplast.

8.2.6 Measurement

Plants adapt to water stress through changes in root:shoot ratio, growth rate and water use efficiency. These are often accompanied

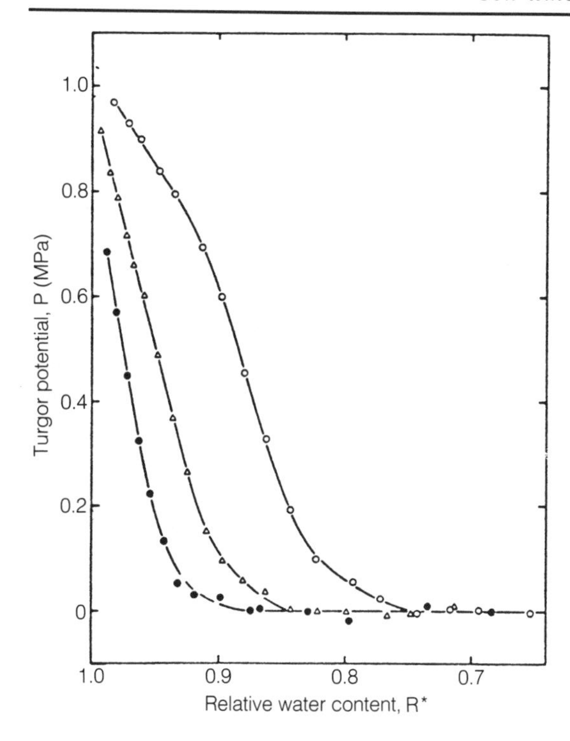

Fig. 8.7. The relationship between turgor potential (P) and relative water content (R*) for two *Dubautia* species and their hydrid combination. ○, *D. ciliolata*; ●, *D. scabra*; △, hybrid. Note the difference between the species in their ability to maintain P at a given R* [27].

by changes in the fundamental properties of their water relations such as the (critical) water potential or relative water content which prompts stomatal closure, diurnal patterns of stomatal conductance, and pre-dawn water potential of leaves/stems, an expression of ψ at the root surface when it is assumed that there is no potential gradient through the plant. In this way, pre-dawn ψ measures the current level of water stress in the plant or crop.

The pressure–volume curve can be used to further understand how changes in the parameters which describe this relationship are related to the capacity of the plant to respond to water stress. For example, differences in the osmotic potential at full turgor, the initial bulk elastic modulus and the apoplastic fraction of water were used to distinguish the tissue water relations of two *Dubautia* species and their hybrid combination [27]. All these parameters can be derived from curves similar to that illustrated in Fig. 8.6.

8.3 SOIL WATER STATUS

8.3.1 Introduction

The maintenance of plant turgor and transpiration from the crop canopy depends on the maintenance of water uptake by roots at the soil-root interface. The more extensive and dense the root system, the more efficiently will these demands be met. As the soil dries however, shrinkage of both the soil and root decreases the soil-root contact. The ultimate effect will be a reduction in water uptake, stomatal closure and a decrease in photosynthesis and biomass production.

As with plant water status, we can consider the water content and water potential as components of soil water status. Soil water content is a measure of the available water. To maximise production, it is important that the available water during the cropping period is sufficient to meet the transpiration demands on that crop. The soil water potential measures the energy status of the soil water and therefore the amount of work required of the plant to absorb this water. Both components are therefore important for an understanding of soil water status and for maximising biomass production through efficient use of irrigation.

8.3.2 Soil water content

Soil water content is simply a means of expressing the quantity of water in the soil. This can be expressed as (i) a weight fraction ($g\,g^{-1}$), or (ii) a volume fraction ($g\,cm^{-3}$). These expressions are related via the soil bulk density, which is given by:

$$\Theta = (\text{weight fraction}) \times (\text{bulk density})$$

where Θ is the volume fraction and bulk density has dimensions of mass per unit volume, in this instance $g\,cm^{-3}$.

The volume fraction (Θ) has an important property since $1.0\,cm^3$ water $= 1.0\,g$. Θ is therefore dimensionless, i.e. $cm^3\,cm^{-3}$. If the volume fraction of water is integrated over depth ($cm^3\,cm^{-2}$), it can be simply expressed in the same terms as rainfall or irrigation, i.e. cm.

The most efficient instrument for measuring available water is the neutron moisture meter. This instrument works on the principle that high energy (fast) neutrons lose energy only on collision with atoms of similar mass, i.e. hydrogen atoms. The low energy (slow) neutrons can be measured and their intensity is proportional to the number of hydrogens in the zone. In soil almost all the hydrogen is present as water. The instrument consists of a neutron source, a slow neutron detector and a means of counting the signals from the detector. Calibration is by means of measurements in media of known water content or by extensive field sampling after uniform readings have been obtained.

As an alternative the weight fraction can be determined by taking soil cores and expressing the results in terms of oven-dry weight at 105°C. The loss in weight relative to oven-dry weight represents the moisture content.

If soil cores are collected carefully so that measurements can be made to determine accurately the volume of soil sampled then simple calculations give the moisture content as a volume fraction (Θ) as well as the soil bulk density (b_ρ):

$$\Theta = \text{loss in weight / sample volume;}$$
$$b_\rho = \text{oven-dry weight / sample volume}$$

b_ρ is an important property of soil linking mass and volume. It enables calculation of the total pore volume in a soil in which water and air are held. Their relationship is given by:

$$\text{pore volume fraction} = 1 - \frac{b_\rho}{\text{particle density}}$$

where particle density is the density of the mineral particles. This is essentally constant and has a value of 2.65. It can be seen from this expression that as b_ρ increases the pore volume decreases, leaving less space for air, water and root growth. This is the case when soil is compacted. Conversely a reduction in b_ρ increases the available space for air and water, and promotes root and plant growth.

Patterns of gross rainfall may alter with climate change to either increase or decrease the available water in soils. If rainfall decreases and/or evaporative demand increases this can be compensated by improving the available water capacity and drainage characteristics of soils. These soil properties can be monitored through soil bulk density and altered, say, by incorporating organic matter into the soil. Organic matter can dramatically improve the structure and available water capacity by decreasing b_ρ and increasing the total pore volume (macro-porosity). This is illustrated by the relationship found for many undisturbed soils:

$$\frac{1}{b_\rho} = \alpha + \frac{\beta I}{100}$$

where α and β are constants representing the bulk volumes of the soil mineral and organic matter (approx. 0.56 and 5.6 respectively) and I is the percentage loss on ignition at 375 ± 5°C [29]. Thus I is a measure of the amount of organic matter in the soil.

8.3.3 Water potential

The major component of water potential (ψ_{soil}) in non-saline soils is the matric potential, τ_{soil}. This force is predominantly influenced by surface tension (capillary) forces originating from the liquid–air interfaces of the finer pores. Capillarity holds water in the soil between potentials of −0.03 and −2.0 MPa, but most of the available water is held at $\psi_{soil} > -0.1$ MPa. In saline soils, there may be a significant osmotic component, π_{soil}, which may reduce the soil water potential by up to −0.2 MPa. As plant membranes are involved in the movement of water into plant roots, the osmotic component can have secondary effects. We can therefore express soil water potential, ψ_{soil}, as:

$$\psi_{soil} = \tau_{soil} + \pi_{soil}$$

The most useful instrument for measuring ψ_{soil} is the tensiometer. This is an airtight, water-filled tube with a porous ceramic tip at the base and a vacuum gauge at the top. The small pores in the wet ceramic tip allow water to move in and out freely but prevent air entry. It can be inserted into soils at selected depths. Water will then move out and create a vacuum which is indicated on the gauge as a measure of ψ_{soil}. This instrument can only be used reliably between 0 and −0.08 MPa but as most established plants will suffer water stress at lower potentials, its main use is for monitoring the need for irrigation requirements and the correct application of irrigation water. ψ_{soil} of drier soils can be measured by thermocouple psychrometry or hygrometry (Section 8.2.3).

8.4 PRACTICAL WORK ON WATER RELATIONS

8.4.1 Objectives

To measure the stomatal conductance, water potential and relative water content of leaves at various heights between the soil surface and the top of the crop. These are related to corresponding environmental factors considered in Chapter 4. Soil water status is measured with a tensiometer and soil cores are removed to estimate volume fraction and (dry) bulk density.

8.4.2 Materials

Grow plants in pots or field plots until a uniform and closed canopy is formed. Select a species which is suitable for your porometer and pressure chamber. For example, ensure that the leaf area sampled by the porometer meets the demands for either a fixed or variable area (which is subsequently measured). The leaves selected for measurement in the pressure chamber should allow a seal to be formed without damaging the material by folding or rolling. Clearly label all leaves which are to be used during the experiment. The following equipment is required:

Transit-time or null-balance porometer, calibrated according to the manufacturer's instructions
Pressure chamber
Leaf punch and screw-top vials
Tensiometer and auger
Soil corer
Forced-draught oven
Balance (to 1 mg)

8.4.3 Procedure

(a) Stomatal conductance (transit-time porometer):
Shade the leaf for one minute before placing the porometer cup so that the stomatal conductance of the lower surface is measured. Keep the porometer cup shaded during measurement. Allow the porometer to go through several cycles until a constant value of the transit time is obtained. Read leaf temperature before removing the porometer cup.

Repeat for the upper surface. Measurements can be made at three levels in the upper, middle and lower canopy so that leaves are sampled which are normally exposed to full sunlight, partial sunlight and shade, respectively. Sample one leaf at each level at a given location and repeat for a total of five locations in the crop.

(b) Stomatal conductance (null-balance porometer):
Set the null-balance point at ambient relative humidity. Place the porometer on the leaf so that the lower surface is measured. When balanced, read the conductance, leaf tem-

perature and quantum flux density. Continue as above for the transit-time porometer.

(c) Leaf water potential:
Place a piece of moist paper towel, a piece of plastic food wrap or a polythene bag around the leaf before cutting the leaf with a razor blade. Take the leaf quickly to the pressure chamber and determine the balance pressure. Sample canopy as for stomatal conductance. N.B. only 1–2 mm of the petiole or blade should project above the chamber seal.

(d) Relative water content:
Punch out 10 leaf disks. If a purpose-built punch is not available, this can be done by pressing the leaf between the face of a cork borer (10–15 mm) and a rubber bung, or alternatively using a leather punch. Place the disks into a weighed screw-top vial and weigh immediately (W_f). Float the disks in distilled water in a Petri dish at 20°C for 3 hours at the light compensation point. Carefully dry between tissues and reweigh (W_t). Dry to constant weight at 70°C (W_d). Do this at various heights in the canopy and at five locations in the crop for each treatment.

(e) Soil moisture content:
Use the soil corer to remove a soil sample. The cylindrical corer may vary such that depths between 2.5 cm and 10 cm can be sampled. Alternatively, sequential samples can be taken to increasing depths. Either quickly sub-sample cores, or place whole cores into a weighed basin and heat in an oven at 105°C for 24 hours. (If the soil is a clay and in large lumps 48 hours may be required.) Cool in a desiccator, since hot soil rapidly picks up moisture from the atmosphere, and when cold weigh quickly.

The loss in weight relative to oven dried weight of soil represents the moisture content. The soil should be sampled at different locations in the crop and in each treatment. Composite samples can be made of cores sampled from a given depth in any treatment if extensive sampling is undertaken.

(f) Soil water potential:
Use the auger to prepare a hole to a selected depth and insert the tensiometer into it. Pack the space around the ceramic tip with fine sand to ensure good contact with the soil. Fill with water and seal with vacuum gauge. Read when the value has stabilized. Place a second tensiometer at a different depth. Depths of 10 cm and 30 cm should be satisfactory for an indication of the gradient of soil water potential.

8.4.4 Calculations

(a) Stomatal conductance (m s^{-1}):
From the transit time and the temperature, calculate stomatal resistance (r_s) using the formula:

$$r_s = \frac{\Delta t}{\text{slope}_t} - r_{cup} \quad (\text{s m}^{-1})$$

where r_{cup} is the cup resistance and slope_t is the slope of the transit time – resistance relationship at the measurement temperature (refer to Section 8.1.5). This step is not necessary for the null-balance porometer.

Calculate stomatal conductance (= 1/ stomatal resistance).

Plot mean stomatal conductance for the five locations against level in the canopy, for each treatment.

(b) Leaf water potential (ψ) and relative water content (R):*
Calculate the leaf water potential in MPa. Plot means as for stomatal conductance. Repeat for relative water content and plot ψ as a function of R*.

Calculate the mean volume fraction of water and bulk density.

8.4.5 Analysis of data

Each student should prepare the profiles (variable plotted against height) of environmental factors and stomatal conductance. Then plot stomatal conductance against the factor(s) which correlate best with it. Write one paragraph giving your interpretation of which factor(s) is controlling stomatal conductance.

Answer the following questions:

1. How could biological factors such as leaf position and leaf age complicate your interpretation of factors controlling stomatal conductance (in two sentences)?
2. Interpret the change in ψ and R* with height and the soil water potential in terms of water flow in the soil-plant-atmosphere continuum. Comment briefly on your values of content and potential for both the plant and the soil.
3. Identify possible errors in the measurements and briefly say how they could be improved.

8.4.6 Suggested additional work

Use data from conductance and leaf area index measurements (see also Chapter 6) to calculate the conductance in each layer of the canopy, and thence (by summing) the canopy conductance. Conductance data can also be combined with measurements of saturation deficit (Chapter 4) and photosynthesis (Chapter 9) to obtain estimates of transpiration and water-use efficiency.

REFERENCES

1. Eamus, D. and P.G. Jarvis (1989) The direct effects of increase in the global atmospheric CO_2 concentration on natural and commercial temperate trees and forests. In: *Advances in Ecological Research* (M. Began *et al.*, eds.), pp. 1–54. Academic Press, New York.
2. Woodward, F.I. (1987) Stomatal numbers are sensitive to increases in CO_2 from pre-industrial levels. *Nature* **327**, 617–618.
3. Burrows, F.J. and F.L. Milthorpe (1976) Stomatal conductance in the control of gas exchange. In: *Water Deficits and Plant Growth*, Vol. 4 (T.T. Kozlowski, ed.), pp. 103–152. Academic Press, New York.
4. Hall, A.E., E.D. Schulze, and O.L. Lange (1976) Current perspectives of steady state stomatal responses to environment. In: *Water and Plant Life* (O.L. Lange, L. Kappen and E.D. Schulze, eds.), pp. 169–188. Springer-Verlag, Berlin.
5. Cowan, I.R. (1977) Stomatal behaviour and environment. *Adv. Bot. Res.* **4**, 117–22.
6. Bartholomew, D., G.S. Campbell, D.C. Davenport, W.L. Ehrler, R.A. Fischer, T.C. Hsaio, E.T. Kanemasu amd H.H. Wiebe (1975) *Measurement of Stomatal Aperture and Diffusive Resistance.* Bull. 809, College of Agric. Res. Centre, Washington State University.
7. Zelitch, I. (1961) Biochemical control of stomatal opening of leaves. *Proc. Natl. Acad. Sci.* **47**, 1423–1433.
8. Rice, J.S., E.M. Glenn and V.L. Quisenberry (1979) A rapid method for obtaining leaf impressions in grasses. *Agronomy J.* **71**, 894–896.
9. Meidner, H. and T.A. Mansfield (1968) *Physiology of Stomata*. McGraw-Hill, London.
10. Meidner, H. and D.W. Sheriff (1976) *Water and Plants*. Blackie, Glasgow.
11. Leverenz, J., J.D. Deans, E.D. Ford, P.G. Jarvis, R. Milne, and D. Whitehead (1982). Systematic spatial variation of stomatal conductance in a Sitka spruce plantation. *J. Applied Ecology* **19**, 835–851.
12. Beadle, C.L., P.G. Jarvis and R.E. Neilson (1979) Leaf conductance as related to xylem water potential and carbon dioxide concentration in Sitka spruce. *Physiol. Plant.* **45**, 158–166.
13. Beardsell, M.F., P.G. Jarvis and B. Davidson (1972) A null-balance diffusion porometer suitable for use with leaves of many shapes. *J. Applied Ecology* **9**, 677–690.
14. Parkinson, K.J. and B.J. Legg (1972) A continuous flow porometer. *J. Applied Ecology* **9**, 669–675.
15. Landsberg, J.J. and M.M. Ludlow (1970) A technique for determining resistance to mass transfer through the boundary layers of plants with complex structure. *J. Applied Ecology* **7**, 187–192.
16. Jarvis, P.G. (1976) The interpretation of the variations in leaf water potential and stomatal conductance found in canopies in the field. *Phil. Trans. Roy. Soc.* B **273**, 543–610.
17. Beadle, C.L., P.G. Jarvis, H. Talbot and R.E. Nielson (1985) Stomatal conductance and photosynthesis in a mature Scots pine forest. II. Dependence on environmental variables of single shoots. *J. Applied Ecology* **22**, 573–586.
18. Beadle, C.L., S.P. Long, S.K. Imbamba, D.O. Hall and R. Olembo (1985) *Photosynthesis in Relation to Plant Production by Terrestrial Ecosystems.* Tycooly/Cassell, London.

19. Slavik, B. (1974) Methods for studying Plant Water Relations Chapman and Hall, London.
20. Barrs, H.D. and P.E. Weatherly (1962) A re-examination of the relative turgidity technique for estimating water deficits in leaves. *Australian J. Biol. Sci.* **15**, 413–428.
21. Koide, R.T., R.H. Robicheaux, S.R. Morse and C.M. Smith (1989) Plant water status, hydraulic resistance and capacitance. In: *Plant Physiological Ecology: Field Methods and Instrumentation* (R.W. Pearcy *et al.*, eds.) pp. 161–183. Chapman and Hall, London and New York.
22. Brown, R.W. and B.P. Van Haveren (1972) *Psychrometry in Water Relations Research.* Utah Agricultural Experiment Station: Logan, Utah, U.S.A.
23. Ritchie, G.A. and T.M. Hinckley (1975) The pressure chamber as an instrument for ecological research. *Adv. Ecol. Res.* **9**, 165–254.
24. Turner, N.C. (1981) Techniques and experimental approaches for the measurement of plant water status. *Plant and Soil* **58**, 339–366.
25. Husken, D., E. Steudle and U. Zimmerman (1978) Pressure probe technique for measuring water relations of cells in higher plants. *Plant Physiol.* **61**, 158–163.
26. Tyree, M.T. and H.T. Hammel (1972) The measurement of the turgor pressure and the water relations of plants by the pressure bomb technique. *J. Exp. Bot.* **23**, 267–282.
27. Robichaux, R.H. (1984) Variations in the tissue water relations of two sympatric Hawaiian Dubautia species and their hybrid combination. *Oecologia* **65**, 45–81.
28. Tyree, M.T. and P.G. Jarvis (1982) Water in tissues and cells. In: *Encyclopaedia of Plant Physiology*, New Series Vol. 12B (O.L. Lange, P.S. Nobel, C.B. Osmond and H. Ziegler, eds.) pp. 35–77. Springer-Verlag, Berlin.
29. Honeysett, J.L. and D.A. Ratkovsky (1989) The use of ignition loss to estimate bulk density of forest soils. *J. Soil. Sci.* **40**, 299–308.

9

Measurement of CO_2 assimilation by plants in the field and the laboratory

S.P. Long and J.-E. Hällgren

9.1 INTRODUCTION

9.1.1 The potential value of CO_2 exchange measurements

The growth of plants and crop stands, in terms of dry weight or carbon gain, has traditionally been measured by sampling, drying, weighing and chemical analysis of the dried material (Chapters 1 and 3). Direct measurement of carbon dioxide (CO_2) uptake provides a complementary approach. Whilst harvesting methods are appropriate in assessing long-term changes, they are unsuitable when interest centres either on short-term carbon gain, i.e. intervals of days, hours or minutes, or on contributions made by individual organs, e.g. the flag leaves of cereals. Measurement of CO_2 uptake provides an alternative and direct method of measuring carbon exchange, with important advantages: it is instantaneous and non-destructive. Furthermore it allows: measurement of the total carbon gain by a plant or stand; separation of the uptake by the different photosynthetic organs of a plant; and separation of photosynthetic gain from respiratory losses. In studying photosynthesis *in vivo*, measurements of O_2 evolution, and of fluorescence and absorption spectroscopy, now provide valuable methods for interpreting limitations and efficiencies. However, only measurement of CO_2 uptake can provide an unambiguous and direct measure of the net rate of photosynthetic carbon assimilation.

Photosynthesis and Production in a Changing Environment: a field and laboratory manual
Edited by D.O. Hall, J.M.O. Scurlock, H.R. Bolhàr-Nordenkampf, R.C. Leegood and S.P. Long
Published in 1993 by Chapman & Hall, London.
ISBN 0 412 42900 4 (HB) and 0 412 42910 1 (PB).

9.1.2 Measures, symbols and units

Many measures, terms and units have been used in the study of plant gas exchange. This chapter has largely adopted those described by von Caemmerer and Farquhar [1]. This system is simpler than that recommended earlier [2], in allowing the direct comparison of fluxes, conductances and derived terms such as quantum efficiency and carboxylation efficiency. The units used are exclusively S.I. Adherence to S.I. and its conventions greatly simplifies the calculation of fluxes and derived terms.

The CO_2 concentration in air may be described by the mole fraction (c). This equals both the partial volume ($cm^3\,m^{-3}$) and the ratio of the partial pressure of CO_2 to the total pressure of the body of air ($Pa\;MPa^{-1}$). Many instruments for CO_2 measurement indicate content as % or as vpm (volumes per million); this is directly proportional to the mole fraction, where $1\,vpm = 1\,cm^3\,m^{-3} = 1\,\mu mol\,mol^{-1}$. The assimilation rate (**A**) is expressed as amount of CO_2 assimilated per unit leaf area and time ($\mu mol\, m^{-2} s^{-1}$). Amount rather than mass is used since this simplifies comparison of the molecular fluxes of different gases and simplifies calculation of quantum efficiency. Mass flux is simply obtained by multiplying **A** by the molecular weight of CO_2, i.e. 44.

The conductance (g) of a diffusion pathway is the ratio of flux to the CO_2 gradient. As the CO_2 concentration gradient (described as a mole fraction) is dimensionless, the dimensions will be those of **A**. Conductance is thus the hypothetical flux when the concentration gradient is unity.

To avoid the use of exponents, sub-multiples of the S.I. base units may be used, but only in the numerator. Thus, $\mu mol\, m^{-2} s^{-1}$ would be correct for **A**, but $mol\, dm^{-2} h^{-1}$ should not be used. Sub-multiples in the denominator add unnecessary complexity to the calculation of derived terms and complicate comparisons [3].

9.1.3 The approach

The majority of CO_2 exchange studies have involved enclosure methods, i.e. enclosure of a leaf, plant or stand of plants in a transparent chamber. The rate of CO_2 assimilation by the material enclosed is determined by measuring the change in the CO_2 concentration of the air flowing across the chamber. This chapter is primarily concerned with techniques based on such enclosure methods, since these have been the most widely used. Alternatively, CO_2 exchange of large areas of vegetation may be measured without enclosure, using micro-meteorological techniques (outlined in Section 9.7). Radio-isotope labelling has provided a further method of assessing the rate of CO_2 assimilation (for an introduction see Field *et al.* [4] and Hällgren [5]). Portable infrared gas analyser (IRGA) systems have now supplanted this technique in field measurements.

9.1.4 Closed systems

In a closed system air is pumped from the chamber enclosing a leaf or plant into an IRGA which continuously records the CO_2 concentration of the system (Fig. 9.1). The air is then recycled back to the chamber. No air leaves the system or enters it from outside. If the leaf enclosed in the chamber

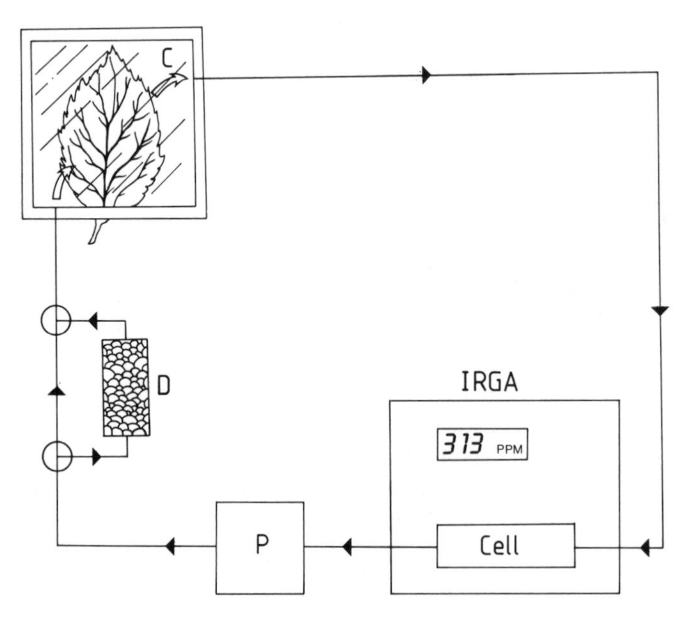

Fig. 9.1. Diagram illustrating air flow in a simple closed system. C, leaf enclosure chamber; D, drier by-pass; IRGA, infra-red gas analyser; P, pump.

is photosynthesising, the CO_2 concentration in the system will decline, and continue to decline until the CO_2 compensation point of photosynthesis (Γ) is reached (see Table 9.1 for definition of symbols). In practice the CO_2 concentration is allowed to drop typically by about 30 μmol mol^{-1} from the ambient level. The rate of CO_2 assimilation is equal to the change in the amount of CO_2 in the system per unit time. Temperature (T) and pressure (p) changes are common and can introduce significant errors which must be corrected for:

$$\mathbf{A} = \frac{[c_1(p_1/T_1) - c_2(p_2/T_2)]V \cdot T_v}{(t_2 - t_1)s \cdot p_v} \qquad (9.1)$$

where c_1, p_1 and T_1 are the mole fractions of CO_2 (μmol mol^{-1}), pressure (kPa) and temperature (K), respectively, of the system at time t_1 (s); c_2, p_2 and T_2 are the corresponding values for time t_2. V is the volume (mol) of the system determined at temperature (T_v) and pressure (p_v).

If humidity is not controlled, transpiration will result in an increase in water vapour concentration and a dilution of the mole fraction of all other gases, including CO_2. However, this will be compensated for by an almost equivalent increase in pressure such that the number of moles of gas per unit volume will remain constant. If the system is not airtight, this increase in pressure will result in a leakage of CO_2. Closed systems are the simplest configurations, being the least demanding of the IRGA and requiring no measurement of flow rate. Such systems do also have important disadvantages.

Recirculation of the air will result in a continuous rise in humidity. A wet humidity trap cannot be used since this will produce a variable volume of liquid water which would represent a sink for CO_2 and complicate the determination of V. Alternatively, a portion of the recirculated air may be passed through a drier, as in the LI-6200 photosynthesis meter (LI-COR Inc., Lincoln, Nebraska, USA). However, this necessitates measurement of flow rate and removes one of the

Table 9.1. List of symbols used in Chapter 9.

Symbol	Parameter	Units
$\mathbf{A}$	net rate of CO_2 uptake per unit leaf area	μmol m^{-2} s^{-1}
$\mathbf{A}_c$	canopy rate per unit of ground area	
$\mathbf{A}_{sat}$	light saturated A	
c	CO_2 concentration in air or mole fraction	μmol mol^{-1}
c_a	in the bulk air	
c_c	at the site of carboxylation	
c_e	at the entrance to an assimilation chamber	
c_i	in the leaf intercellular air space	
c_o	at the outlet from an assimilation chamber	
c_p	specific heat of air or specific gases	J kg^{-1} K^{-1}
ce	carboxylation efficiency	mol m^{-2} s^{-1}
$\mathbf{E}$	transpiration per unit leaf area	mol m^{-2} s^{-1}
F	net downward CO_2 flux per unit ground area	μmol m^{-2} s^{-1}
F_{soil}	net soil efflux of CO_2 per unit ground area	μmol m^{-2} s^{-1}
g	conductance to CO_2 diffusion	mol m^{-2} s^{-1}
g_b	boundary layer	
g_s	stomatal	
g_l	total, i.e. boundary layer and stomatal	
g′	conductance to H_2O diffusion	mol m^{-2} s^{-1}
g'_b	boundary layer	
g'_s	stomatal	
g'_l	total, i.e. boundary layer and stomatal	
J_c	rate of absorption of CO_2 by a leaf	μmol s^{-1}
$J_{c,i}$	J_c at an absorbed photon flux of $\mathbf{Q}_{abs,i}$	
$J_{c,0}$	J_c in darkness	
J_Q	rate of absorption of visible photons by a leaf	μmol s^{-1}
$J_{Q,i}$	J_Q at an absorbed photon flux of $\mathbf{Q}_{abs,i}$	
k	von Karman constant (0.41)	dimensionless

Table 9.1. *Continued*

Symbol	Parameter	Units
k_Q	value of **Q** at which **A** = $\mathbf{A}_{sat}/2$	$\mu mol\ m^{-2}\ s^{-1}$
k_w	radiation extinction coefficient at wavelength w	dimensionless
l	radiation path-length	m
M	molar concentration of a gas	$mol\ m^{-3}$
o	O_2 concentration in air or mole fraction	$\mu mol\ mol^{-1}$
o_i	in the leaf intercellular air space	
p	atmospheric pressure	kPa
p_o	standard at sea level (101.325 kPa)	
Q	photon flux	$\mu mol\ m^{-2}\ s^{-1}$
$\mathbf{Q}_{abs}$	**Q** absorbed by the leaf	
$\mathbf{R}_d$	dark respiration rate	$\mu mol\ m^{-2}\ s^{-1}$
RH	relative humidity	dimensionless
s	projected surface area	m^2
s_a	total surface area of a standard absorber	m^2
s_s	internal surface area of integrating sphere	m^2
t	time, elapsed time	s
T	temperature	K
T_a	air	
T_l	leaf	
T_o	standard (273.13 K)	
u	mole flow of air, wind speed	$mol\ s^{-1}$, $m\ s^{-1}$
u_c	mole flow of CO_2	
u_v	volumetric flow of air/gas	$m^3\ s^{-1}$
u_w	mass flow of air/gas	$kg\ s^{-1}$
u*	friction velocity	$m\ s^{-1}$

Table 9.1. *Continued*

Symbol	Parameter	Units
V	the volume of a closed system	mol
w	water vapour concentration in air	$mmol\ mol^{-1}$
w_a	in the bulk air	
w_e	at the entrance to an assimilation chamber	
w_i	at the mesophyll cell surface	
w_o	at the outlet from an assimilation chamber	
$w_{s[T]}$	saturated air at temperature T	
z	distance, height above surface	m
α	absorptivity, absorptance of radiation	dimensionless
α_a	absorptance of integrating sphere walls	
α_l	leaf absorptance	
α_s	absorptance of standard absorber	
α_w	absorptivity at wavelength w	
Γ	CO_2 compensation point of photosynthesis	$\mu mol\ mol^{-1}$
Θ	convexity of the response of **A** to **Q**	dimensionless
ϕ	quantum yield of photosynthesis calculated from the incident photon flux (**Q**)	dimensionless
ϕ_{abs}	quantum yield of photosynthesis calculated from the absorbed photon flux ($\mathbf{Q}_{abs}$)	dimensionless

key advantages of a closed system. Errors resulting from CO_2 adsorption/desorption to and permeation through tubing and chamber walls will be amplified by continuous re-circulation. A theoretical objection to closed systems is that since the CO_2 concentration is changing, **A** cannot reach steady-state and the measured value might not be a true reflection of the rate which would be obtained at a constant CO_2 concentration. Oscillations in CO_2 assimilation, such as those produced by stomatal cycling at low humidities or those produced through feed-back effects in carbon metabolism, may occur at a lower frequency than the period required for a measurement; thus cyclic variations in **A**, which would be apparent in a system capable of continuously monitoring **A** at constant c_a, would appear as random noise in measurements made in a closed system.

The simplicity of closed systems makes these the most easily adapted for field use. In the LI-COR LI-6200 Photosynthesis System, all components, except for the leaf chamber, are contained in one unit which can be carried and operated by a single person. Different types and sizes of leaf chamber are available which enable measurements on different shapes of leaves and shoots (Appendix A).

9.1.5 Semi-closed or null-balance systems

Semi-closed systems, also known as null-balance or compensating systems, are a variation on the closed system in which c_a is maintained at a constant value. The IRGA is used as a null-balance instrument which controls a flow of CO_2 into the system at a rate equivalent to the rate of uptake by the leaf (Fig. 9.2). In practice, when CO_2 is removed by the photosynthesising leaf a decrease in c_a sensed by the IRGA switches on a supply of CO_2 to the system. This supply is then maintained at a rate just sufficient to keep c_a constant:

$$\mathbf{A} = \frac{u_c}{s} \tag{9.2}$$

To determine **A** in a semi-closed system it is necessary that the IRGA is calibrated in absolute mode and that the rate of addition of CO_2 is known with great accuracy, since error in the estimation of **A** will be directly proportional to and primarily dependent upon the measurement of u_c. The humidity

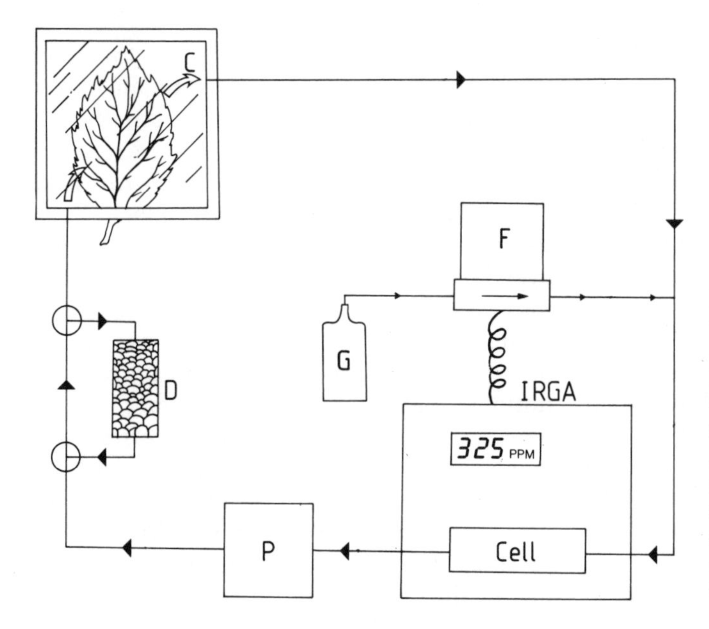

Fig. 9.2. Diagram illustrating air flow in a simple semi-closed system. C, leaf enclosure chamber; D, drier by-pass; F, thermal mass flowmeter and controller; G, cylinder of compressed CO_2; IRGA, infra-red gas analyser and P, pump. Photosynthesis by the enclosed leaf causes a decrease in system CO_2 concentration which is sensed at the IRGA and opens a controlled and metered flow (F) of CO_2 into the system.

within the system should either be maintained at a constant level or monitored so that the CO_2 concentration may be corrected to that of dry air. Assimilation rate will otherwise be overestimated, since the mole fraction of CO_2 within the system will decline not only as a result of photosynthesis, but also as a result of transpiration. In a semi-closed system, transpiration will increase humidity and thus dilute the concentrations of the other gases in enclosed air, including CO_2. A further requirement of the IRGA (less important in closed systems) is absence of zero drift, i.e. long-term stability, since any drift will produce a systematic error in the estimate of **A**. Addition of CO_2 to semi-closed systems may be achieved with electronic flow controllers [6] which allow injection of CO_2 into the system at a constant rate.

The advantage of semi-closed systems over closed systems is that c_a is maintained at a constant level so that **A** is determined at steady-state. Errors arising from permeation or leakage of CO_2 will be a constant rather than an accumulating error, as in closed systems. A further advantage is that **A**, and (if humidity control and measurement are included) **E**, may be studied at a range of values of c_a and w_a, simply by changing the set-point values in the system. The same measurements with an open system would require a complex air-conditioning system. Two practical disadvantages of semi-closed systems should be noted. Firstly, only one chamber can be monitored by one IRGA; thus such systems can be expensive relative to the number of measurements that are made. Secondly, rapid transient changes in **A**, such as those arising from sudden changes in light level, cannot be resolved easily. If the recirculation of air is slower than the transient change, a damped oscillation in the system will be set up which is a function of the system and not the plant material. To monitor transient changes, the volume of the system would need to be minimised and the air rapidly circulated.

9.1.6 Open or differential systems

In an open or differential system there is a net flow of air through the system with no addition of CO_2 to compensate for photosynthesis (Fig. 9.3). The IRGA is used to measure the difference in CO_2 concentration between the chamber entrance (c_e) and outlet (c_o). Where the air streams are dried prior to entering the IRGA, the assimilation rate will approximate to:

$$\mathbf{A} = \frac{u(c_e - c_o)}{s} \tag{9.3}$$

If the air is not dried before entering the IRGA the effects of water vapour on the measurements of CO_2 flux must be considered. Increase in the water vapour content of the air as it passes over the leaf will affect calculation of **A** in the above equation in two ways:

(1) Increase in the water vapour mole fraction must decrease the mole fraction of all other gases. Thus, c_o will be less than c_e simply because of transpiration. If the change in humidity is determined then dilution of c may be accounted for in calculating **A**:

$$\mathbf{A} = \frac{u(c_e - c_o)}{s} \times \frac{(1 - w_o)}{(1 - w_e)} \tag{9.4}$$

(2) Sensitivity of CO_2 IRGAs to water vapour may cause an underestimate of c_o. This may be accounted for if both the response to water vapour and the increase in water vapour ($w_o - w_e$) are known. Many CO_2 IRGAs incorporate optical filters which decrease response to water vapour. However, the presence of water vapour will alter the spectral composition of infrared radiation in the cells and may alter the response to CO_2 in a complex manner dependent on concentrations, path-lengths, source spectra and detector sensitivity. Since these errors can be substantial and not easily predicted [7] it is probably simplest to avoid the uncertainties by routinely drying the IRGA inlet air streams. This may be achieved by passing the air through columns of calcium chloride

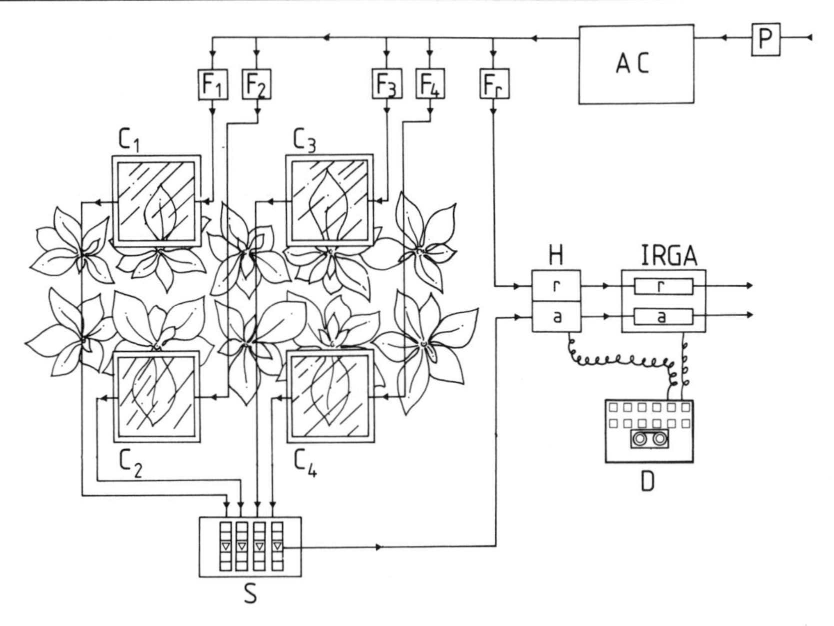

Fig. 9.3. Diagram illustrating air flow in a multichamber open system. Outside air is drawn into the system by a pump (P) and then passed into an air conditioning system (AC) to control humidity and gas concentrations. Flow to a reference and several leaf chambers is controlled by individual mass flow-meters and flow controllers (F). Air from the leaf chambers (C) enters a sample selector (S) which passes gas from each chamber in sequence to a differential hygrometer (H) and differential infra-red gas analyser (IRGA). Change in water vapour and CO_2 concentrations across each leaf is determined by comparison to a reference (r) gas stream. Typically, the large amounts of data gathered by such a system would be sent to a data-logger (D).

and magnesium perchlorate in series (Day, 1985). Silica gel should not be used for this purpose as it can exchange CO_2. In the LI-6262 (LI-COR), CO_2 and water vapour are measured simultaneously within the same optical path allowing direct correction for water vapour interference via the instruments microprocessor.

In an open system it is necessary that: a) the IRGA is calibrated in differential mode; b) the changes in both humidity and CO_2 across the leaf are known; c) the flow rate of air (u) through the leaf chamber is constant and accurately known; and d) the leaf surface area is determined accurately. The main disadvantages of such a system are the initial expense, in particular, the requirement for an air-conditioning system and for an IRGA which can accurately sense small differences in CO_2 mole fraction between two air streams, i.e. of the order of $\leqslant 1\ \mu mol\ mol^{-1}$. However, there are advantages to such a system:

1. By use of a switching device, **A** can be simultaneously determined for a number of chambers.
2. The CO_2, O_2 and water vapour concentrations around the leaf can easily be manipulated.
3. By linking a water vapour IRGA or electrical humidity sensor in series with the CO_2 IRGA, transpiration and photosynthetic CO_2 assimilation can be measured simultaneously for several chambers.

4. Transient changes in gas exchange with step changes in the microclimate of the leaf may be monitored easily.

Commercial portable open systems are listed in Appendix A.

9.2 INFRARED GAS ANALYSIS

9.2.1 Principles

Infrared (IR) gas analysis of CO_2 is the most widespread contemporary method of determining CO_2 concentrations in the measurement of photosynthetic and respiratory rates of plants. The popularity of the method stems from the reliability, accuracy and simplicity of this technique compared to others, coupled with recent miniaturisation of IRGAs. To determine CO_2 exchange for a leaf area of about $10\,cm^2$ in an open or semi-closed system the instrument should be capable of resolving a CO_2 mole fraction of $0.1-1\,\mu mol\,mol^{-1}$ (=0.1–1.0 vpm) against the normal mean atmospheric concentration of about 354 $\mu mol\,mol^{-1}$. Although many IRGAs designed for laboratory operation will meet this specification, there are few truly portable instruments capable of such resolution (Appendix A). This section examines the principles of IR gas analysis.

Hetero-atomic gas molecules absorb radiation at specific sub-millimetre IR wavebands, each gas having a characteristic absorption spectrum. Gas molecules consisting of two identical atoms (e.g. O_2, N_2) do not absorb IR radiation, and thus do not interfere with determination of the concentration of hetero-atomic molecules [8]. IR gas analysis has been used for the measurement of a wide range of hetero-atomic gas molecules, including CO_2, H_2O, NH_3, CO, N_2O, NO and gaseous hydrocarbons. The major absorption band of CO_2 is at 4.25 μm with secondary peaks at 2.66, 2.77 and 14.99 μm. The only hetero-atomic gas normally present in air with an absorption spectrum overlapping that of CO_2 is water vapour (both molecules absorb IR in the 2.7 μm region). Since water vapour is usually present in air at much higher concentrations than CO_2, this interference is significant, but may be overcome simply by drying the air. Absorption bands are in fact made up of a series of absorption lines corresponding to the rotational states of CO_2. Absorption of radiation by CO_2 at any one wavelength follows the Beer–Lambert Law [2], and thus depends on the radiation path length through the measuring cell and the molar concentration of CO_2:

$$\alpha_w = 1 - \exp(-l \cdot M \cdot k_w) \qquad (9.5)$$

Most CO_2 IRGAs use broad-band radiation and total absorption will therefore be determined by integrating over all the absorption lines in the 4.25 μm band. Since k_w will differ between specific wavelengths, the spectral distribution of energy will change with passage of the broad-band radiation through the sample, the more strongly absorbed wavelengths being depleted more rapidly than the more weakly absorbed wavelengths.

9.2.2 Configurations

An IRGA consists of three basic parts; an IR source, a gas cell and a detector (Fig. 9.4). CO_2 in the gas cell will decrease the radiation reaching the detector, causing a decrease in detector output signal. For true differential measurements, two parallel chambers are needed; the detector must be capable of measuring the difference in the amounts of radiation in the two cells.

(a) Source:

Two types of IRGA may be recognised: dispersive (DIR) and non-dispersive (NDIR). In DIR instruments, the source radiation is passed through a monochromator, the selected narrow band of radiation being passed through the cell. Thus the sample may be scanned over a range of wavelengths, and absorption by several hetero-atomic

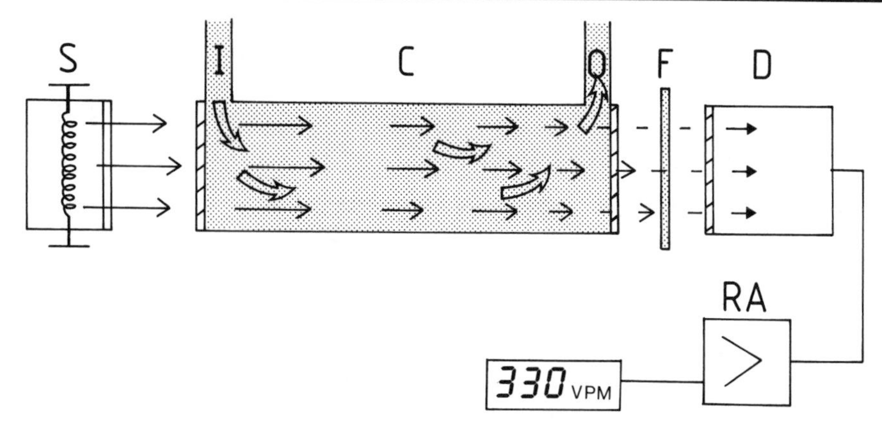

Fig. 9.4. Generalised layout of a simple infra-red CO_2 analyser. Infra-red radiation from a source (S) is passed through a gas cell (C), with an inlet (I) and outlet (O), which allows a continuous flow of the gas being analysed. The infra-red radiation leaving the cell may be filtered, typically with a 4.3 μm bandpass filter, (F) before reaching the detector (D). The detector signal will be rectified and amplified (RA) before display. Any increase in the concentration of the IR absorbing gas in the cell will result in a fall in the detector signal.

species measured. However, where concern centres on one molecular species (e.g. CO_2) this is unnecessary. Most instruments currently used in crop physiology are NDIR, that is, they use the broad-band radiation emitted by the source.

The IR source is typically a spiral of nichrome alloy or tungsten, heated to about 600–800°C (dull red glow) through a low voltage circuit. The coil may be coated to reduce sublimation which will otherwise contaminate windows and reflective surfaces [8]. The delicate spiral of metal must be firmly mounted to minimise movement in response to vibration which would otherwise cause random noise in the detector signal. Often the source is embedded in a transparent ceramic material to prevent any movement. Care must be taken to check that the ceramic casing does not fracture.

In dual beam instruments parallel beams of IR must be produced. This is achieved by the use of two sources connected in series in the same circuit or by use of a single source split between the two parallel cells with reflectors. The latter method avoids the problem of differential aging between two sources.

The source radiation is 'chopped', either mechanically with a rotating shutter or electronically by pulsing the electrical supply to the source. Mechanical shutters are inherently sensitive to vibration. This sensitivity and the resulting noise may be reduced by increased chopping speeds or eliminated by solid-state (electronic) chopping. For example, the LI-6251/6252/6262 (LI-COR), and the LCA3 (ADC) analysers use solid-state chopping, eliminating all moving parts from the IRGA optical bench.

(b) Cells:
Most IRGAs used for photosynthesis measurement are dual beam, passing equal amounts of radiation into two parallel cells, termed the analysis and the reference cell (Figs. 9.5 and 9.6). The analysis cell is a through-fall cell, i.e. there is a continuous flow of the sample gas through the cell. Reference cells may be factory-sealed (e.g. with CO_2-free air), or may also be through-fall. The latter configuration ensures greater flexibility in the use of the instrument. Cells therefore contain a gas inlet and an outlet,

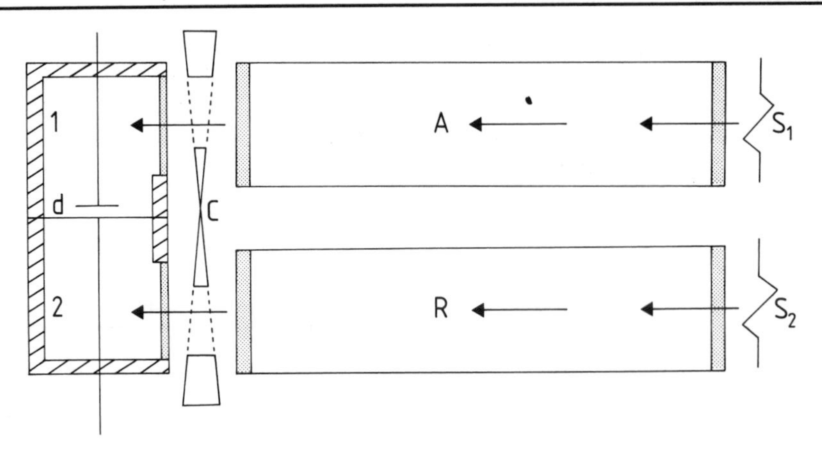

Fig. 9.5. An example layout of a double beam IRGA with Luft absorption cells of the detector (d) in parallel. Radiation from both sources (S_1 and S_2) is chopped simultaneously (C) so that radiation from both the analysis (A) and reference (R) cells reach the corresponding detector cells simultaneously.

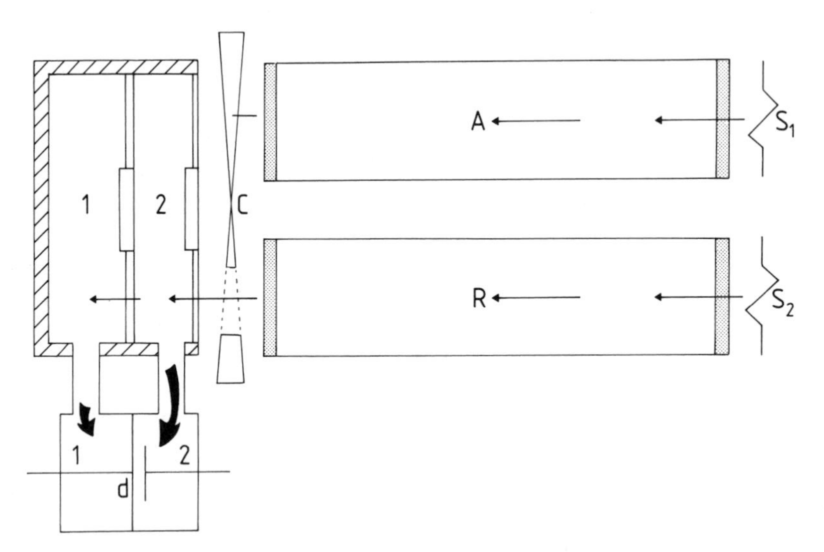

Fig. 9.6. An example layout of a double beam IRGA with Luft absorption cells (1 and 2) of the detector (d) in series. Radiation from the sources (S_1 and S_2) is chopped alternatively (C) so that radiation from first the analysis (A) and then the reference (R) cell is received alternately by both detector cells.

windows of an IR transmitting material such as calcium fluoride, and a highly reflective inner lining. To maximise transmission, cell inner surfaces are commonly gold-plated. The need for this highly reflective coating would be removed if imaging optics were used to produce a beam of parallel radiation passing through the centre of the cells. However, very few instruments use imaging optics. By reference to Equation 9.5 it may be seen that sensitivity will increase with increase in the path length of the cell. In most current laboratory analysers a cell length of about 250 mm is commonly employed for the $\mu mol\,mol^{-1}$ range [8]. This is a limitation to miniaturisation for field applications, although it could be overcome by the use of folded optical paths.

The use of a split cell, i.e. an analysis cell which is split into two lengths in series by a window, provides a simple means for varying cell length to provide two broad ranges of measurement sensitivity. In the Series 225 (ADC) instruments the analysis cell is normally split to provide a short and long cell, representing respectively 5% and 95% of the total path length. Thus by passing the sample gas through the short cell, and CO_2-free air through the long cell, the instrument may be switched from a 0–50 μmol mol^{-1} range to a 0–1000 μmol mol^{-1} range, both of 1% precision.

Some of the recent developments in portable analysers have been in single cell instruments. In the LCA3 (ADC) a reference is provided by alternating the sample with CO_2-free air; thus a comparison is made in time rather than in space, as in dual beam instruments.

(c) Detectors:

The most common type of detector has been the Luft type which operates on the principle of positive filtration, i.e. it absorbs IR in the CO_2 absorption bands. This is achieved by filling the detector with CO_2. The detector is divided into two chambers separated by a thin diaphragm of copper-beryllium, aluminium or gold, which forms one electrode of a diaphragm condenser. The chambers may be arranged in parallel or series configurations (Figs. 9.5 and 9.6). The principle of detector operation will be described by considering the parallel configuration. Here, radiation passing through the reference cell enters one chamber and radiation passing through the analysis cell enters the other. Both chambers will absorb radiation in the CO_2 absorbing bands, the amount available for absorption being inversely proportional to the amounts absorbed within the cells. The chopped radiation will thus cause periodic pressure changes in the detector with simultaneous vibration of the membrane. The amplitude of vibration is determined by the pressure difference between the two chambers, which in turn is determined by the CO_2 concentration difference between the analysis and reference cells. Change in the amplitude of vibration of the membrane produces a change in the condenser capacity which is inversely proportional to the voltage change across the condenser [9]. Where the detector chambers are arranged in series, radiation is chopped such that it passes alternately through reference and sample cells. The radiation passing through the cell then passes into the front and rear absorption chambers. The gas in the front chamber absorbs radiation primarily in the centre of the wavebands, leaving the 'tails' for absorption in the rear chamber. The rear chamber is made deeper such that pressure pulses will balance in the zero position. The in-series configuration, by comparison with the parallel configuration, leads to less cross-sensitivity to other gases and a more stable zero [2,8].

The Luft-type detector provides a very sensitive method for detecting small CO_2 differences, but it has two serious limitations to its use in field studies:

1. The minimum practical size of these detectors is too great to allow the manufacture of miniaturised or even readily portable instruments.
2. It is prone to 'microphony', i.e. signal noise arising from spurious vibrations of the diaphragm. Microphony may be reduced by incorporating a capillary between the two absorption chambers, the bore being selected to allow pressure equilibration between the chambers at frequencies below the chopper frequency vibration [2]. Phase-selective rectification also reduces sensitivity to vibration.

Two developments have helped to overcome these limitations. In the Binos II (Leybold–Heraeus) the Luft principle of two absorption chambers filled with the absorbing gas (CO_2) is retained, but they are

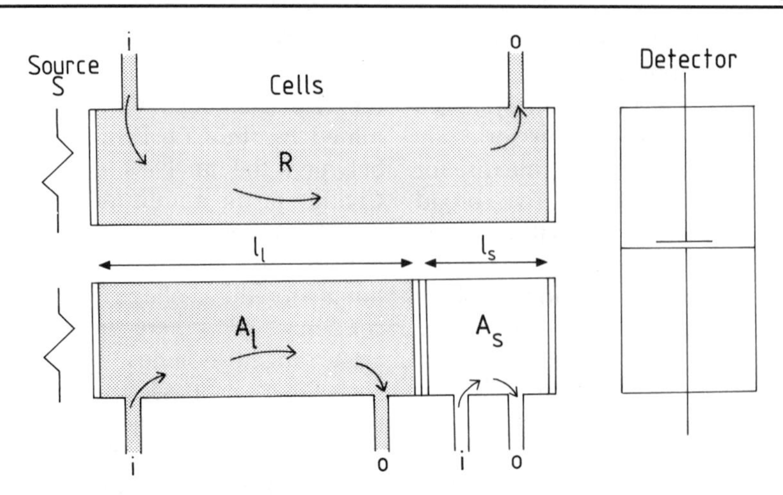

Fig. 9.7. Double beam IRGA with a split analysis cell for differential calibration. To calibrate, air of known CO_2 mole fraction (c) is passed through the reference cell (R) and long section of the analysis cell (A_l), whilst CO_2-free air is passed through the short section of the analysis cell (A_s). The change in detector signal will be the same as that which would be obtained by reducing the CO_2 concentration in the whole of the analysis cell by $c.l_s/(l_s + l_l)$; where l_s and l_l are the optical path lengths of A_s and A_l respectively.

separated by a rigid diaphragm and the gas volumes linked via a tube which incorporates a mass flow sensor, so that a measurement of flow replaces measurement of diaphragm condenser capacitance. This eliminates the problem of vibration and allows the manufacture of smaller detectors.

A development of even greater importance is the solid-state detector. Generally, these are broad-band pyroelectric detectors. They are internally polarised and produce a voltage proportional to temperature change, i.e. when they receive a pulse of IR [9]. These detectors are therefore very sensitive to ambient temperature change and cannot simply replace the Luft detector. However, this limitation has been overcome in the LCA3 and RF instruments from ADC by 'gas-chopping', i.e. chopping only the CO_2 absorbing wavebands of the source. The LCA3 has a single sample cell, with a single solid-state radiation detector. In this instrument radiation from the source passes through the sample gas in the cell and into the detector via a narrow band-pass thin-film filter (to isolate the 4.3 μm absorption band). For absolute measurements the instrument pumps sample gas and CO_2-free air, in alternation, through the cell for periods of 2 seconds each (Fig. 9.7). The signal from the amount of energy received in one half cycle is stored and compared with the amount received in the next half cycle. The difference in energy reaching the detector between half cycles is thus directly proportional to the quantity of CO_2 in the sample gas. This technique of gas alternation or gas-chopping of the radiation beam allows the manufacture of sensitive miniaturised instruments, ideal for field use and for use in remote locations. The greatest practical limitation is that the instrument cannot be used easily within a semi-closed system, since CO_2-free air will enter the system as the analyser switches between gas streams. An alternative approach has been adopted in the solid-state LI-6251/6252/6262 IRGAs from LI-COR. Here conventional dual beam systems are used, but the detector temperature is lowered and controlled by a Peltier block. Controlled cooling of the detector has the added advantage of increasing sensitivity, since the maximum sensitivity of pyroelectric sensors is achieved at sub-zero temperatures [9,10]. These dual beam solid-state instruments can now provide sensitivities comparable to those of the much larger instruments incorporating Luft detectors (Appendix A).

9.2.3 Calibration

Although the IRGA constructions described allow highly sensitive and continuous monitoring of CO_2 concentration, most IRGAs will require frequent calibration. Instruments vary considerably in the frequency with which they require recalibration. Initially, therefore, it is advisable to calibrate daily. However, if no significant shift in calibration settings are found, then a longer interval could be employed. Regular calibration is also a useful diagnostic tool. The development of chronic faults, such as a slowly leaking Luft detector or contamination of cell windows, will be seen as a repeated need to increase amplifier gain to regain the set-points upon each calibration. The minimum requirement for reliable calibration is a source of CO_2-free air and a source of air containing a precisely known concentration of CO_2 in the range to be analysed, preferably this should be contained in a cylinder with stainless inner surfaces to avoid CO_2 adsorption. Alternatively, scale point or calibration mixtures may be made by mixing known volumes of pure CO_2 and N_2, either with precision gas mixing pumps (SA27 or M300 series, H. Wösthoff, Bochum, Germany), high precision mass flow meters [6], or gas syringes (LI-6000-01, LI-COR). With mass flow meters a good accuracy and precision can be achieved, and the flow range is less restricted than with gas mixing pumps. The gas syringe method is relatively simple and especially useful for field use or in places where calibration gas of known CO_2 concentration is difficult to obtain. Some IRGAs provide a built-in 'CO_2-free' air supply which pumps air through a column of CO_2 absorbent material, e.g. soda lime. It is advisable to check the efficiency of these supplies by comparison with high purity nitrogen.

(a) Principles:
Calibration is the process of adjusting an IRGAs signal output, either an electrical output signal or a meter deflection, to the quantities and units required for calculation of CO_2 exchange. The measure given on the output meters of most commercial IRGAs is the partial volume (synonymous with mixing ratio), with units of vpm (volumes per million), the S.I. equivalent being $cm^3\,m^{-3}$. This measure is also directly equivalent to mole fraction in $\mu mol\,mol^{-1}$. However, radiation absorption is not directly dependent on partial volume or mole fraction, but on the molar concentration, i.e. amount per unit volume ($mol\,m^{-3}$). Essentially the instrument measures the quantity of absorbing molecules per unit volume irrespective of the amount of non-absorbing background matter. Because mixtures are commonly made by volumetric mixing, manufacturers of compressed gases provide scale-point gases, i.e. calibration mixtures specified in terms of the volume fraction (vpm = $cm^3\,m^{-3}$). Volume fraction has the advantage of constancy irrespective of temperature and pressure variation. The volume fraction may be converted to molar concentrations for calibration purposes:

$$M = \frac{c \cdot T_o \cdot p}{22.4T \cdot p_o} \times 10^{-3} \qquad (9.6)$$

The IRGA could be calibrated in units of volume or mole fraction, but the calibration will only be valid for the temperature conditions of calibration. Ambient pressure changes will produce a small error in such a calibration, typically ±2%.

However, since the gas must flow through the cell, there will be a pressure difference proportional to the rate of flow. It is therefore important to calibrate an IRGA at the flow rate which will be used in taking measurements. An *in situ* calibration should be made in the field at the prevailing conditions. A greater problem in field use will be the effect of ambient temperature variation on molar concentration. By reference to Equation 9.6 it may be shown that an IRGA calibrated against a scale-point gas of $300\,cm^3\,m^{-3}$ will underestimate the true volume and mole fractions by 6.8% at 0°C and overestimate by 6.9% at 40°C.

(b) Absolute calibration:
Where an IRGA is used to determine the CO_2 concentration of an air sample, it is calibrated in absolute mode, i.e. the CO_2-free air provides the reference. To calibrate, CO_2-free air is passed through the analysis cell, and in dual beam instruments through the reference cell as well. Zero is set on the output meter by adjusting the zero shutter. Air samples of known CO_2 concentrations are then passed through the analysis cell, starting with the highest concentration.

The electrical output of the IRGA is not necessarily linearly related to the CO_2 concentration, so a range of points spanning the range of mole fractions to be analysed are required. These may be obtained by using a single calibration cylinder which provides the highest scale point, and obtaining lower scale points by dilution of the mixture with CO_2-free air using a precision gas mixing pump, mass flow meters, or through a CO_2 absorber within a gas diluter (e.g. GD600, ADC).

(c) Differential calibration:
Where an IRGA is used to determine a change in CO_2 concentration, for example the difference in CO_2 concentration in an air stream before and after it has passed over a leaf, the analyser is calibrated in differential mode. In this mode it is possible to detect very small changes in CO_2 mole fraction, down to $0.1\,\mu mol\,mol^{-1}$, with several of the larger dual beam instruments. Precise calibration requires that the analysis and reference cells are filled with air of known, but only slightly different, CO_2 concentrations. In practice this can only be achieved by a small dilution of the calibration gas using a precision gas mixing pump or gas diluter. Here the calibration gas flow is split such that one stream passes through the reference cell while the other is precisely diluted with CO_2-free air and passed through the analysis cell. An alternative method of differential calibration is possible for instruments in which the analysis cell is split (Fig. 9.7) [11].

In the Series 225/3 (ADC) the analysis tube is split into two lengths, typically a long cell representing 95% of the path-length and a short cell representing the remaining 5%. To set zero, the air stream to be used in the experiment (of previously determined absolute CO_2 concentration) is passed through both the reference and analysis tube, and zero is then set on the output meter. At atmospheric concentrations of CO_2, the amount of IR absorbed in the analysis tube is only a small fraction of the total IR in the CO_2-absorbing bands which pass through the tube; thus removal of CO_2 from the short cell is optically equivalent to reducing the CO_2 concentration over the whole length of the analysis tube by 5%. The change in the electrical output of the IRGA meter produced by this operation represents the change that a 5% depletion in CO_2 concentration would produce. This provides a simple, yet highly accurate means of calibration [11].

9.3 THE MEASUREMENT AND CONTROL OF GAS FLOW

9.3.1 Flowmetering

Flow measurement is fundamental to the measurement of CO_2 exchange between plants and the flowing atmosphere of an enclosing chamber. In an open gas-exchange system, **A** is proportional to the product of the difference in CO_2 mole fraction and the flow rate of air across the leaf (Equation 9.3). In semi-closed systems a constant CO_2 mole fraction is maintained by an inflow of CO_2 into the system, the rate of this flow being the measure of CO_2 uptake by the plant (Equation 9.2). Because of the physical principles employed, most commercial flowmeters provide an estimate of either volumetric flow (u_v; $m^3 s^{-1}$) or mass flow (u_w; $kg\, s^{-1}$) and not the mole flow (u; $mol\, s^{-1}$) required for the direct calculation of **A**. Direct measurement of either u or u_w is clearly not possible since it is not practical to collect and weigh gas [6]. Direct measurement of u_v is possible and forms the basis of the calibration techniques considered later. Even these methods are unsuitable for use within a gas exchange system. Mole flow rate may be determined from volumetric flow rates by Equation 9.7a and from mass flow rates by Equation 9.7b:

$$u = \frac{u_v \cdot T_o \cdot p}{22 \cdot 4T \cdot p_o} \qquad (9.7a)$$

$$u = u_w/m \qquad (9.7b)$$

where m is the molecular weight of gas ($g\, mol^{-1}$) (CO_2 = 44.01; H_2O = 18.02; dry air = 28.97)

To determine flow rate the experimenter must resort to the measurement of some physical effect arising from the motion of the gas in the tube. Three of these effects are widely used in measuring flow:

1. consequent mechanical effects, such as the rate of rotation of a rotor mounted in the stream;
2. pressure changes; and
3. the rate of heat transfer from a heated body in the air stream [12].

This section covers only the major techniques which have application in the study of plant gas exchange. The range of volume flow rates measured in gas exchange studies vary from 1 $\mu mol\, s^{-1}$, for CO_2 being fed into a semi-closed system for the measurement of photosynthesis by small leaves, to 100 $mol\, s^{-1}$ (u_v = approx. 2 $dm^3 s^{-1}$), for air in open systems used for the measurement of CO_2 exchange by whole plants or swards in enclosures such as open-top chambers.

9.3.2 Variable-area flowmeters

These have been the most widely used instruments for flow rate measurement, but are now supplanted in many applications by thermal mass flowmeters. Their low cost, simplicity and simple visual indication of flow rate suggest that variable-area flowmeters will continue to be used as secondary flow rate indicators in gas exchange systems.

Variable-area flowmeters consist of a transparent graduated tube with a slightly tapered bore, in which the diameter decreases downwards and the gas flow to be measured passes upwards. A float of diameter slightly less than the minimum bore of the tube is forced by the flow of gas up the tube to the point where its weight is balanced by the force of gas flowing past it (Fig. 9.8a). In a constant volume flow the ball shape is inherently liable to sudden fluctuations in position within the tube, and may 'chatter' against the side of the tube. One solution to this problem is the use of a conical float with angled grooves which cause rotation around the vertical axis, giving the float central stability [6]. Precision depends on tube length, float shape and manufacturing tolerances of the glass tube. Instruments are calibrated by the manufacturer for a given gas, at a specified pressure and temperature. Such calibrations will only change if deposits of dirt are allowed to form in the tube or on the float, or if the tube or float surfaces become damaged or corroded. These instruments are particularly sensitive to ambient temperature and pressure fluctuations. For example, if a variable-area flowmeter was factory calibrated at 15°C and 101.3 kPa, normal sea-level atmospheric pressure fluctuations from 88 to 108 kPa could produce an error of +7.3 to −3.3% in indicated flow rates. In the field, ambient temperature variations between 0 and 40°C could produce errors of +2.7% to −4.0%.

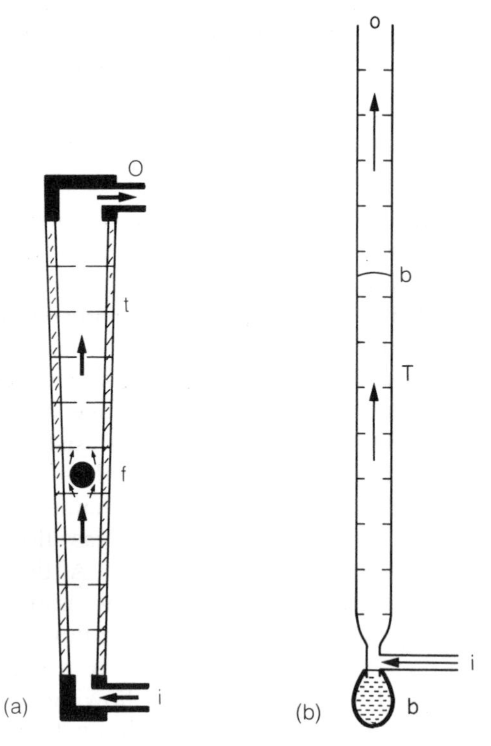

Fig. 9.8. (a) A variable area flowmeter. Air from the inlet (i) forces the float (f) to rise in the tapered graduated glass tube (t) to an equilibrium position. The outlet (o) of these meters is necessarily at the top.
(b) A soap-film flowmeter. Inlet (i) gas passes through a soap solution contained in a compressible bulb (B). The films or bubbles (b) generated rise up in a graduated tube (T) to the outlet (o). Time taken for the bubble to rise through a known volume of tube provides an accurate measure of flow.

Calibration of these flowmeters for the range of working temperatures is therefore necessary, unless the manufacturer provides temperature correction graphs [6].

Since, in most gas-exchange systems, flowmeters are inserted immediately upstream of the assimilation chamber, gas in the outlet of the flowmeter will be above atmospheric pressure. When used in this way either the actual outlet pressure should be measured so that the equivalent flow rate at atmospheric pressure can be calculated, or the flowmeter should be recalibrated *in situ* [6].

The major disadvantages of variable-area flowmeters are that: (i) they must be mounted perfectly upright; (ii) subjective errors in assessing the float position relative to the tube graduations are difficult to avoid (especially if the float position is unstable); and (iii) precision is inherently low ($> \pm 2\%$) except in the longest tubes. A further practical problem is that the slightest amount of moisture in the tube may cause the float to stick. Furthermore, the design is not well suited to the production of an electrical output and hence computerised recording. Alternative mechanical methods of flow measurement include turbine meters and pressure drop measurement across a tube constriction [6,12].

9.3.3 Thermal mass flowmeters

The application of thermal mass flowmeters to the measurement of photosynthetic gas fluxes has increased the potential accuracy of determination of assimilation rate. They consist of a sensor tube which carries the flow, or a constant fraction of the flow. This tube is heated such that the temperature distribution with no flow is symmetrical about its mid-point (Fig. 9.9). Two temperature sensors, typically thermocouples or platinum resistance thermometers, are situated equidistant from the mid-point, one on either side. With no gas flowing, the temperature at both sensors will be equal. Gas flow will transfer heat downstream. The temperature at the sensor downstream of the mid-point (T_d) will be higher than the temperature of the upstream sensor (T_u). The magnitude of the temperature difference will be a function of the flow rate [13]:

$$u_w = \frac{(T_d - T_u) \cdot N}{c_p \cdot H \cdot K_s} \tag{9.8}$$

where H = injected power ($J\,s^{-1}$);
K_s = a system related constant (dimensionless) [13];
N = Correction factor, dependent on the molecular structure of the gas (dimensionless) [13].

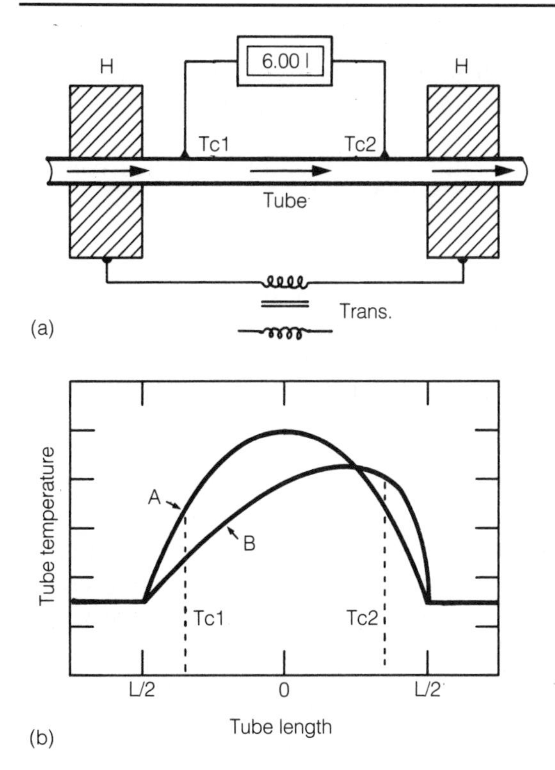

Fig. 9.9. A thermal mass flowmeter. (a) A capillary tube is heated uniformly by a transformer. Thermocouples (Tc1 and Tc2) compare the tube temperature at points equidistant from the centre of the tube. (b) Temperature distribution within the tube. Line A illustrates temperature with no flow; here the temperature at Tc1 equals that at Tc2. Line B illustrates temperature with a small flow of gas through the tube. The moving gas transfers heat upstream such that the temperature at Tc2 is greater than at Tc1, this difference being a function of mass flow rate.

In many commercial units (e.g. MKS Instruments, Burlington, MA, USA; Teledyne Hastings-Radyst, Hampton, VA, USA) T is measured in a tube shunt which takes only a fraction of the total flow. The relation between T and u_w is complex, differing between instruments and flow conditions [13].

Thermal mass flowmeters are normally factory calibrated for one gas, but single correction factors (N), dependent on the molecular structure of the gas and its specific heat capacity, can be used to recalculate the flow rate of other gases. Both manufacturers and independent assessors suggest precisions of ±0.5–1.0% of maximum flow rate for thermal mass flowmeters at a given temperature, depending on design. By comparison to variable-area flowmeters, performance suffers less from fluctuations in pressure ($\pm 0.003\%\ kPa^{-1}$) and from fluctuations in temperature, which may be $\pm 0.1\%\ K^{-1}$ between 5 and 43°C [6]. This source of error may be reduced by controlling the temperature of the flowmeter or by including a compensating circuit, which measures the temperature of a duplicate or reference sensor tube with zero flow rate (Appendix A).

9.3.4 Flowmeter calibration

It is frequently necessary to recalibrate a flowmeter for use with different gases or under different operating conditions. It is also advisable to re-check calibration at regular intervals, especially in the field. For the flow rate ranges used in gas-exchange systems, soap-film flowmeters provide a simple means of calibration, although there are other calibration methods [4,6].

Soap-film meters (Fig. 9.8b) are operated by introducing soap solution over the air inlet to the measuring tube, where a soap bubble or film forms across the tube and is forced upwards through the vertical tube by the flow of the gas. The time taken for the soap film to travel between two points separated by a known volume is recorded and provides a direct measure of volume flow rate (u_v). Such a flowmeter may be simply constructed by adding a T-junction to the base of an accurate burette. By using different diameter tubes a range of flow rates from $10\ mm^3\ s^{-1}$ to $100\ cm^3 s^{-1}$ may be measured with an accuracy of ±0.25%, decreasing to ±1% at $1\ dm^3\ s^{-1}$. However, if the movement of the soap-film is monitored by eye and timed with a stop-watch, timing precision is unlikely to be better than ±0.1 s. Thus to obtain optimum accuracy with this instrument, the combination of tube length and diameter should be such that the time required for the passage of the film between the fixed points exceeds 20 s. The average of several readings should be used. Timing accuracy may be improved by the use of photoelectric detectors which trigger an electric timer. For larger volume flow rates ($>10\,cm^3\,s^{-1}$) it is more practical and accurate to use a wet gas meter for calibration [6].

9.3.5 Flow control

Although its importance is less easily quantified, flow rate control is an essential consideration in gas-exchange studies. Fluctuation in flow rate will decrease the resolution with which both flow rate and gas concentration difference may be measured, may cause fluctuations in gas composition if different gases are being mixed or humidified, and may introduce both random and systematic errors into calculations of flux [6].

(a) Critical flow orifices:
A particular problem in mixing gases, either in the supply to gas-exchange systems or for the calibration of IRGAs and other instruments measuring gas concentration, is that the flow rate of each gas will be influenced by slight changes in downstream pressure. These may be produced by small changes in the pipework and connections, or by changes in the flow rate of a second gas used in generating the mixtures. If a constriction is introduced into the gas stream such that the gas reaches sonic velocity through the constriction, velocity and hence volumetric flow rate will be maintained within very close limits. Thus for any given constriction, once the sonic velocity is reached and provided that the ratio of upstream to downstream pressure is sufficient to maintain sonic velocity then volumetric flow rate will be insensitive to changes in downstream pressure. However, mass flow rate may be varied simply by altering the upstream pressure and hence density of the gas. Constrictions may be small nozzles, venturis or orifices. It has been shown that a range of orifice sizes would provide a fixed range of constant volumetric flow rates, provided that a constant upstream pressure is maintained [14]. This principle is utilised in the GD600 and WG600 (ADC) instruments for the generation of a range of controlled concentrations of CO_2 and water vapour [6].

(b) Electronic flow control:
An accurate control of flow rate may be obtained by linking the electrical output of a flowmeter to an electronic flow rate control valve, such that the flow rate is continually monitored and automatically adjusted to maintain a pre-set value. A wide range of electronic flow rate controllers is available commercially, many of which have been patented (Appendix A). The selection of a suitable valve depends on the pressure drop, range of flow rates and response time required.

The two established methods of electrical control are solenoid and servo-driven valves. In solenoid valves, the valve seat is typically connected to the armature and is lifted away from an orifice as increasing current is supplied to the solenoid. In servo-driven systems the valve is operated by a small stepping motor. Both of these methods can give good precision. More recent developments are thermal expansion valves and piezo-electric crystal valves.

Thermal expansion valves consist of a small thin-walled tube with a ball welded to one end which rests in the gas pipe. The tube contains a small resistance heating element and heat-transfer fluid. When a voltage is applied to this element the tube expands and moves the ball which thus controls the flow rate. This design has the advantage of having no moving seals, virtually no moving parts (total travel approx. 0.1 mm) and no friction; it may thus be expected to be both precise and reliable. However, the device cannot completely stop the flow; thus a solenoid valve must be added to the line if it is necessary to interrupt flow during experiments. The slow response time of 6–10 s would be a serious limitation in semi-closed systems if the experimental objectives are the study or recording of non-steady-state changes in gas exchange, e.g. in kinetic studies of the effects of dark–light transitions on CO_2 assimilation [6]. An alternative which overcomes this time limitation is a proportioning magnet, allowing a variable occlusion of the gas pipe, as used in the Type 1259A (MKS Instruments).

Piezo-electric crystal valves consist of a viton seal cemented to a piezo-electric crystal which has the property of flexing in response to an applied electrical potential. In the resting position the valve is closed, but when a voltage is applied the flexing of the crystal opens the valve by an amount proportional to the voltage. The particular advantage of this design is its very rapid response time of 2 ms [6].

9.4 LEAF AREA DETERMINATION

To determine the assimilation rate (**A**) of a leaf an accurate measure of its surface area (s) is a prerequisite. Leaves do not always present simple flat surfaces, but will often be twisted, crinkled, or finely divided and may be hemispherical, diamond or ellipsoidal in cross-section. Therefore the determination of surface area of the leaf may be complex and can cause significant error in the calculated rate of assimilation. By convention, projected area (one side) of the leaf is used in the calculation of **A**.

The simplest way to determine leaf area is to place the leaf flat on a piece of paper and trace its outline. If graph paper is used, the area can be counted. Alternatively, the image of the leaf can be cut out, dried and then weighed on an accurate balance, and the area calculated from the weight/area ratio for the same paper. Photocopiers may also be used to obtain a photographic image, which is similarly cut out and weighed. A problem in making such copies is that of flattening the leaf to get the true projected area. This is best achieved by pressing the leaf between a sheet of transparent plastic foil and one of white paper in order to obtain good contrast.

Commercially available leaf area meters are both portable (e.g. LI-3000A, LI-COR; LAM-100, ADC) and laboratory based (e.g. AMS, Delta-T Devices, Cambridge, UK; LI-3100, LI-COR; S1700, Skye Instruments, Llandridnod Wells, UK: Appendix A). These operate on a range of principles. In the simpler field instruments the leaves are passed between an array of light-emitting diodes and light sensors, and the total area is estimated from the occlusion of light by the leaf. The portable instruments have the great advantage that the area can be measured whilst the leaf is still attached to the plant, avoiding destructive harvesting and allowing repeated measurements to be made throughout growth. The portion of the leaf enclosed in a leaf chamber can thus be measured. To obtain a more accurate estimate of the leaf area the same leaf may be measured several times to yield an average. Such instruments are accurate (commonly 1 mm^2 resolution) for large simple leaves. Small and/or narrow leaves, such as very fine grass leaves or conifer needles, are more difficult to determine accurately because of the high edge to area ratio. Many of the laboratory-based area measuring systems can have a higher resolution and accuracy, which allows a small object to be enlarged for measurement. Commonly these systems consist of a TV camera which views the object on a light box (e.g. AMS, Delta-T; S1700, Skye Instruments: Appendix A). The video image is analysed by a microprocessor or microcomputer to determine the projected leaf area. Variegated or diseased portions of the leaves can sometimes be determined separately, providing that there is sufficient contrast. A conveyor belt system (e.g. LI-3050A) can be used to make the determination of areas more rapid and efficient, if many objects are to be measured. The resolution depends largely on the quality of the camera used.

9.5 CHAMBER CONDITIONS AND CONSTRUCTION

9.5.1 Principles

The environment of the enclosed leaf or plant is influenced by the design of the chamber and the effectiveness of the air-conditioning system which determines the composition of the air supplied to the chamber. Chamber design will be determined by the objectives of the study, the prevailing climatic conditions, and the size and shape of the material to be enclosed. Two broad groups of objectives may be identified in the study of CO_2 assimilation by crops.

(a) Measurement under prevailing field conditions:
Where the objective is to determine assimilation rate under field conditions at any one instant, conditions within the chamber must be close to those outside. Two approaches have been employed to achieve this objective: 1) to design the chamber such that its effect upon leaf environment is minimal; 2) to monitor outside conditions and control chamber internal conditions so that they follow those outside. The former approach is clearly the simpler, in that the chamber requires no ancillary control equipment and is thus more readily portable. Chambers supplied with the LI6200 system (LI-COR) and the LCA3 system (ADC) follow this approach. This is most effective if the leaf is enclosed for a brief period of time, i.e. $<1-2$ minutes, the effect and extent of modification of the environment increasing with time. In the other approach an efficient air conditioning system allowing cooling and heating, with a short response time, is a prerequisite. A double-walled chamber can be cooled by air or water so that the air temperature inside the chamber equals the outside air temperature. The tubing carrying air to the chamber in the field can be heated or cooled to effect control of the air temperature within the chamber. An advantage of an efficient air-conditioning system is that not only may the temperature of the chamber be controlled to track outside temperatures, but it may also be set to other temperatures to determine the temperature responses *in situ*. Similarly, humidity, CO_2 and (by addition of a portable light source) photon flux may be manipulated, e.g. various versions of the Minicuvette System (H. Walz, Effetrich, Germany) [4]. The two approaches are suited to different purposes, the first through increased portability allowing sampling of large numbers of different leaves and the second through its provision of environmental control, allows prolonged study of individual leaves, and also allows provision for measurement under controlled conditions; see below. Selecting an appropriate system for field studies therefore requires a careful prior definition of objectives.

(b) Measurement under controlled conditions:
Where comparisons between treatments or genotypes are required, e.g. screening of varieties in crop breeding, control of some aspects of the leaf chamber environment will be necessary. It is necessary to provide control of those aspects of microclimate which will strongly influence CO_2 assimilation so that valid comparisons are possible. This may be achieved, for example, by controlling the leaf temperature at a fixed value for all measurements or by addition of saturating artificial light. Control of the chamber environment is also necessary if the study aims to establish the basis of genotype or treatment differences, through light and CO_2 response curves (Section 9.6).

9.5.2 Chamber design

As discussed above, the objective of the study is the primary determinant of chamber design. Regardless of objective, a further primary requirement is that the environment within the chamber be as homogeneous as possible. Gradients of temperature, CO_2 and water vapour across and along the leaf should be minimised so that assimilation rate will be determined for a well-defined microclimate. The degree of temperature and gaseous homogeneity within the chamber depends upon boundary layer conductance to heat and gaseous transfers; this should be maximised. Chamber design must take this prerequisite into account with respect to the tissue under investigation, since a chamber designed for a leaf or leaves of one species may be unsuitable for many others.

9.5.3 Boundary layer conditions

The boundary layer conductance (g_b) will determine the homogeneity of the gaseous and thermal microenvironment of the leaf. It is maximised by obtaining a high rate of air movement around the leaf. This can be achieved either by stirring the air vigorously with a fan within the chamber or by rapidly recirculating the air with a pump outside the chamber. Even when a ventilation technique is used, care should still be taken in design to avoid the creation of pockets of still air. These can occur where a leaf is in close proximity to the chamber wall, and in the corners of a rectangular chamber. In the absence of stirring or recirculation, g_b depends on the net rate of flow of air through the chamber and its pattern of movement across the leaf. However, the velocity of air around the leaf may be accelerated by reducing chamber volume and by even spacing of gas inlet and outlet ports. Unstirred chambers of this design have been described [15]. It is not possible to predict g_b exactly; it will depend not only on chamber design, but on the size, shape and surface properties of the leaf. It is therefore advisable to determine g_b empirically for any new chamber or species.

Boundary layer conductance to water vapour transfer (g'_b) for simple and flat leaves may be determined by placing a wet filter paper replica of the leaf into the chamber and measuring the rate of water vapour efflux (**E**), leaf temperature (T_l) and the ambient humidity of the chamber (w_a); this will equal w_o in a stirred chamber and approximately $(w_e + w_o)/2$ in an unstirred chamber:

$$g'_b = \frac{\mathbf{E}}{w_{s[T]}/p - w_a/p} \qquad (9.9)$$

The temperature of the filter paper replica may be determined by small wire thermocouples appressed to the lower surface. Where the leaves and shoots being studied have complex shapes, an alternative may be to coat the surface with plaster of Paris and then wet this surface.

Parkinson [16] provides an alternative approach using dry leaf replicas. Here the temperature of the replica is determined from its energy balance so avoiding the uncertainties of thermocouple measurement. Alternatively, g_b may be determined from the heat exchange either of a replica or from actual leaves [2].

If boundary layer conditions are to be similar in magnitude to those for leaves in the open, then chambers should be designed to produce a g'_b of $>5\,mol\,m^{-2}\,s^{-1}$ [2].

9.5.4 Temperature

From a physiological standpoint it is leaf temperature rather than air temperature that is of interest. Control of leaf temperature is greatly facilitated by minimising the thermal radiation which reaches the leaf. Heat-reflecting glass ('hot mirror') above the chamber can intercept much of the incoming IR, e.g. Wide Band Hot Mirror (OCLI, Santa Rosa, CA, USA). The use of a long wave IR transmitting window in the chamber, such as polypropylene film (Propafilm C, ICI Plastics, Runcorn, UK), is useful in preventing a 'greenhouse effect' within the chamber. Good chamber ventilation (as indicated by a large g_b) keeps leaf-air temperature gradients to a minimum and minimises temperature gradients along and across the leaf. Leaf temperature is most commonly controlled by regulating the temperature of the ambient air in the chamber. This can be achieved by jacketing the chamber so that coolant can be circulated over the chamber walls, by inserting cooling coils inside the chamber, or by building Peltier modules into the chamber walls. The thermal conductivity of aluminium is $205\,W\,m^{-1}\,K^{-1}$ compared to approx. $0.2\,W\,m^{-1}\,K^{-1}$ for acrylic plastics, i.e. 1000 times greater. An aluminium walled chamber will therefore be far less prone to heating above the ambient temperature. The speed of heat dissipation may be further enhanced by the addition of cooling fins on the chamber undersurface.

9.5.5 Photon flux

Usually, the light required will be that which would have been incident on the leaf in the absence of the chamber. The chamber window must therefore be as transparent as possible. Few window materials are perfect transmitters. Acrylic plastic has a transmission of 0.92 when new. Scratches and smears on plastic windows can produce a surprisingly large reduction in photon flux density (**Q**) at points on the leaf surface, so it is important to have a supply of replacement windows. This is easier if thin film windows such a 'Propafilm' are used, since they may be simply sealed with double-sided tape, making rapid replacement of a damaged window feasible in the field.

Where a saturating light level is required a supplementary light must be provided above the chamber. An important criterion in the selection of light sources used for controlled environment CO_2 exchange studies is similarity to natural daylight. Xenon-arc lamps provide a close match, but they have a high heat (near infra-red radiation) output. A combination of high-pressure sodium and mercury lamps also gives a good spectral match to daylight, without the same level of heat output. Highly efficient short-arc lamps of good wavelength characteristics are available (Osram Sales Corporation, General Electric Company, London, UK) but they are much more expensive than the less efficient and longer arc lamps. Studio spot lamps of this type are used in the television and film industries and are available commercially. Powerful lamps up to 5 kW can illuminate with photon flux densities equivalent to full sunlight. Small spot lamps of the type used in store window displays (GE 85 W multi-mirror spot lamp, Phillips Lighting, Croydon, UK) can be used to illuminate small areas of flat leaves. A 500 or 100 W mercury-halide lamp (HQI, Wotan Lamps Ltd., London, UK), a hemispherical mirror and an inexpensive acrylic Fresnel lens (Ealing Beck Ltd., Watford, UK) of the appropriate focal length and diameter can be used to evenly illuminate a $5 \times 5 \times 10$ cm space within a chamber [17]. Photon flux can be varied at the level of the chamber, ideally by placing neutral density filters (Schott Glaswerke, Mainz, Germany) above the chamber, although sheets of white muslin may be used. A limitation of white muslin and similar materials is the simultaneous change in directional distribution of light with the change in photon flux density. Inexpensive wire screens can be used, but care should be taken that interference lines will not result in uneven illumination. Decrease in the voltage supply to lamps has also been used to vary photon flux, but change in the supply voltage to lamps will not only alter the quantity but also the spectral composition of the emitted radiation.

Most leaf chambers are designed to receive direct radiation on the upper surface. If the base of the chamber is painted with optically black paint then the radiation conditions of the leaf can be defined precisely. Finally, the leaf should be held in the horizontal plane in such a chamber, if all parts of the surface are to receive the same photon flux. In large chambers this may be achieved by placing the leaf between two coarse meshes of fine transparent nylon. By comparison to leaves in the natural environment this direct illumination from above may be atypical and because of the poor distribution of light which it creates may suggest an atypically low efficiency. Under conditions of low photon flux, e.g. around dawn or dusk, under heavy cloud conditions, or below the upper layer of dense canopies, most radiation will be diffuse with the result that the upper and lower surfaces of a leaf will receive diffuse radiation. Only by using a perfectly spherical chamber with reflective walls and light entering via an effective diffuser could the radiation supplied to the leaf be totally diffuse and the quantity at all surfaces of the leaf be exactly defined. The advantage of such chambers is that they may be used to measure total light absorption by the leaf simultaneously with gas exchange, so that the true quantum efficiency can be determined [18].

Manufacturers of chambers for field and laboratory include ADC, LI-COR, PP Systems and H. Walz (Appendix A).

9.5.6 Materials

Materials used in the construction of chambers, gas connections between system components, and air-conditioning systems have a major influence on the effectiveness of the system and the accuracy of determinations of gaseous fluxes [19]. Permeation of CO_2 and water vapour between the surrounding air and that enclosed in the system will produce errors in determination of rates of CO_2 assimilation and water movement in all systems, although the error is likely to be most pronounced in a closed system since recirculation will cause an amplification of error in flux calculations. Adsorption and desorption of CO_2 and water vapour from internal surfaces will also produce errors in calculations of steady-state CO_2 uptake and transpiration rates in closed systems. Strictly, this problem is avoided in open systems, since steady-state fluxes here cannot be achieved until the whole system is in equilibrium: thus adsorption must be balanced by desorption. Materials with an ability to absorb significant quantities of CO_2 or water vapour will greatly affect the apparent responses of CO_2 uptake and transpiration rates to changes in microclimate. In addition, the characteristics of non-steady-state changes (e.g. the induction of CO_2 assimilation following a dark–light transition) will be altered since the system will impose a lag in response in addition to any lag in the response of the leaf itself. Ideally, then, all systems should be constructed from materials which do not adsorb, absorb or allow permeation of either CO_2 or water vapour. Many materials not only adsorb, but also absorb, water vapour. Since CO_2 is soluble in water (36.5 mol m^{-3} at 20°C) any material which absorbs water vapour will also absorb CO_2. The amount of CO_2 absorbed will be strongly dependent on temperature and humidity. Any sudden change in temperature, such as a sunfleck falling on the tubing, could suddenly release the absorbed CO_2 and produce a spurious flux estimate. A slow change in temperature, such as gradual warming throughout the day, may produce a systematic error which could go undetected.

(a) Glass:

Glass chambers are near ideal as far as permeability, absorption and light penetration are concerned. Dixon and Grace [20] found that glass was the least water-retentive material in a comparison of materials used in chamber construction. The surface absorption of water by soda glass is 0.05 g m^{-2} after 24 h at 293 K. A skillful glassblower will be able to form almost any shape of glass chamber. However, quartz glass must be used in the field and in the laboratory if natural light or UV-light effects are to be studied and this will increase costs considerably. The disadvantage of glass is the obvious risk of breakage, and the technical difficulties of inserting fans, thermocouples, etc.

(b) Metals:

Metals have near-zero permeability to gases. However, if a metal is reactive it may adsorb or absorb gases. Non-stainless steels absorb water vapour strongly. Rusting and other forms of corrosion produce the additional problems of surface roughness. A rough surface will contain many microcavities, trapping still air and greatly increasing the time taken for the surfaces to come into equilibrium with air in the system. The problem of surface reactivity is avoided in good-quality stainless steels, though even here the material should not be assumed perfect. Contamination of surfaces with greases and oils is common, and if the surfaces have been roughly worked they will contain surface microcavities. Stainless steel surfaces should be washed in a degreasing agent and polished to improve surface smoothness. A practical disadvantage in chamber construction is that stainless steels are difficult to work. Copper, aluminium and alloys based on either of these are easier to work, but are more likely to be reactive. In particular, copper absorbs and adsorbs CO_2 strongly. This problem may be overcome by chrome plating which will provide a smooth and largely unreactive surface. Aluminium is easily worked and provides heat conduction properties only slightly inferior to copper. It will rapidly oxidise and absorb water because the oxides possess surface hydroxyl groups which form hydrogen bonds with water [19]. This problem is removed if the more inert alloys of aluminium are used, e.g. duralinium. Stainless steel tubing is difficult to fit, lacking flexibility, and therefore only convenient for permanent connections. Copper is easier to work, but should be avoided because of its reactivity both with water and CO_2. Semi-rigid butyl rubber tubing with a thin stainless steel or polyethylene-coated aluminium liner provides, in theory, an ideal tubing (e.g. Dekaron Tubing, Eaton Corp., Dayton, OH, USA) since it combines the flexibility of plastics with the good surface and permeability properties of metals.

(c) Plastics:

Although few plastics approach the excellent low permeabilities and adsorptivities of some metals, their use in some parts of the system is unavoidable and invariably more convenient. Since at least some part of the assimilation chamber must be transparent, a plastic or glass window is essential. Plastics are generally easier to work, and the wide availability of adhesives for joining plastics can mean that mechanical workshop facilities are not essential for chamber construction. Most plastics can also be vacuum-formed so that almost any chamber design can be made. There is a wide variety of types of flexible plastic tubing and connectors which simplify system construction. Properties of some plastics and rubbers are listed in Table 9.2. These figures are intended only as a guide. Properties vary considerably depending on density and method of manufacture.

Acrylic plastics ('Perspex' or 'Plexiglass') have been used widely in the construction of chambers. They transmit 92% of light uniformly in the wavelengths 400–800 µm [19]. They are also light, easy to cut and simple to bond. However, water adsorption is high and thermal conductivity low. This second property makes the temperature of an acrylic plastic chamber difficult to control, unless a large heat exchanger is added. Water and CO_2 adsorption represent a very significant problem. Bloom *et al.* [19] showed that rates of CO_2 and water vapour adsorption in such a chamber are large enough to create errors in excess of 50% for estimates of both **A** and **E**; significant wall fluxes of both CO_2 and water vapour were apparent even four hours after changing the chamber temperature. Absorption by acrylic plastics may be reduced by coating the internal surfaces with a material of lower water permeability. A

Table 9.2. Properties of plastics and rubbers used for some common types of tubing in gas exchange systems*

Material	Trade names	Permeability† (nmol mm s^{-1} m^{-2} Pa^{-1}) CO_2	H_2O	Water absorption (mmol kg^{-1} day^{-1})
Plastics				
Polyethylene (low density)		1.5	42.0	<5
Polyethylene (high density)		0.6	5.4	0
Polyprolylene		0.6	22.2	<2
Polyvinyl chloride (soft)	Tygon	0.7	–	128
Polyvinyl chloride (hard)		0.04	55.5	–
Polyamide 6	Nylon 6	0.03	560.	5300
Polyamide 12	Nylon 12	0	72.2	140
Polycarbonate		2.2	560.	–
Polytetrafluoroethylene	Teflon PTFE	0.3	7.	<10
Fluorinated ethylpropylene	Teflon FEP	1.7	10.	<2
Polytrifluorochloroethylene	Plaskon CTFE	0.02	–	0
Rubbers				
Polyisoprene	Natural rubber	23.	–	270
Polychloroprene	Neoprene	4.5	–	270
Poly(dimethylbutadiene)	Methyl rubber	1.3	–	170

* Permeabilities of plastics calculated from von Oberbach (1975). Data on rubbers and absorption of water vapour after Bloom *et al.* (1980).
† Permeability expressed as the product of net amount diffusing and wall thickness per unit time, surface area and applied pressure.

PTFE coating provides one solution to this problem.

For tubing, reference to Table 9.2 shows that CTFE has excellent properties with respect to both water absorption and CO_2 permeability. PTFE has a very low water absorptivity, whilst Nylon 12 has a low CO_2 permeability. It must be appreciated that the values given in Table 9.2 are only the mid-points of ranges. Different batches of plastics from the same manufacturer can vary. Plastics vary considerably depending in particular on the quality of the tubing. Poorly manufactured tubing may have small holes, making properties such as water adsorption irrelevant. Temperature will also have a marked effect on permeability; in PTFE this increases five-fold between 23°C and 35°C. The density of the tubing also has an important effect: the higher density polyethylene and nylon have lower permeabilities and adsorptivities. Soft polyvinyl chlorides, which are widely used in gas exchange systems, adsorb significant amounts of water vapour (0.25% of their weight) and consequently CO_2. They have the further disadvantage that volatile plasticisers are released slowly; these not only support microbial growth but have IR absorption spectra which coincide with CO_2 [19].

9.5.7 Air conditioning

An essential part of any system is a means of controlling the concentration of gases entering the chamber, particularly CO_2, O_2 and water vapour. It is also necessary in most environments to control the concentration of air pollutants such as ozone, sulphur dioxide and nitrogen oxides entering the chamber. The level of air pollutants can be kept very low by filtering the air through columns with active charcoal and, e.g. Purafil™ (Purafil II, Purafil Inc, Norcross, CA, USA).

(a) CO_2 mole fraction:
Since the concentration of CO_2 is often limiting to **A**, its precise control is essential. A more practical problem is that rapid fluctuations in the CO_2 mole fraction of air supplied to the chamber will be apparent as random noise in the determination of CO_2 differential with an IRGA, so decreasing the accuracy of determination of assimilation rate. In a closed system CO_2 cannot be controlled, whereas in a semi-closed system CO_2 concentration is controlled by definition. However, in open systems, control of CO_2 concentration can become far more complex. Where only the atmospheric CO_2 level is required (ca. 354 μmol mol^{-1} of CO_2 in air), air from outside the system can be used. However, the inlet must be distant from any source of CO_2 (chimneys, combustion engines, people). In practice an inlet at a height of 4 m above the ground will usually prove adequate; however temperature inversions can result in large changes in c_a even at 4 m. Even when distant from obvious sources of CO_2 pollution, small fluctuations in atmospheric CO_2 concentration occur. These can be damped by passing the air intake through two or three large containers linked in series; the actual volume required depends on the flow rate into the system. Until the last decade most of the units that have been described for field use were large and transportable only within a mobile laboratory [4]. The ASU (ADC) air-supply system provides one portable alternative, although its maximum delivery flow of about 10 cm^3 s^{-1} limits its use to single leaves or small plant chambers. It consists of a lightweight telescopic mast, extensible to 4 m. Air is drawn down the mast by a pump capable of continuous operation for more than 5 h using an internal 12 V rechargeable battery. Either the whole or a portion of the air stream may be passed through two columns, which may contain a drying agent and/or a CO_2 absorber. By varying the flow through these columns a range of CO_2 and humidities, up to atmospheric mole fractions, may be generated in the chamber inlet stream.

An alternative method of controlling CO_2 concentration is to provide the leaf with air from compressed gas cylinders. A range of CO_2 concentrations may be generated from one cylinder by use of critical flow orifices, to divert a precise portion of the flow through a CO_2 absorber or to mix gas from a second cylinder of CO_2-free air [6]. Mass flow controllers coupled with cylinder supplies allow the generation of a range of CO_2 and O_2 mole fractions, as in the Minicuvette System (H. Walz), although with an obvious decrease in portability.

(b) Humidity control:
Water vapour concentration has an important influence on stomatal opening. The water vapour pressure deficits (VPD) that can be withstood, without detriment to the CO_2 uptake rate (**A**), vary from species to species. In many mesophytes stomatal closure, sufficient to depress **A**, begins at a VPD of >1.0 kPa. Therefore, care should be taken that VPD does not become inhibitory within the gas exchange system. In the laboratory, humidity may be controlled by bubbling the air through water at a known temperature. A more efficient system is to pass the air first through water well above the required dew-point and then to bubble the air through water at the required temperature. This second bubbler acts as a condenser. Similarly, the wet air stream can also be passed over a surface in a tube submerged in a temperature-controlled water bath. Alternatively, water may be replaced by hydrated salt crystals with a high equilibrium water vapour pressure (e.g. ferrous sulphate crystals [21]). In many instances in field studies, the humidity required will be that of the ambient air; thus if there are no materials within the system which absorb water vapour the humidity of air supplied to the chamber will be that required. However, transpiration by the enclosed leaf will raise the actual chamber humidity; in a well-stirred chamber the increase could be very significant. It may therefore be necessary to partially dry the air

before it enters the chamber, to compensate for the humidity rise resulting from enclosure. Silica gel is commonly used for drying and has the advantage of easy regeneration. However, it does absorb CO_2 and is not appropriate where CO_2 concentration is being varied. Magnesium perchlorate or zinc chloride provide effective driers which will equilibrate more rapidly with any change in CO_2 mole fraction [2]. If dry air from cylinders is used an efficient humidification system is essential. The design of a humidifier-condenser for use with field and laboratory systems is given in Field *et al.* [4].

9.6 ANALYSIS OF GAS EXCHANGE MEASUREMENTS

Studies of CO_2 exchange can provide an *in vivo* probe of limitations to photosynthetic carbon assimilation, allowing quantitative assessments of the effects of environmental variables on different steps in the diffusion pathway. From simultaneous measurements of CO_2 and water vapour fluxes it is possible to separate stomatal limitations from limitations within the mesophyll, and to separate effects on the light-limiting and CO_2-limiting phases of photosynthesis.

9.6.1 Resistance analogies

These have been applied extensively to gas exchange measurements in analyses of the effects of environmental variables on assimilation rate and on the fluxes of other gases from leaves [2,22]. A net influx of CO_2 to the leaf occurs in photosynthesis because a diffusion gradient exists between the atmosphere and the sites of photosynthetic CO_2 assimilation within the mesophyll. By applying Fick's Law of diffusion it may be shown that the net flux of a gas in a one-dimensional diffusion pathway is proportional to the ratio of the concentration difference to the diffusion resistance across that pathway. Thus, in photosynthetic CO_2 assimilation and transpiration:

$$\mathbf{A} = (c_a - c_c)/\Sigma r \tag{9.10}$$

$$\mathbf{E} = (w_i - w_a)/\Sigma r' \tag{9.11}$$

where Σr and $\Sigma r'$ are the total resistances to transfer of CO_2 and water vapour, respectively, across these gradients.

These equations are analogous with Ohm's Law, such that $(c_a - c_c)$ and $(w_i - w_a)$ are analogous to the potential difference and **A** and **E** are analogous to the electrical current. The diffusion pathway into the leaf may be divided into a number of discrete stages, each analogous to a resistor in an electrical circuit. However, most recent studies have preferred to use conductance (g), i.e. the reciprocal of resistance [4,22]. The major advantage of conductance, as an expression of limitations to gas exchange, is that it is directly proportional to flux (i.e. the terms **A** and **E**), so its interpretation is simpler. In earlier studies, conductance was expressed in units of $m\,s^{-1}$. These units were determined by those used to express the flux and concentration gradient, the latter being described in the older literature by the amount of gas per unit volume. An alternative method of expressing conductance results from expression of the gradient as the difference in mole fraction of the gas. Since moles are the units of numerator and denominator, the dimensions of conductance will be those of flux.

9.6.2 The gaseous diffusion pathway

(a) Gas phase conductance and transpiration:
To reach the leaf, CO_2 must first diffuse through the boundary layer. Boundary layer conductance (g_b) is a function of the aerodynamic properties of the leaf and leaf chamber, wind speed and turbulence. The empirical determination of g'_b was described earlier (Section 9.5.3). Under field conditions the boundary layer conductance for many crop plants and mesophytes will be at least

an order of magnitude greater than the maximum stomatal conductance (g_s) [22].

Since the stomata are considered to be the dominant limitation to diffusion of CO_2 in the gas phase, this conductance is often referred to as the stomatal conductance: however gas phase conductance of the leaf (g_l') provides a less ambiguous term. Here, the term stomatal conductance is used to refer to diffusion from the internal air space to the leaf outer surface. By measuring the flux of water vapour and the atmospheric humidity, and by assuming that the humidity of air at the mesophyll-internal air space interface is saturated at the leaf temperature, g_l' may be calculated:

$$g_l' = \mathbf{E}/(w_{s[T]} - w_a) \quad (9.12)$$

where $w_{s[T]}$ at the leaf temperature is assumed to equal w_i.

The conductance to CO_2 through this pathway (g_l) has then been assumed to be g_l' divided by the ratio of the binary diffusivities of water vapour/air and CO_2/air, with 1.6 as the accepted value for this ratio [4]. However, the ratio in the boundary layer will be lower because molecular transfer will be by both diffusion (dependent on molecular size) and turbulent transfer (independent of molecular size). A ratio of 1.37 has been suggested appropriate to the boundary layer [2]. The gas phase conductance to CO_2 will then be given by:

$$1/g_l = 1.61/g_s' + 1.37/g_b' \quad (9.13)$$

where:

$$g_s' = g_l' \cdot g_b'/(g_b' - g_l') \quad (9.14)$$

From considerations of diffusion alone it may be assumed that mean CO_2 mole fraction at the mesophyll cell wall-internal air space interface (c_i) will be:

$$c_i = c_a - \mathbf{A}/g_l \quad (9.15)$$

The volumetric net efflux of gas, predominantly water vapour, from a photosynthesising leaf will normally exceed the total influx so that a pressure gradient will exist, driving a mass flow of gases including CO_2 out of the leaf and so depressing c_i relative to c_a. This will occur even in the absence of any consumption of CO_2 within the mesophyll. Thus a correction to Equation 9.15 which takes account of the transpiration rate is necessary [1]:

$$c_i = \frac{(g_l - \mathbf{E}/2) \cdot c_a - \mathbf{A}}{g_l + \mathbf{E}/2} \quad (9.16)$$

Calculations of c_i assume that conductance is uniform across the leaf. Following application of abscisic acid or severe water stress, closure of stomata in patches has been observed. This will affect the calculation of c_i [23]. Such patchy closure will be of greatest significance where the internal air spaces of the leaf are not continuous, and in particular, within the leaves of dicotyledonous species where net venation may isolate pockets of mesophyll [10,23]. A mathematical approach to examining the potential significance of this patchy closure of stomata on c_i is provided by Cheeseman [24].

(b) Feedback control of stomatal conductance by CO_2 assimilation:

Earlier studies used the resistance analogy approach to determine the relative importance of stomatal limitations. By assuming c_c to be zero, Σr could be assumed to represent the total diffusion resistance to CO_2 assimilation (Equation 9.10). Thus, the relative limitation imposed by the stomata and cuticle would be $1/(g_s \cdot \Sigma r)$. To be correct, the response of **A** to c at all points in the diffusion pathway must be linear. The available evidence suggests that this condition would never be satisfied [22]. The response of **A** to c_i shows a hyperbolic or asymptotic response (Fig. 9.10); thus when **A** approaches saturation its value will be independent of g_s. The resistance analogy also ignores the fact that stomatal aperture and the capacity of the mesophyll for CO_2 assimilation may be closely linked. Stomata may balance the need for the leaf to allow the entry of CO_2 for

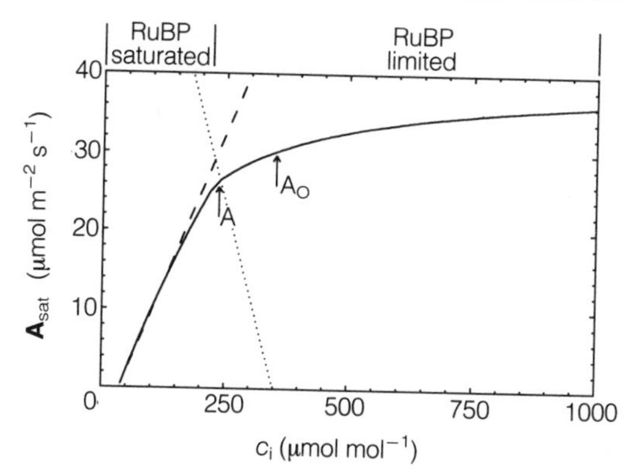

Fig. 9.10. A generalised response of the light saturated CO_2 assimilation rate ($\mathbf{A}_{sat}$) to leaf internal CO_2 mole fraction (c_i) as described by equation 9.18. $\mathbf{A}_{sat}$ on the initial slope of the curve is limited by the carboxylation efficiency (dashed line) and is therefore RuBP saturated. Beyond the inflection of the curve, $\mathbf{A}_{sat}$ is assumed to be limited by the potential rate of regeneration of RuBP and is therefore RuBP limited. Point A represents the operating point, i.e. the rate of CO_2 assimilation that would be achieved given an external CO_2 concentration (c_a) of 354 μmol mol^{-1} and the actual stomatal conductance; in this example g_s = 0.235 mol m^{-2} s^{-1}. Point A_o illustrates the $\mathbf{A}_{sat}$ that would be achieved if there was no stomatal restriction, i.e. $g_s = \infty$ and $c_i = c_a$. The dotted line indicates the 'supply function', the c_i obtained for a given g_s with varying $\mathbf{A}_{sat}$.

photosynthesis whilst limiting the transpiratory loss of water vapour [25]. This compromise is effective if conditions favouring rapid CO_2 assimilation tend to raise g_s, such that $\delta\mathbf{E}/\delta\mathbf{A}$ remains constant [25]. Wong *et al.* [26] hypothesised that rather than g_s influencing the rate of CO_2 assimilation, the capacity of the mesophyll to fix CO_2 influences g_s. An alternative method of assessing stomatal limitation is therefore necessary, and is provided by the response of **A** to c_i.

9.6.3 The A/c_i response

Here **A** is plotted against CO_2 concentration or mole fraction at the mesophyll cell surface (c_i); c_i is determined according to Equation 9.16. The plot illustrates the response of assimilation rate to CO_2 concentration in the absence of stomatal limitations. The response has had two important applications: (1) as an alternative method of separating stomatal from mesophyll limitations; and (2) in separating *in vivo* carboxylation from electron transport limitations within the mesophyll.

(a) Separation of stomatal and mesophyll limitations:
Farquhar and Sharkey [24] developed a simple method of separating stomatal and mesophyll limitations using the **A**/c_i response. Assimilation rate (**A**), measured at the normal atmospheric CO_2 concentration (c_a = 354 μmol mol^{-1}), is subtracted from $\mathbf{A}_o$, the rate which would occur if there was no stomatal limitations (i.e. the value of **A** interpolated from the response curve at c_i = 350 μ mol mol^{-1}; Fig. 9.10). The relative limitations (l) which the stomata impose, may then be calculated:

$$l = (\mathbf{A}_o - \mathbf{A})/\mathbf{A}_o \qquad (9.17)$$

Stomatal limitation (l) is the proportionate decrease in **A** that may be attributed to the stomata. The method has the advantage, when calculated graphically, that it makes no assumptions about the shape of the response of **A** to c_i.

(b) Separation of limitations due to carboxylation and to capacity for regeneration of RuBP:
From a steady-state model of photosynthetic carbon metabolism, it has been suggested that the **A**/c_i response should consist of two phases, an initial linear response where the efficiency of carboxylation (i.e. amount of active ribulose-1,5-bisphosphate carboxylase-oxygenase, Rubisco) determines the slope $\delta\mathbf{A}/\delta c_i$, followed by an inflection to a slower rise where $\delta\mathbf{A}/\delta c_i$ approaches zero because **A** is limited by the supply of substrate (ribulose-1,5-bisphosphate, RuBP) for carboxylation [27] (Fig. 9.10). Following the model of Farquhar *et al.* [27] and a subsequent modification [28] it may be stated

that **A** is related to c_i in the following manner:

$$\mathbf{A} = \left[1 - \frac{o_i}{2\tau \cdot c_i}\right] \cdot \min\{W_c, W_j, W_p\} - \mathbf{R_d} \tag{9.18}$$

where o_i = O_2 concentration in the leaf intercellular air space ($\mu mol\, mol^{-1}$)
min = selection of the minimum value between {and};
W_c = Rubisco-limited rate of carboxylation ($\mu mol\, m^{-2}\, s^{-1}$);
W_j = electron transport limited rate of regeneration of RuBP ($\mu mol\, m^{-2}\, s^{-1}$);
W_p = inorganic phosphate limited rate of regeneration of RuBP ($\mu mol\, m^{-2}\, s^{-1}$);
$\mathbf{R_d}$ = dark respiration rate ($\mu mol\, m^{-2}\, s^{-1}$);
τ = specificity factor for Rubisco carboxylation (dimensionless).

W_c is assumed to obey Michaelis–Menten kinetics:

$$W_c = \frac{V_{c,max} \cdot c_i}{c_i + k_c(1 + o_i/k_o)} \tag{9.19}$$

where $V_{c,max}$ = maximum rate of carboxylation ($\mu mol\, m^{-2}\, s^{-1}$);
k_c, k_o = Michaelis constants for CO_2 and O_2, respectively ($\mu mol\, mol^{-1}$ and $mmol\, mol^{-1}$, respectively).

At low values of c_i, **A** will be limited by W_c (Equation 9.18), since capacity for regeneration of RuBP will be in excess here. Since the affinities of Rubisco for CO_2 and O_2 are considered to vary little between terrestrial C_3 species, the main determinant of **A** at low c_i will be $V_{c,max}$ which will be directly dependent on the quantity of active Rubisco *in vivo*. The initial slope of the **A**/c_i response, or carboxylation efficiency (ce; Fig. 9.10), therefore provides an *in vivo* measure of the activity of Rubisco ($mol\, m^{-2}\, s^{-1}$). Subsequent studies have shown close agreement between the initial slope of the **A**/c_i curve, or ce, predicted by the method of Farquhar *et al.* [27] and the extractable activity of Rubisco [22]. This initial phase of the **A**/c_i response is therefore considered to be limited in C_3 species by the activity of Rubisco and saturated by RuBP (Fig. 9.10).

Beyond the inflection of the **A**/c_i response (Fig. 9.10), **A** is constrained by the capacity of the leaf to regenerate RuBP for carboxylation, and is therefore limited by the concentration of RuBP (Fig. 9.10). Two factors may limit **A** at these higher c_i values: (1) W_j, the potential rate of non-cyclic electron transport, which is a function of the photon flux, size of the apparatus for light capture, and efficiency of energy transduction on the photosynthetic membrane [27]; and (2) W_p, the rate of regeneration of RuBP that may be supported by the available inorganic phosphate [28]. The latter may be particularly important

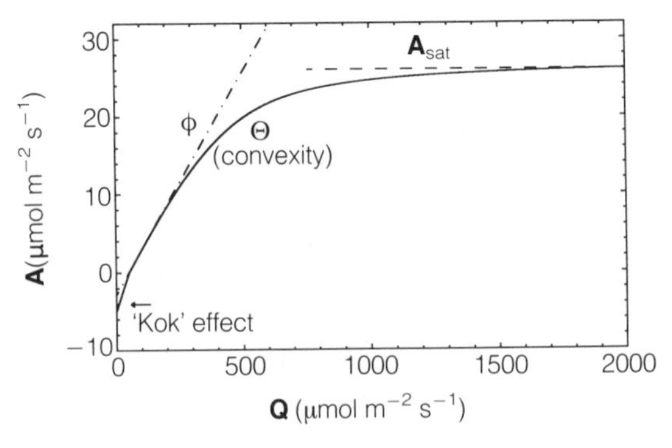

Fig. 9.11. An idealised response of CO_2 assimilation rate (**A**) to photon flux density (**Q**). This illustrates the four possible phases of the curve: (1) an initial rapid increase in **A** with increase in **Q** below the light-compensation point, termed the 'Kok' effect; (2) a longer linear phase in which $\delta\mathbf{A}/\delta\mathbf{Q}$ is considered to represent the maximum quantum yield of photosynthesis (ϕ); (3) a transition from light limited to light saturated photosynthesis described by the convexity of the curve (Θ; equation 9.21); and (4) the light saturated rate of **A** ($\mathbf{A_{sat}}$).

under conditions which lead to an accumulation of soluble carbohydrates in the leaf, e.g. rising atmospheric CO_2 concentrations [28]. The upper portion of the $\mathbf{A}/c_i$ response therefore provides an *in vivo* measure of the maximum capacity of the leaf to regenerate RuBP at that point in time. Beyond the inflection of the $\mathbf{A}/c_i$ response an increase in **A**, albeit at a lower $\delta\mathbf{A}/\delta c_i$, is apparent if W_j is limiting, even though **A** is not limited here by carboxylation efficiency. This results from the continued increase in c_i, whilst o_i remains constant (assumed to be $209\,\mathrm{mmol\,mol^{-1}}$), since the proportion of carbon lost in photorespiration will continue to decline, as predicted by Equation 9.18.

9.6.4 Light response

The response of **A** to photon flux (**Q**) describes a curve or curvilinear progression, consisting of several phases (Fig. 9.11): a respiration rate in the dark and an increase with light through the light compensation point (Fig. 9.11). The so-called Kok effect [8] may be observed as an initial rapid increase in **A** with **Q** up to the light compensation point. This is assumed to result from a progressive suppression of dark respiration. Beyond this point a linear response of **A** to **Q** persists to between 50 and $200\,\mu\mathrm{mol\,m^{-2}\,s^{-1}}$ [29,30]. This slope is assumed to correspond to the maximum apparent quantum yield of photosynthesis (ϕ). Beyond this linear phase a progressive rate of bending (convexity = Θ) in the slope of the light response curve ($\delta\mathbf{A}/\delta\mathbf{Q}$) (Fig. 9.11) is observed, until a plateau is reached beyond the light saturation point of photosynthesis. This value of **A** is termed the light-saturated rate of CO_2 uptake ($\mathbf{A}_{sat}$) (Fig. 9.11). In some cases the response curve may approximate to a rectangular hyperbola:

$$\mathbf{A} = \frac{\mathbf{A}_{sat}\cdot\mathbf{Q}}{k_Q + \mathbf{Q}} \qquad (9.20)$$

Here convexity (Θ) is zero. If the curve showed an ideal Blackman response, then Θ would be unity. In reality, Θ lies between these two extremes and the curve may be expressed by a quadratic equation, after Prioul and Chartier [31]:

$$\mathbf{A} = \frac{\mathbf{Q}\cdot\phi + \mathbf{A}_{sat} - [(\phi\cdot\mathbf{Q} + \mathbf{A}_{sat})^2 - 4\phi\cdot\mathbf{A}_{sat}\cdot\Theta]^{0.5}}{2\Theta} - R_d \qquad (9.21)$$

A can be expressed on any basis without affecting the value of Θ, and Θ is independent of differences in absorbed light. Thus Equation 9.21 can be used to compare values for Θ for light response curves for a wide range of plant material. It follows from Equation 9.21, that Θ will determine the extent to which Φ may influence **A** beyond the initial slope of the light response curve of photosynthesis and the extent to which $\mathbf{A}_{sat}$ may influence **A** at photon fluxes below light saturation.

(a) Light saturated rate of CO_2 assimilation ($\mathbf{A}_{sat}$):

Net photosynthesis responds hyperbolically to **Q** as light becomes of decreasing importance as a limiting factor. Individual leaves of many C_3 plants are unable to use additional light above **Q** = ca. $500\text{–}1000\,\mu\mathrm{mol\,m^{-2}\,s^{-1}}$, or roughly 25–50% of full sunlight; this is not true of C_4 plants, which in general fail to saturate even at full sunlight. The light-saturated assimilation rate ($\mathbf{A}_{sat}$) may be considered as a measure of the photosynthetic capacity of the leaf. In contrast with ϕ, $\mathbf{A}_{sat}$ varies markedly both within and between C_4 and C_3 species. Maximum rates of photosynthesis of C_4 plants usually exceed values for C_3 plants, whilst rates for C_4 grasses ($32\text{–}66\,\mu\mathrm{mol\,m^{-2}\,s^{-1}}$) are the highest recorded [10,22]. $\mathbf{A}_{sat}$ varies with almost all environmental variables which influence photosynthesis. It also varies markedly with preconditioning, leaf age and ontogeny. Considerable variation in $\mathbf{A}_{sat}$ may be found within one genotype, or even between leaves within the same plant.

(b) Quantum yield and its determination:
Quantum yield (ϕ) is the efficiency of light utilisation by photosynthesis, the number of moles of CO_2 fixed per mole quanta absorbed by a leaf. Since light becomes of less importance as a factor limiting photosynthesis with increasing **Q**, the maximum quantum yield can only be measured at low **Q**, when photosynthesis is strictly light limited and proportional to **Q** (Fig. 9.11). In practice this is measured at quantum fluxes from ca. 20 to 150 μmol m^{-2} s^{-1}.

The initial slope of the light response curve (ϕ) may be described as the apparent maximum quantum yield. The qualification 'apparent' is important since the estimate is based on incident and not absorbed light; if reflected and transmitted light are taken into account, the true maximum quantum yield (ϕ_{abs}) may be obtained. ϕ_{abs} may be determined in two ways. The first is to obtain ϕ from the initial slope of a light response curve determined in a conventional leaf chamber. Subsequently, the absorptance (α_l) of the leaf is determined, e.g. following the procedure of Rackham and Wilson [32], and ϕ_{abs} estimated:

$$\phi_{abs} = \phi/\alpha_l \qquad (9.22)$$

This assumes that α_l is constant and the same under the conditions of illumination within the leaf chamber and in the separate integrating sphere in which absorptance is determined. As noted earlier, this method also uses direct illumination of the leaf from above, when at the low photon fluxes relevant to the determination of ϕ, light in the natural environment would be largely diffuse. An alternative approach is to determine CO_2 in a transparent leaf chamber incorporated into an Ulbricht integrating sphere [18,30,33]. This allows the simultaneous measurement of light and CO_2 uptake.

To determine quantum yield within an integrating sphere leaf chamber the rates of CO_2 and photon absorption must be determined. For each discrete light level (i) the flux of CO_2 into the leaf $J_{c,i}$ may be determined as:

$$J_{c,i} = (c_e - c_o) \cdot u \qquad (9.23)$$

Note that leaf surface area is not required and thus the flux is irrespective of area. To determine the rate of photon absorption (J_Q) it is first necessary to determine the absorptance of the empty sphere. This should be done at regular intervals. To determine sphere absorptance (α_s), the empty sphere is illuminated and the photon flux on the wall recorded ($\mathbf{Q}_{empty}$). A black absorber is placed in the centre of the chamber in the orientation assumed by the leaves placed in the chamber, and the photon flux on the wall again recorded ($\mathbf{Q}_{absorber}$). The black absorber may be a paper leaf replica of known area painted with optically black paint. Sphere absorptance is given by:

$$\alpha_s = \alpha_a \cdot s_a/[s_s(\mathbf{Q}_{empty} - \mathbf{Q}_{absorber})] \qquad (9.24)$$

Once α_s is known then the rate of photon absorption by enclosed leaves or other photosynthetic organs may be determined, for any light level:

$$J_{Q,i} = s_s \cdot \alpha_s \cdot (\mathbf{Q}_{empty,i} - \mathbf{Q}_{sample,i}) \qquad (9.25)$$

ϕ_{abs} is obtained by determining the slope of the response of $J_{c,i}$ to $J_{Q,i}$ by finding the best-fitting straight line by the least-squares method:

$$J_{c,i} = J_{c,0} + \phi \cdot J_{Q,i} \qquad (9.26)$$

9.7 MICROMETEOROLOGICAL OR CHAMBERLESS TECHNIQUES

Photosynthesis removes CO_2 from the atmosphere around leaves and a plant canopy. However, this CO_2 is usually replenished by air movement and gas diffusion, atmospheric processes which transport CO_2 from regions of high concentration to the regions of low concentration around the photosynthesising canopy. The reverse may occur at night. Just as the product of the CO_2 difference and flow through a chamber can yield CO_2 uptake for a leaf in an open system, information of the CO_2 concentration gradient and transport across that gradient can be used to calculate the net rate of CO_2 uptake by an unenclosed stand of vegetation in the field.

Meteorological methods have been widely applied for estimation of water and heat fluxes from crop canopies. These principles can also be applied to the study of CO_2 transfer. The physical principles underlying micro-meteorological measurements have been reviewed comprehensively elsewhere [34–36].

The two most widely used methods, the *aerodynamic* and *eddy correlation* or *eddy covariance* methods, are outlined below.

9.7.1 Aerodynamic method

In this method CO_2 concentration (c_a), windspeed (u) and temperature must be measured at the same intervals within the boundary layer to establish their respective gradients with height (z) above the canopy $\delta c_a/\delta z$ and $\delta u/\delta z$. It is assumed, by analogy with molecular diffusion, that the flux (F) towards the canopy is proportional to the vertical gradient of mean CO_2 concentration, $\delta c_a/\delta z$ (averaged over several minutes, i.e. 30–60 minutes). Typically, these measurements will be made by placing the sensors on a mast over the canopy.

The flux density equation can be written:

$$F = -(\delta c_a/\delta u)u^{*2} \quad (9.27)$$

where u^* is the friction velocity:

$$u^* = k\delta u/\delta[\ln(z)] \quad (9.28)$$

These equations are developed for stable neutral conditions, i.e. where the potential temperature (actual temperature corrected for the dry adiabatic lapse rate with height, z) remains constant with height. Under neutral stable conditions, an estimate of flux may also be obtained from measurements at just two points over the canopy [35]:

$$F = \rho \cdot c_\rho \cdot k^2 \cdot \frac{(u_1 - u_2)(c_{a,1} - c_{a,2})}{\{\ln[(z_2 - d)/(z_1 - d)]\}^2} \quad (9.29)$$

where d = zero plane displacement
ρ = density of air

This has the advantage of simplifying the amount of equipment required, but the disadvantage that estimates are sensitive to error in a single instrument and local irregularities of the site. Under non-neutral conditions it is necessary to know the profile of T, as well as u, to estimate u^*. Equations developed to deal with these conditions are described by Monteith and Unsworth [35].

9.7.2 Eddy covariance

The eddy covariance method provides a direct method of measuring CO_2 exchange, which eliminates the need for time-averaging used in the aerodynamic method, because it quantifies the transport associated with each pulse of the turbulent flow of air above the canopy. Wind directly above the canopy may move in any direction, including up and down, the upward pulses transporting air depleted of CO_2 from a photosynthesising canopy, the downward pulses transporting CO_2 into the canopy. Over a period of time this turbulent flow can be expected to result in a net transfer of CO_2 from the atmosphere into a photosynthesising canopy. Rapidly responding instruments (e.g. sonic anemometers and an open-path CO_2 IRGA) may be used to measure turbulent fluctuations in vertical wind velocity and CO_2 concentration associated with individual eddies of air. Net flux is equal to the mean covariance between fluctuations in vertical wind velocity and the concentration of CO_2, which is given by the time-average of the product of the two measurements:

$$F = \overline{u'c_a} \quad (9.30)$$

Thus, if u' and c_a are independent, then $F = 0$.

9.7.3 Limitations of meteorological methods

(a) Relating fluxes to canopy photosynthesis:
In the above methods a flux of CO_2 into the canopy (F) is derived. Although canopy photosynthesis (and respiration at night) is assumed to be the major sink and source of CO_2, it is not a direct measure of $\mathbf{A}_c$. Soil respiration will tend to decrease the flux by adding CO_2 at the base of the canopy. If the efflux of CO_2 from the soil (F_{soil}) is determined independently, e.g. by open-system determination of efflux into inverted cups, then an estimate of canopy CO_2 uptake ($\mathbf{A}_c$) may be obtained:

$$\mathbf{A}_c \approx F + F_{soil} \tag{9.31}$$

(b) Relative merits and demerits of the methods:
All meteorological methods suffer two potential weaknesses. Firstly, the techniques concern exchange from an area of vegetation which is difficult to define and varies with the prevailing weather. Secondly, no method provides a direct measurement of photosynthesis, but rather of whole community CO_2 exchange [10]. A large area (i.e. several hectares) of the vegetation under consideration is required to ensure sufficient fetch, such that the observed vertical profiles result from the properties of the canopy under consideration, rather than adjacent areas. Theoretically, smaller areas may be adequate for the eddy covariance method, when measurements are made close to the canopy surface.

Despite its theoretical attractions, the eddy covariance method has rarely been used over a period of sufficient duration to advance understanding of community carbon exchange. Limitations to its use close to vegetation have generally been the availability of reliable and durable sensors for CO_2 and wind velocity. Eddy sizes typically grow with height above a surface, such that sensors operating between 0.1 and 10 Hz may be adequate for use several metres above a rough forest canopy, whilst a frequency response of 1 mHz may be needed for measurements close to a smooth surface [35]. However, the development of sonic anemometers [37] and open-path analysers [8], together with the decreasing costs of digital recording would all seem to make eddy covariance an increasingly powerful and more widely feasible technique.

The aerodynamic method is simple for neutral stability, but becomes unreliable at low wind speeds when anemometers may stall. Empirical corrections for other stability conditions appear effective over short uniform crops, but appear seriously in error over tall rough crops and forests [35].

9.8 CONCLUSION

Measurement of CO_2 assimilation provides not only a sensitive probe for determining the short-term growth of both crops and wild plants, but also a method for separating the bases of effects of the environment or genotype differences on photosynthetic capacity. Recent developments in miniaturisation (especially with respect to IRGAs) mean that this technique, once confined to the laboratory, may now be applied routinely to plants in the field over a wide range of spatial scales.

9.9 EXPERIMENTAL WORK

9.9.1 Photosynthetic gas exchange in the laboratory

An experiment which is used in the UNEP Training Courses, and can be completed by a group of students within a half-day, is the determination of light utilisation efficiency by a leaf in a range of photon flux densities.

(a) Objectives:
The objectives are to illustrate the operation, construction and calibration of an IRGA; the construction of simple closed and open gas

exchange systems; absolute and differential calibration and measurement of CO_2 concentrations with an IRGA; calculation of assimilation rate (**A**) and apparent quantum efficiency; and sources of error in gas analysis studies.

(b) Equipment required:

1. CO_2 IRGA suitable for operation in both absolute and differential mode, connected to a pen recorder or data logger.
2. Temperature-controlled leaf chamber (e.g. Leaf Section Chamber, ADC, or Parkinson Laboratory Chamber, PP Systems).
3. Cooled artificial light source capable of providing a photon flux density of 1800 $\mu mol\,m^{-2}\,s^{-1}$ at the chamber surface (e.g. 200 W slide projector).
4. Neutral density filters (or sheets of white muslin or similar material of various thicknesses).
5. Cylinder of calibration gas and, ideally, a gas diluter (e.g. GD600, ADC) or a gas-mixing pump (e.g. SA27f, Wösthoff).
6. A means of determining leaf area.
7. Supply of CO_2-free air or pure N_2.
8. Thermocouples suitable for leaf temperature measurement.
9. Quantum sensor and meter.
10. Graph paper and calculators, and a PC-compatible microcomputer with the GASEX software (Appendix E).

(c) Plant material:

Leaves ranging from fronds of *Cocos nucifera* to annual grasses such as *Digitaria sanguinalis* have been used in past UNEP Training Courses. Mature leaves of 3-week-old *Vigna sinensis* and 5-week-old *Amaranthus edulis* have proved the most successful, although almost any healthy leaves are suitable (except for succulents and other CAM species, which may show nocturnal stomatal opening). *V. sinensis* and *A. edulis* can be used in the same leaf chambers, and provide a comparison between a C_3 and a C_4 crop, respectively. These plants must be grown under unstressed conditions, e.g. in well-watered and fertilised potting compost (e.g. John Innes No. 2) or in sand, watered daily with a complete nutrient solution.

(d) Procedure:

The group of students (ideally no more than five, if they are to obtain 'hands-on' experience) is shown the construction of the IRGA, and the principle of its operation is explained. Calibration of the IRGA is then demonstrated in absolute mode using the calibration cylinder and, if available, a gas diluter or gas-mixing pump.

The zero and gain settings are then deliberately upset, and the group asked to recalibrate the IRGA. Pumping of laboratory air through the analysis cell demonstrates the high and fluctuating CO_2 concentrations encountered in laboratories, and emphasises the need to collect air for open systems at some distance from the laboratory. Students now measure the area of the leaf or leaf segment to be placed in the chamber with a leaf area meter. Alternatively, trace its outline onto graph paper, cut out, weigh, and calculate the area from the weight/area ratio for the paper. The ADC leaf section chamber encloses an area of $10\,cm^2$, so measurement is not required if the leaf fills the chamber.

Seal the chamber and construct a simple closed system. Check for leaks by breathing onto individual joints and seals. Exhaled breath contains about $50\,000\,\mu mol\,mol^{-1}$ CO_2, so even small leaks are quickly detected by the system. It is instructive to deliberately leave one or two joints loose.

Once it has been made gas-tight, record the rate of CO_2 depletion in the system. Note the time taken for CO_2 concentration to decrease from about 380 to $340\,\mu mol\,mol^{-1}$. Calculate **A** (Equation 9.4) whilst waiting for the CO_2 concentration to fall to the compensation point (an estimate of the system volume is required).

Reconnect the chamber to make an open system, with a second empty chamber as a reference. First, measure the reduction in CO_2 concentration caused by the leaf, using the IRGA in absolute mode. Compare the CO_2 concentrations of the leaf and reference streams by passing each in turn through the analysis cell of the IRGA; at best, this difference will be no more than 1–2% of full-scale deflection for most IRGAs. Now calibrate the instrument in differential mode. Connect the leaf and reference chambers to the analysis and reference cells of the IRGA, respectively. This amplifies the difference between the two gas streams, and demonstrates the value of differential gas analysis. Recalculate **A** using Equation 9.3. The value calculated for the open system will probably be higher than that obtained for the closed system with the same leaf, since a steady-state could not be obtained for the latter.

Reduce the photon flux density (**Q**) in the leaf chamber by interposing neutral density filters between the light source and the leaf. Once the leaf has reached a new steady state, **Q** may be further reduced. Use the filters to give 10–20 values of **Q**, from about 2000 $\mu mol\ m^{-2}\ s^{-1}$ to zero. Since the most rapid changes in **A** occur at low **Q**, make sure that the low light levels are well represented. As **Q** decreases, the leaf will cool; chamber air temperatures must therefore be increased to keep leaf temperature constant.

(e) Analysis of results:
See Section 9.9.3 for a listing of equations for calculations that may be made from the gas exchange data and Appendix E for instructions on the use of the GASEX software. Plot the light response curve (**A** against **Q**). Since both are calculated in $\mu mol\ m^{-2}\ s^{-1}$, efficiency of light conversion or apparent quantum yield may be calculated (ϕ = **A**/**Q**) and plotted against **Q**. This will show how efficiency decreases with increase in **Q**. Such graphs plotted for *V. sinensis* (C_3) and *A. edulis* (C_4) provide a good comparison for the UNEP Training Courses. At 20°C *V. sinensis* shows higher efficiencies at low light levels, but *A. edulis* has a higher efficiency than *V. sinensis* at light levels close to full sunlight. At temperatures above 30°C, *A. edulis* is more efficient at all light levels.

(f) Suggested additional work:
Spectral filters may be employed to show the effects of light quality on quantum efficiency and light energy conversion efficiency. The effect of high water vapour pressure deficit may be illustrated by drying the air with a column of dry calcium chloride granules before it enters the leaf chamber.

9.9.2 Photosynthetic gas exchange in the field

(a) Objective:
This experiment analyses CO_2 assimilation in leaves at various positions in a C_4 (*Zea mays*) or a C_3 canopy (*Vigna sinensis*). CO_2 assimilation is studied in relation to **Q** at these positions, and leaf stomatal conductance, transpiration rate and leaf temperature are measured at the same time.

(b) Equipment:

1. ADC LCA3 portable IRGA or LI-COR 6200 portable system.
2. ADC Parkinson broad-leaf chamber or LI-COR 6200 chambers.

(c) Plant material:
Stands of *Z. mays* and *V. sinensis* approaching flowering. However, any healthy plant with leaves larger than about 2 × 2 cm could be used.

(d) Procedure:
The construction of the portable IRGA and its principles of operation are explained before field work commences. Students should work in groups of 3–4 to gain 'hands-on' experience with the IRGA, the leaf chamber and the data handling software. Air and leaf temperature are measured separately, and all readings taken down in a notebook (even if the recorded data are stored in the logger).

Select a stand, and decide where the measurements are going to be taken. Aim for the minimum disturbance of the plants and their microenvironment (photon flux density etc.). Check the analyser and leaf chamber and all connections, as well as light sensor, humidity sensor, flow control, battery, silica gel, soda lime, etc., before the start of the measurements. If an open system is used, erect the air sampling mast and measure the ambient CO_2 concentration. If a closed system is used, be sure that air of ambient CO_2 concentration enters the chamber and that the relative humidity (and flow rate) are set correctly.

Insert a leaf into the chamber, and check for leaks; this is very critical with the closed system. Deliberately create a leak around the leaf seal to illustrate the effect of this on calculated values. Measure, record and store your measured data. It is often necessary to record an average of several readings. Take several measurements (at least 10) at each position in the canopy.

(e) Calculation and analysis of results:
See Section 9.9.3 for a listing of equations for calculations that may be made from the gas exchange data and Appendix E for instructions on the use of the GASEX software.

9.9.3 Calculations for field and laboratory gas exchange experiments

(1) Convert volumetric or mass flow to mole flow or air: follow Equation 9.7a given earlier.
(2) Calculate leaf area in m^2. The ADC broad-leaf chamber has an area of $6.25\,cm^2 = 6.25 \times 10^{-4}\,m^2$. The alternative ADC narrow-leaf chamber (intended for grass leaves) has an area of $10\,cm^3$. If the leaf does not fill the chamber, its area must be estimated separately. For narrow leaves this can be done by measuring length and width of the part enclosed in the chamber. The area of leaf exposed in the LI-COR chambers can be pre-set. In other cases the measurements should be performed as described earlier.
(3) Calculate transpiration rate, **E**:

$$E = \frac{f}{s} \cdot \frac{(w_o - w_e)}{(p - w_o)} \quad (9.32)$$

w_o and w_e are calculated from the saturation humidity ($w_{s[T]}$) at the measured leaf temperature (T), given the relative humidity:

$$w_o = w_{s(T)} \cdot RH/100 \quad (9.33)$$

$w_{s[T]}$ may be obtained from standard humidity tables or calculated approximately from:

$$w_{s[T]} = (0.61121e^{\{17.502T/(240.97 + T)\}})/p \quad (9.34)$$

where T is temperature (°C)
(4) Calculate assimilation rate, **A**: follow Equation 9.4 given earlier.
(5) Calculate total conductance:

$$g_l' = \frac{E}{w_{s[T]} - w_o} \quad (9.35)$$

g_s' may be calculated if g_b' is known (Equation 9.14).
(6) Calculate the internal CO_2 concentration: follow Equation 9.16 given earlier

DISCLAIMER

Inclusion of the name of a manufacturer or commercial instrument in this chapter is not a recommendation by the authors, nor a guarantee of the suitability of the equipment. The authors have been neither employees nor paid consultants of any of the manufacturers or suppliers named within the text.

REFERENCES

1. von Caemmerer, S. and G.D. Farquhar (1981) Some relationships between the biochemistry of photosynthesis and the gas exchange of leaves. *Planta* **153**, 376–387.
2. Sestak, Z., J. Catsky and P.G. Jarvis (1971) *Plant Photosynthetic Production: Manual of Methods*. Dr. W. Junk, The Hague, Netherlands.

3. Incoll, L.D., S.P. Long and M.R. Ashmore (1977) S.I. units in publications in Plant Science. *Current Advances in Plant Sciences* **27**, 331–43.
4. Field, C.B., J.T. Ball and J.A. Berry (1989) Photosynthesis: principles and field techniques, in *Plant Physiological Ecology: Field Methods and Instrumentation* (J.W. Pearcy, J. Ehleringer, H.A. Mooney and P.W. Rundel, eds.). Chapman and Hall, London, pp. 208–53.
5. Hällgren, J.-E. (1982) Field photosynthesis monitoring with $^{14}CO_2$, in *Techniques in Bioproductivity and Photosynthesis* (J. Coombs and D.O. Hall, eds.). Pergamon Press, Oxford, pp. 36–44.
6. Long, S.P. and C.R. Ireland (1985) The measurement and control of air and gas flow rates for the determination of gaseous exchanges of living organisms, in *Instrumentation for Environmental Physiology* (B. Marshall and F.I. Woodward, eds.), Cambridge University Press, Cambridge, pp. 123–138.
7. Bunce, J.A. and D.A. Ward (1985) Errors in differential infrared carbon dioxide analysis resulting from water vapour. *Photosynthesis Research* **6**, 289–94.
8. Jarvis, P.G. and A.P. Sandford (1985) The measurement of carbon dioxide in air, in *Instrumentation for Environmental Physiology* (B. Marshall and F.I. Woodward, eds.), Cambridge University Press, Cambridge, pp. 29–58.
9. Hill, D.W. and T. Powell (1968) *Non-dispersive Infra-red Gas Analysis*. Adam Hilger, London.
10. Long, S.P. (1989) Gas exchange of plants in the field, in *Toward a More Exact Ecology*, 30th Symposium of the British Ecological Society (P.J. Grubb and J.B. Whittaker, eds.). Blackwell, Oxford, pp. 33–62.
11. Parkinson, K.J. and B.J. Legg (1971) A new method for calibrating infra-red gas analysers. *J. Physics E. Scientific Instruments* **4**, 598–600.
12. Brain, T.J.S. and R.W.W. Scott (1982) Survey of pipeline flowmeters. *J. Physics E. Scientific Instruments* **15**, 967–80.
13. Widmer, A.E., R. Fehlmann and W. Rehwald (1982) A calibration system for calorimetric mass flow devices. *J. Physics E. Scientific Instruments* **15**, 213–220.
14. Parkinson, K.J. and W. Day (1979) Use of orifices to control the flow rate of gases. *J. Applied Ecology* **16**, 623–32.
15. Harris, G.C., J.K. Cheeseborough and D.A. Walker (1983) Measurement of gas exchange in leaf discs. *Plant Physiology* **71**, 102–7.
16. Parkinson, K.J. (1985) A simple method for determining the boundary layer resistance in leaf cuvettes. *Plant Cell and Environment* **8**, 223–6.
17. Leverenz, J.W. and P.G. Jarvis (1979) Photosynthesis in sitka spruce. *J. Applied Ecology* **16**, 919–32.
18. Ireland, C.R., S.P. Long and N.R. Baker (1989) An integrated portable apparatus for the simultaneous field measurement of photosynthetic CO_2 and water vapour exchange, light adsorption and chlorophyll fluorescence of attached leaves. *Plant Cell and Environment* **12**, 947–58.
19. Bloom, A., H.A. Mooney, O. Björkman and J.A. Berry (1980) Materials and methods for carbon dioxide and water exchange analysis. *Plant Cell and Environment* **3**, 371–6.
20. Dixon, M. and J. Grace (1982) Water uptake by some chamber materials. *Plant Cell and Environment* **5**, 323–7.
21. Day, W. (1985) Water vapour measurement and control, in *Instrumentation for Environmental Physiology* (B. Marshall and F.I. Woodward, eds.). Cambridge University Press, Cambridge, pp. 59–78.
22. Long, S.P. (1985) Leaf gas exchange, in *Photosynthetic Mechanisms and the Environment* (J. Barber and N.R. Baker, eds.), Elsevier, Amsterdam, pp. 453–500.
23. Downton, W.J.S., B.R. Loveys and W.J.R. Grant, (1988) Stomatal closure fully accounts for the inhibition of photosynthesis by abscisic acid. *New Phytologist* **108**, 263–6.
24. Cheeseman, J. (1991) PATCHY: simulating and visualizing the effects of stomatal patchiness on photosynthetic CO_2 exchange studies. *Plant Cell and Environment* **14**, 593–601.
25. Farquhar, G.D. and T.D. Sharkey (1982) Stomatal conductance and photosynthesis. *Ann. Rev. Plant Physiology* **33**, 317–45.
26. Wong, S.C., I.R. Cowan and G.D. Farquhar (1979) Stomatal conductance correlates with photosynthetic capacity. *Nature* **282**, 424–426.
27. Farquhar, G.D., S. von Caemmerer and J.A. Berry (1980) A biochemical model of photosynthetic (CO_2) assimilation in leaves of C_3 species. *Planta* **149**, 78–90.
28. Harley, P.C., R.B. Thomas, J.F. Reynolds and B.R. Strain (1992) Modelling photosynthesis of cotton grown in elevated CO_2. *Plant Cell and Environment* **15**, 271–282.
29. Björkman, O. and B. Demmig (1987) Photon yield of O_2 evolution and chlorophyll fluorescence characteristics at 77K among vascular plants of diverse origins. *Planta* **170**, 489–504.
30. Long, S.P. and B.G. Drake (1991) Effect of the long-term elevation of CO_2 concentration in the field on the quantum yield of photo-

synthesis of the C_3 sedge, *Scirpus olneyi*. *Plant Physiology* **96**, 221–6.
31. Prioul, J.L. and P. Chartier (1977) Partitioning of transfer and carboxylation components of intracellular resistance to photosynthetic CO_2 fixation: A critical analysis of the methods used. *Annals of Botany* **41**, 789–800.
32. Rackham, O. and J. Wilson (1967) Integrating sphere, in *The Measurement of Environmental Factors in Terrestrial Ecology* (R.M. Wadsworth, ed.). Blackwell, Oxford, pp. 259–63.
33. Öquist, G., J.-E. Hällgren and L. Brunes (1978) An apparatus for measuring photosynthetic quantum yields and quanta absorption spectra of intact plants. *Plant Cell and Environment* **1**, 21–27.
34. Baldomcchi, D.D., B.B. Hicks and T.P. Meyers (1988) Measuring biosphere-atmosphere exchanges of biologically related gases with micrometeorological methods. *Ecology* **69**, 1331–1340.
35. Monteith, J.L. and M.H. Unsworth (1990) *Principles of Environmental Physics*, 2nd edn. Arnold, London.
36. Nobel, P.S. (1991) *Physicochemical and Environmental Plant Physiology*. Academic Press, San Diego.
37. Grace, J. (1989) Measurement of wind speed near vegetation, in *Plant Physiological Ecology: Field Methods and Instrumentation* (J.W. Pearcy, J. Ehleringer, H.A. Mooney and P.W. Rundel, eds.). Chapman and Hall, London, pp. 57–74.
38. von Oberbach, K. (1975) *Kunstoff-Kennwerte für Kunstrukteure*. Carl Hanser, Munich.

10

Polarographic measurement of oxygen

D.A. Walker

10.1 THE OXYGEN ELECTRODE

10.1.1 Introduction

Photosynthesis by isolated organelles, protoplasts and cells, prepared as described in Chapter 17, cannot easily be followed by infra-red gas analysis. $^{14}CO_2$ fixation can provide invaluable information about products, but it does not lend itself to continuous measurement. For these reasons, routine measurements of photosynthesis are usually carried out with an oxygen electrode, which can easily detect changes of the order of 0.01 µmol in the oxygen content of a suspension. If necessary, $^{14}CO_2$ fixation can be followed simultaneously in the same vessel by withdrawing samples with a micro-syringe at appropriate time intervals. These are injected immediately into HCl, and can be determined with a Geiger or scintillation counter, or subjected to chromatography.

An oxygen electrode (Fig. 10.1) is a special form of electrochemical cell, in which a current is generated that is proportional to the activity of oxygen present in a solution. The release of gaseous hydrogen and oxygen at the surface of wires submerged in water and connected to the negative and positive poles, respectively, of a battery was demonstrated in 1800 by Nicholson and Carlisle. About 100 years later, Danneel showed that the reverse reaction was possible; i.e. that gaseous dissolved oxygen could be electrolysed to the hydroxyl ion, the ionic component of water. Furthermore, Danneel demonstrated that the current flow during this electrolytic reaction was proportional to the concentration of gaseous dissolved oxygen [2].

This is the basis of the polarographic measurement of oxygen. When the negative pole of a battery is attached to an electrode (usually platinum, but silver, gold or lead can also be used), electrons (e^-) will tend to move from the battery to this metal surface, making it negatively charged (a cathode). Similarly, when a second electrode (usually silver) is attached to the positive pole of the battery, electrons will tend to move towards the battery *from* the electrode, making it also positively charged (an anode). When the circuit is completed by a film of electrolyte (a solution such as KCl, which ionises to K^+ and Cl^-), the application of this polarising voltage (i.e. a voltage which creates new positive and negative poles) causes reduction of oxygen at the cathode, and initiation of a flow of electrons or electric current (Fig. 10.3). Traditionally, this current has been converted to a voltage signal and registered on a pen recorder. More recently it has been possible to display this signal on a computer screen, using an analogue-to-digital converter.

Photosynthesis and Production in a Changing Environment: a field and laboratory manual
Edited by D.O. Hall, J.M.O. Scurlock, H.R. Bolhàr-Nordenkampf, R.C. Leegood and S.P. Long
Published in 1993 by Chapman & Hall, London.
ISBN 0 412 42900 4 (HB) and 0 412 42910 1 (PB).

10.1.2 Aqueous-phase measurements

Figure 10.1 shows an apparatus designed for measuring oxygen in the aqueous phase. The sensor itself constitutes the floor of a small chamber used for this purpose. If no oxygen is present, the reaction at the platinum electrode will not occur and no current should flow. The platinum and reference electrodes are shielded from the solution by a thin membrane which is permeable to oxygen but impermeable to most poisons for platinum. The membrane is kept in place by a rubber 'O'-ring. Since the platinum cathode consumes oxygen (Fig. 10.3), it is necessary to stir the solution at about 500–600 rpm with a small magnetic follower or 'flea', in order to maintain a stable oxygen gradient across the membrane.

As described above, a polarizing voltage is applied to the oxygen electrode from a battery (Fig. 10.2). If the electrode output at a given oxygen activity is plotted against polarizing voltage, it is found that a plateau exists between about 0.4 V and 0.8 V (platinum negative with respect to the reference electrode). If a polarizing voltage near the centre of the plateau is chosen (−0.7 V), then the current output from the oxygen electrode is linearly proportional to the oxygen in solution.

Much of the account which follows relates to the *Hansatech* oxygen electrode (Hansatech Ltd., King's Lynn, UK), but there are in addition a great many commercially available oxygen sensors based on the original Clark principle [1]. Some of these, particularly those designed for medical use (e.g. *Drager*) have reached a high degree of sophistication [2,4]. However, the *Hansatech* electrode is the only one which was specifically designed for photosynthetic measurements. It remains the most modestly priced and reliable system incorporating all of the features which favour use in photosynthetic work. These include total illumination of the reaction chamber, a high degree of heat dissipation and infrared filtration (which is important when high photon flux densities are applied) and stirring at the cathode surface which makes for a very responsive system. Furthermore, there are now variants of the standard *Hansatech DW2* chamber which facilitate measurements of fluorescence, etc.

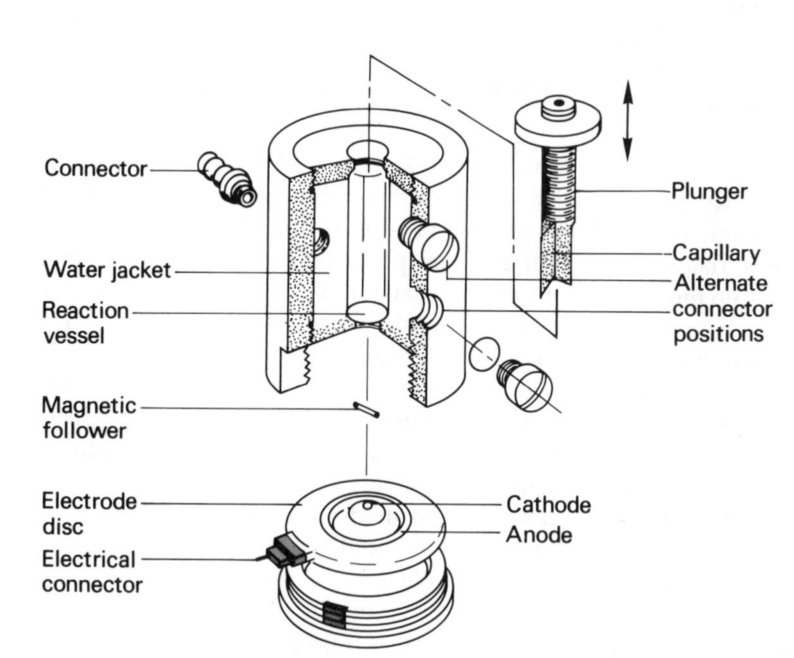

Fig. 10.1. The Hansatech DW Oxygen Electrode Unit is a high-precision laboratory instrument ideally suited for measuring the kinetics of oxygen uptake or evolution [11]. It is a Clark-type oxygen electrode based on a design by Delieu and Walker. It can be obtained from Hansatech Ltd., Hardwick Industrial Estate, King's Lynn, Norfolk PE30 4NE, U.K. Details of the instrument and its use may be found in Walker [11].

10.1.3 Characteristics of the oxygen electrode

Membranes:
Many different types of membrane have been used including collodion, cellophane, polythene, nylon, teflon, mylar, silicone rubber and 'cling-film'. Oxygen permeability of the membrane will depend on both composition and thickness, and the thinnest membrane that can reasonably be used is about 12.5 μm. The most commonly used membranes are polythene and Teflon in thicknesses of 12.5–25 μm. Thinner membranes will give increased response speed at the expense of increased fragility.

Output Current:
The current generated by an oxygen electrode is governed by the amount of oxygen reaching the platinum cathode, and by the cathode area. Thus for the given oxygen tension and membrane, a large platinum disc will give a large output current, but will consume measurable amounts of oxygen from the solution. A large cathode will require better stirring to maintain equilibrium of oxygen between the solution and the electrode, compared with a smaller cathode. For a given set of conditions and membrane type, a large cathode will give a faster response time than a small cathode.

Response Time:
The 90% response time of an oxygen electrode is defined as the length of time required for the electrode to make a 90% response to an instantaneous change in oxygen tension in the solution. Factors affecting the response time have already been mentioned, namely the thinness and composition of the membrane, the size of the cathode and the rate of stirring.

Using modern electrode designs and a 0.5–1.0 mm cathode, a response time of 1–10 seconds is normal at 25°C. A very rough check on the response time can be made by switching off the stirrer, when the electrode output will fall because oxygen diffusion across the membrane is not fast enough to balance the oxygen consumption at the cathode. After the output has fallen by about three recorder units, the stirrer can be switched back on at the same speed as before, and the time taken to recover 90% of the former signal can be measured.

Electrode Leakage or Residual Current:
In theory, a perfect electrode will give zero output when the oxygen tension in the solution is zero. In practice, small faults in the sealing of the platinum electrode into the insulator can occur; these will cause some leakage (residual) current to flow in the absence of oxygen. This can be tested for by setting up the electrode to read 100% on air-saturated water, and then adding a few crystals of sodium hydrosulphite (dithionite) ($Na_2S_2O_4$) which will chemically remove all the oxygen in the solution. Any remaining signal will then be caused by electrode leakage current. It is advisable to test electrodes regularly in this way. A large residual current (more than about 10% of the air value) will suggest that the electrode needs cleaning or that it has become unserviceable. Caution: dithionite is corrosive, so the electrode chamber should be thoroughly washed three or four times with water before further use.

Exclusion of Oxygen:
The oxygen electrode measures the rate of oxygen uptake from the solution. It is essential for this purpose that oxygen does not enter the reaction vessel at an appreciable rate. This is particularly important when monitoring reactions in which the oxygen concentration changes slowly. Atmospheric oxygen is normally excluded from the reaction vessel by a close-fitting plastic plunger with a small hole (1 mm) through which additions can be made with an 'auto-zero' type pipette or a Hamilton syringe. It is essential that no bubbles are allowed to remain under the plunger when the electrode is in use, as these will cause rapid reoxygenation of the

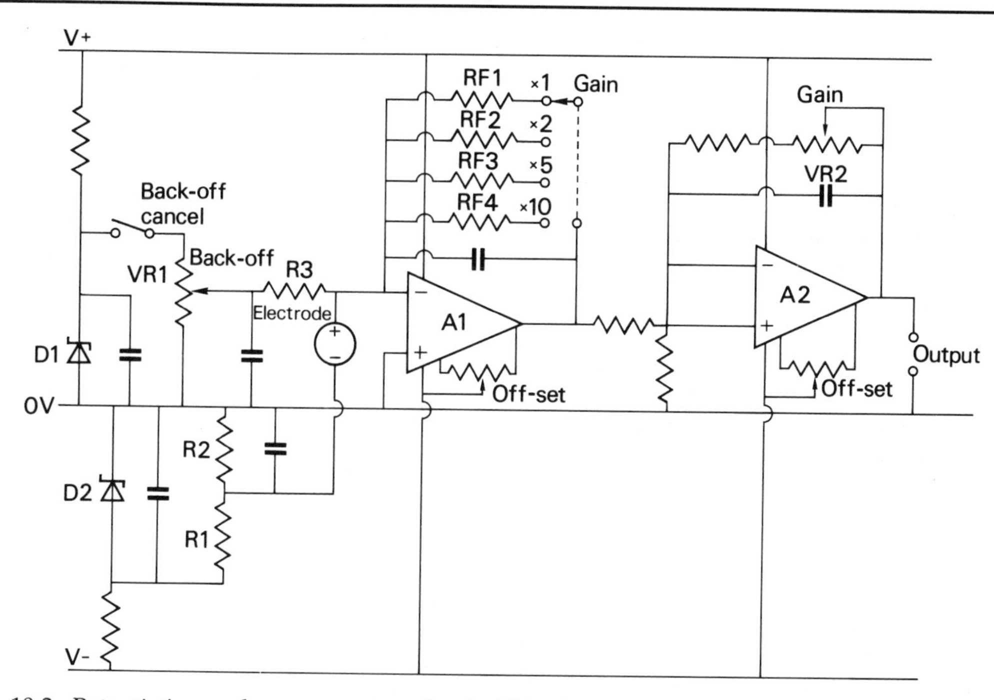

Fig. 10.2. Potentiating and zero suppress circuit. This circuit is used to provide a stable potentiating voltage of approximately 700 mV to the oxygen electrode. This is derived from the voltage reference diode D2 through the potential divider (R1, R2). The amplifier A1 forms a current-to-voltage converter which converts the sum of the currents through the electrode and R3 into a voltage. The conversion factor is determined by the feedback resistors RF1–RF4, which are selected by a gain switch. Amplifier A2 forms a non-inverting voltage amplifier, the gain of which is varied by variable resistor VR2. A back-off current is generated using VR1 to adjust the voltage across R3. This current, which opposes that produced by the oxygen electrode, can be used to 'null' the residual current produced by the electrode at zero oxygen activity. (Reproduced by kind permission of Hansatech Ltd.).

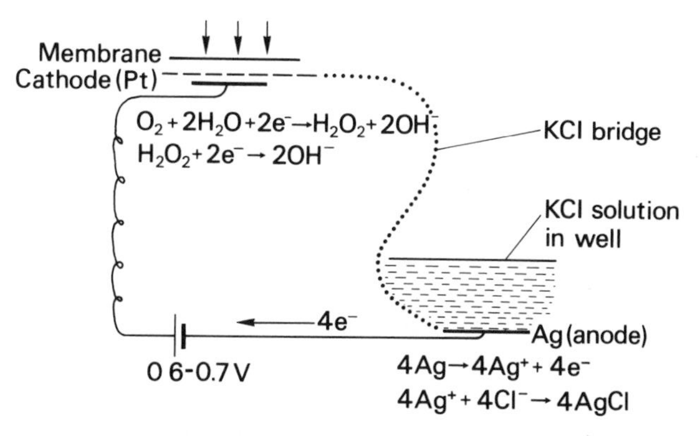

Fig. 10.3. Diagrammatic representation of electrode reactions. When the potentiating voltage is applied across the two electrodes the platinum becomes negative and the silver becomes positive. Oxygen diffusing through the membrane is reduced at the platinum surface and a current flows through the circuit (which is completed by the KCl bridge). The silver is oxidized and silver chloride is deposited. The current is stoichiometrically related to the oxygen reduced.

medium. The liquid in the chamber should therefore enter the hole in the plunger.

10.1.4 Calibration of the aqueous-phase system

It is normal to fill the oxygen electrode with air-saturated water and then use the output gain control to set the meter or the recorder to a convenient value. It should be noted that a marked drift of output occurs for the first 2 or 3 minutes, while equilibration takes place. Following the addition of dithionite (see above) to establish the residual current (i.e. the output that the electrode would give in the complete absence of oxygen) the electrical output should be readjusted so that the difference between this oxygen-free or 'nitrogen' value and the 'air' value is a convenient multiple of the number of micromoles of oxygen in solution. For example, air-saturated water at 20°C contains 0.28 μmol O_2 ml^{-1}. If you are using 2 ml of solution, it may be convenient to set the difference between the air and nitrogen values at 56 units (on the *Hansatech* control box) or 56 divisions on the recorder paper (where full scale deflection is 100 divisions). Thus a change of 0.1 μmol O_2 correspond to 10 divisions or units (since $56 = 2 \times 0.28 \times 100$). For other volumes and other temperatures the setting should be adjusted accordingly; the solubility of oxygen in water at various temperatures is shown in Table 10.1.

If absolute calibration is required, note that the solubility of oxygen is influenced by the presence of other compounds in solution. Absolute calibraton is best achieved by generating known amounts of oxygen in solution, e.g. by adding aliquots of standardised hydrogen peroxide to a dilute solution of catalase. For most purposes the effect of other solutes is small and may be disregarded.

Table 10.1. Solubility of oxygen in water with temperature

Temp. (°C)	O_2 (ppm)	O_2(μmol ml^{-1})
0	14.16	0.442
5	12.37	0.386
10	10.92	0.341
15	9.76	0.305
20	8.84	0.276
25	8.11	0.253
30	7.52	0.230
35	7.02	0.210

General procedure for setting up a liquid-phase oxygen electrode:

1. Assemble oxygen electrode with membrane, etc. and switch on stirrer at desired rate.
2. Switch on recorder, and set range to 1.0 V. Adjust recorder zero with control box switched off. Do not make any further adjustments to recorder zero.
3. Fill electrode with air-saturated water (a source of air bubbles such as an aquarium pump is useful, but do not bubble too vigorously otherwise water may be supersaturated with oxygen). Switch on control box, with polarising voltage set to −0.7 V (this is controlled automatically in the *Hansatech* control box). Set the 'output' or 'sensitivity' gain control to minimum (in the case of the *Hansatech* system, set the range to '×1' and the gain control knob fully anticlockwise). 'Back-off' controls (coarse and fine) should be set to minimum (i.e. fully clockwise on the *Hansatech* control box, so that switching between 'back-off' and 'cancel' produces no effect).
4. Adjust the 'output' gain control so that the recorder shows 90% of full scale deflection (the digital display on the *Hansatech* control box will correspondingly read about 900 mV).
5. Add a few crystals of dithionite. Oxygen concentration should immediately decrease and approach zero on the recorder scale (and the *Hansatech* digital display). Note that fresh dithionite should be used otherwise oxygen in the cell may not be reduced fully. Do not leave dithionite in contact with the electrode membrane for more than 1–2 minutes.

6. Wash the electrode with three or four changes of air-saturated water. The recorder pen (and the *Hansatech* digital signal) should now return to the original 90% mark. If this does not happen, wash the electrode further to remove all traces of dithionite.
7. The electrode is now ready for studying oxygen uptake (either in the light or dark). To study oxygen evolution, use the back-off control to adjust the recorder pen to about 10% of full scale deflection before starting the reaction. If only small changes in oxygen concentration are being studied, the *Hansatech* gain control may be set to '×2', '×5' or '×10' to increase the sensitivity.

10.2 THE LEAF DISC ELECTRODE

10.2.1 Introduction

Photosynthesis by whole leaves is usually followed by infra-red gas analysis (Chapter 9). However, the polarographic detection of oxygen with the leaf disc electrode [1] is a cheap alternative which can provide useful additional information (for example, the one-to-one relationship between oxygen evolved and carbon dioxide fixed does not *always* hold). The leaf disc electrode can also be used for measurements which are not readily undertaken with a conventional infrared gas analyser (IRGA). It is ideal for maximal rate measurements because it can be used under conditions of saturating CO_2. Similarly, rate vs. light intensity measurements ('quantum requirement') can be carried out very easily. This account is based on the Hansatech LD2 leaf disc electrode, which is light, robust and portable. This instrument can obtain its power from a car battery, enabling it to be used under field conditions. As with certain other electronic equipment, operation from a direct current source may be preferable even in the laboratory, since it ensures freedom from errors introduced by the 'spikes' which are present in most mains electricity supplies.

If a leaf is enclosed in a chamber, provided with CO_2 (or bicarbonate as a source of CO_2) and then illuminated, oxygen will be evolved. In the Hansatech LD2 (Fig. 10.4), a leaf disc is used, and CO_2 provided usually in the form of sodium bicarbonate (which dissociates in solution):

$$NaHCO_3 \rightarrow NaOH + CO_2$$

The oxygen which accumulates in the gas-phase during photosynthesis is then detected, polarographically, by a 'Clark-type' electrode.

10.2.2 Calibration

In a closed system, the amount of oxygen is governed by the partial pressure of O_2 and the volume of the chamber. Thus 1.0 ml (1000 µl) of atmospheric air contains 210 µl O_2. If 1.0 ml of the space in a 5.0 ml chamber is occupied by a leaf disc, the increase in signal which follows the introduction of a given volume of air into the chamber will be correspondingly greater than in the absence of a leaf. For most purposes, therefore, the recorder 'excursions' following the introduction (or removal) of small volumes of air into the chamber (using a gas-tight syringe) can be taken as the basis of calibration, and all that remains to be done is some simple arithmetic (Section 10.2.3).

Using a 2 ml gas-tight syringe, introduce successive 200 µl aliquots of air into the leaf disc chamber through one of the two gas vents. (The other should be kept closed during this procedure.) As each volume of air is pushed into the chamber, the reading on the recorder should rise quickly to a new level and stay there (Fig. 10.5).

Care should be taken to avoid over-tightening the threaded base of the chamber which pushes the electrode disc against its sealing O-ring. Once an effective seal is established, excessive tightening can constrict the layer of electrolyte between the anode and the cathode, leading to loss of linearity of response. It should also be noted that if this

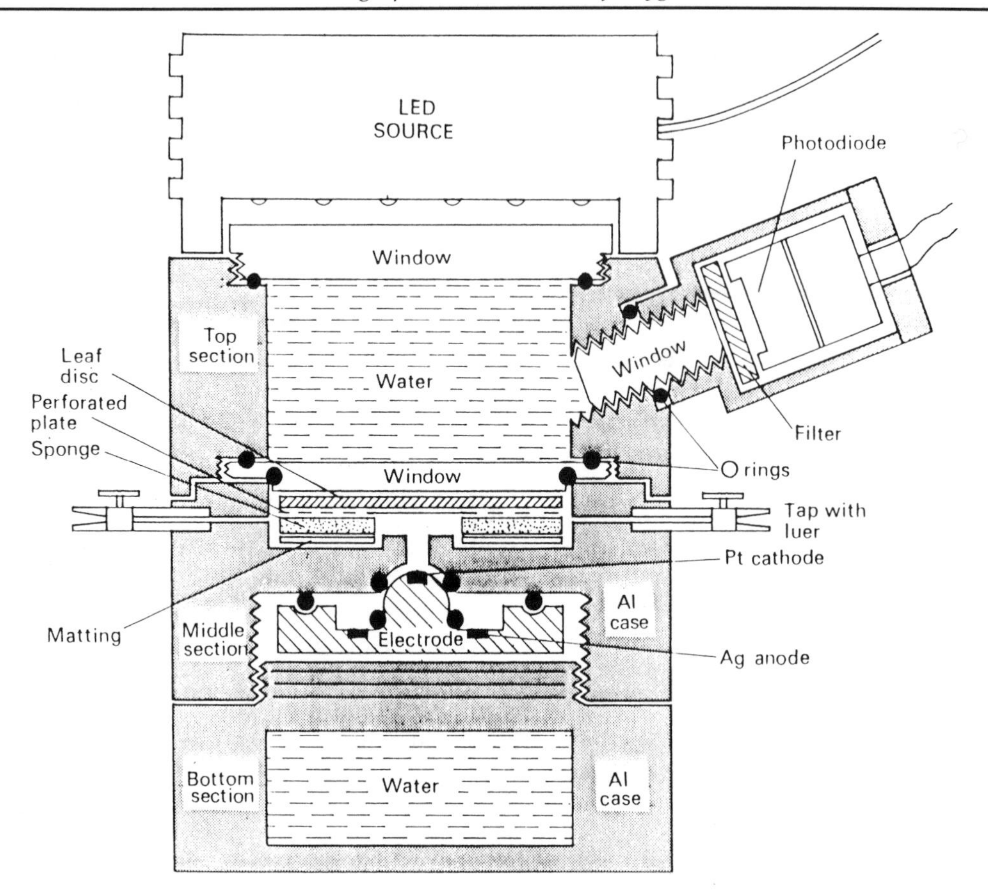

Fig. 10.4. Schematic diagram of gas phase oxygen electrode and fluorescence probe. The leaf disc or leaf pieces are supported on a stainless steel mesh in the chamber which is located in the middle section of the apparatus. The O_2 sensor (Clark-type electrode) lies beneath the leaf chamber with its Pt cathode exposed to the atmosphere within it. The leaf tissue is pressed lightly against the temperature-controlled roof of the chamber by a foam rubber disc which also separates it from the carbonate/bicarbonate buffer carried on capillary matting. The leaf is illuminated through this window which also allows fluorescence to reach a probe (inserted at an angle of 40°) where it is monitored by a photodiode. Actinic light is delivered to the top of the apparatus, through a fibre optic, as shown, or directly from an appropriate light source such as the Hansatech LS1 or LS2. The photodiode is protected from the actinic light by a suitable filter or filters. The clips which draw the top section onto the middle section (so that the roof of the leaf chamber is sealed against an O-ring) and the tubes which carry temperature-controlled water to the top and bottom sections are not shown. The taps (with luers) are for calibration and adjustment of the gas phase.

procedure is followed with a leaf already in position in the chamber, the reading will decline with time because of dark respiration. Even with an empty and properly sealed chamber there will be a small decline with time as oxygen in the chamber is consumed in the cathode reaction. In normal use this will be negligible, but the actual size of all signals will, of course, vary with the degree of pre-amplification used and the channel on the recorder which has been selected. If a slight leak is suspected, the chamber should be subjected to negative pressure by attaching an empty syringe to the chamber and withdrawing aliquots of air. Clearly the excursions of the recorder will now be negative

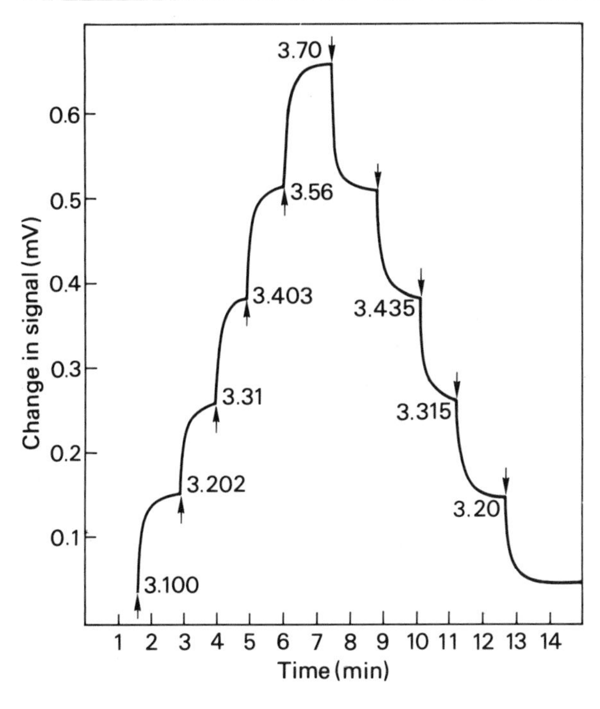

Fig. 10.5. Recorder trace excursions produced by introducing and removing aliquots of air. A 1.0 ml gas-tight syringe with its plunger fully withdrawn (to the 1.0 ml position) was attached to an open tap on the leaf chamber and the other tap closed. The figure illustrates the responses which were observed as the plunger was then pressed inwards in 200 μl stages and finally withdrawn again in a similar fashion. Note that each successive addition produces a larger signal (for explanation see text).

and a leak will be indicated by an increase in signal rather than the converse.

It will be seen that the initial rise is fast, but decreasing as the final level is approached. This is because the rate of diffusion of oxygen to the detector will become limiting as the difference in concentration between the atmosphere in the chamber and that at the electrode surface approaches equilibrium. If equilibrium is not approached within about one minute (i.e. if the response is sluggish) the electrode may require cleaning. If the signal rises rapidly but then falls equally quickly, the chamber is leaking; it should be checked that the other gas tap is closed, that the electrode disc is pressed securely against its O-ring, and that the clips which hold the two parts of the chamber together are closed properly. If the response seems too slow for the experiments intended, it can be speeded up by using a thinner membrane, by cutting a hole in the 'spacer' immediately above the cathode, or by dispensing with the spacer altogether. If no spacer is used, departure from linearity may sometimes result (see above).

10.2.3 Calculations relating to volume and calibration

At S.T.P. (Standard Temperature and Pressure) the volume of oxygen in 1.0 ml of atmospheric air (containing 21% O_2) is 210 μl. Since one mole of gas occupies 22.414 litres (and therefore 1.0 μmol occupies 22.414 μl), this volume of oxygen contains 210/22.414 or 9.37 μmol. At any other temperature (T), the amount of oxygen can be obtained by multiplying by 273/(273 + T); thus at 20°C the corresponding value is 8.73 μmol. Because the oxygen electrode measures concentration (or, more strictly, activity) it is also necessary, for purposes such as this, to know the effective volume of the chamber itself. This, of course, is variable because the leaf disc and any other material enclosed within the fixed volume of the chamber will occupy a significant space. However, the increase in signal (i.e. the increase in the electrical output) from the electrode circuit can be used to calculate both the effective volume of the chamber and the linearity of the electrode response. Suppose, for example, that the initial electrode output (R_1) was 1.5 mV and that, after the plunger was fully depressed, this reading had increased to 1.8 mV (R_2). The effective volume (v) of the chamber can be calculated from the relationship:

$$v = \frac{R_1}{R_2 - R_1} = \frac{1.5}{1.8 - 1.5} = 5\,\text{ml}$$

The general form of this equation:

$$v = \frac{R_1 - [R_2 \times (1 - \text{volume injected})]}{R_2 - R_1}$$

can also be used if sufficiently large excursions result from the introduction of volumes of air smaller than 1.0 ml.

Having assembled and calibrated the apparatus you are now presumably anxious to see that it will record oxygen evolution from an illuminated leaf. At this stage it must be remembered that photosynthesis requires CO_2, and that atmospheric air contains only about 350 parts per million. This means that the CO_2 content of the closed chamber will not sustain rapid photosynthesis for long. For example, 4 ml of air contains approximately 14 µl CO_2. Under favourable conditions, a healthy spinach leaf would have little difficulty in maintaining a rate of CO_2 fixation in excess of 240 µmol h^{-1} (mg Chl)$^{-1}$; or for a disc of 10 cm^2 (containing, say, 500 µg chlorophyll) about 2 µmol min^{-1}. At this rate, the CO_2 consumption would be $2 \times 22.4 = 44.8$ µl min^{-1}; enough to deplete the chamber of CO_2 in less than 20 s.

10.2.4 Experiments: does it work with a leaf?

Using the cutter provided, cut a disc of area 10 cm^2 from a broad leaf such as spinach. Place it on damp (not wet) capillary matting supported by the stainless steel grid provided (use the grid with the unperforated centre which prevents light shining directly on to the cathode). This is best supported, in turn, by a sponge disc and a second stainless steel grid. Close the chamber with the clips but leave both taps open. Attach a tube to the tap and breathe heavily into the chamber. Close both taps. You should have succeeded in replacing air (21% O_2, 0.035% CO_2) with expired air (say, 14% O_2, 5% CO_2).

Offset the signal on the recorder using 'back-off' until the trace occupies the lower 10% of the chart. Record the dark respiration for two or three minutes (until this reaches a steady O_2 consumption as the temperature within the chamber adjusts to that of the circulated water). Switch on the light and allow the leaf to photosynthesize at high irradiance. If the leaf disc has been taken from the dark or from low light you will observe an initial lag or induction period (Fig. 10.6). Thereafter the rate should become linear; if illumination is interrupted by 30–60 s of darkness, you should observe enhanced oxygen uptake and little or no lag following re-illumination. Gradually the rate of oxygen evolution in the light will decline. If you then open both taps and displace the atmosphere in the chamber with more expired air you will observe an abrupt fall in the recorded reading as the oxygen content is returned to about 14% and (possibly after some temperature re-equilibration) a return to a higher rate of photosynthesis.

10.2.5 Measurement of photosynthetic efficiency

The leaf disc electrode is particularly useful for measuring quantum yield in saturating CO_2. In principle, this is achieved by plotting the rate of oxygen evolution against photon flux density (PFD). Quantum yield is then derived from the slope of the initial linear part of this relationship. With a fully computerised version of this apparatus, even the changes in PFD can be automated and rate vs. PFD relationships can be established in as little as 10 minutes. In addition to quantum

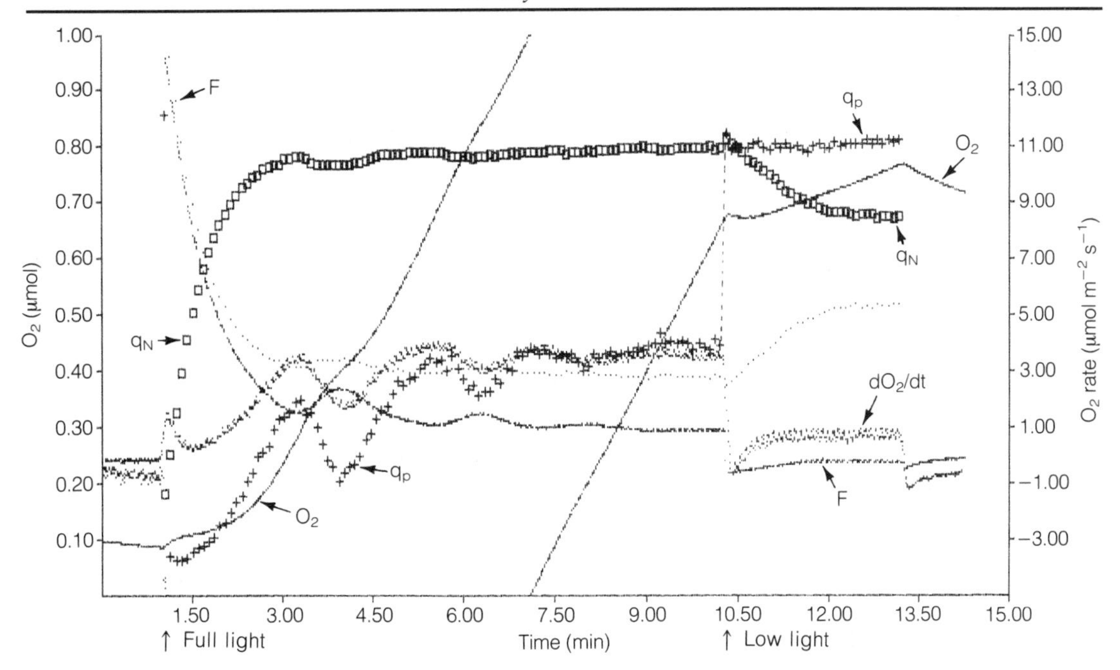

Fig. 10.6. Simultaneous measurements of fluorescence and oxygen made with the leaf disc electrode and the *Walz PAM 101* (H. Walz, Effeltrich, Germany).

The leaf disc electrode can be fitted with a photodiode which enables chlorophyll *a* fluorescence (Chapter 12) and oxygen evolution to be measured simultaneously. Microcomputer software is also available (Hansatech Ltd., King's Lynn, UK) which permits the change in O_2 to be recorded as the rate function dO_2/dt. Thus three of the five traces illustrated here could have been recorded with the Hansatech apparatus alone. The combination of the leaf disc electrode with the *Walz PAM 101* [6] allows the resolution of fluorescence (F) into photochemical (q_P) and non-photochemical quenching (q_N); this is the only commercially available apparatus which enables simultaneous fluorescence and oxygen measurement.

This experiment was carried out at high photon flux density (PFD) (850 µmol m^{-2} s^{-1}) and in saturating CO_2. After an initial lag or induction period, during which q_N (a measure of ΔpH) rises in advance of q_P (a measure of the reduction status of Q_A), dampening oscillations ensue. There is a broadly parallel but phase-shifted relationship between q_P, q_N and O_2, and a broadly anti-parallel relationship between q_P, q_N and O_2 on the one hand and fluorescence on the other. The dots mark the upper limits of the fluorescence excursion caused by the saturating light phases which are used in the quenching analysis. The decrease in PFD to 83 µmol m^{-2} s^{-1} at the time indicated abolishes oscillations but induces a new lag before O_2 rises to its new steady-state value.

More complete descriptions of such phenomena and procedures are available [5–12].

yield, these will give the values for dark respiration, the light compensation point and the percentage of incident energy used in photosynthetic carbon assimilation within the range provided by the incident light source [10,11].

In practice, illumination by an array of light-emitting diodes is adequate for most purposes (Fig. 10.4). For example, the data shown in Fig. 10.7 were derived in this way. However, it should be borne in mind that the lack of uniformity with this light source can cause problems, particularly if small pieces of leaf or pine needles are used. In such circumstances the Bjørkmann lamp [11] may be used, but this does not lend itself to com-

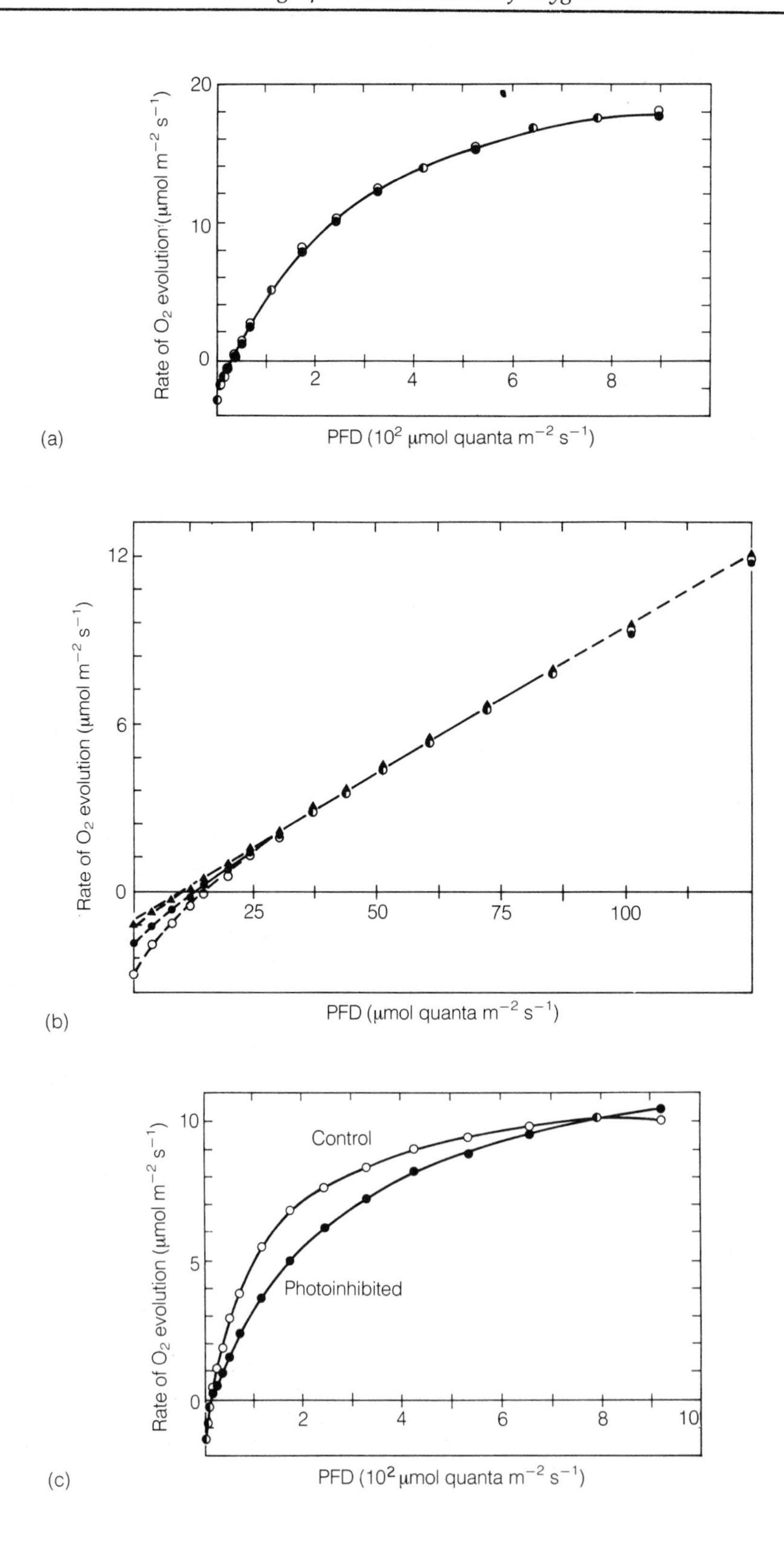

Rate of O2 evolution (μmol m−2 s−1)
20
10
0
2
4
6
8
(a)
PFD (10² μmol quanta m−2 s−1)
Rate of O2 evolution (μmol m−2 s−1)
12
6
0
25
50
75
100
(b)
PFD (μmol quanta m−2 s−1)
Rate of O2 evolution (μmol m−2 s−1)
10
5
0
Control
Photoinhibited
2
4
6
8
10
(c)
PFD (10² μmol quanta m−2 s−1)

puterisation. Accurate measurements of quantum yield with the light-emitting diode array are possible if a diffusing screen is placed immediately above the top of the chamber window, and if PFD is restricted to the range 0–100 μmol $m^{-2} s^{-1}$.

10.2.6 Combining oxygen measurement with fluorescence analysis

The leaf disc electrode chamber has a facility which permits simultaneous measurement of chlorophyll *a* fluorescence using a photodiode sensor [11] (see also Chapter 12). Although fluorescence *per se* is undoubtedly informative, much more information may be derived from quenching analysis involving the apparatus and procedures employed by Schreiber [3]. Minor modifications to the chamber and appropriate computer software, all available from Hansatech Ltd., permit the leaf disc electrode to be combined with the *Schreiber* apparatus (H. Walz, Effeltrich, Germany). At present, these are the only commercially available systems which permit quenching analysis to be combined with oxygen measurements. With a fully computerised version, 'real time' signals of the rate of oxygen exchange (dO_2/dt), photochemical quenching (q_P) and non-photochemical quenching (q_{NP}) can be recorded simultaneously with changes in oxygen and chlorophyll *a* fluorescence (Fig. 10.6). Using such a combination, Seaton and Walker [7,8] recently confirmed and extended the results of Genty *et al.* [3], and showed that, at least under some circumstances, many C_3, C_4 and CAM species follow a single curvilinear relationship between easily measured aspects of fluorescence emission and quantum yield (Fig. 10.8). In such circumstances, this relationship allows the actual quantum yield at a given light intensity to be derived *by fluorescence measurement alone*. The rate of photosynthesis can then be derived by simple arithmetic. For example, in Fig. 10.7b, the photosynthetic rate at a PFD of 100 μmol quanta $m^{-2} s^{-1}$ is seen to be *approximately* 10 μmol O_2 $m^{-2} s^{-1}$. This would imply a quantum yield of 10/100 = 0.1. Conversely, if it is known that the quantum yield is 0.1, it

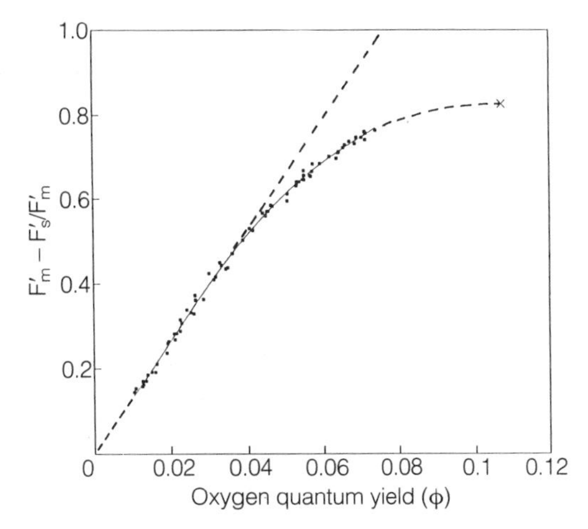

Fig. 10.8. Steady-state relationship between the fluorescence parameter ($F'_m - F'_s/F'_m$) and quantum yield (ϕ) for unstressed plants (see also Chapter 12). Chlorophyll fluorescence (i.e. $F'_m - F'_s/F'_m$) was measured according to Genty *et al.* [3] and Schreiber *et al.* [6]. Data are from a number of C_3, C_4 and CAM species. This relationship allows the rate of photosynthesis to be derived from fluorescence measurements alone, provided that the incident PFD is known.

Fig. 10.7. Automated measurement of rate as a function of light intensity using the leaf disc electrode (from Walker [10]; reproduced by permission).
(a) Reproductivity over a range of photon flux densities (PFDs). Leaf disc of *Vitis vinifera* (cv. Rhine Reisling) assayed twice (○,●) in quick succession.
(b) Reproductivity at low PFDs showing departure from linearity consequent upon increased O_2 uptake resulting from pre-illunination at high PFDs. Measurements were made consecutively following 5 minutes pre-illumination at 12.5 (▲), 125 (●) and 900 (○) μmol $m^{-2} s^{-1}$.
(c) Early stages of photoinhibition in *Helleborus niger*. Rate versus PFD before (○) and after (●) 2.5 h exposure to 1000 μmol $m^{-2} s^{-1}$ at 8°C.

follows that the photosynthetic rate at a PFD of 100 μmol quanta $m^{-2} s^{-1}$ is 10 μmol $O_2 m^{-2} s^{-1}$.

The computer programme 'LeafDisc', written by J. McCauley and G. Seaton in collaboration with the author, allows the leaf disc electrode and associated equipment to be used in conjunction with an IBM-compatible microcomputer. Copies of the programme are available from Hansatech Ltd., Hardwick Industrial Estate, King's Lynn, Norfolk PE30 4NE, U.K.

10.3 SUMMARY

The above account is a brief introduction to the use of aqueous-phase and gas-phase measurements of oxygen and their relation to photosynthesis by isolated organelles and leaves. All of these matters and many related topics are discussed in more detail in the manual listed below [11].

REFERENCES

1. Clark, L.C. (1956) Monitor and control of blood and tissue oxygen tension. *Transactions of the American Society for Artificial Internal Organs* **2**, 42.
2. Fatt, I. (1982) Theory of operation and its application in biology, medicine and technology. In: *Polarographic Oxygen Sensors.* Robert E. Krieger Publ., Malabar, Florida. pp. 1–278.
3. Genty, B., J.M. Briantais and B.R. Barker (1989) The relationship between the quantum yield of photosynthetic electron transport and quenching of chlorophyll fluorescence. *Biochim. Biophys. Acta* **990**, 87–92.
4. Gnaiger, E. and H. Forstner (1983) Aquatic and physiological applications. In: *Polarographic Oxygen Sensors.* Springer-Verlag, Berlin and New York. pp. 1–370.
5. Pinto, M., K.J. Bailey, G.G.R. Seaton and D.A. Walker (1991) Excursions in quenching regulation and philosophy. In: *Molecular Approaches to Compartmentation and Metabolic Regulation,* Proc. 14th Ann. Symp. Am. Soc. Plant Physiol., Univ. California, Riverside, Jan. 1991 (A. Huang and Lincoln Taiz, eds.).
6. Schreiber, U., U. Schliwa and W. Bilger (1986) Continuous recording of photochemical and non-photochemical chlorophyll fluorescence quenching with a new type of modulation fluorimeter. *Photosynthesis Research* **10**, 51–62.
7. Seaton, G.G.R. and D.A. Walker (1990) The use of chlorophyll fluorescence to predict CO_2 fixation during photosynthetic oscillations. *Proc. Roy. Soc. Lond. B* **241**, 59–64.
8. Seaton, G.G.R. and D.A. Walker (1991) Measuring photosynthesis by measuring chlorophyll fluorescence. In: *Trends in Photosynthesis Research* (J. Barber, H. Medrano and M.G. Guerrero, eds.) Intercept Ltd., UK.
9. Stokes, D., D.A. Walker, C. Grof and G.G.R. Seaton (1990) Light enhanced dark respiration. In: *Perspectives in Biochemical and Genetic Regulation of Photosynthesis* (I. Zelitch, ed.). Alan Liss Publ., New York. pp. 319–338.
10. Walker, D.A. (1990) Automated measurement of leaf photosynthetic O_2 evolution as a function of photon flux density. *Phil. Trans. Roy. Soc. Lond. B* **323**, 313–326.
11. Walker, D.A. (1990) *Use of the Oxygen Electrode and Fluorescence Probes in Simple Measurements of Photosynthesis.* 2nd edn, 2nd impression. Oxygraphics Ltd., Brighton/ Packard Publishing, Chichester. 203 pp.
12. Walker, D.A. and K.J. Bailey (1992) Changes in fluorescence quenching brought about by feeding dithiothreitol to leaves. *Plant Physiol.* **99**, 124–129.

11

Carbon isotope discrimination

H. Griffiths

11.1 INTRODUCTION

The stable isotopes of carbon occur in relatively constant proportions throughout the global carbon pool, with 98.89% comprising ^{12}C, and ^{13}C making up 1.11%. However, small but significant variations occur in the proportion of the heavy isotope ^{13}C incorporated into organic and inorganic matter. These depend on the extent of discrimination against (or in favour of) the heavy isotope during equilibrium reactions (e.g. dissolution) or kinetic processes such as irreversible enzyme reactions (particularly CO_2 fixation). When investigating the carbon isotope composition of plant material in relation to photosynthetic characteristics, it is important to know the species of inorganic carbon being fixed (whether CO_2 or HCO_3^-). In addition, it is also essential to identify the isotopic signature (i.e. proportion of ^{13}C:^{12}C) of the inorganic carbon prior to fixation (i.e. 'source' carbon).

Discrimination against ^{13}C during the photosynthetic incorporation of inorganic carbon is tempered by the characteristics of carboxylating enzymes and by diffusion limitation. Fractionation against $^{13}CO_2$ occurs because of the physical and chemical properties and altered metabolic transformation of the heavy isotope, leading to a smaller proportion of ^{13}C (as opposed to ^{12}C) being incorporated into organic material during photosynthesis. This depletion in ^{13}C, as compared to the isotopic signature of source inorganic carbon, is measured on CO_2 prepared and purified from plant organic material and analysed with an isotope ratio mass spectrometer. In view of the small differences in natural ^{13}C abundance which must be measured, results have traditionally been expressed as a carbon isotope ratio ($\delta^{13}C$) with units of parts per thousand (‰), with the differential notation comparing the ratio of the 45/44 masses for a sample with that of a defined standard (see below). Atmospheric CO_2 may also be collected and analysed either so as to verify 'source-CO_2' contributions or as part of on-line fractionation studies which measure discrimination instantaneously during gas exchange.

Initially, applications predominantly involved the analysis of the distribution of photosynthetic pathways (C_3, C_4, CAM), since C_4 and obligate CAM plants were less depleted in ^{13}C than C_3 plants [13,14,16]. However, with the development of models which described specific fractionation processes during CO_2 uptake and assimilation for C_3, C_4 and CAM plants [6,7,15], it is now possible to quantify carboxylation and diffusion limitation in terms of the regulation of

Photosynthesis and Production in a Changing Environment: a field and laboratory manual
Edited by D.O. Hall, J.M.O. Scurlock, H.R. Bolhàr-Nordenkampf, R.C. Leegood and S.P. Long
Published in 1993 by Chapman & Hall, London.
ISBN 0 412 42900 4 (HB) and 0 412 42910 1 (PB).

internal CO_2 concentration. The relationship between carbon isotope discrimination and water use efficiency can thus be used to select for improved crop productivity in arid habitats [6,8,21].

Applications involving stable isotopes range from cellular to ecosystem processes and are not confined to carbon: it is now increasingly apparent that integrated analyses using a combination of other stable isotopes, such as deuterium (δD), oxygen ($\delta^{18}O$), nitrogen ($\delta^{15}N$) and sulphur ($\delta^{34}S$) hold great promise for the future. Potential research areas include hydraulic properties of catchments and water use by plants ($\delta^{13}C$, δD, $\delta^{18}O$), carbon and nitrogen use in agroforestry ($\delta^{13}C$, $\delta^{15}N$), modelling ecosystems and food chains ($\delta^{13}C$, $\delta^{15}N$, $\delta^{34}S$) and analysing long-term changes in vegetation and climate ($\delta^{13}C$, $\delta^{18}O$, δD) [4,9,17,18,21].

However, we still do not fully understand the fractionation processes for all of these isotopes as compared to carbon [9]. In view of the exponential increase in applications involving stable isotopes [9,17,18,20,21], the approach adopted for carbon in this chapter will be to consider recent developments in the theoretical understanding of discrimination processes and their terminology, together with details of sample collection, purification and analysis, and finally applications in agriculture and the natural environment.

11.2 THEORY OF CARBON ISOTOPE FRACTIONATION

11.2.1 Derivation of carbon isotope ratio and discrimination nomenclature

Differences in absolute $^{13}C:^{12}C$ composition of two samples are usually small and, with mass spectrometric stability prone to variation, individual analyses are always compared directly against a defined standard. For instance, where two samples have absolute $^{13}C:^{12}C$ abundances of 1.111% and 1.095%, these would be compared with the PDB carbonate standard (1.124%), using the differential notation which is common to all stable isotope studies:

$$\delta(‰) = \frac{R_{sample} - R_{standard}}{R_{standard}} \times 1000 \qquad (11.1)$$

where R = 45/44 for carbon in CO_2. Thus R is the molar abundance ratio of sample or standard, and δ is expressed per mil (‰), with the standard set arbitrarily at zero. For the examples using carbon given above, the carbon isotope ratio of plant material would be derived from:

$$\left(\frac{R_p}{R_s} - 1\right) \times 1000 \qquad (11.2)$$

where R_p and R_s refer to plant organic material and PDB standard, respectively. The above samples are thus depleted in ^{13}C by −11.6‰ and −25.8‰ compared with the PDB standard, and could represent C_4 (or CAM) and C_3 plant material, although the isotopic constituents of source air (R_a) should also be considered (see below). A typical distribution for $\delta^{13}C$ in C_3 and CAM plants is shown in Fig. 11.1. In order to simplify the interpretation of isotope effects, some workers using carbon isotopes prefer to adopt a different expression for fractionation derived from standard, sample and source isotope composition [6]. Thus discrimination, Δ, is *positive* in plant material depleted in ^{13}C and the factor 1000 used in the derivation of 'per mil' (Equation 11.1) is not always used. Using this terminology we may calculate Δ for the examples used above. Given that δ_a (source air) will be calculated from $R_a/R_s - 1$, and δ_p (plant material) calculated from $R_p/R_s - 1$, overall discrimination, Δ, is defined as:

$$\Delta = \frac{\delta_a - \delta_p}{1 + \delta_p} \qquad (11.3)$$

where R_a (source) = 0.01115 in the year 1990 (i.e. $\delta^{13}C$ = −8‰ or δ_a = 0.008)
R_p = 0.01111 or 0.01095 (i.e. δ_p = 0.0116 or 0.0258, respectively)

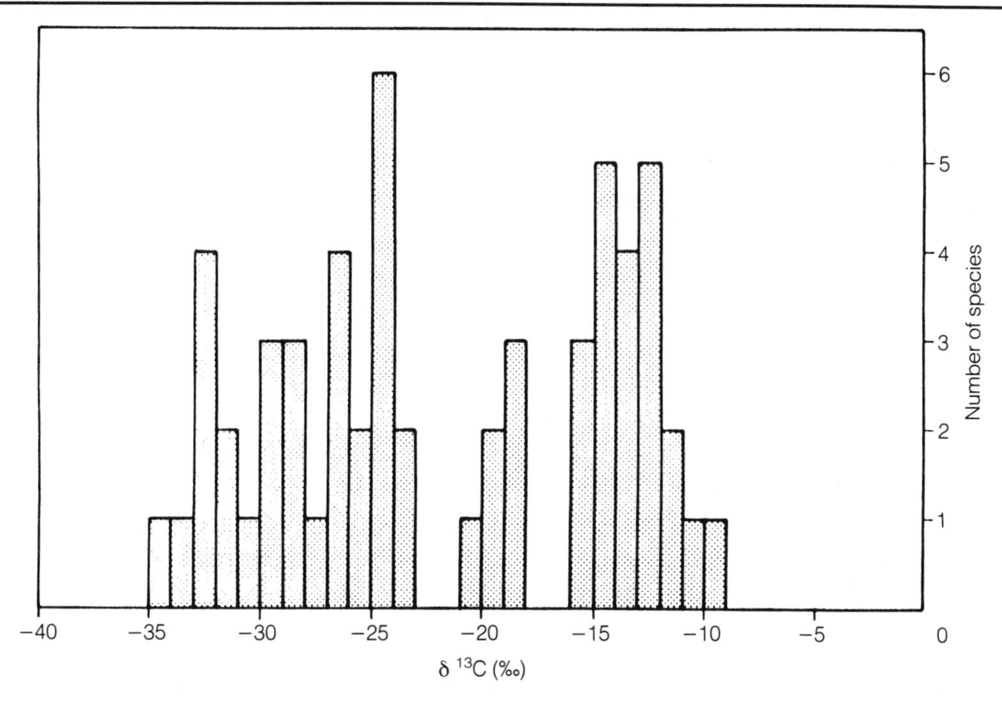

Fig. 11.1. Occurrence of C_3 and CAM within the Bromeliaceae of Trinidad. Only two of the 59 species were not available for analysis, with samples taken either from fresh material or from herbarium specimens. The intermediate group between 'C_4-like' and C_3 ranges is typical for CAM, and indicates potential C_3-CAM intermediates.

Δ = 0.0036 (3.6‰: i.e. C_4-like) and 0.0183 (18.3‰: i.e. C_3) for each sample.

The use of this nomenclature enables the magnitude of Δ to be directly proportional to the extent of fractionation for any biological process, and any experimental effects of source CO_2 are corrected automatically (such as the use of respired CO_2 already depleted in $^{13}CO_2$ by plants either experimentally or in natural habitats) [6].

11.2.2 Magnitude of isotope effects: kinetic and equilibrium reactions

The potentially wide range of δ_{source} for each isotope, shown in Table 11.1, contrasts with the smaller isotope fractionation (Δ) processes associated with incorporation into organic material. Isotope effects or fractionation factors (α) reflect the separate rate constants for reactions involving normal and enriched substrate molecules, where $\alpha = R_r/R_p$, i.e. the molar abundance ratio of reactant (r) and product (p). Kinetic effects (reversible and irreversible) result in discrimination because of the higher mass of the heavy isotope, since molecules containing heavier isotopes will react or diffuse more slowly and form stronger bonds as compared with the lighter isotope. For carbon this can be represented by the ratio of the rate constants, whereby $\alpha = k^{12}/k^{13}$, and since $\Delta = \alpha - 1$, α and Δ have been measured (or estimated) for nearly all reactions involved in the uptake and assimilation of CO_2 [6,14,21]. Thus for diffusion of CO_2 in air (reversible), α = 1.0044 and Δ = 4.4‰, and for discrimination by ribulose bisphosphate carboxylase (Rubisco: irreversible effect), α = 1.029 and Δ = 29‰.

Table 11.1. Variations of carbon isotope ratio ($\delta^{13}C$) and discrimination (Δ) in plants.

Isotope	Natural abundance (%)	Standard (‰)	Range of source δ (‰)	Range of δ_p in plants (‰)		Major factors regulating discrimination (Δ)	Range of discrimination (Δ‰)	Secondary discrimination processes
^{13}C	1.11	PDB carbonate	−8 to (CO_2) −30	C_3	−23 to −36	Rubisco/diffusion	15 to 28	lipid depleted vs. carbohydrate, 4 to 5‰; trophic level reflects source C signature
				C_4 + CAM	−10 to −18	Rubisco/diffusion/ leakage	2 to 10	
^{12}C	98.89	0		$C_3 - C_4$, C_3 − CAM intermediates	−18 to −36	PEPc	10 to 28	
			0 to (HCO_3) −11	aquatic	−10 to −50	Rubisco/boundary layers/CO_2 concentrating mechanism?	2 to 28	

Several models were developed, and have been refined by G.D. Farquhar and co-workers to describe the balance between the diffusion fractionation and Rubisco fractionation in terms of the co-limitation of C_3 photosynthesis (see Equations 11.4 and 11.5 below) [6,7,8].

Equilibrium reaction rate constants, where $\alpha = K^{12}/K^{13}$, result in generally smaller fractionation since kinetic effects are then at a thermodynamic balance: the predominant example is the distribution of ^{13}C and ^{12}C in air and water, such that the net effect of dissolution and hydration is normally a Δ of −8‰ (i.e. enrichment in ^{13}C). This isotope effect varies with temperature, and should be corrected for aquatic plants growing under natural conditions [12,19]. In terrestrial C_4 and CAM plants, the use of HCO_3^- as substrate by phosphoenol pyruvate carboxylase (PEPc) results in a net discrimination (including diffusion of dissolved CO_2) of −5.7‰, since PEPc shows only a small discrimination against $H^{13}CO_3^-$ (2‰).

11.2.3 Origin and magnitude of fractionation processes

Discrimination during photosynthesis results from the combined effects of diffusion and carboxylation limitation. In theory, if diffusion were the sole limiting factor, the isotope signature of the plant material would tend towards 4.4‰ (i.e. $\delta^{13}C$ = −12.4‰ including source CO_2); in contrast, were inherent Rubisco fractionation to be the predominating factor (i.e. stomatal conductance not limiting), Δ would tend towards 29‰, with $\delta^{13}C$ values approaching −37‰. The range of $\delta^{13}C$ values encountered in plant material (Fig. 11.1) may then be described in terms of Δ for each process (Table 11.1). The average C_3 carbon isotope ratio of −28‰ suggests that carboxylation resistance is twice as limiting as diffusion, implying that a molecule of CO_2 is twice as likely to diffuse back out of a leaf as to be carboxylated. Thus we may relate the diffusion limitation imposed by stomatal conductance (a = 4.4‰), to the extent of discrimination expressed by Rubisco (b = 27‰) in C_3 plants in terms of the ratio of c_i/c_a (internal:external CO_2 partial pressure) by [6,7]:

$$\delta_p = \delta_a + a\frac{c_a - c_i}{c_a} + b\frac{c_i}{c_a} \qquad (11.4)$$

and

$$\Delta = a + (b - a)(c_i/c_a) \qquad (11.5)$$

It should be noted that the value of b normally adopted is 27‰, to account for the contribution that PEPc makes to CO_2 fixation even in C_3 plants. The theoretical relationship between Δ and c_i/c_a is shown in Fig. 11.4a, with the values typically found for C_3 and C_4 plants. Since water use efficiency (WUE) is defined as the molar ratio of CO_2 uptake (**A**) to transpiration (**E**), calculated in

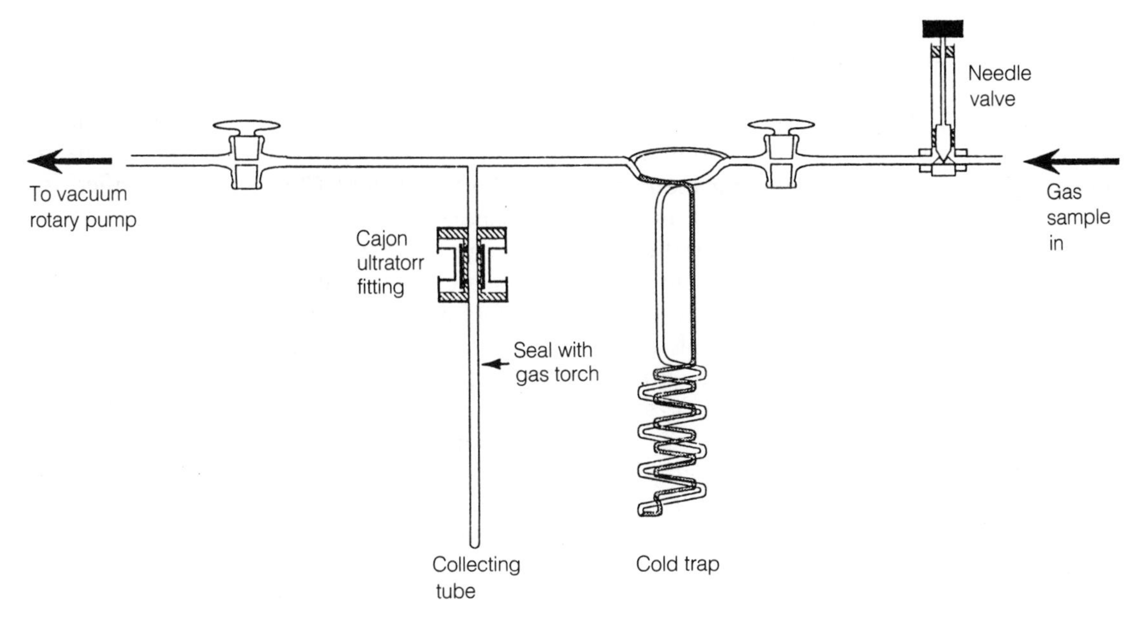

Fig. 11.2. A simple preparation line which may be used to collect gaseous CO_2 samples in the field and laboratory. CO_2 may be collected downstream from an infra-red gas analayser following determination of assimilation or ambient CO_2 concentration. The needle valve is used to regulate the flow from the atmosphere to the vacuum system at a flow rate of 300 ml min^{-1}, with CO_2 collected by freezing in the liquid nitrogen trap over a 10–15 minute period. The needle valve is then closed (or isolated), and following evacuation of the preparation line (to 1 Pa or 10^{-2} torr), the sample may be cryodistilled into the side-arm tubing, and sealed with a gas torch to form a sample vial. This may be transported to the laboratory where the CO_2 should be repurified through a conventional line to remove N_2O.

relation to stomatal conductance and leaf-air vapour pressure deficit (Δw), we see that c_i is also related to WUE:

$$\frac{\mathbf{A}}{\mathbf{E}} = \frac{(c_a - c_i)}{1.6\,\Delta w} \qquad (11.6)$$

The isotope discrimination in C_3 plant material is therefore related to WUE, with Δ greater for plants where diffusion is less limiting (Fig. 11.4b). It has so far been possible to evaluate WUE in the selection of crop genotypes, because Δ is proportional to stomatal conductance and hence inversely related to WUE (Section 11.5.1).

Finally, it should be noted that secondary fractionation occurs in the synthesis of lipids during the decarboxylation of pyruvic acid, with lipids being depleted in ^{13}C by 4–5‰ with respect to carbohydrates [13].

11.3 SAMPLE COLLECTION AND ANALYSIS

11.3.1 Sample preparation and purification

Although many studies to date have used leaf material, other plant components are often analysed as these may have a distinct isotopic signature. In view of the relationship between c_i/c_a, Δ and environmental variables it is important to standardise sampling procedures to use comparable plant material from natural sites or experimental plots. It is generally acceptable to oven-dry fresh tissue at 70°C, so that organic materials are not lost or modified, although if material can be frozen on collection it is preferable to freeze-dry samples. Only a small amount of plant material (<5 mg) is required for most combustion systems, but it must be finely ground (<0.1 mm particle size) to ensure homogeneity and to improve combustion characteristics.

If samples are being analysed by a central facility, then samples are usually combusted *in situ* using one of the following methods: combustion under oxygen and cycling of products through oxidation and reduction columns *in vacuo*; automatic combustion trains such as the VG ISOPREP (VG Isotech, U.K.); automated elemental analyser coupled to a mass spectrometer (e.g. *ANCA-MS*; Europa Scientific, U.K.); batch processing of silica-glass vials in a muffle furnace.

For the latter technique, the sample is mixed with CuO in a silica-glass tube which can be linked to a vacuum line using *Cajon Ultra-torr* (Swagelok, Ohio, U.S.A.) fittings and sealed with a gas-torch under vacuum. Following combustion at 850°C for 4–8 h the vials can be broken in a vacuum line using *Cajon Flexible Tubing* (Swagelok, Ohio, U.S.A.) and re-purified by cryodistillation through an oxidation column (with MnO_2, CuO and silver foil at 850°C) and reduction column (reduced copper at 550°C). Dry-ice/ethanol traps before and after these columns ensure the removal of water, and the CO_2 can be drawn *in vacuo* into a collecting vial using liquid nitrogen once residual gases have been pumped away. It is preferable to analyse samples soon after combustion to prevent any reaction between H_2O, CO_2 and copper and also to minimise contamination through high-vacuum stopcocks.

Atmospheric 'source' CO_2 may be collected using a preparation line similar to that used in 'on-line' measurements [5,10], whereby CO_2 is frozen out of air bled into a vacuum system through a needle valve at about $5\,cm^3\,s^{-1}$, with a liquid nitrogen cold trap sufficient to remove the CO_2 passing through the line (Fig. 11.2). Such a technique can be used in the field, if a generator and rotary pump are available, with samples frozen and sealed into pyrex tubes which may then be transported to the laboratory for repurification and analysis (Fig. 11.2; see batch processing above). It is important to remove N_2O from atmospheric CO_2 samples by passage over a reduction column, since the N_2O has the same mass as $^{12}CO_2$. Alternatively, pre-evacuated vessels may be used to collect sufficient CO_2 to be re-purified in the laboratory. For aquatic systems, source inorganic carbon may be precipitated as carbonate using a saturated solution of $SrCl_2$-NH_4OH or $BaCl_2$-NaOH [1,19].

11.3.2 Mass spectrometry

The isotope ratio mass spectrometer used for CO_2 analysis is a high-precision, high-vacuum machine. It requires maintenance and operation by a qualified technician, with typical running costs of US$ 50,000 per annum to cover spare parts, consumables and technical support. Such machines are capable of analysing a range of low molecular weight, gaseous samples encompassing those isotopes used in biological systems (H/D, C, O, N, S: Section 11.1) [4,9,18,21], although it should be noted that a separate configuration is required for δD analysis.

Sample and standard gas are introduced via a matched inlet system, with inlet volumes balanced by means of metal bellows (Fig. 11.3). Secondary standards such as NBS graphite at −28.1‰ (calibrated against the original PDB carbonate) are available from the National Bureau of Standards, Washington DC, USA, and are routinely used to calibrate a 'working' standard gas for day-to-day use. Gases are introduced into the ultra-high vacuum of the mass spectrometer (10^{-6} Pa or 10^{-8} torr) via crimped capillaries, with sample and standard alternately selected by means of a change-over valve (Fig. 11.3). The source consists of a filament which produces electrons to ionise the sample, with the positive ions then accelerated through focusing slits into the flight tube where, depending on the mass-to-charge ratio (m/e), each group of ions are deflected and separated by a magnetic field through a 90° or 120° sector to a series of

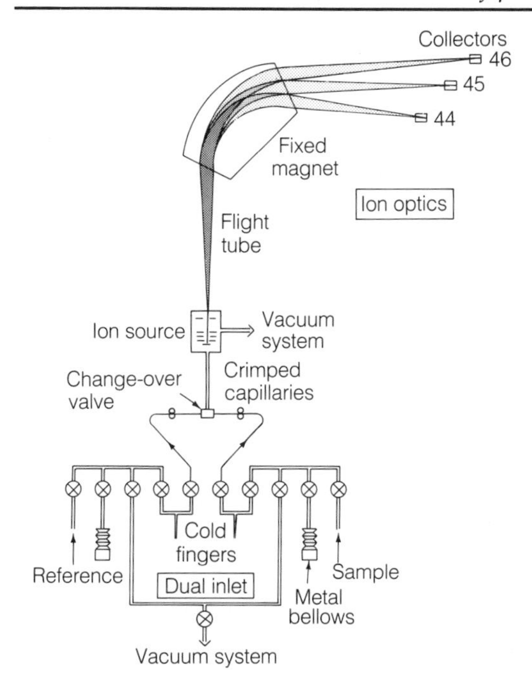

Fig. 11.3. Important features of a dual-inlet, triple collector mass spectrometer. Details are described in Section 11.3.2.

collectors. The impacts are converted by Faraday cups into a voltage and subsequently into a frequency (Fig. 11.3).

Modern mass spectrometers may use three collectors for each of the 44, 45 and 46 m/e peaks, with the contribution of ^{18}O to the 46 m/e analysed so as to calculate the ^{17}O in the 45 m/e. Alternatively, dual collector machines analyse firstly 44 vs. 45 m/e and then 46 vs. (44 + 45) m/e in the major collector. While the precision of any given mass spectrometer is often quoted to 0.01‰, it should be noted that errors in sample preparation and sample variability result in an overall precision of some 0.2–0.5‰. Results are expressed routinely as a $2\sigma_{10}$, which is the standard error of 10 measurements of the 44/45 m/e for sample vs. standard.

Automated analyses may allow increased throughput of samples in two ways: firstly, a manifold with electronic inlet valves may allow analysis of up to 36 samples; secondly, a CHN elemental analyser may be interfaced with the mass spectrometer, allowing batches of up to 100 samples to be analysed in a potentially portable unit (*ANCA-MS*; Europa Scientific, U.K.).

11.4 EVALUATION OF PHOTOSYNTHETIC PATHWAYS

11.4.1 Genotypic and environmental regulation of discrimination in C_3 plants

Given the interrelationship between carbon isotope discrimination (Δ) and c_i/c_a (Equations 11.4 and 11.5), the relationship between Δ, stomatal conductance and water use efficiency (WUE) can be determined for a variety of plants (Equation 11.6; Fig. 11.4). When stomatal conductance is low in relation to Rubisco capacity, Δ will tend to be regulated more by diffusion (a = 4.4‰). Under these conditions c_i will be small, WUE will be high and Δ reduced (i.e. $\delta^{13}C$ is less negative, enriched in ^{13}C; Fig. 11.4). When stomatal conductance (and hence c_i) is high, Δ will tend to be dominated more by the inherent fractionation expressed by Rubisco (b = 27‰); when WUE is low, Δ is high with organic material more depleted in ^{13}C. These observations have profound implications for the interpretation of genotypic and environmental expression of discrimination. Firstly, with carboxylation more limiting than diffusion, plant breeders should attempt to improve Rubisco efficiency so that CO_2 uptake will increase without affecting WUE. Secondly, carbon isotope discrimination should identify genotypes and cultivars with differing WUE.

It should be noted that compared to the wide range of $\delta^{13}C$ values found for C_3 plants (Table 11.1), the interactions between Δ, c_i/c_a and WUE only usually result in changes of some 3–4‰ (Fig. 11.4b). However, there have now been many studies which have shown the validity of relating Δ to diffusion/carboxylation limitation and have demonstrated the effectiveness of Δ in the selection

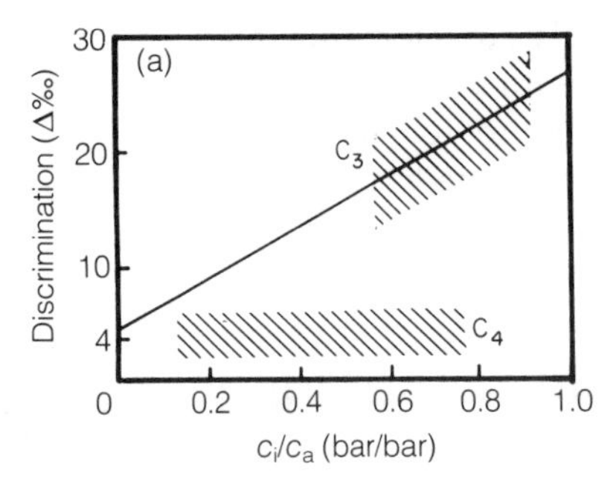

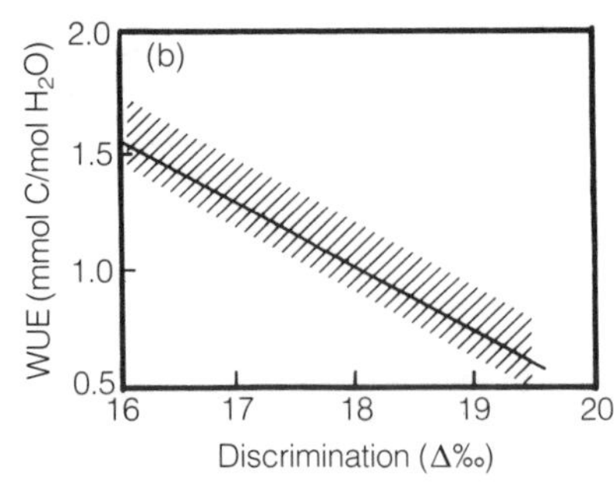

Fig. 11.4. The relationship between carbon isotope discrimination, c_i/c_a and water use efficiency (WUE): (a) the continuous line shows the relationship between Δ and c_i/c_a when a = 4.4‰ and b = 27‰ (Equation 11.4). Data for C_3 plants, with c_i/c_a measured by gas exchange, is normally found close to this line with values of c_i/c_a ranging from 0.6–0.8 (upper hatched area). Data for C_4 plants (lower hatched area) show that Δ is largely independent of c_i/c_a; (b) a typical inverse relationship between WUE and Δ, as predicted from Equations 11.5 and 11.6.

of improved WUE [6,8,21]. This has been shown for a number of cultivars of wheat, barley, peanut, cotton, sunflower and tomato in laboratory studies and field trials (see typical distribution of data in Fig. 11.4b). Ultimately, it has been shown that the relationship between Δ and WUE is heritable, and thus carbon isotope discrimination may be used to screen for WUE genotypically. Having identified suitable cultivars (with low Δ and hence high WUE), these may then be crossed with conventional high-yielding varieties.

However, such studies must be careful to exclude those variations which are related to microclimate within a crop canopy or variations in water supply. Although there have been few studies of crop canopies, a gradient of $\delta^{13}C$ is often found within temperate and tropical forest canopies, with leaf material in the understorey often depleted in ^{13}C by some 2–5‰ [6,22]. This is attributed to a combination of factors: firstly, re-fixation of CO_2 derived from respiration or from degrading leaf litter which is already depleted in ^{13}C (i.e. variations in δ_a, source CO_2); secondly, light limitation within the canopy may result in increased c_i and hence alters δ_p. Given that both CO_2 partial pressure (c_a) and isotope composition (δ_a) are likely to vary dynamically, temporally and vertically within any canopy, such observations emphasise the need for measurement of source CO_2 [22].

Alternatively, any increase in leaf-air vapour pressure deficit or temperature will affect c_i and hence Δ, and it should be noted that interactions with atmospheric pollutants may also affect c_i/c_a. Additionally, soil water supply, soil strength and salinity may vary within an experimental location and on a seasonal basis; water limitation will reduce c_i and affect the Δ of new growth subsequent to the imposition of stress [6].

While the Δ of plant material reflects the integrated WUE, there may be variations between leaf, stem and roots which reflect both environmental interactions and carbon assimilate partitioning [6]. In order to relate day-to-day variations in Δ and environment, it is possible to extract and analyse starch and sugar pools in relation to the daily weighted c_i/c_a [2]. Alternatively, the 'on-line' discrimination technique measures fractionation directly in terms of gas exchange [3,5,10]. The isotope composition of CO_2 which has passed over a leaf surface is analysed by this method, with the ^{13}C enrichment corresponding to the extent of discrimination being expressed internally (see Fig. 11.2).

11.4.2 Discrimination in C_4 plants

The average ^{13}C value of C_4 plants (−14‰) is characteristically different from C_3 plants (Table 11.1), but is also more depleted in ^{13}C than is predicted from theoretical analyses of carboxylation and diffusion limitation [6]. The low discrimination shown by PEPc (2‰) effectively provides an irreversible carboxylation step prior to the CO_2 being refixed by Rubisco in the bundle sheath. Diffusion limitation should result in more negative $\delta^{13}C$ values in C_4 plants [6,14]. The difference between observed and theoretical values of discrimination is attributed to leakage of CO_2 from the bundle sheath, allowing a greater discrimination to be expressed by Rubisco. Higher 'leakiness' is related to lower quantum yields in the different categories of C_4 plants.

Application of carbon isotope discrimination analysis to C_4 plants is of more use in evaluating the interactions between anatomy and the co-ordination of mesophyll and bundle sheath function [6], although genotypic variations between cultivars are as yet poorly characterised [14]. However, discrimination studies have led to a better understanding of C_3–C_4 intermediates, where it is now thought that refixation of photorespiratory CO_2 by Rubisco in the bundle sheath results in 'overcycling' or refixation by Rubisco with repeated discrimination being expressed [3]. Such observations have been confirmed by analysis of 'on-line' instantaneous discrimination.

11.4.3 Discrimination by CAM plants

Although the dark fixation by CAM plants is analogous to that in C_4 plants, carbon isotope discrimination is tempered by the extent of daytime CO_2 uptake with direct carboxylation by Rubisco [6,14,16]. In general terms, $\delta^{13}C$ values show a broader range than for C_4 plants (Table 11.1; Fig. 11.1) although plants with constitutive CAM are always distinct from C_3 plants. Analysis of carboxylation and diffusion limitation during dark fixation of CO_2 have shown that c_i/c_a may vary during the dark period, although analysis of newly fixed CO_2 (decarboxylated from the extracted pool of malic acid) suggested that both equally limit CO_2 fixation [15]. At dawn, during the period that stomata close, carboxylation may be mediated by both PEPc and Rubisco. However, increasing c_i leads to complete closure of stomata; during the decarboxylation of malic acid and refixation by Rubisco, leakage may occur allowing Rubisco discrimination to be expressed. During the late afternoon, when c_i declines, stomata may re-open and CO_2 fixation occurs directly via Rubisco, although PEPc activity increases towards dusk. This complex mixture of carboxylation has been modelled [6], and also quantified directly using the 'on-line' fractionation technique [10]. Here, the combined discrimination weighted according to the magnitude of CO_2 uptake during each phase of CAM correlated well with the Δ measured for organic material.

In terms of genetic variation, the occurrence of CAM has been shown to be derived in the Crassulaceae, with the extent of dark fixation associated with smaller Δ. It is important to note that where possible, studies on the distribution of CAM within plant populations should not be determined solely from carbon isotope discrimination. Analysis of the dawn-dusk variations in titratable acidity is a useful adjunct, particularly in the case of C_3–CAM intermediates where direct carboxylation by Rubisco may predominate through much of the growing season.

11.4.4 Discrimination in aquatic plants

Variations in source inorganic species should be quantified in terms of CO_2 or HCO_3^-, and the extent of allochthonous and terrigenous respiratory inputs (which may lead to δ_a being depleted by as much as −30‰; Table 11.1). In aquatic habitats, HCO_3^- may be the inorganic species accumulated by plants, with δ_p reflecting equilibrium fractionation in favour of ^{13}C as well as any variations in source inorganic carbon isotope composition [11,19].

Carbon isotope discrimination may then be used to estimate the effects of boundary layers in aquatic plants. Modification of Equation 11.4, replacing δ_a with the $\delta^{13}C$ of dissolved CO_2 at the given water temperature and correcting for dissolved CO_2 diffusion (0.7‰), enables diffusion limitation to be determined. From measurements of photosynthetic characteristics, values of c_i/c_a may then be used to calculate the thickness of the unstirred layer [19], which is likely to be a much greater limitation for aquatic plants in view of the 10^4 slower diffusivity of CO_2 in water as compared with air. A number of CO_2 concentrating mechanisms have evolved in freshwater and marine habitats to counter this limitation, such as CAM in isoetids, C_4-like mechanisms and biophysical CO_2/HCO_3^- accumulating mechanisms [16,18]. When analysing aquatic plant discrimination a wide range of Δ may be demonstrated (Table 11.1) and careful evaluation of photosynthetic characteristics and source isotope composition is required [11,19].

11.5 CARBON ISOTOPE DISCRIMINATION: ECOLOGICAL APPLICATIONS

11.5.1 Photosynthetic variations in ecosystems

While many applications have been mentioned above, it is important to note the wide-ranging uses for carbon isotope discrimination studies. While many surveys have been made from herbarium material showing the distribution of C_3, C_4 and CAM within a given plant population [16], it is also possible to examine changes in the natural vegetation on a seasonal basis. For C_3 plants, the variations in discrimination have already been described for a variety of environmental conditions (Section 11.4.1), but interactions between C_3 and C_4 pathways are also important for natural vegetation, crops and weeds [21]. While the altitudinal zonation of C_3 and C_4 plants has been shown for several tropical mountain habitats, so has the decreasing proportion of C_4 species been shown for northern latitudes [16]. Within a given habitat, the phenology of C_3 and C_4 grasses may alter according to water supply (with C_4 predominating during the dry season). Indeed, carbon isotope discrimination provides a succinct measure of belowground competition between C_3 and C_4 root systems [4].

There is now interest in reconstructing climate change from analysis of tree rings, although further analysis is required before we may relate differences within a canopy to the material deposited in the stem [9,21]. Measurements of atmospheric CO_2 isotope composition have shown a decline in δ_a in the last 40 years, reflecting anthropogenic inputs mainly from fossil fuels. Whether genotypic or phenotypic variations in WUE by natural vegetation will occur in response to any climate change has yet to be determined. However, carbon isotope discrimination analyses, similar to those carried out for crop plants, could be a useful adjunct. Several programmes are currently being developed using stable isotopes to evaluate the WUE of 'multi-purpose' tree varieties, so as to identify suitable clones for re-afforestation in arid regions.

11.5.2 Analysis of food webs

Carbon isotopes are useful markers for the origin and distribution of primary production within ecosystems, with analyses of food webs and trophic levels carried out in both terrestrial and aquatic environments [9,17,20,21]. Animals usually have a similar isotopic signature to the predominant dietary material, and so analysis of selectivity for C_3 and C_4 vegetation has been related to C_3 and

C_4 distribution for a number of herbivores (ranging from grasshoppers to elephants!) [21]. In aquatic ecosystems, the discrimination expressed in particulate organic matter (mainly phytoplankton) can be used both to model mixing within and between coastal and oceanic waters as well as analysing trophic levels [17,20].

11.6 CONCLUSIONS

When initiating any investigation using carbon isotope discrimination it is important to take into consideration environmental and physiological factors which may alter Δ within any particular habitat [9]. While analysis of Δ is a useful adjunct to many studies of gas exchange, water use efficiency and the integration of photosynthetic processes, care should be taken to standardise sampling procedures and sample preparation. In view of the temporal variations in c_a and c_i, and spatial variations in water supply and other environmental factors affecting c_i/c_a, scaling-up of individual plant processes to the crop or natural canopy level should be attempted with caution. Mass spectrometry facilities are expensive to maintain, but the semi-automation offered by some systems will enable the large throughput of samples necessary to validate statistically the field observations. For a limited number of samples, however, it is often better to approach a recognised central facility rather than trying to establish and maintain an individual mass spectrometer.

ADDRESSES OF EQUIPMENT MANUFACTURERS

Dual-inlet isotope ratio mass spectrometers:
VG Isotech Ltd., Aston Way, Middlewich, Cheshire CW10 0HT, U.K.
Finnegan MAT GmbH, Barkhausenstrasse 2, D-2800 Bremen 14, Germany.

Automatic carbon and nitrogen analyser:
Europa Scientific Ltd., Europa House, Electra Way, Crewe CW1 1ZA, U.K.

General vacuum preparation line equipment:
Swagelok Companies, Solon, Ohio 44139, USA.

REFERENCES

1. Bishop, P.K. (1990) Precipitation of dissolved carbonate species from natural waters for $\delta^{13}C$ analysis – a critical appraisal. *Chemical Geology (Isotope Geoscience Section)* **80**, 251–259.
2. Brugnoli, E., K.T. Hubick, S. von Caemmerer, S.C. Wong and G.D. Farquhar (1989) Correlation between the carbon isotope discrimination in leaf starch and sugars of C_3 plants and the ratio of intercellular and atmospheric partial pressures of carbon dioxide. *Plant Physiol.* **88**, 1418–1424.
3. von Caemmerer, S. and K.T. Hubick (1989) Variation in short-term carbon isotope discrimination among C_3, C_4 and C_3-C_4 intermediate species. *Planta* **178**, 475–481.
4. Ehleringer, J.R. and C.B. Osmond (1989) Stable isotopes. In: *Plant Physiological Ecology: Field Methods and Instrumentation* (eds. R.W. Pearcy, J.R. Ehleringer, H.A. Mooney and P.W. Rundel) pp. 281–299. Chapman and Hall, London and New York.
5. Evans, J.R., T.D. Sharkey, J.A. Berry and G.D. Farquhar (1986) Isotope discrimination measured concurrently with gas exchange to investigate CO_2 diffusion of leaves in higher plants. *Australian J. Plant Physiol.* **9**, 121–137.
6. Farquhar, G.D., J.R. Ehleringer and K.T. Hubick (1989) Carbon isotope discrimination and photosynthesis. *Ann. Rev. Pl. Physiol. Pl. Mol. Biol.* **40**, 503–537.
7. Farquhar, G.D., M.H. O'Leary and J.A. Berry (1982) On the relationship between carbon isotope discrimination and the intercellular carbon dioxide concentration in leaves. *Australian J. Plant Physiol.* **11**, 539–552.
8. Farquhar, G.D. and R.A. Richards (1984) Isotopic composition of plant carbon correlates with water use efficiency of wheat genotypes. *Australian J. Plant Physiol.* **11**, 539–552.
9. Griffiths, H. (1991) Applications of stable isotope technology in physiological ecology. *Functional Ecology* **5**, 254–269.
10. Griffiths, H., M.S.J. Broadmeadow, A.M. Borland and C.S. Hetherington (1990) Short-term changes in carbon isotope discrimination identify transitions between C_3 and C_4

carboxylation during crassulacean acid metabolism. *Planta* **181**, 604–610.
11. Kerby, N.W. and J.A. Raven (1987) Transport and fixation of inorganic carbon by marine algae. *Adv. Bot. Research* **11**, 71–123.
12. Mook, W.G., J.C. Bommerson and W.H. Staverman (1974) Carbon isotope fractionations between dissolved bicarbonate and gaseous carbon dioxide. *Earth Planetary Science Letters* **22**, 169–176.
13. O'Leary, M.H. (1981) Carbon isotope fractionation in plants. *Phytochemistry* **20**, 553–567.
14. O'Leary, M.H. (1988) Carbon isotopes in photosynthesis. *BioScience*, **38** 325–336.
15. O'Leary, M.H. and C.B. Osmond (1980) Diffusional contribution to carbon isotope fractionation during dark CO_2 fixation in CAM plants. *Plant Physiol.* **66**, 931–934.
16. Osmond, C.B., K. Winter and H. Ziegler (1982) Functional significance of different pathways of CO_2 fixation in photosynthesis. In: *Encyclopaedia of Plant Physiology*, Vol. 12B; Physiological Plant Ecology II (eds. O.L. Lange, P.S. Nobel, C.B. Osmond and H. Ziegler). pp. 479–547. Springer Verlag, New York.
17. Peterson, B.J. and B. Fry (1987) Stable isotopes in ecosystem studies. *Ann. Rev. Ecology and Systematics* **18**, 293–320.
18. Raven, J.A. (1987) The applications of mass spectrometry to biochemical and physiological studies. In: *The Biochemistry of Plants* Vol. 13 (ed. D.D. Davies) pp. 127–180. Academic Press, New York.
19. Raven, J.A., J. Beardall and H. Griffiths (1982) Inorganic C-sources of *Lemanea, Cladophora* and *Ranunculus* in a fast-flowing stream: measurements of gas exchange and of carbon isotope ratio and of their ecological implications. *Oecologia* **53**, 68–78.
20. Rounick, J.S. and M.J. Winterbourn (1986) Stable carbon isotopes and carbon flow in ecosystems. *BioScience* **36**, 171–177.
21. Rundel, P.W., J.R. Ehleringer and K.A. Nagy (1989) *Stable Isotopes in Ecological Research*. Springer-Verlag, New York.
22. Sternberg, L.D.S.L.O. (1989) A model to estimate carbon dioxide recycling in forests using $^{13}C/^{12}C$ ratios and concentrations of ambient carbon dioxide. *Agricultural and Forest Meteorology* **48**, 163–173.

12

Chlorophyll fluorescence as a tool in photosynthesis research

H.R. Bolhàr-Nordenkampf and G. Öquist

12.1 INTRODUCTION

The introduction of fluorescence microscopy in the 1920s opened up the practical investigation of chlorophyll fluorescence. By microscopic study of thin sections it was shown that all photosynthetic tissues emitted red fluorescence when exposed to UV-A or blue light. Kautsky and Hirsch [13] were the first to report the pattern of fluorescence induction when dark-adapted leaves were exposed to light. It was assumed that the kinetics of fluorescence emission during induction reflected the initial phases of the induction of photosynthesis. Later research has proved this assumption correct and chlorophyll fluorescence is today used as a sensitive, *in vivo* probe of photosynthetic function, both in the field and in the laboratory. Today, new instruments provide ready means for the measurement of chlorophyll fluorescence *in vivo*. However, data must be interpreted with caution.

The aim of this chapter is to give a short introduction to (i) the relationship between chlorophyll fluorescence and photosynthesis, (ii) the use of chlorophyll fluorescence in stress physiology and (iii) instrumentation now available. More extensive accounts of chlorophyll fluorescence and its applications are available [6,10,14–18,22].

12.2 FLUORESCENCE EMISSION

Light energy is absorbed by chlorophyll molecules for photosynthesis. However, portions of the absorbed light are always lost as heat or by re-emission as fluorescence (Fig. 12.1). Since these decay processes of excited chlorophyll are competitive, changes in the photosynthetic rate and/or in dissipative heat emission will cause complementary changes in the intensity of the emitted fluorescence.

A photon of red light (670 nm) contains enough energy for a chlorophyll molecule to reach the first excited state, termed singlet 1. The excited chlorophyll molecule is stable for less than 10^{-8} s. During this short period, charge separation within the reaction centre takes place, comprising the primary photochemical step of photosynthesis. If charge separation does not take place the absorbed light energy is released as heat and/or fluorescence when the excited electron of the molecule returns to ground level (Fig. 12.2). The relationship between the three decay processes (Figs. 12.1. and 12.2), photochemistry (P), fluorescence (F) and heat (radiationless deactivation, D) is best expressed mathemat-

Photosynthesis and Production in a Changing Environment: a field and laboratory manual
Edited by D.O. Hall, J.M.O. Scurlock, H.R. Bolhàr-Nordenkampf, R.C. Leegood and S.P. Long
Published in 1993 by Chapman & Hall, London.
ISBN 0 412 42900 4 (HB) and 0 412 42910 1 (PB).

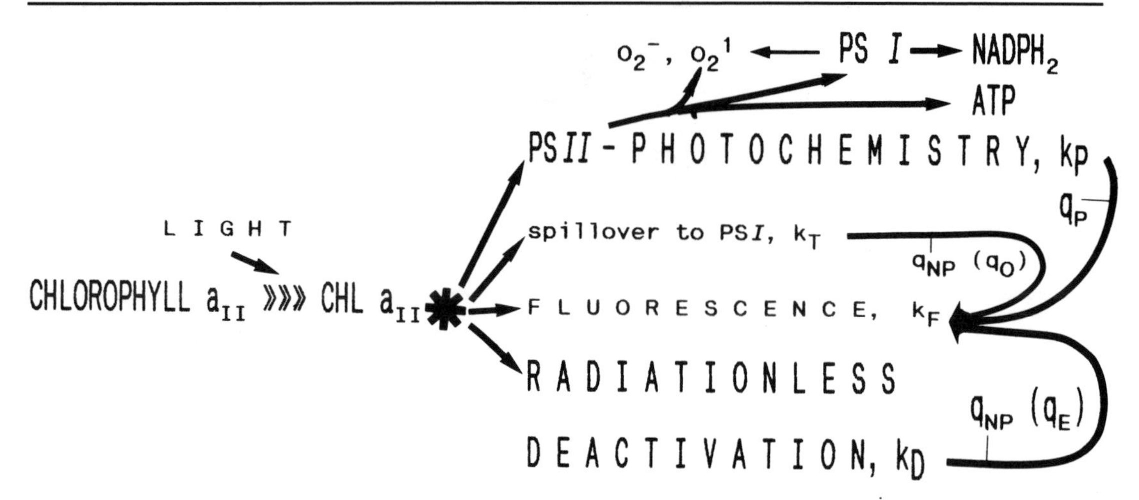

Fig. 12.1. Deactivation scheme of excited chlorophyll *a* (P_{680}) in Photosystem II (PS II). O_2^-, oxygen radical, 'Mehler' reaction at PS I; O_2^1, suggested formation of singlet oxygen at PS II; k, rates of de-excitation: k_P, via photochemistry; k_T, via spillover to PS I; k_F, via fluorescence; k_D, via heat; q_P, photochemical quenching; q_{NP}, non-photochemical quenching; q_0, F_0 quenching; q_E, energy quenching.

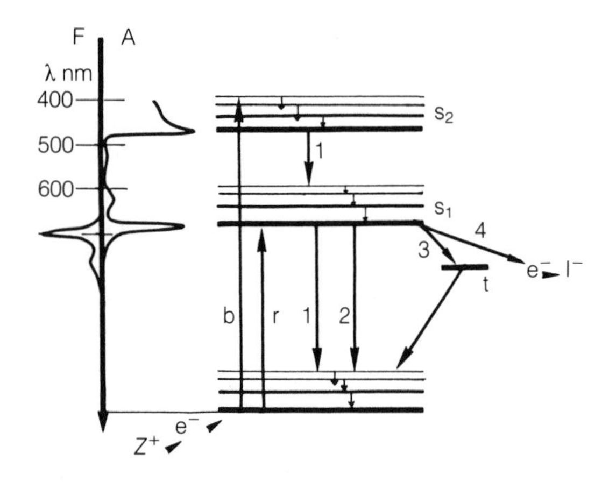

Fig. 12.2. Energy level diagram for chlorophyll: F, fluorescence spectrum; A, absorbance spectrum; b, blue photon; r, red photon; 1, internal conversion (k_D); 2, fluorescence (k_F); 3, inter-system crossing; 4, charge separation (k_P); I^-, intermediate e^- acceptor of PS II; s_1, s_2, singlet states; t, triplet state; Z^+, electron donation from water splitting system.

ically using rate constants: de-excitation via photochemistry, k_P; via fluorescence, k_F; and via heat, k_D. The total number of de-excitations per second is:

$$(k_P + k_F + k_D)n,$$

where n is the number of chlorophyll molecules. As essentially all fluorescence at room temperature originates from chlorophyll molecules associated with Photosystem II (PS II), the yield (Φ) of PS II fluorescence can be defined under constant illumination conditions as:

$$\Phi F = k_F/(k_P + k_F + k_D).$$

Under low light conditions, when high quantum yields can be achieved, about 97% of the absorbed photons are used in photochemistry, 2.5% are transformed to heat and 0.5% are re-emitted as red fluorescence light. If all PS II reaction centres are closed, 95–97% of the absorbed energy may be deactivated via heat and 2.5–5.0% via fluorescence.

Chlorophyll fluorescence is red because the energy difference between the ground level and the first singlet level equals the energy of a photon of red light; i.e. approximately the reverse of what happens when chlorophyll absorbs red light. However, the red fluorescence emission peak, which is around 685 nm *in vivo*, is shifted towards a longer

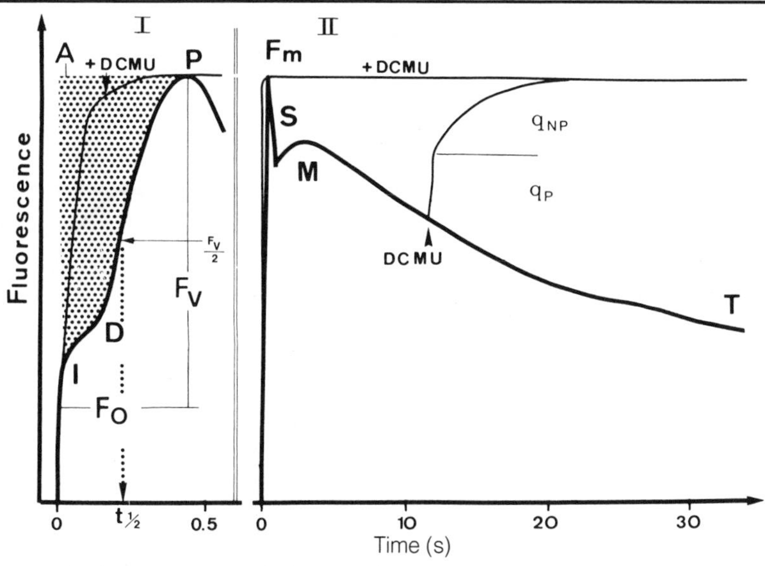

Fig. 12.3. Characteristic fluorescence induction kinetics or Kautsky curve as recorded from a dark-adapted *Chlorella* suspension. I, fast kinetics; II, slow kinetics; A, area above the curve; F_0, minimal initial fluorescence (dark) at open PS II reaction centres; O, origin: I, inflection or intermediate level; D, dip or plateau; P, peak level; F_v, variable fluorescence; $t_{1/2}$, half rise time from F_0 to F_m; F_m, maximal fluorescence (dark) at closed PS II reaction centres; S, slope; M, secondary maximum; T, terminal level; DCMU, 3-(3,4-dichlorophenyl)-1,1-dimethylurea; q_{NP}, non-photochemical quenching; q_P, photochemical quenching. The curves were measured with a time-resolving fluorimeter (*PSM, BioMonitor, Sweden*).

wavelength than the red chlorophyll absorption peak. This phenomenon is called Stoke's shift and results from a loss of energy as heat when fluorescence is emitted (Fig. 12.2).

12.3 CHLOROPHYLL FLUORESCENCE IN RELATION TO PHOTOSYNTHESIS

The fluorescence induction curve is sometimes called the *Kautsky curve* (Fig. 12.3). The different phases are denoted O I D P S M T. The fluorescence rise from O to P occurs during the first second of illumination and is called the 'fast phase'. The 'slow phase' follows after P and it may take several minutes before the terminal phase, T, is reached. The fast phase is related to primary processes of PS II, whereas the slow phase is mainly related to interactions between processes in the thylakoid membranes and metabolic processes in the stroma, primarily carbon metabolism.

12.3.1 Induction: fast kinetics

Following illumination of dark-adapted photosynthetic tissues, there is an immediate rise in fluorescence to a so-called minimal level F_0. This is attained when light has been absorbed by the chlorophyll antenna, but before the excitons have been trapped by the reaction centres of PS II. This state lasts only for picoseconds to nanoseconds, and it is therefore difficult to resolve the real value of F_0 with conventional electromagnetic shutters or by using illumination with ultra-bright light-emitting diodes. The problem can, however, be solved by using a mathematical method to extrapolate to the true F_0. The F_0 level can also be obtained if the intensity of the exciting light is low enough

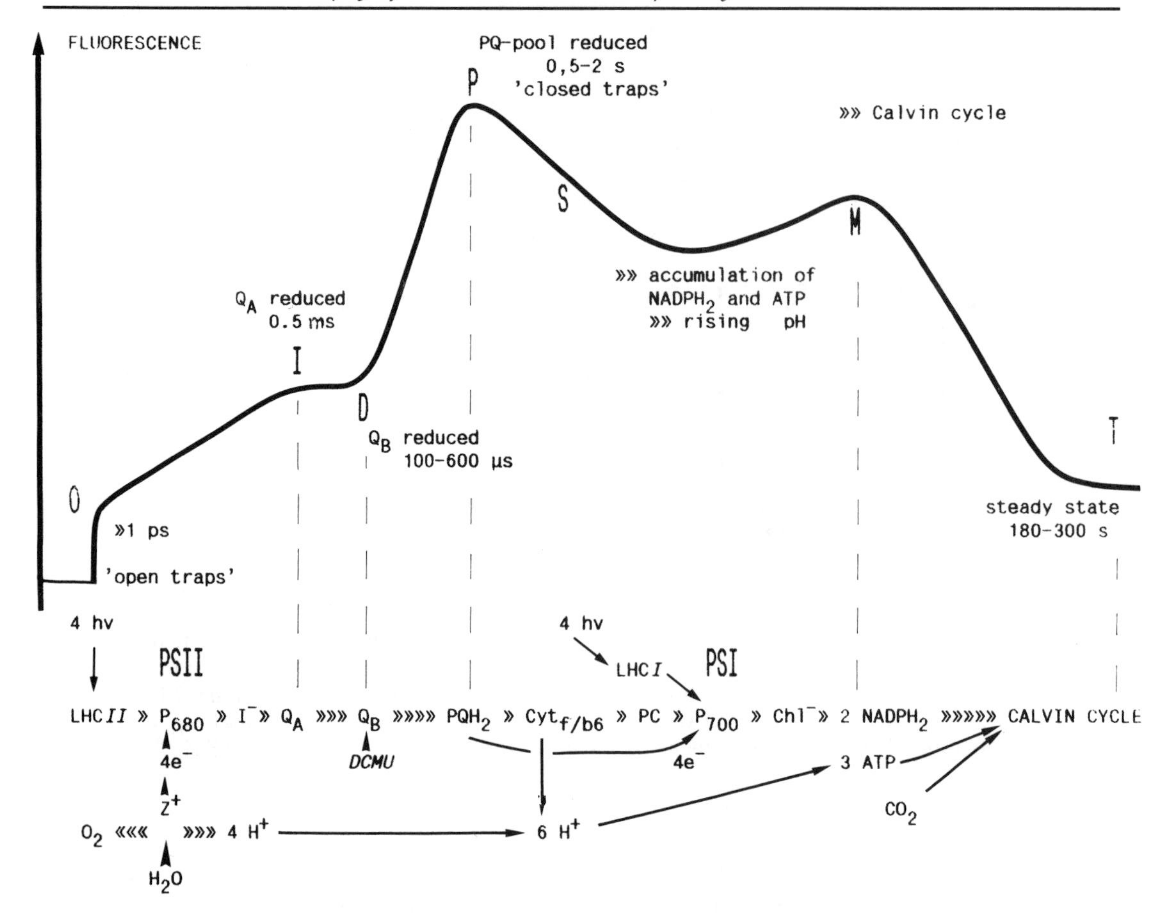

Fig. 12.4. Chlorophyll fluorescence induction curve aligned with the reactions in the electron transport chain: O I D P S M T, successive fluorescence levels during the time course of chlorophyll *a* fluorescence; hv, photon; LHC II,I, light-harvesting complexes of Photosystems II and I (PS II, PS I); P_{680} and P_{700}, reaction centres of PS II and PS I; e^-, electron; Z^+, electron donation from water splitting system; H^+, proton; I, intermediate acceptor, pheophytin *a*; Q_A, primary plastoquinone acceptor of PS II; Q_B, secondary plastoquinone (two electron) acceptor of PS II; PQH_2, reduced plastoquinone; $Cyt_{f/b6}$, cytochrome f/b_6 complex; PC, plastocyanin; Chl^-, intermediate electron acceptor of PS I, presumably a chlorophyll *a*.

to prevent detectable photochemistry from occurring. This method is used in modulated fluorescence measuring systems (Section 12.3.3). However, in both time-resolved and modulated systems, the key feature of F_0 is the same, i.e. it is measured when all reaction centres are open for primary photochemistry ('open traps'). At F_0 both the potential for photochemical use of excitation energy and therefore for photochemical quenching (q_P) of fluorescence, are maximal. The F_0 signal provides a valuable reference for all other fluorescence parameters of the induction curve.

Upon illumination with a sufficiently strong light, fluorescence increases from F_0 via an intermediate level (I), and often a dip (D), to a peak level (P) (Figs. 12.3 and 12.4). This rise reflects a gradual increase in the yield of chlorophyll fluorescence, as the rate of photochemistry concurrently declines (Fig. 12.5). The reason for this decline is that

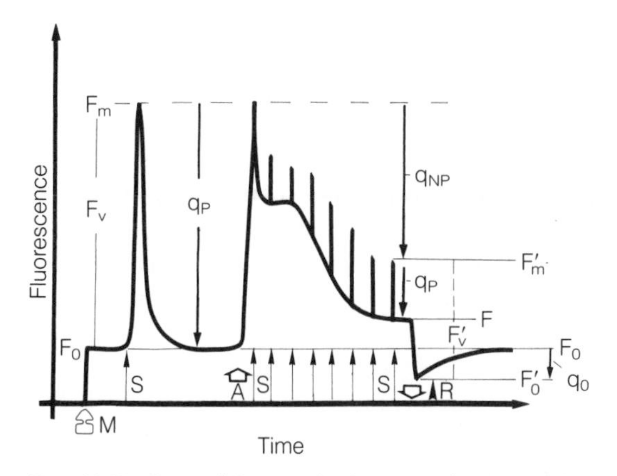

Fig. 12.5. Quenching analysis using the saturation pulse method. The curve was recorded by a modulated fluorimeter (MFMS, Hansatech) with a dark-adapted castor bean leaf. M, weak modulated measuring beam ($2\,\mu mol\,m^{-2}\,s^{-1}$ at 583 nm); S, saturating light pulse ($8500\,\mu mol\,m^{-2}\,s^{-1}$, white light, applied for 500 ms); A, continuous actinic light ($150\,\mu mol\,m^{-2}\,s^{-1}$, white light) on simultaneously with a saturating pulse; R, far-red light on ($7\,\mu mol\,m^{-2}\,s^{-1}$, >700 nm) and actinic light off. Along the curve characteristic points of fluorescence yield are marked, i.e. dark adapted: F_0, F_m, F_v; with actinic light added: F'_m, F'_v, F'_0; F, level of fluorescence at any time of the induction curve; quenching coefficients: q_P, photochemical quenching; q_{NP}, non-photochemical quenching; q_O, 'F_0-quenching' (see also Van Kooten and Snel [27]).

$$q_P = \frac{F'_m - F}{F'_m - F'_0} \qquad q_{NP} = 1 - \frac{F'_m - F'_0}{F_m - F_0}$$

the acceptor Q_A becomes increasingly reduced as the 'traps' close. At I, the electrons start to become transferred via Q_B to the plastoquinone pool, so that a transient reoxidation of Q_A occurs. After the dip D, there is a sharp rise in fluorescence towards the peak P. If the actinic light is strong enough, then all reaction centres become closed at P ('closed traps'; $k_P = 0$; see below) and the maximum level of fluorescence (F_{max} or F_m) is attained, with no photochemistry taking place.

A useful parameter derived from the fast kinetics of the fluorescence induction curve is the so-called variable fluorescence (F_v), which equals the fluorescence increase from F_0 to F_m. The ratio F_v/F_m has a typical range of 0.75–0.85 and can be shown to be proportional to the quantum yield of photochemistry by calculation of rate constants for competing decay reactions at F_0 and F_m. To consider the de-excitation by energy transfer from PS II to Photosystem I (PS I) the rate constant k_T is introduced.

$$\Phi F_0 = k_F/(k_F + k_D + k_T + k_P)$$
$$\Phi F_m = k_F/(k_F + k_D + k_T) \qquad k_P = 0$$
$$\Phi F_v = \Phi F_m - \Phi F_0$$
$$\Phi F_v/\Phi F_m = k_P/(k_F + k_D + k_T + k_P)$$
$$\Phi P = F_v/F_m = (F_m - F_0)/F_m$$

Björkman and Demmig [4] measured F_v/F_m ratios for a large number of vascular plant species. They found that at 77 K healthy leaves show an uniform ratio of 0.832 ± 0.004 (Section 12.3.4). Empirically, the photochemical efficiency of PS II relates approximately linearly to the quantum yield of overall photosynthesis.

Recently, new measurements of fluorescence life times reveal that several changing rate constants are involved in the deactivation process to heat [16]. These findings seem to show that the relationship between F_v/F_m and ΦP is merely empirical, and certainly more complex than previously thought. The demonstration of at least two populations of PS II centres (α, β) and possibly variable cooperation between centres provide some evidence that there is no unique linear relationship between F_v/F_m and ΦP. Nevertheless, this correlation is highly reproducible with respect to photoinhibition.

The area above the curve (A; Fig. 12.3) between F_0 and F_m is proportional to the pool size of the electron acceptors on the reducing side of PS II. If electron transfer is blocked by DCMU at the Q_B site (Figs. 12.3 and 12.4), the area above the fluorescence curve will strongly decrease because no re-oxidation of Q_A can occur. The latest microprocessor-based systems are able to determine the area over the curve directly (*PEA, Hansatech Ltd., Kings Lynn, UK*), but the half-rise time ($t_{1/2}$)

for the rise from F_0 to F_m is an alternative simple indicator for estimating the size of the plastoquinone pool. Shade leaves with large chlorophyll antenna and a small plastoquinone pool exhibit a shorter $t_{1/2}$ (and thus a smaller area above the curve) compared with sun leaves, which have a small antenna size and a large plastoquinone pool [1].

Chlorophyll fluorescence kinetics can also be used to demonstrate heterogeneity of PS II. Under low incident light, the rise from F_0 to I reflects the reduction of Q_A in centres which are not connected to Q_B and PQ ('inactive centres') [16]. Analysis of the area increment A_t (the area at time t divided by the total area A; Fig. 12.3) above the fluorescence induction curve in the presence of DCMU reveals an initial fast non-exponential phase and a slow exponential phase. The nature of PS II heterogeneity is very complex and still under debate, but PS II_α (in the granal region) is generally characterized by a large light-harvesting complex (LHC) antenna, whereas PS II_β (in the stromal region) is characterized by a small LHC antenna [11].

12.3.2 Induction: slow kinetics

Following the P peak, fluorescence decreases through the S-phase as Q_A becomes partially oxidised through electron transfer to PS I via the cytochrome *f*/*b*$_6$ complex and plastocyanin (Fig. 12.4). However, the fluorescence quenching following the P peak is only partially explained by a re-oxidation of Q_A, i.e. by an increased photochemical quenching. During the S phase following the P peak, electron transport generates a proton gradient which is gradually built up across the thylakoid membrane causing its energization. Besides being the driving force for ATP synthesis, this pH gradient also causes additional quenching of the fluorescence, often called energy-dependent quenching, q_E. This type of non-photochemical quenching (q_{NP}) of fluorescence occurs via an increased deactivation through heat formation. Thus the interpretation of the slow phase of the fluorescence induction curve is complicated by a disproportionate change in heat de-excitation of chlorophyll (k_D). Since the Calvin cycle consumes ATP and NADPH.H, carbon assimilation can affect the slow fluorescence kinetics through both photochemical and non-photochemical quenching [3,8].

The M peak reflects the induction of increased rates of carbon fixation, and the height of the M peak depends on the relative effects of photochemical and non-photochemical quenching. These two types of fluorescence quenching can be determined by the use of modulated fluorescence (Fig. 12.5). However, a simple but useful parameter to describe the state of fluorescence quenching under steady-state photosynthesis is the 'fluorescence decline ratio', which is thought to be determined by Calvin Cycle activity [5]:

$$F_{dr} = F_m - F_T/F_m.$$

In summary, during the slow induction phase as well as during steady-state photosynthesis, we must consider two types of fluorescence quenching processes in order to understand the relationship between photosynthesis and fluorescence. Photochemical quenching (q_P) is determined by photochemistry so that photochemical quenching decreases in proportion to the closure of the reaction centres, i.e. reduction of Q_A. Non-photochemical quenching (q_{NP}) is mainly determined by de-excitation through heat generation, and its largest contributor is the energy-dependent quenching created by the pH gradient across the thylakoids. Minor components of non-photochemical quenching are controlled by the amount of excitation energy transferred to PS I (Figs. 12.4 and 12.5).

The effect of DCMU:
If DCMU is added in the middle of the slow fluorescence induction phase of, for example,

an unicellular green alga such as *Chlorella*, it is found that fluorescence increases abruptly, then more slowly until it approaches the fluorescence level at the P peak (sometimes even lower because Q_A^- + Q_BDCMU is quenching the fluorescence [3,8] [16]; Fig. 12.3). During fast kinetics of induction DCMU causes rapid relaxation of photochemical quenching (q_P). From this we would expect an immediate increase of fluorescence to reach, or exceed, the P-peak. The reason why this does not occur is that algae (and higher plants) under the course of the slow induction phase have built up additional quenching processes (q_{NP}). This type of fluorescence decrease results from conformational changes probably caused by generation of the pH gradient which drives ATP synthesis (q_E), and increasing deactivation via heat (k_D).

12.3.3 Modulated light fluorescence

For the quantitative separation of fluorescence quenching, modulated light fluorimeters are used to facilitate the analysis. Like a stroboscopic light, the measuring light is switched on and off with frequencies from 1 to 100 kHz. In a dark-adapted leaf, this modulated light beam of very low intensity induces such a low rate of excitons transferred to PS II reaction centres that reoxidation processes are fast enough (<1.3 ms) to keep all reaction centres open for any incoming exciton, and no variable fluorescence is generated. The weak initial rise in fluorescence is assumed to be the F_0 level. By using an amplifier which only monitors the fluorescence emitted at the frequency and phase of the modulated beam, light of any wavelength can be used to induce a Kautsky curve or to close all reaction centres by a saturating light pulse (Fig. 12.5).

If no changes in chloroplast arrangement and ultra-structure occur, the modulated beam will always 'see' the same collective of chlorophyll molecules, and therefore only the changes in deexcitation of these chlorophyll molecules will be monitored in fluorescence measurements. The addition of a brief strong light pulse to the modulated beam will close all PS II reaction centres in the observed collective and will give rise to the maximal fluorescence, F_m. The variable fluorescence, F_v, almost equals the maximal possible fluorescence quenching and therefore at F_0 it equals also the photochemical quenching. If, during the generation of the Kautsky curve by 'actinic' light, saturating light pulses are applied, maximal fluorescence (F'_m) will not reach F_m, indicating that non-photochemical quenching mechanisms (which means the de-excitation via heat) have been developed. When the actinic light is switched off after several minutes the F'_0 value of the light-adapted state will be recorded which is generally lower than the initial F_0. By the application of saturating light pulses the coefficients of the two quenching mechanisms may be calculated (see also Fig. 12.5).

12.3.4 Low temperature fluorescence

The influence on chlorophyll fluorescence of electron transport capacity (limited, for example, by Calvin Cycle activity or any other enzymatically determined reaction) can be avoided if measurements are performed at the temperature of liquid nitrogen (77 K) so that only primary photochemical reactions will be measured [15,20].

If a leaf is dark adapted prior to freezing, the time dependent fluorescence yield of PS II will depend only on the redox state of Q_A, i.e. on photochemical quenching. All other temperature-dependent processes causing non-photochemical quenching are inhibited. For example, if a leaf is frozen in darkness and subsequently exposed to light, the F_0 typical of 'open' reaction centres will first be resolved followed by a rise to F_m as Q_A becomes reduced and the reaction centres 'close' (Fig. 12.4). Hence 77 K fluorescence induction kinetics have become an important tool for the determination of the photo-

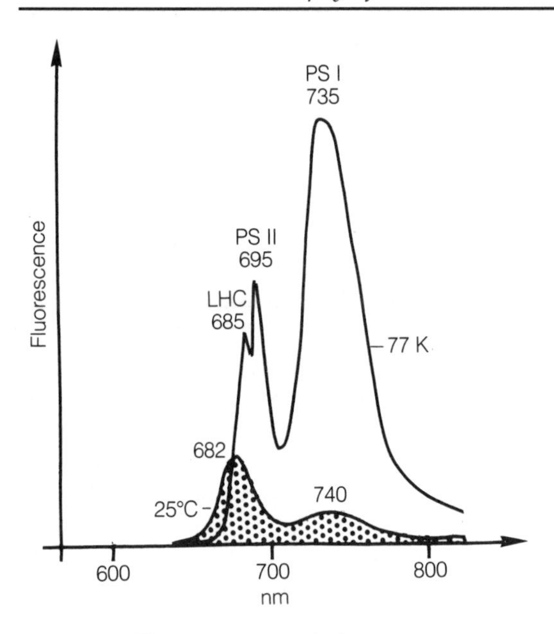

Fig. 12.6. Fluorescence emission spectra at room temperature (25 C) and at 77 K of intact isolated spinach chloroplasts (after Krause and Weis [15]).

chemical efficiency of PS II as expressed by the ratio of F_v/F_m.

Fluorescence spectroscopy at 77 K has also frequently been used to study energy distribution between the two photosystems [7]. This is because both PS II and PS I exhibit distinct emission bands at 77 K, 695 and 735 nm, respectively (Fig. 12.6). The variable part of PS I fluorescence observed at 735 nm is suggested to result from excitation energy transfer from PS II to PS I ('spillover') and reflects closure of PS II reaction centres, i.e. reduction of Q_A.

12.3.5 Dark adaptation

In darkness, q_{NP} will reverse completely, albeit relatively slowly. If NADPH.H is seen as the final product of non-cyclic electron transport, then its redox state will affect the capacity for electron transport and photochemistry, hence q_P and indirectly q_E. However, this redox state will also be affected by the rate of consumption of NADPH.H in the carbon cycle, and in the other stromal reductive reactions of photosynthetic carbon and nitrogen metabolism. Similarly, the rate of ATP consumption will affect capacity for consuming the transthylakoid pH gradient and thus affect q_E. It is clear that if we want to use fluorescence to determine the maximal photochemical efficiency of PS II expressed as F_v/F_m, we must dark adapt the leaf prior to measurement. A dark adaptation period of 30 minutes is usually enough to reverse all non-photochemical fluorescence quenching provided that photoinhibition of photosynthesis (Chapter 9) is not involved [21]. The two quenching components, q_P and q_{NP}, can be easily resolved by use of modulated fluorescence (Fig. 12.5) [15,18,22].

12.4 FLUORESCENCE AS A SENSITIVE INDICATOR OF STRESS

It is well documented that the function of thylakoid membranes is sensitive to several environmental stresses. PS II, including the water oxidising step, appears to be particularly sensitive to a number of stress factors such as high temperatures, chilling [12,24], freezing, drought and excessive radiation [21].

Since all these stresses affect the function of PS II, directly or indirectly, fluorescence can be used as a tool, not only in revealing stress response mechanisms but also in quantifying the stress response under laboratory and field conditions. A very good correlation has been found between photoinhibition of photosynthesis, as induced by excessive excitation, and a decrease of the F_v/F_m ratio. This ratio is therefore a very useful measure of photoinhibition [14,21]. Photoinhibition can be observed as a secondary stress response to moderate, or even low light if the photosynthetic process as a whole is restricted by some other stress factor, particularly low temperature (Fig. 12.7).

Furthermore, the frost-killing temperature for pine can be indicated by chlorophyll flu-

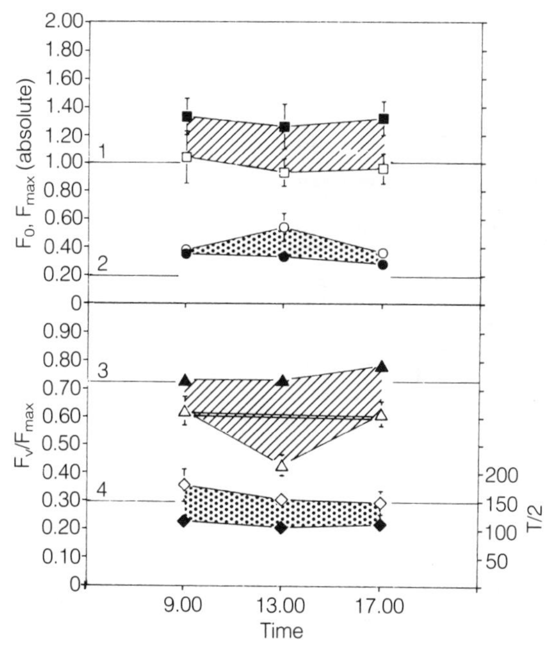

Fig. 12.7. Chlorophyll fluorescence induction of french bean leaves on a day with full sunshine. Closed symbols, artificially shaded plants; open symbols, sun-exposed plants. The area between the two graphs is related to the degree of photoinhibition. Lines 1 and 2, normalised values of F_m (1) and F_0 (2) (F_m level normally equals 5 times F_0 level); line 3, threshold value of 0.725 for F_v/F_m; line 4, mean of all measured $t_{1/2}$ (T/2) values. The area below the parallel line in the F_v/F_m graph represents fast relaxing photoinhibition [5].

orescence, because of direct or indirect damage to primary photochemistry [25]. In studies of forest decline, and more generally in studies of effects of air pollution on plants and photosynthesis, chlorophyll fluorescence has become an important tool, not only in screening for effects but also in elucidating the nature of damage to photosynthesis (Fig. 12.8) [5,6].

The slow fluorescence kinetics have also been related to stresses such as drought. One site for inhibition of photosynthesis by low water potential appears to be the decrease in carbon reduction cycle activity caused by stomatal closure. Since this is an effect on the 'energy-consuming' part of photosynthesis the following scenario may be suggested.

Inhibition of CO_2 fixation will reduce the consumption of ATP and NADPH.H, causing an increased pH gradient and a more reduced electron transport chain. An increased pH gradient will cause decreased fluorescence through increased q_{NP}, whereas an increased level of reduced electron acceptor in PS II, Q_A, will cause increased fluorescence through decreased q_P. The resulting fluorescence change observed upon water stress will be determined by the relative importance of the change in these two quenching processes.

A good example for the demonstration of the relationship between fluorescence, q_P and q_{NP} is provided in studies of photosynthesis induction at different levels of CO_2 for dark-adapted leaves [2,22,29].

12.4.1 Fluorescence as a screening tool

Chlorophyll fluorescence kinetics have been used in screening for plant individuals in certain acclimatised states or with certain stress tolerance. Sunblad and co-workers [26] demonstrated that the slow fluorescence kinetics phase contains information that can be correlated to the degree of frost hardiness in Scots pine, and that fluorescence kinetics may be used as a predictive tool to screen plants for state of cold hardiness. This observation is an example that chlorophyll fluorescence properly calibrated can, even if we do not know the mechanistic background, be used as a screening test to select individual plants with changed stress tolerance. Chlorophyll fluorescence has similarly been used to select for high tolerance to chilling and freezing temperatures.

12.5 FLUORIMETERS

Two types of fluorimeter are available (Fig. 12.9): time-resolving systems and modulated systems.

12.5.1 Time-resolving systems

These instruments record only 'Kautsky' curves to measure F_0 and F_m for calculation of

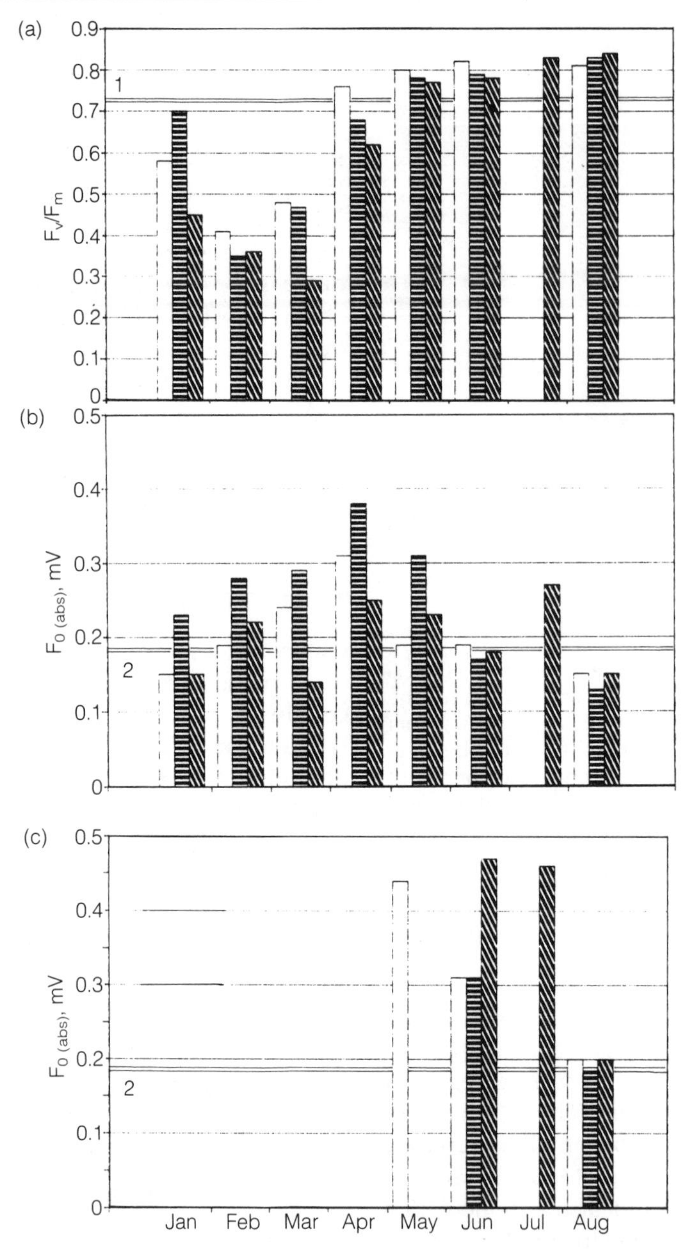

Fig. 12.8. Annual variation in (a) variable fluorescence (F_v/F_m), and (b) and (c) in initial fluorescence (F_0), due to winter stress, ozone and ontogenesis. Data gained from spruce in alpine regions using a time-resolving fluorimeter (*PSM, BioMonitor, Sweden*). □ Site 1, 700 m; ▤, Site 2, 1000 m; ▧ Site 3, 1500 m. (a) Needle set 1987: the decline of F_v/F_m in February (freezing temperatures) and March (low temperatures and relatively high light) is a typical winter stress phenomenon. (b) Needle set 1987: F_0 values from January to April can be correlated with the recorded ozone concentrations especially at site 2. (c) Needle set 1988: during the development of the new needle set the F_0 value is high as well. Line 1, threshold value for F_v/F_m of 0.725: line 2, mean of all measured F_0 values.

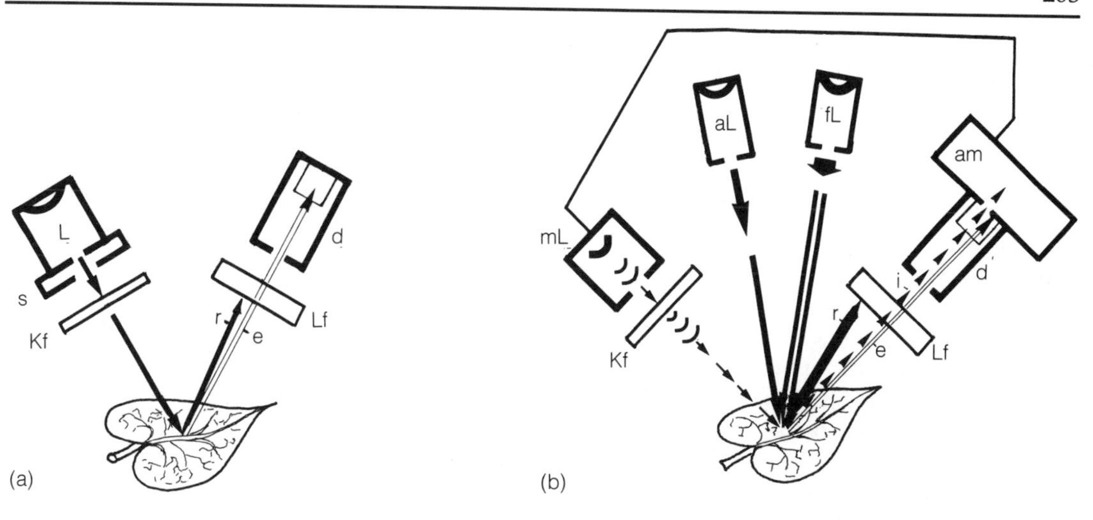

Fig. 12.9. Generic construction of different types of fluorimeter. (a) Non-modulated (time-resolving) system. (b) Modulated system. L, light source (white halogen or red LEDs); mL, modulated (pulsed) light (red or yellow LEDs); aL, 'actinic' white light; fL, saturating flash of white light; d, detector; am, 'log in' or 'window' amplifier synchronised with the pulsing light source; Kf, short-pass filter (e.g. *Schott BG38*, <660 nm); Lf, long-pass and band filter (e.g. 690, 710, 730 nm); r, reflected white light; e, emitted fluorescent light; i, modulated (pulsed) fluorescent light.

F_v/F_m and $t_{1/2}$ (Fig. 12.3; e.g. *Plant Stress Meter (PSM), BioMonitor AB, Umea, Sweden; Plant Efficiency Analyser (PEA), Hansatech Ltd, King's Lynn, (UK)* [19]. The leaves have to be dark adapted for at least 30 minutes and the excitation light must be passed through a cut-off filter which allows only the passage of light with wavelengths less than 620 nm. The photodiode measures the fluorescence light via a band filter of 695 nm. As the definition of F_v depends on the accurate determination of F_0 the resolution (or calculation) of this signal is a potential source of error. A second problem might be caused by use of an excitation light which is insufficient to reduce all acceptors at F_m. Due to the differences in penetration of blue, yellow or red light into the leaf tissue the signal from different instruments can vary markedly [28]. The bandwidth of the filters used in fluorimeters has important implications for the accurate estimation of F_0. If F_0 and F_v are measured at 690 nm, the absoute values are strongly influenced by re-absorption of the red fluorescence light by chlorophyll, whereas measurement of F_0 at wavelengths above 700 nm gives rise to significant errors caused by substantial contribution of fluorescence from PS I [9].

12.5.2 Modulated systems

The problems mentioned above can be avoided partly by using a modulated fluorimeter (Section 12.3.3; e.g. *PAM Fluorimeter 101, Walz, Effeltrich, Germany; Modulated Fluorescence Measuring System (MFMS), Hansatech, King's Lynn, UK*). These allow the measurement of continuous fluorescence signals from a leaf exposed to light of any wave-length (see also Fig. 12.5). This is achieved by using a weak modulated light source in conjunction with a fluorescence system, which monitors only the fluorescence emitted at the frequency and phase of the modulated light [18]. To generate a Kautsky curve a second light source (around $150\,\mu mol\,m^{-2}\,s^{-1}$) is needed. To generate trap closure at any point on the induction curve, high intensity flashes (5–

20,000 μmol $m^{-2} s^{-1}$) from a third light source are used for execution of the so-called quenching analysis.

12.5.3 Applications

The choice of fluorimeter for a research project depends upon the questions to be answered. For most routine field work, estimation of changes in photosynthetic capacity, i.e. the efficiency of energy capture by open PS II centres (F_v/F_m) will be sufficient, using a portable battery-operated non-modulated system.

If the research involves studies on energy de-activation in PS II reaction centres (photochemical and non-photochemical quenching of fluorescence), then modulated fluorescence systems provide additional information. Changes in the quantum yield of non-cyclic electron transport (ϕ_e) are thought to have particular application. The efficiency of excitation capture by open PS II reaction centres (F_v/F_m), multiplied by the concentration of open PS II reaction centres (q_P), indicates the yield of non-cyclic electron transport *in vivo* ($F'_m - F/F'_m$) (Fig. 12.5). ϕ_e shows a linear to curvilinear correlation with both the quantum yield of oxygen evolution and the quantum yield of CO_2 fixation over a wide range of light intensities and CO_2 concentrations [18,23].

12.6 EXPERIMENTS

12.6.1 Laboratory experiments

(i) Red fluorescence can be demonstrated with a concentrated chlorophyll solution. Spinach leaves (5 g) are ground with 5 ml quartz sand in 20 ml of an acetone-based buffered extraction medium (80% acetone, 19.8% H_2O, 0.2% NH_3 or $MgCO_3$). The resulting green paste is diluted with 80 ml extraction medium and filtered carefully, preferably using a vacuum flask. Quantitative extraction is possible if performed under low light (5–10 μmol $m^{-2} s^{-1}$) and at low temperature (4°C) to avoid formation of pheophytin. The best results are obtained by grinding the leaves in liquid nitrogen. A round-bottomed flask of the chlorophyll solution will be seen to emit scattered red light when illuminated by a narrow beam of white light.

(ii) Basic fluorescence phenomena are measured using dark-adapted leaves of barley, maize or wheat. Alternatively, algae or intact chloroplasts of spinach or pea are suitable.

(a) Measure the fluorescence induction curve in the presence of an uncoupler (NH_4Cl) and/or electron acceptor (methyl viologen) (Chapter 18). The leaves can be pre-incubated for one hour to allow the additives to penetrate the cells. Alternatively, rub the leaves gently with carborundum or strip off the epidermis. If a fibre-optic system is used protect the fibre from chemicals by putting a piece of transparent plastic between the fibre and the leaf. Compare the fluorescence parameters F_0, F_m, F_v, F_v/F_m, $t_{1/2}$ obtained with treated and untreated leaves.

(b) Repeat the experiment with the addition of DCMU (20 μmol).

(iii) Demonstration of photochemical (q_P) and non-photochemical (q_{NP}) quenching with isolated intact chloroplasts (spinach or pea), protoplasts or unicellular green algae (Fig. 12.3). A fluorescence induction curve is recorded until steady-state fluorescence is reached during the T-phase. Then add DCMU (20 μmol) and follow the fast (q_P) and slow (q_{NP}) phases of the fluorescence increase.

(iv) Demonstration of loss of F_v due to photoinhibition. For this experiment, a shade leaf of any plant material can be used. Leaves kept at room temperature (control) or on ice are illuminated by a strong light source (1000–2000 μmol $m^{-2} s^{-1}$) for at least 30 min. To reduce the heat emitted from the lamp a water filter should be used. After dark adaptation for another 30 minutes, fluorescence parameters are determined. Data should be compared with controls kept in the dark, and controls kept under non-inhibitory low light.

(v) Chilling stress (Fig. 12.8) can be detected in darkness and light using leaves of chilling-sensitive (maize) and chilling-resistant plants (wheat). The leaves of maize and wheat are kept on ice in darkness or light (about 400 $\mu mol\, m^{-2} s^{-1}$) for 12 hours. Measure fluorescence parameters at both room temperature and 77 K. The results of both measurements should be compared to reveal the site of chilling damage in darkness and light.

(vi) Relationship between F_v/F_m and quantum yield of photosynthesis. Such measurements can be performed with leaves of any plant. To induce photoinhibition a strong light source (2000 $\mu mol\, m^{-2} s^{-1}$) is needed. To determine the quantum yield, equipment for measuring CO_2 uptake under increasing light intensities must be available. Photoinhibition will increase with the length of time at high irradiance. During this period, several measurements of F_v/F_m and the quantum yield may be performed. Plot the quantum yield as a function of F_v/F_m.

12.6.2 Field experiments

(i) Photoinhibition (Fig. 12.7). On a sunny day with high irradiance many plants will show a pronounced decline in photosynthetic rates, especially around midday. This decline can in part be attributed to photoinhibition. Several plants (bean, eucalypts, castor bean, maize, maple, spinach, spruce, wheat) may be selected. Measurements of fluorescence parameters are made in the morning, at midday and during the late afternoon. Shade some of the plants, or for potted plants, move some of them to a shady place. A comparison of well-watered and water-stressed plants gives impressive results. As photoinhibition develops first on the upper, sun-exposed surface of a leaf, measurements of the lower and upper surface can give quite different results.

ACKNOWLEDGEMENT

We are grateful to Dr. Christa Critchley, University of Queensland, Australia, for comments during the revision of the manuscript.

REFERENCES

1. Anderson J.M., W.S. Chow and D.J. Goodchild (1988) Thylakoid membrane organisation in sun/shade acclimation. In: *Ecology of Photosynthesis in Sun and Shade* (Evans, J.R., von Caemerer, S. and Adams III, W.W., eds.) pp 11–26. CSIRO, Melbourne.
2. Baker N.R., T.M. East and S.P. Long (1983) Chilling damage to photosynthesis in young *Zea mays* leaves. *J. Exp. Bot.* **139**, 189–197.
3. Baker N.R. and P. Horton (1988) Chlorophyll fluorescence quenching during photoinhibition. In: *Photoinhibition* (Kyle, D.J., Osmond C.B. and Arntzen, C.J., eds.) pp 145–168. Elsevier Science Publishers.
4. Björkman, O. and B. Demmig (1987) Photon yield of O_2 evolution and chlorophyll fluorescence characteristics at 77K among vascular plants of diverse origins. *Planta* **170**, 489–504.
5. Bolhàr-Nordenkampf, H.R. and E.G. Lechner (1988) Winter stress and chlorophyll fluorescence in Norway spruce (*Picea abies* (L.) Karst). In: *Applications of Chlorophyll Fluorescence* (Lichtenthaler, H.K., ed.) pp. 173–180. Kluwer Academic Publ., Dordrecht.
6. Bolhàr-Nordenkampf, H.R., S.P. Long, N.R. Baker, G. Öquist, U. Schreiber and E.G. Lechner (1989) Chlorophyll fluorescence as a probe of the photosynthetic competence of leaves in the field: a review of current instrumentation. *Functional Ecol.* **3**, 497–514.
7. Butler, W.L. (1978) Energy distribution in the photochemical apparatus of photosynthesis. *Ann. Rev. Plant Physiol.* **29**, 345– 378.
8. Genty, B., J-M. Briantais, and N.R. Baker (1989) The relationship between the quantum yield of photosynthetic electron transport and photochemical quenching of chlorophyll fluorescence. *Biochim. Biophys. Acta* **990**, 87–92.
9. Genty, B., J. Wonders and N.R. Baker (1990) Non-photochemical quenching of F_0 in leaves

is emission wavelength dependent: consequences for quenching analysis and its interpretation. *Photosynth. Research* **26**, 133–139.

10. Govindjee and P.A. Jursinic (1979) Photosynthesis and fast change in light emissions by green plants. In: *Photochemical and Photobiological Reviews*, Vol 4. (Smith, K.C., ed.) Plenum Press, New York and London.
11. Guenther, J.E. and A. Melis (1990) The physiological significance of Photosystem II heterogeneity in chloroplasts. *Photosynth. Research* **23**, 105–109.
12. Havaux, M. and R. Lannoye (1984) Effects of chilling temperatures on prompt and delayed chlorophyll fluorescence in maize and barley leaves. *Photosynthetica* **18**, 117–127.
13. Kautsky, H. and A. Hirsch (1934) Chlorophyl-lfluoreszenz und Kohlensäureassimilation. Das Fluoreszenzverhalten grüner Pflanzen. *Biochem. Zeitschrift* **274**, 423–434.
14. Krause, G.H. (1988) Photoinhibition of photosynthesis. An evaluation of damaging and protective mechanisms. *Physiologia Plantarum* **74**, 566–574.
15. Krause G.H. and E. Weis (1984) Chlorophyll fluorescence as a tool in plant physiology.II. Interpretation of fluorescence signals. *Photosynth. Research* **5**, 139–157.
16. Krause G.H. and E. Weis (1991) Chlorophyll fluorescence and photosynthesis: the basics. *Ann. Rev. Plant Physiol. Plant Mol. Biol.* **42**, 313–349.
17. Lavorel, J. and A.L. Etienne (1977) *In vivo* chlorophyll fluorescence. In: *Primary Processes in Photosynthesis* (Barber, J., ed.) pp. 203–268. Elsevier, Amsterdam.
18. Ögren, E. and N.R. Baker (1985) Evaluation of a technique for the measurement of chlorophyll fluorescence from leaves exposed to continuous light. *Plant Cell and Environment* **8**, 539–547.
19. Öquist, G. and R. Wass (1988) A portable microprocessor operated instrument for measuring chlorophyll fluorescence kinetics in stress physiology. *Physiologia Plantarum* **73**, 211–217.
20. Powles, S.B. and O. Björkmann (1982) Photoinhibition of photosynthesis: effect on chlorophyll fluorescence at 77 K in intact leaves and in chloroplast membranes of *Nerium oleander*. *Planta* **156**, 97–107.
21. Powles, S.B. (1984) Photoinhibition of photosynthesis induced by visible light. *Ann. Rev. Plant Physiol.* **35**, 14–55.
22. Schreiber, U., W. Schliwa and U. Bilger (1986) Continuous recording of photochemical and non-photochemical chlorophyll fluorescence quenching with a new type of modulation fluorimeter. *Photosynth. Research* **10**, 51–62.
23. Seaton, G.G.R. and D.A. Walker (1990) Chlorophyll fluorescence as a measure of photosynthetic carbon assimilation. *Proc. Roy. Soc. Lond. B* **242**, 29–35.
24. Smillie, R.M. (1979) The useful chloroplast: A new approach for investigating chilling stress in plants. In: *Low Temperature Stress in Crop Plants* (Lyons, J.M., Graham, D. and Raison, J.K. eds.). pp. 187–202. Academic Press. New York.
25. Strand, M. and G. Öquist (1985) Inhibition of photosynthesis by freezing temperatures and high light levels in cold acclimated seedlings of Scots pine (*Pinus sylvestris*). I. Effects on the light limited and light saturated rates of CO2 assimilation. *Physiologia Plantarum* **64**, 425–430.
26. Sunblad, L-G., M. Sjostrom, G. Malmberg and G. Öquist (1990) Prediction of frost hardiness in seedlings of Scots pine (*Pinus sylvestris*) using multivariate analysis of chlorophyll fluorescence and luminescence kinetics. *Canadian J. Forest Research* **20**, 592–597.
27. Van Kooten, O. and J.F.H. Snel (1990) The use of chlorophyll fluorescence nomenclature in plant stress physiology. *Photosynth. Research* **25**, 147–150.
28. Vogelmann, T.C., J.F. Bornman and S.A. Josserand (1989) Photosynthetic light gradients and spectral regime within leaves of *Medicago sativa*. *Phil. Trans. R. Soc. Lond. B* **323**, 411–421.
29. Walker, D.A., M.N. Sivak, R.T. Prinsley and J.K. Cheesborough (1983) Simultaneous measurements of oscillations in oxygen evolution and chlorophyll fluorescence in leaf pieces. *Plant Physiol.* **73**, 542–549.

13

Modelling of solar irradiance, leaf energy budget and canopy photosynthesis

I.N. Forseth and J.M. Norman

13.1 INTRODUCTION

The elements of climate (radiation, temperature, wind and moisture) determine to a large extent the type of vegetation and development of soil in a biome [25]. The driving force behind the Earth's climate is radiation from the sun. Incident solar radiation on the Earth's surface produces differential temperature regimes that, when coupled with the Earth's rotation around its axis and orbit around the sun, generate wind patterns and ocean currents. These air and water movements in turn influence global patterns of rainfall and heat distribution, generating the diurnal, seasonal and latitudinal changes in climate that occur on the Earth's surface. In addition to its effects on climate, solar energy is the ultimate source of energy for all life on Earth. Thus, the ability to calculate solar irradiance on a surface, whether it is plant, animal, or non-living is a valuable tool in physiological ecology. The first part of this chapter presents relationships that may be used to calculate solar radiation on idealized surfaces under clear sky conditions.

All organisms interact with their physical environment through energy exchange processes. For plants, these energy exchange processes are central to determining leaf temperatures, rates of photosynthetic carbon gain and rates of water loss [13]. Since plants are sessile, there are restrictions on their ability to alter their microhabitat and the surrounding energy environment. Thus, mechanisms to modulate energy exchange between the leaf and its microhabitat are intrinsic to successful plant growth and reproduction. Leaf orientation (angle from the horizontal and azimuth) will determine the amount of solar radiation intercepted by the leaf. Leaf surface characters will determine the proportion of this incident radiation which is absorbed. Finally, leaf size, shape, and the degree of stomatal opening interact to determine how this absorbed radiation will be partitioned among re-radiative, convective, and latent heat exchange avenues. The second part of this chapter explores relationships among leaf properties and leaf energy exchange with the environment, while the third section investigates the connection between leaf characteristics, water loss and carbon gain by adopting a phenomenological approach of combining

Photosynthesis and Production in a Changing Environment: a field and laboratory manual
Edited by D.O. Hall, J.M.O. Scurlock, H.R. Bolhàr-Nordenkampf, R.C. Leegood and S.P. Long
Published in 1993 by Chapman & Hall, London.
ISBN 0 412 42900 4 (HB) and 0 412 42910 1 (PB).

photosynthetic responses to light and temperature with leaf energy budget models.

Although single leaf level analyses can provide many insights into plant adaptation to the environment, canopy level processes are most important in analysing crop and biome-level productivity. The final section of this chapter demonstrates a method which utilizes leaf characteristics to calculate canopy conductance and photosynthesis in response to solar radiation. This approach may be used to study the effects of canopy structural characteristics, such as mean leaf inclination and leaf area index, on canopy photosynthetic carbon gain and water loss.

13.2 CALCULATION OF SOLAR IRRADIANCE

Before we examine the relationships that determine the level of incident solar radiation on a surface, we must define some terms. One of the most important of these is the *solar constant*. The solar constant is the amount of radiation falling on a surface perpendicular to the sun's rays immediately outside of the Earth's atmosphere. It varies with season and sunspot activity, but a good approximation of its value is $1360\,W\,m^{-2}$ [15]. *Declination* (δ) is the tilt of the Earth on its axis. This value is usually defined in relation to the northern hemisphere, where declination is 0° on the spring and autumnal equinoxes (21st March, 22nd September), 23.5° on the summer solstice (22nd June), and −23.5° on the winter solstice (22nd December).

The amount of solar energy striking a surface on the Earth is dependent upon the orientation of the surface, relative to the position of the sun, and on atmospheric transmissivity. The position of the sun in the sky is determined by latitude, time of year, and time of day. Atmospheric transmissivity is a function of altitude, cloudiness, and the amount of particulates (pollutants, dust and water vapour) in the air. The following equations can be used to calculate the sun's position in the sky. First, solar zenith angle (θ), the angle of the sun from the vertical can be calculated from:

$$\cos\theta = \sin(\Lambda)\sin(\delta) + \cos(\Lambda)\cos(\delta)\cos(15(T - T_{SN})) \tag{13.1a}$$

where Λ = latitude;
δ = declination;
T = solar time;
T_{SN} = solar noon.

Declination may be estimated from:

$$\delta = -23.5\cos[360(D_j + 10)/365]$$

where D_j is the Julian date. Next, we may determine what compass direction the sun is at, termed the solar azimuth (η_s) from:

$$\sin(\eta_s) = \cos(\delta)\sin(15[T - T_{SN}])/\sin(\Lambda) \tag{13.1b}$$

The relationship $180 + \eta_s$ may be used to convert azimuth to compass directions (0° = N, 90° = E, 180° = S, 270° = W). After determining the sun's zenith angle and azimuth, we may calculate solar radiation, Φ_{SP} ($W\,m^{-2}$) perpendicular to the sun's rays, after it has been attenuated by passage through the atmosphere [2]:

$$\Phi_{SP} = 1360 \times \tau_A^{[(p/p_0/\cos(\theta)]} \tag{13.2a}$$

where τ_A = atmospheric transmissivity (usually 0.6–0.8 for clear skies);
p = atmospheric pressure (Pa);
p_0 = atmospheric pressure at sea level.

A similar relationship could be used to calculate the energy in the photosynthetically active portion of the solar spectrum (400–700 nm), by using a value of $600\,W\,m^{-2}$ for the solar constant [26]. Equation 13.2a is an estimate of direct solar radiation only. We must also calculate diffuse solar radiation, $\Phi_{S(DIFF)}$. This can be approximated for clear sky conditions by [2]:

$$\Phi_{S(DIFF)} = 0.5 \times 1360 \times \{1 - t^{[(p/p_0)/\cos(\theta)]}\} \times \cos(\theta) \tag{13.2b}$$

The amount of reflected solar radiation, $\Phi_{S(REF)}$ may be approximated by:

$$\Phi_{S(REF)} = (1 + a)(\Phi_{SP} + \Phi_{S(DIFF)})$$

where a is the albedo or reflectance of the surrounding substrates.

Finally, we must calculate how much of the direct solar radiation is incident on a surface that is not perpendicular to the sun's direct rays. To do this, we calculate a term called the *cosine of the angle of incidence* ($\cos \iota$). This is defined as the cosine of the angle between the surface and a perpendicular to the sun's direct beam. It is equivalent to the fraction of the direct beam the surface actually intercepts [11].

$$\cos\iota = \cos(\alpha)\cos(\theta) + [\sin(\alpha)\sin(\theta)\cos(\eta_s - \eta_l)] \quad (13.3)$$

where α is the angle of the surface from the horizontal, and η_l is the surface azimuth. The product of $\cos\iota$ and Φ_{SP} is the amount of direct solar radiation on any surface, at any time of year, at any time of the day. This value may then be added to $\Phi_{S(DIFF)}$ and $\Phi_{S(REF)}$ to provide an estimate of total incident solar radiation, I_s. It should be emphasized that the equations presented here are applicable only to single, suspended leaves under clear sky conditions. Diffuse and reflected radiation components will vary markedly from these relationships for leaves within canopies, or under cloudy skies [20, 23,26]. Equations 13.1–13.3 may be used to illustrate the effects of season (Fig. 13.1a), surface angle (Fig. 13.1b), and surface azimuth (Fig. 13.1c) on the interception of direct solar radiation.

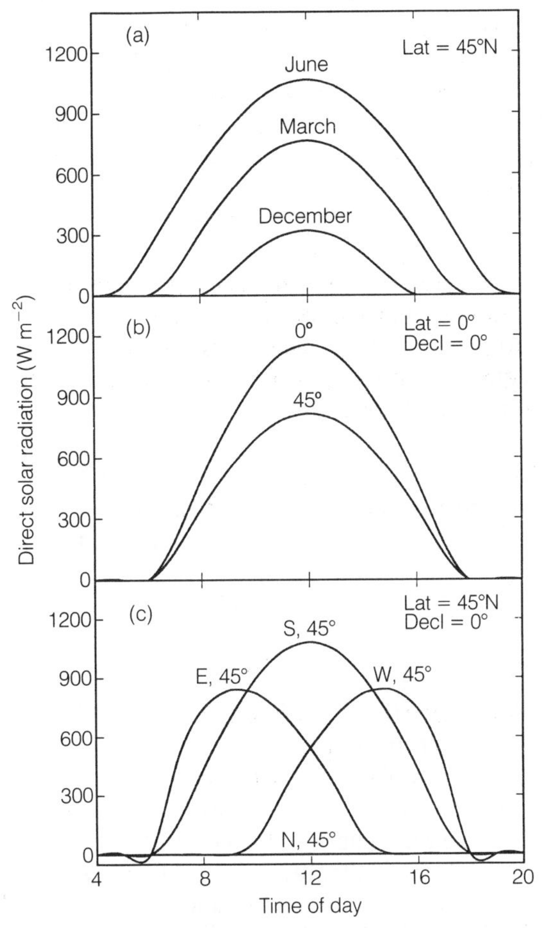

Fig. 13.1. Calculated direct solar radiation levels incident on a variety of surfaces under clear sky conditions. Equations used in the calculations are presented in the text.
(a) The effect of declination is illustrated by calculating direct solar radiation incident on a horizontal slope at a latitude of 45°N on the summer solstice (June), spring equinox (March) and winter solstice (December).
(b) The effect of angle from the horizontal is illustrated by calculating direct solar radiation on a horizontal surface and a surface at 45° from the horizontal at a latitude of 0° on the spring equinox.
(c) The effect of azimuth is illustrated by calculating direct solar radiation on a surface which is at an angle of 45° from the horizontal pointing north, south, east and west at a latitude of 45°N on the spring equinox.

13.3 LEAF RADIANT ENERGY BALANCE

The early work of Gates [12] pioneered leaf energy budget investigations in plant ecology. Although the interaction between leaves and their radiant energy environment can be extremely complex, a simplified approach that serves to introduce basic concepts governing leaf energy exchange is now fairly well-established [7]. This approach starts with the premise that at equilibrium, i.e. when leaf temperature is constant, all energy inputs into a leaf are balanced by energy outputs:

$$\text{energy}_{in} = \text{energy}_{out} \tag{13.4a}$$

When either of these terms change, leaf temperature will change (increase or decrease) until an equilibrium is re-established. The equation can be expanded to include different kinds of energy exchange:

$$\text{absorbed radiation} = R + C + LE + M + G \tag{13.4b}$$

where absorbed radiation = both short-wave (solar) and long-wave (infrared) electro-magnetic radiation inputs;

R = re-radiation of long-wave radiation from the leaf
C = sensible heat exchange between the leaf and the surrounding air
LE = latent energy exchange that occurs when water evaporates from or dew forms on the leaf surface
M = metabolic heat used in photosynthetic or catabolic functions
G = heat energy stored in tissues.

For most situations in nature, the M and G components in this expression are relatively small. Therefore, leaf energy budget can be expressed by:

$$A[\Phi_{S(DIR)} + \Phi_{S(DIFF)} + \Phi_{S(REF)}] + \epsilon[\Phi_{L(UP)} + \Phi_{L(DN)}] = 2[\epsilon\sigma(T_l + 273)^4] + [c_p\rho(T_l - T_a)]/r_{a,H} + ([0.622\rho\lambda/p](e_{s,T_l} - e_a))/(r_s + r_{a,H_2O}) \tag{13.4c}$$

where A = leaf absorptance to solar radiation;
ε = emissivity of the leaf to infrared radiation;
$\Phi_{L(UP)}$ = infrared radiation incident on the leaf from below (soil);
$\Phi_{L(DN)}$ = infrared radiation incident on the leaf from above (sky);
σ = Stefan–Boltzmann constant ($5.67 \times 10^{-8}\,W\,m^{-2}K^{-1}$);
T_l = leaf temperature (°C);
c_p = specific heat of air;
ρ = density of air;
T_a = air temperature (°C);
$r_{a,H}$ = mean leaf boundary layer resistance to heat transfer (both sides of the leaf);
λ = latent heat of vaporisation of water;
p = atmospheric pressure;
e_{s,T_l} = saturated vapour pressure at leaf temperature;
e_a = water vapour pressure of ambient air;
r_s = mean stomatal resistance to water vapour (both sides of the leaf);
r_{a,H_2O} = mean leaf boundary layer resistance to water vapour.

With the advent of the microcomputer, the ability to evaluate individual leaf properties for their impact on leaf temperature and water loss using this equation has improved dramatically. This is done by an iterative procedure, started by setting leaf temperature equal to air temperature, computing each term in the equation, and then comparing the values of energy_{in} to energy_{out}. Depending on the result, leaf temperature is either raised or lowered by small increments and the process is repeated until $\text{energy}_{in} = \text{energy}_{out}$.

In order to use a minimum of inputs into such a program, incident solar radiation may be calculated from Equations 13.1–13.3. Infrared radiation from the sky and soil may be calculated from the Stefan–Boltzmann relationship given in Chapter 4, $\Phi_L = \varepsilon\sigma(T_s + 273)^4$, where ε is the emittance of the radiating surface, and T_s is the temperature (°C) of the surface. For calculation of $\Phi_{L(DN)}$, the emittance of the atmosphere (ε_a) may be estimated using $\varepsilon_a = 0.72 + 0.005\,T_a$ [2].

There are several key morphological and physiological factors that influence the rates at which energy is exchanged between leaves

Table 13.1. Leaf coupling factors and their relationship to different avenues of energy exchange (Adapted from Collier *et al.* [4] and Ehleringer [7]).

Energy exchange process	Key environmental variables	Coupling factor	Key organism properties	Organism response
Short-wave radiation	Φ_s	A	Leaf surface characteristics	T_l, **A**, **E**
	$\Phi_{S,DIR}$	cos ι	Leaf angle, azimuth	T_l, **A**, **E**
Long-wave radiation	Φ_L	ε	Molecular composition	T_l
Convective heat exchange	wind, T_a	$r_{a,H}$	Leaf size, leaf shape	T_l, **E**
Latent heat exchange	wind	r_{a,H_2O}	Leaf size, leaf shape	T_l, **E**
	e	r_s	Stomatal density, stomatal opening, cuticular resistance	T_l, **A**, **E**

and their environment. These properties are termed leaf coupling factors (Table 13.1) [4,7]. Because of their effects on convective and latent heat exchange, two of the most important coupling factors are stomatal (r_s) and boundary layer (r_a) resistances. Stomatal resistance is a function of the aperture and density of stomatal pores in the leaf epi-. dermis. Virtually all of the carbon fixed in photosynthesis and the water lost in transpiration must pass through the stomata. Thus, the regulation of stomatal opening by the plant determines the balance between carbon gain and water loss. This balance is crucial to a terrestrial existence, where plants constantly face the rigors of a dessicating environment. Techniques to determine stomatal conductance (the inverse of resistance) are outlined in Chapter 8.

Boundary layer resistance is a function of the thin layer of still air next to a leaf surface. The exchange of mass and energy occurs primarily by diffusion through this layer. Since diffusion is a relatively slow process, the thickness of this boundary layer will be directly proportional to the rate with which heat is exchanged between the leaf and the surrounding air. Boundary layer resistance is dependent upon leaf size, shape, and wind speed. A relationship which estimates a mean leaf (both sides) $r_{A,H}$ under a combination of free and forced convection is given in Jones [14]; $r_{a,H} = 151 \sqrt{(w/v)}$, where $r_{a,H}$ is in s m^{-1}, w is the leaf width (m), and v is the wind speed ($m\,s^{-1}$). Boundary layer resistance to water vapour, r_{a,H_2O} can be approximated by multiplying $r_{a,H}$ by 0.96.

Other leaf coupling factors in Table 13.1 are involved with the absorptance or interception of either short-wave or long-wave radiation. Emissivity (ε) is dependent upon the molecular composition of a substance and is generally unresponsive to environmental and plant factors. Because plant leaves have many different kinds of atoms making up their leaf surfaces, ε is quite high (0.95–0.97) [13] and plant leaves behave nearly like black bodies (a black body absorbs all energy incident on it) in the far infrared wavelengths.

In contrast to ε, differences in leaf absorptance to solar radiation (A) do occur, both within and among species [5,6,9]. Absorptance to solar radiation is determined by leaf surface characteristics such as pubescence (hairs) or waxes. There are examples of leaves altering these characteristics in response to temperature and water availability, by either producing new leaves, or altering the surface of old leaves [9]. The effects of changes in leaf absorptance on leaf radiant energy balance and temperature can be quite large. Typical values for absorptance of green leaves range between 0.4 and 0.5 for radiation in the total solar spectrum, but some white hairy leaves may reduce this value to 0.1.

Changes in leaf absorptance are not the only mechanism for reducing the amount of solar radiation received by the leaf. An alternative seen in many species is to manipulate incident solar radiation by changing leaf angle or orienting the leaf away from the sun (Equation 13.3) [10,16].

Figure 13.2 illustrates how the relationship between leaf temperature and leaf width is altered by changes in r_s, leaf angle and leaf absorptance. Table 13.2 provides additional information on how the balance among different energy exchange avenues, i.e. reradiation, convection and latent heat exchange, was affected by leaf width, leaf angle, leaf absorptance and r_s.

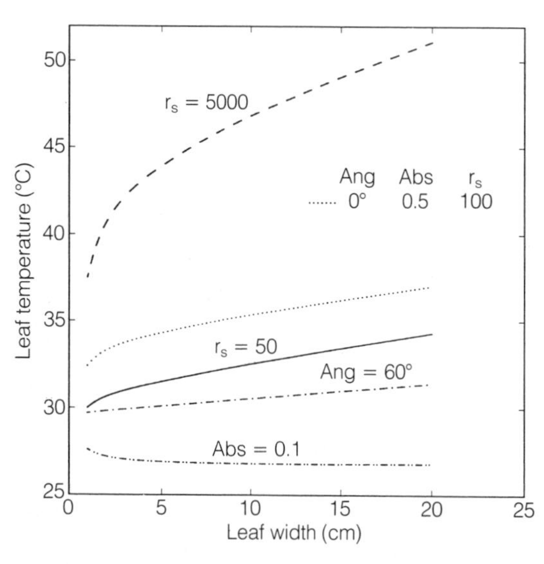

Fig. 13.2. The results of an energy budget simulation are presented as a plot of leaf temperature vs. leaf width. The basic conditions were a leaf angle (Ang) of 0°, leaf absorptance to solar radiation (Abs) of 0.5, and a stomatal resistance to water vapour [r_s] of 100 s m^{-1}. Simulations were then conducted by varying each of these parameters one at a time. The variable altered is listed above the line presenting the results of each simulation. For this simulation, the sun was assumed to be directly overhead, the sky was clear, air temperature was 30°C, soil temperature was 40°C, wind speed was 1.0 m s^{-1} and relative humidity was 30%.

13.4 A PHENOMENOLOGICAL MODEL OF LEAF PHOTOSYNTHETIC RESPONSE TO LIGHT AND TEMPERATURE

Two of the most important factors controlling photosynthetic carbon gain in plants are light and temperature. Generalized photosynthetic responses to these two variables are presented in Chapter 9. The model presented here is based upon these generalized responses, and has been parameterized for current global conditions of CO_2 concentration. A more complex, mechanism-based approach, which incorporates the effects of CO_2 concentration on light-saturated rates of photosynthesis ($\mathbf{A}_{sat}$) and photosynthetic quantum yield (ϕ) is developed in Chapter 14. We can approximate the photosynthetic response to light with the relationship:

$$\mathbf{A} = \phi \times \mathbf{Q}_{abs}/(1 + (\phi^2 \times \mathbf{Q}_{abs}^2/\mathbf{A}_{sat}^2))^{1/2} \qquad (13.5a)$$

where $\mathbf{A}$ = photosynthetic CO_2 fixation;
ϕ = quantum yield;
$\mathbf{Q}_{abs}$ = photosynthetic photon flux (PPF) absorbed by the leaf;
$\mathbf{A}_{sat}$ = light saturated photosynthetic rate.

Quantum yield, ϕ, is the slope of the initial, linear phase of the photosynthetic light response curve and is a measure of the efficiency of light utilization. For C_3 plants a typical value is 0.05 μmol CO_2 μmol^{-1} PPF [8]. $\mathbf{A}_{sat}$, the light-saturated rate of photosynthesis, will vary with the habitat and physiological status of the plant. For example, high-light-adapted plants will have significantly higher values for $\mathbf{A}_{sat}$ than will shade-adapted plants. Furthermore, changes in $\mathbf{A}_{sat}$ are usually associated with changes in the light level at which photosaturation occurs (i.e. higher values of $\mathbf{A}_{sat}$ = higher light saturation points).

Photosynthetic response to temperature usually shows reduced rates at low temperatures, higher rates at intermediate temperatures, and reduced rates again at higher

Table 13.2. The results of a computer simulation of leaf radiant energy exchange. Conditions for the simulation included clear skies, with the sun directly over the leaf, $T_a = 30°$, relative humidity = 40%, and a wind speed of $1.0\,m\,s^{-1}$. Initial leaf characteristics were; width = 1 cm, angle = 0°, absorptance = 0.5, $r_s = 100\,s\,m^{-1}$. These characters were varied individually for each row in the table.

Altered character	Absorbed solar radiation ($W\,m^{-2}$)	Leaf temperature (°C)	Re-radiation of infrared radiation ($W\,m^{-2}$)	Convective heat exchange ($W\,m^{-2}$)	Latent heat exchange ($W\,m^{-2}$)
None	755	32.4	959	188	488
Width (20 cm)	755	37.0	1018	122	487
Angle (60°)	408	29.7	925	−23	382
A (0.1)	151	27.6	900	−188	309
r_s (5000)	755	37.5	1025	587	17

temperatures (Chapter 9). This function can be described by fitting an equation of the appropriate form to measured photosynthetic temperature responses. A general equation suitable for many C_3 temperate species is:

$$p = (-0.004(T_l - T_{opt})^2) + (0.00023(|T_l - T_{opt}|)) + 1 \quad (13.5b)$$

where p = the proportion of $\mathbf{A}_{sat}$ (0–1);
T_l = leaf temperature (°C);
T_{opt} = photosynthetic temperature optima.

Again, a model predicting photosynthetic response to the environment should have as few inputs as possible. The solar radiation and leaf energy budget models developed earlier provide us with a means to calculate absorbed radiation and leaf temperature using only a few inputs. However, in order to calculate leaf temperature using the energy budget equation, we require an estimate of stomatal resistance. One approach to estimating values of stomatal resistance relies on an Ohm's Law analogy for photosynthesis:

$$\mathbf{A} = (c - c_{leaf})/(r_{s,CO_2} + r_{a,CO_2}) \quad (13.5c)$$

where c = CO_2 concentration in the air;
c_{leaf} = CO_2 concentration inside the leaf;
r_{s,CO_2} = stomatal resistance to CO_2;
r_{a,CO_2} = boundary layer resistance to CO_2.

Resistances to CO_2 are calculated by multiplying the appropriate resistances to H_2O vapour by 1.65 [14].

The relationships in Equations 13.5a–c were used to construct a phenomenological model of daily carbon gain and water loss for a single leaf suspended above the ground. First, Equations 13.1–13.3 were used to calculate incident radiation on a leaf. This incident radiation is then converted to an equivalent amount of photosynthetic photon flux (PPF) by assuming that PPF comprises 45.4% of the total solar radiation. A number of studies have shown that under clear sky conditions, and for solar zenith angles up to 80°, PPF is a relatively constant fraction of total solar radiation, with values ranging from 45 to 50% [17,22,24,26]. Photosynthesis (**A**) is calculated, using this PPF value, from Equation 13.5a. This value for **A** is substituted into Equation 13.5c, assuming an intercellular CO_2 concentration of $220\,\mu l\,l^{-1}$ [27] and a value for c of $350\,\mu l\,l^{-1}$, and the equation is solved for r_{s,H_2O}. The energy budget equation (Equation 13.4c) is then used to calculate T_l, which is used in Equation 13.5b to calculate p. The value p is then multiplied by the original value of **A** from equation 13.5a to yield a modified value for photosynthesis ($\mathbf{A}_{MOD}$).

If this modified value is different from the original **A**, a new r_{s,H_2O} is calculated using $\mathbf{A}_{MOD}$ in Equation 13.5c, and a new T_l is then calculated. The whole process is reiterated in this fashion until T_l, $\mathbf{A}_{MOD}$ and r_{s,H_2O} converge onto stable solutions.

The results of a simulation of photosynthetic response to light and temperature are presented in Fig. 13.3. This particular analysis suggests that as leaf size increases, photosynthetic carbon gain may decrease due to excess leaf temperatures (Fig. 13.3a). Although total daily transpiration (**E**) is shown to decrease (due to the model's linkage of r_{s,H_2O} and **A**; Fig. 13.3b), the relatively greater decrease in **A** causes a rise in the mean daily transpiration ratio (**E**/**A**) as leaf size increases (Fig. 13.3c). Steep leaf angles, by reducing midday solar radiation interception and T_l, were predicted to eliminate the disadvantage of large leaf sizes in terms of both **A** and **E**/**A** in this environment (Fig. 13.3a–c).

Although this type of simple, non-mechanistic model is limited in its predictive applications, it can be useful in illustrating potential impacts of leaf coupling factors (such as absorptance and orientation) on diurnal patterns of leaf temperature, carbon gain and water loss. Leaf photosynthesis in various environments may be examined by manipulating environmental inputs such as

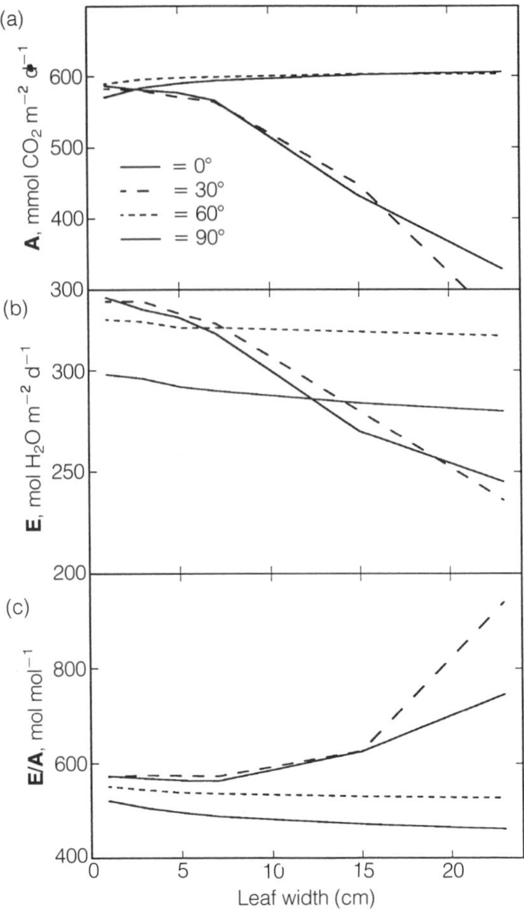

Fig. 13.3. The results of a phenomenological model combining leaf energy budget equations with photosynthetic responses to light and temperature. The simulation assumed clear sky conditions and used environmental parameters for a desert environment located at 30°N latitude listed in Table 13.3. All leaves were assumed to be stationary throughout the day, with a south-eastern azimuth. The equations used for photosynthetic response to temperature and incident light are given in the text. (a) Calculated daily carbon gain versus leaf width for leaves oriented at 0°, 30°, 60°, and 90° from the horizontal. (b) Calculated daily transpirational water loss versus leaf width for leaves oriented at 0°, 30°, 60°, and 90° from the horizontal. (c) Calculated daily mean transpiration ratio plotted versus leaf width for leaves oriented at 0°, 30°, 60°, and 90° from the horizontal.

Table 13.3. Various environmental parameters used to differentiate climates in a phenomenological model of photosynthetic response to light and temperature.

	Desert environment	Tropical environment
Max. air temp. (°C)	37	26
Max. soil temp. (°C)	42	21
Albedo	0.3	0.1
Relative humidity (%)	10	70
Wind speed ($m\,s^{-1}$)	2	1
Latitude	30	0
Declination	0	0
Transmissivity	0.85	0.7

relative humidity, wind speed and ambient temperatures. Two such examples of the effects of contrasting environments are presented in Table 13.3.

13.5 CROP PHOTOSYNTHESIS AND CONDUCTANCE FROM LEAF MEASUREMENTS

13.5.1 Theory

If leaf photosynthetic rate (**A**) and stomatal conductance (g_s) are known as a function of PPF incident on the leaf, then canopy photosynthetic rate ($\mathbf{A}_c$) and canopy conductance (g_c) can be estimated from some simple measurements above and below the canopy. The basic assumptions involved in this simplified procedure are that light is the dominant factor, the photosynthetic and stomatal dependence on light is independent of depth in the canopy, and foliage is randomly distributed in space. Examples of leaf measurements required for this approach are shown in Figs. 13.4 and 13.5. Alternatively, the dependencies of photosynthetic rate and stomatal conductance on light can be fitted using Equation 13.5a.

The following light measurements are necessary to extend leaf measurements to canopy estimates of conductance and photosynthetic rate: (1) total ($\mathbf{Q}_{tot}$) and diffuse ($\mathbf{Q}_{diff}$) PPF above the canopy in the horizontal plane; (2) average PPF on a horizontal plane in the canopy ($\mathbf{Q}_i$) using a ceptometer (*Decagon Inc., Pullman, WA, USA*) or line quantum sensor (*LI-COR Inc., Lincoln, NE, USA*), and (3) zenith angle of the sun (θ).

The procedure for estimating canopy conductance and photosynthetic rate from single leaf rates involves dividing the canopy into sunlit and shaded leaves and summing up the contributions separately. Thus we need to know the amount of leaf area that is sunlit and the amount shaded, and we also need to estimate the mean illumination levels on both sunlit and shaded leaves. This is necessary

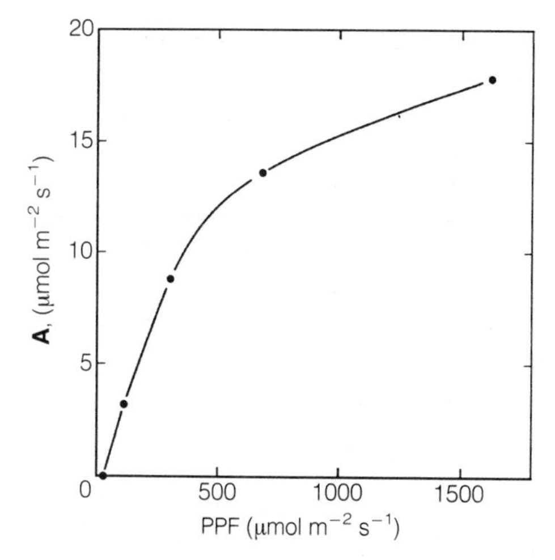

Fig. 13.4. Photosynthetic rate (**A**) of a wheat leaf as a function of incident photosynthetic photon flux (PPF) at 340 μl l^{-1} of ambient CO_2 concentration and 27°C. The symbols represent measurements with a gas exchange system (*LI-6200, LI-COR, Inc., Lincoln, NE, USA*) and the curve is a least squares fit to a rectangular parabola.

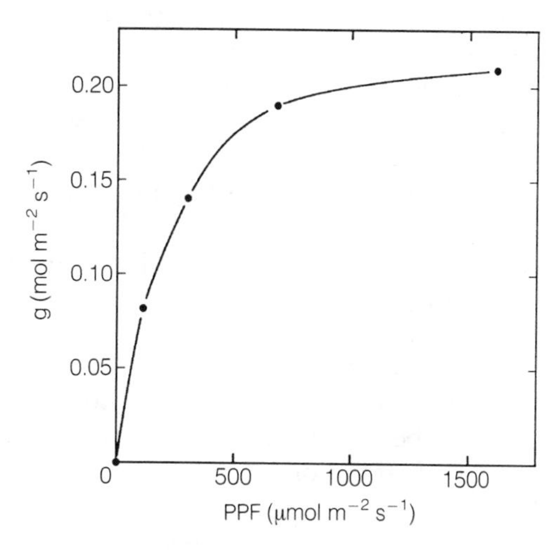

Fig. 13.5. Stomatal conductance (g) of a wheat leaf as a function of incident PPF. See legend to Fig. 13.4 for other details.

because of the nonlinear dependence of leaf conductance and photosynthetic rate on light (Figs. 13.4, 13.5; Chapter 9).

The canopy conductance based on a unit of ground area (g_c) can be estimated from the sum of contributions of sunlit and shaded leaves:

$$g_c = g_{sun}F_{sun} + g_{shade}(F - F_{sun}) \qquad (13.6)$$

where g_{sun} and g_{shade} are obtained from Fig. 13.5 using the mean values of PPF flux densities for each leaf type. F is the leaf area index for the canopy, and F_{sun} is the sunlit leaf area index. Likewise, the canopy photosynthetic rate ($\mathbf{A}_c$) per unit ground area can be estimated from:

$$\mathbf{A}_c = \mathbf{A}_{sun}F_{sun} + \mathbf{A}_{shade}(F - F_{sun}) \qquad (13.7)$$

where $\mathbf{A}_{sun}$ and $\mathbf{A}_{shade}$ are obtained from Fig. 13.4 using the mean values of PPF flux densities for each leaf type. The sunlit leaf area index can be calculated if leaves are assumed to be randomly distributed:

$$F_{sun} = [1 - \exp(-kF/\cos\theta)]\cos\theta/k \qquad (13.8)$$

where k is the foliar absorption coefficient of light in the canopy, which is dependent upon the leaf inclination angle distribution.

The leaf inclination angle distribution can be adequately described by a single parameter, if leaves are assumed to be distributed symmetrically about the azimuth. Several different parameters have been used, for example: (1) the mean inclination angle of the leaf plane from the horizontal ($\bar{\alpha}$), (2) the inclination index (v) of Ross [23], or (3) the mean ratio of horizontal to vertical projections of leaves (x) defined by Campbell [3]. The relation between these three are tabulated in Table 13.4. The extinction coefficient can be calculated from the ellipsoidal leaf angle distribution [3]:

$$k = \frac{(x^2 + \tan^2)^{1/2}\cos\theta}{x + 1.744(x + 1.182)^{-0.773}} \qquad (13.9)$$

where θ is the sun zenith angle and x is the leaf angle distribution parameter (mean ratio of horizontal to vertical projections of leaves). Table 13.5 contains values of k and F_{sun} for a range of leaf inclination angles and sun zenith angles. Clearly, the leaf area index exposed to direct sunlight can vary widely with canopy architecture and sun zenith angle. In fact, a canopy with a leaf area index of 1.0 can have more sunlit leaf area at midday than a canopy with a leaf area index of 3.0 has in the morning or evening. Table

Table 13.4. Comparison of single-parameter methods for describing leaf inclination angle distributions (see text for explanation).

x	v	$\bar{\alpha}$
0.0	−1.000	90.0 (vertical)
0.1	−0.767	84.2
0.2	−0.614	82.0
0.3	−0.518	78.7
0.4	−0.418	75.3
0.5	−0.320	71.9
0.6	−0.249	68.7
0.7	−0.179	65.5
0.8	−0.113	62.6
0.9	−0.053	59.8
1.0	0.0	57.1 (spherical)
1.5	0.212	46.2
2.0	0.365	38.2
2.5	0.462	32.2
4.0	0.649	21.1
6.0	0.764	13.8
8.0	0.832	10.3
10.0	0.868	8.6
∞	+1.000	0.0 (horizontal)

Table 13.5. Canopy foliar absorption coefficient (k), mean leaf-sun angle (d), and sunlit leaf area index (F_{sun})* for several canopy types (x) and several solar zenith angles (θ).

x	θ (°)	k	d(°)	F_{sun} (F = 1.0)	F_{sun} (F = 3.0)
0.4	15	0.283	73.6	0.87	2.00
	45	0.463	62.4	0.73	1.31
	75	0.590	53.8	0.39	0.44
1.0	15	0.500	59.7	0.78	1.52
	45	0.500	59.7	0.72	1.24
	75	0.500	59.7	0.44	0.51
4.0	15	0.856	31.1	0.66	1.05
	45	0.644	49.9	0.66	1.03
	75	0.313	71.8	0.58	0.80

*F_{sun} is calculated from two values of leaf area index (F).

Table 13.6. Foliage inclination index for various plant canopies [23].

Species	Conditions	v
Corn	Young	+0.2
	Tasseling	−0.1 to +0.2
Strawberry	Fruiting	+0.5
White clover	Flowering	+0.4 to +0.6
Sunflower	Flower formation	+0.3 to +0.4
Horsebeans	Flowering	+0.3 to +0.4
Potato	Flowering	+0.3 to +0.4
Lucerne	50–60 cm	+0.2
Sorghum	Leading	+0.2
Barley	Shooting-milk ripeness	+0.1
Wheat	–	0.0
Rye	–	−0.1 to +0.1
Perennial ryegrass	Various ages	−0.3 to +0.4

13.6 is a list of some species with their inclination indexes from Ross [23].

The canopy conductance (photosynthetic rate) depends on a weighted sum of sunlit and shaded leaf conductance (photosynthetic rate) represented by Equations 13.6 and 13.7. To obtain the mean leaf conductance (photosynthetic rate), we must estimate the average incident PPF for both sunlit and shaded leaves.

The average PPF received by all shaded leaves ($\mathbf{Q}_{shade}$) in a canopy was estimated by Norman [19], assuming a spherical leaf angle distribution. Although this will depend on the leaf angle distribution, at the present time no simple equation is available for predicting the mean shaded illumination on leaves as a function of leaf inclination. Based on the data in Table 13.6, we expect the results from spherical-distribution calculations to be reasonable for many canopies. Therefore, we will use the equation from Norman [19]:

$$\mathbf{Q}_{shade} = \mathbf{Q}_{diff} \exp[-0.5\ F^{0.7}] + \mathbf{Q}_{scat} \quad (13.10)$$

$$\mathbf{Q}_{scat} = 0.07\ \mathbf{Q}_{dir}\ (1.1 - 0.1\ F) \exp(-\cos\theta) \quad (13.11)$$

where $\mathbf{Q}_{diff}$ and $\mathbf{Q}_{dir}$ are the incident sky diffuse and direct solar beam PPF on a horizontal plane above the canopy, respectively, and $\mathbf{Q}_{scat}$ represents the direct beam scattered by leaves in the canopy. Since our measurements include $\mathbf{Q}_{tot}$ and $\mathbf{Q}_{diff}$, $\mathbf{Q}_{dir} = \mathbf{Q}_{tot} - \mathbf{Q}_{diff}$.

The PPF received by sunlit leaves is the sum of that from the direct beam along with the diffuse given by Equations 13.10 and 13.11:

$$\mathbf{Q}_{sun} = \mathbf{Q}_{dir}^{(\cos d/\cos\theta)} + \mathbf{Q}_{shade} \quad (13.12)$$

where d is the mean leaf–sun angle (zero when sunbeam is perpendicular to leaf), given by:

$$d = \cos^{-1}(k) \quad (13.13)$$

Some values of d are contained in Table 13.5.

13.6 EXAMPLE CALCULATION

13.6.1 Measurements

$\mathbf{Q}_{tot} = 1680\ \mu mol\ m^{-2}\ s^{-1}$
$\mathbf{Q}_{diff} = 504\ \mu mol\ m^{-2}\ s^{-1}$
$\mathbf{Q}_{i} = 134\ \mu mol\ m^{-2}\ s^{-1}$
$\mathbf{Q}_{dir} = 1680 - 504 = 1176\ \mu mol\ m^{-2}\ s^{-1}$
$\theta = 36.9°$ ($\cos\theta = 0.800$)
time = 1230 hours
$\bar{\alpha} = 57°$
$F = 4.8$
$\mathbf{Q}_{scat} = 0.07(1176)[1.1 - 0.1(4.8)] \exp(-0.8) = 22.9\ \mu mol\ m^{-2}\ s^{-1}$
$\mathbf{Q}_{shade} = 504 \exp[-0.5(4.8)^{0.7}] + 22.9 = 135\ \mu mol\ m^{-2}\ s^{-1}$
$\mathbf{Q}_{sun} = 1176^{(0.50/0.800)} + 135 = 870\ \mu mol\ m^{-2}\ s^{-1}$

From Table 13.4, $x = 1.0$

From Table 13.5, $k = 0.50$
$d = \cos^{-1}(0.50) = 60°$
$F_{sun} = [1 - \exp(-0.500[4.8/0.800])] \times 0.800/0.500 = 1.52$
$F - F_{sun} = 4.8 - 1.52 = 3.28$

13.6.2. Leaf conductance

From Fig. 13.5:

$$g_{sun} = 0.194\,\text{mol}\,H_2O\,m^{-2}\,s^{-1}$$
$$g_{shade} = 0.093\,\text{mol}\,H_2O\,m^{-2}\,s^{-1}$$

13.6.3. Leaf photosynthetic rate

From Fig. 13.4:

$$\mathbf{A}_{sun} = 15.7\,\mu\text{mol}\,CO_2\,m^{-2}\,s^{-1}$$
$$\mathbf{A}_{shade} = 4.4\,\mu\text{mol}\,CO_2\,m^{-2}\,s^{-1}$$

13.6.4. Canopy conductance

$$g_c = 0.194(1.5) + 0.093(3.3)$$
$$= 0.60\,\text{mol}\,H_2O\,m^{-2}\,\text{(ground area)}\,s^{-1}$$

13.6.5. Canopy photosynthetic rate

$$\mathbf{A}_c = 15.7(1.5) + 4.4(3.3)$$
$$= 38\,\mu\text{mol}\,CO_2\,m^{-2}\,\text{(ground area)}\,s^{-1}$$

13.7 INTERPRETATION OF RESULTS

The canopy conductance value may be used to estimate transpirational water vapour flux from the canopy by multiplying it by the water vapour mole fraction gradient, $(e_{s,Tl} - e_a)/p$. This value may then be used in the Penman–Monteith equation [18] to estimate the ratio of actual transpiration to potential evapotranspiration. This method does not account for any soil evaporation or evaporation of liquid water on the surface of leaves (dew or rainfall interception), or for any canopy boundary layer effects. Equations for potential evapotranspiration can be obtained from Brutsaert [1].

The canopy light use efficiency, which is the number of moles of CO_2 fixed per moles of PPF quanta intercepted (IPPF), can be estimated from the light measurements and the estimates of canopy photosynthetic rate:

$$\phi_c = \frac{\mathbf{A}_c}{\mathbf{Q}_{tot} - \mathbf{Q}_i} = \frac{38}{1680 - 134}$$
$$= 0.0246\,\frac{\text{mole } CO_2}{\text{mole quanta}}$$

This translates to about 40 quanta per CO_2 molecule, or about one-fifth the maximum theoretical limit of efficiency of the photosynthetic system. Another way of interpreting this rate is that it is equivalent to $1.2\,g\,CO_2\,mol^{-1}$ IPPF. Converting this to energy units would involve multiplying the moles of photons in each wavelength by their energy content. Assuming that yellow light (wavelength of 570 nm) is representative of the photosynthetically active radiation (PAR) band, a value of $210\,kJ\,mol^{-1}$ PPF may be used. This would result in a value of $5.7\,g\,CO_2\,MJ^{-1}$ IPAR. Assuming that 6 CO_2 molecules are fixed for every glucose produced, and using a value of 0.651 g plant tissue produced per g glucose [21], this CO_2 fixation rate could be converted to a dry matter accumulation rate. With molecular weights for CO_2 and glucose of 44 and 180, respectively, this dry matter accumulation rate would approximate to 2.5 g dry matter MJ^{-1} IPAR for a typical grass crop.

ACKNOWLEDGEMENTS

The authors wish to acknowledge the assistance of Dr. Dayle McDermitt, Dr. Steve Roemer and Larry Middendorf of LI-COR Inc. in collection of the gas exchange data on wheat, and L.M. Rosa for help in preparing the manuscript.

REFERENCES

1. Brutsaert, W. (1982) *Evapotranspiration into the Atmosphere: Theory, History and Applications*. D. Reidel Publishing Co., Boston, Mass. 299 pp.
2. Campbell, G.S. (1977) *An Introduction to Environmental Biophysics*. Springer-Verlag, New York.
3. Campbell, G.S. (1986) Extinction coefficients for radiation in plant canopies calculated using an ellipsoidal inclination angle distribution. *Agricultural and Forest Meteorology* **36**, 317–321.
4. Collier, B.D., G.W. Cox, A.W. Johnson and P.C. Miller (1973) *Dynamic Ecology*. Prentice-Hall, Englewood Cliffs, New Jersey.

5. Ehleringer, J.R. (1981) Leaf absorptances of Mohave and Sonoran Desert plants. *Oecologia* **49**, 366–370.
6. Ehleringer, J.R. (1988) Changes in leaf characteristics in species along elevational gradients in the Wasatch Front, Utah. *American J. Bot.* **75**, 680–689.
7. Ehleringer, J.R. (1989) Temperature and energy budgets. In: *Plant Physiological Ecology: Field methods and instrumentation.* (R.W. Pearcy, J.R. Ehleringer, H.A. Mooney, and P.W. Rundel, eds.). pp. 117–135. Chapman and Hall, London and New York.
8. Ehleringer, J.R. and O. Björkman (1977) Quantum yields for CO_2 uptake in C_3 and C_4 plants: dependence on temperature, CO_2, and O_2 concentration. *Plant Physiol.* **59**, 86–90.
9. Ehleringer, J.R. and O. Björkman (1978) Pubescence and leaf spectral characteristics in a desert shrub, *Encelia farinosa*. *Oecologia* **36**, 151–162.
10. Ehleringer, J.R. and I.N. Forseth (1980) Solar tracking by plants. *Science* **210**, 1094–1098.
11. Forseth, I.N. and J.R. Ehleringer (1980) Solar tracking response to drought in a desert annual. *Oecologia* **44**, 159–163.
12. Gates, D.M. (1962) *Energy Exchange in the Biosphere.* Harper and Row, New York.
13. Gates, D.M. (1980) *Biophysical Ecology.* Springer-Verlag, New York.
14. Jones, H.G. (1983) *Plants and Microclimate.* Cambridge University Press.
15. List, R.J. (1968) *Smithsonian Meteorological Tables*, 6th edn. Smithsonian Misc. Coll. 114. Smithsonian Press, Washington, D.C.
16. McMillen, C.G. and J.H. McClendon (1979) Leaf angle: an adaptive feature of sun and shade leaves. *Bot. Gazette* **140**, 437–442.
17. Moon, P. (1940) Proposed standard solar radiation curves for engineering use. *J. Franklin Institute* **230**, 583–618.
18. Monteith, J.L. (1965) Evaporation and environment. *SEB Symposia* **19**, 205–234.
19. Norman, J.M. (1982) Simulation of microclimates. In: *Biometeorology and Integrated Pest Management.* (J.L. Hatfield and I.J. Thomason, eds.). pp. 65–99. Academic Press, New York.
20. Norman, J.M. and G.S. Campbell (1989) Canopy structure. In: *Plant Physiological Ecology: Field methods and instrumentation.* (R.W. Pearcy, J.R. Ehleringer, H.A. Mooney, and P.W. Rundel, eds.). pp. 301–325. Chapman and Hall, London and New York.
21. Penning de Vries, F.W.T., A.H.M. Brunsting and H.H. van Laar (1974) Products, requirements and efficiency of biosynthesis: a quantitative approach. *J. Theoretical Biology* **45**, 339–377.
22. Pereira, A.R., E.C. Machado and M.B.P. de Camargo (1982) Solar radiation regime in three cassava (*Manihot esculenta* Crantz) canopies. *Agricultural Meteorology* **26**, 1–10.
23. Ross, J. (1981) *The Radiation Regime and Architecture of Plant Stands.* Dr. W. Junk, The Hague, Netherlands.
24. Szeicz, G. (1974) Solar radiation for plant growth. *J. Applied Ecology* **19**, 617–636.
25. Walter, H. (1979) *Vegetation of the Earth and Ecological Systems of the Geo-Biosphere.* 2nd ed. Springer-Verlag, New York.
26. Weiss, A. and J.M. Norman (1985) Partitioning solar radiation into direct and diffuse, visible and near-infrared components. *Agricultural and Forest Meteorology* **34**, 205–213.
27. Wong, S.C., I.R. Cowan and G.D. Farquhar (1979) Stomatal conductance correlates with photosynthetic capacity. *Nature* **282**, 424–426.

14

Modelling of canopy carbon and water balance

R.E. McMurtrie

14.1 INTRODUCTION

Understanding how ecosystems function is essential to their management and manipulation. But it is important that the understanding be sought in an organized and systematic way. Several eco-physiological processes are now relatively well understood. Examples are photosynthesis, respiration, leaf energy balance and turbulent exchange within canopies, all of which have benefited from advances made recently in fundamental theory and instrumentation. The advances, many of which are described in chapters of this volume, have enhanced our capacity to predict the response of individual plant processes to changes in micrometeorological conditions. A means is required however to integrate up from the micro-environmental scale to explain and predict the response of communities of plants. This requires integration over space and time, and sound formulations of the linkages between interacting processes, tasks which can only be achieved using relatively complex process-based models. Such models provide frameworks for research aimed at understanding how plants grow and predicting their response to environmental change. The range of available models of plant stand bioproductivity has been surveyed by Ågren *et al.* [1].

This chapter outlines a mechanistic model of the growth of tree stands, the BIOMASS model [9,10,11,12]. The model simulates carbon and water balance of trees in relation to their nutrition. The structure of the model is described in Section 14.2, followed by a discussion of its use in the synthesis of experimental data compiled in a multi-disciplinary research project on the growth of *Pinus radiata* D. Don near Canberra, Australia. The final section assesses what has been learned from the experience of using the model as a research framework.

14.2 STRUCTURE OF THE MODEL

The growth model BIOMASS is intended to be general. It is process-based or mechanistic, consisting of a series of sub-models which describe the operation of the various physiological processes involved in the growth of trees. The model describes the growth of stands and simulated outputs are expressed on a unit ground area basis. The carbon balance model, depicted on the right-hand side of Fig. 14.1, consists of sub-models for processes of radiation interception, net photosynthesis, respiration, carbon partitioning

Photosynthesis and Production in a Changing Environment: a field and laboratory manual
Edited by D.O. Hall, J.M.O. Scurlock, H.R. Bolhàr-Nordenkampf, R.C. Leegood and S.P. Long
Published in 1993 by Chapman & Hall, London.
ISBN 0 412 42900 4 (HB) and 0 412 42910 1 (PB).

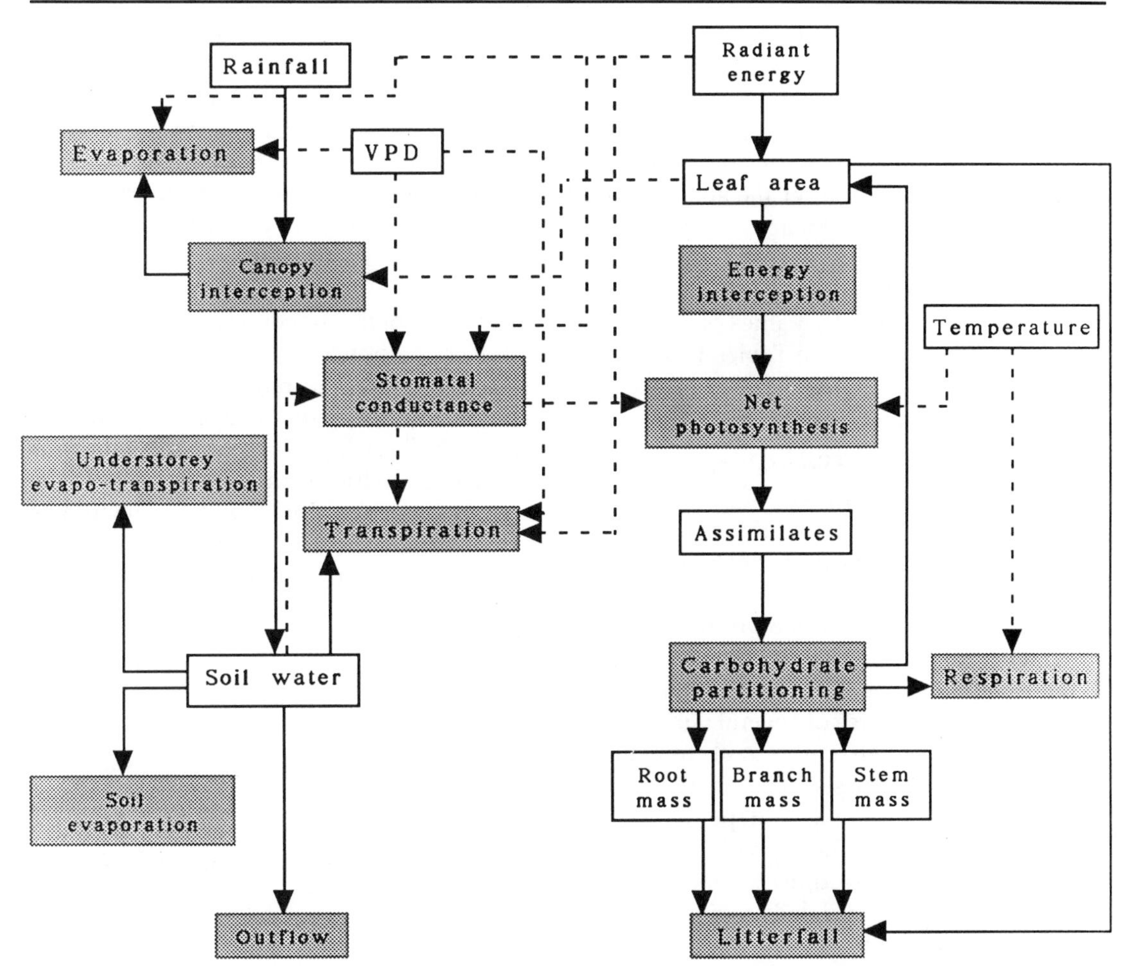

Fig. 14.1. Schematic representation of the BIOMASS model. The carbon and water balance models are displayed on the right-hand and left-hand sides, respectively. Unshaded boxes denote meteorological and state variables, shaded boxes denote processes, solid arrows denote transfers of matter and energy and broken arrows denote influences.

and litter fall. The water balance model on the left-hand side of Fig. 14.1 considers processes of canopy interception of rainfall, evaporation from wet canopies, transpiration which depends on stomatal and canopy conductance, understorey evaporation, soil evaporation, and moisture loss from the rooting zone through run-off and drainage. Biomass components considered are stem, branch, foliage and root. Simulations are performed with a daily time resolution, requiring input of standard meteorological-station data and site/crop-specific information such as latitude, longitude, tree density, soil water characteristics, rooting depth, physical characteristics of crop and physiological parameters of the tree species. BIOMASS consists of a series of equations, some processes being described by a single equation, others by several. The equations are largely based on established theories of plant physiological processes and soil water dynamics. Linkages between processes are illustrated schematically in Fig. 14.1.

The BIOMASS model has been implemented using IBM PC-compatible microcomputers. To run a simulation of BIOMASS for a particular site, a user must provide input files of meteorological data and of parameter values. The latter file specifies the values of all coefficients appearing in the model's equations, as well as initial dry weights of stem, branch, foliage and root and initial soil water contents. Values of several of the parameters are given in Tables 14.1–14.5.

14.2.1 Meteorological inputs

The minimum meteorological data set required is daily precipitation, and maximum and minimum air temperatures. Other meteorological data used by the model, e.g. daily total short-wave radiant energy, can be read from files or generated from formulae provided within the simulation package. While the model operates on a daily time step, the diurnal course of several variables is required to calculate daily total transpiration and photosynthesis. This is necessary because of the non-linear dependence of these processes on meteorological conditions. Details of how diurnal patterns are generated from standard daily meteorological data are given in Table 14.1.

14.2.2 Carbon balance

Tree crown shape is represented by either of two geometrical constructions, a cone or ellipsoid of revolution, with all crowns assumed identical in dimensions. The plant community is represented by a randomly spaced array of trees with a correction employed so that simulated and measured total projected ground cover agree. Total ground cover may vary with stand age. The foliage is divided into three horizontal layers, and model users are able to specify different photosynthetic parameters for each layer.

Canopy net photosynthesis:
Interception of both direct and diffuse radiation depends upon canopy architecture. The sunlit leaf area index for a given solar zenith angle is derived for each canopy layer [12] with a modification to account for the fraction of beam radiation passing through gaps in the canopy. The fraction of the sunlit leaf area photosynthesising at the irradiance saturated rate ($\mathbf{A}_{sat}$) is estimated under the assumption of a spherical leaf angle distribution (extinction coefficient, k = 0.5). The contribution of this fraction to daily net photosynthesis is derived by integration over the daylight period. The rate of photosynthesis of the remainder of the foliage (shade foliage plus sunlit foliage below light saturation) is proportional to the absorbed energy flux. The constant of proportionality, β, is related to the quantum yield (ϕ_{abs}). The calculation of absorbed sky diffuse radiation assumes that incident radiation intensity is uniform over the sky. Equations used to simulate canopy photosynthesis are given in Table 14.2.

The model of canopy photosynthesis in BIOMASS is less demanding computationally than models with detailed descriptions of the distribution of radiation within canopies and it requires input of fewer parameters and less meteorological data. Analysis by Norman [13] gives endorsement to the approach of dividing the canopy into sunlit and shaded foliage in simulating canopy photosynthesis.

Values of β and $\mathbf{A}_{sat}$ are required by BIOMASS, and the user can supply these either directly as input parameters derived from an empirical photosynthetic light response curve [12] or indirectly by specifying the parameters of a mechanistic model of photosynthesis by leaves of C_3 plants [4,11] (Table 14.4). Dependences of photosynthesis on temperature and foliar nitrogen concentration are explicit in the model. Nutrient concentrations are not simulated variables, but are supplied by the user as model inputs.

Respiration:
The carbon balance of the canopy is updated daily, with growth occurring when assimilation is surplus to tissue maintenance requirements. Maintenance respiration rates are

Table 14.1. Equations used to estimate the diurnal course of weather variables from input of daily values of total short-wave radiation (K_s, $MJ\,m^{-2}\,d^{-1}$), maximum and minimum air temperature (T_{max} and T_{min}, °C). Symbols are: D_j, year day (Julian date) with $D_j = 1$ on 1st July; e_a, absolute humidity of air (kPa); S_o, solar constant (= $1360\,W\,m^{-2}$); t, time of day with t = 0 at sunrise; θ, solar zenith angle; Λ, latitude; and τ_A, atmospheric transmittance under clear sky conditions. Six-point Gaussian quadrature is employed in BIOMASS for evaluation of diurnal and other integrals [5]. Further reading on the derivations of relationships below may be found in McMurtrie *et al.* [12]

Declination (radians):

$$\delta = 0.1303\,\pi\cos(2\pi[D_j + 10]/365)$$

Daylength (hour):

$$\Delta = 24\cos^{-1}(-\tan[\Lambda\pi/180]\tan\delta)/\pi$$

Daytime pattern of air temperature:

$$T(t) = 0.5\,(T_{min} + T_{max} - (T_{max} - T_{min})\cos(1.5\pi t/\Delta))$$

Average daytime temperature:

$$\bar{T} = 0.5\,(T_{max} + T_{min}) + (T_{max} - T_{min})/(3\pi)$$

Extra-terrestrial radiation ($MJ\,m^{-2}\,day^{-1}$):

$$K_o = 0.0864\,S_o\,(\delta/24\sin[\Lambda\pi/180]\sin\delta + 1/\pi\cos[\Lambda\pi/180]\cos\delta\sin[\Delta\pi/24])$$

Diffuse sky short-wave radiation ($MJ\,m^{-2}\,d^{-1}$):

$$K_{sd} = fK_s$$

where
$$f = \begin{cases} 1 & \text{if } K_s/K_o \leqslant 0.07 \\ 1 - 2.3(K_s/K_o - 0.07)^2 & \text{if } 0.07 < K_s/K_o \leqslant 0.35 \\ 1.33 - 1.46\,K_s/K_o & \text{if } 0.35 < K_s/K_o \leqslant 0.75 \\ 0.23 & \text{if } K_s/K_o > 0.75 \end{cases}$$

Incident beam radiation ($MJ\,m^{-2}\,d^{-1}$):

$$K_{sb} = K_s - K_{sd}$$

Diurnal course of incident diffuse short-wave radiation ($MJ\,m^{-2}\,d^{-1}$):

$$\phi_d = \frac{K_{sd}\cos\theta}{\int_0^{\Delta}\cos\theta\,dt}$$

Diurnal course of incident photosynthetic photon flux density (μmol quanta $m^{-2}\,s^{-1}$) under clear sky conditions:

$$\phi_b = 2.3\,S_o\cos\theta\,\tau_A^{1/\cos\theta}$$

Saturation vapour pressure of air (kPa):

$$e_s = 0.61078\exp(17.269\,T)/(T + 237.3)$$

Saturation vapour deficit of air (kPa):

$$\Delta e = e_s - e_a$$

Relative humidity (%):

$$h = 100\,(e_a/e_s)$$

Table 14.2. Equations used to calculate daily total canopy photosynthesis for a spatially uniform canopy consisting of a single canopy layer with 100% ground cover. Symbols as follows: $\mathbf{A}_{sat}$, light-saturated rate of net photosynthesis at ambient CO_2 (mol CO_2 m^{-2} s^{-1}); f_d, fraction of intercepted diffuse radiation incident on non-saturated foliage; k, radiation extinction coefficient (= 0.5 for spherically symmetric leaf angle distribution, R_{di}, non-photorespiratory day and dark respiration (kg C ha^{-1} day^{-1}); β, initial slope of assumed photosynthetic light response; ρ_1 and τ_1, reflectance and transmittance of foliage to photosynthetically active radiation. Other symbols as defined in Table 14.1

Daily net photosynthesis by light-saturated sunlit foliage (kg C ha^{-1} d^{-1}):

$$P_{n1} = f_\delta \int_0^\Delta L_1{}^*(t)\, A_{sat}(t)\, dt - R_{d1}$$

Sunlit leaf area index:

$$L_0{}^* = (1 - \exp(-kL^*/\cos\theta))\cos\theta/k$$

Light-saturated sunlit leaf area index, assuming spherically symmetric leaf angle distribution:

$$L_1{}^* = L_0{}^* (1 - \mathbf{A}_{max}/(\beta\phi_b(1 - \rho_1 - \tau_1)))$$

Sunshine duration:

$$f_s = (1 - K_{sd}/K_s)/0.77$$

Daily net photosynthesis by non-saturated sunlit foliage (kg C ha^{-1} d^{-1}):

$$P_{n2} = \beta\,(1 - \rho_1 - \tau_1) \int_0^\Delta Q_2(t)\, dt - R_{d2}$$

Photosynthetically active radiation intercepted at time t by non-saturated sunlit foliage (MJ m^{-2} d^{-1}):

$$Q_2(t) = Q_b(t)\, f_b\,(t)$$

Photosynthetically active radiation intercepted at time t (MJ m^{-2} d^{-1}):

$$Q_b(t) = 0.5\, K_{sb} \frac{(1 - \exp(-kL^*/\cos\theta))\cos\theta\, \tau_A{}^{1/\cos\theta}}{\int_0^\Delta \cos\theta\, \tau_A{}^{1/\cos\theta}\, dt}$$

Fraction of beam incident on non-saturated foliage:

$$f_b(t) = \min\,(1, (\mathbf{A}_{sat}/(\beta\phi_b(1 - \rho_1 - \tau_1)))^2)$$

Daily net photosynthesis resulting from interception of diffuse radiation (kg C ha^{-1} d^{-1}):

$$P_{n3} = \beta\,(1 - \rho_1 - \tau_1) \int_0^\Delta Q_3\, f_d\, dt - R_{d3}$$

Diffuse photosynthetically active radiation intercepted (MJ m^{-2} d^{-1}):

Q_3 = intercepted sky diffuse radiation + intercepted scattered radiation

Daily canopy net photosynthesis (kg C ha^{-1} d^{-1}):

$$P_n = P_{n1} + P_{n2} + P_{n3}$$

expressed in terms of per unit surface area for stem, branch and foliage and per unit dry weight for roots. Rates depend on temperature (T), and daily total respiration is calculated by integration over the diurnal temperature cycle (Table 14.3). Respiration also occurs in the conversion of assimilate to structural tissue. This component, termed growth respiration, is assumed to consume 30% of assimilate available after subtraction of maintenance respiration [7]. Net primary production (NPP) represents daily net photosynthesis less daily total maintenance and growth respiration. Dry matter production

Table 14.3. Equations for net primary production and biomass allocation. Symbols as follow: a_j, b_j, constants relating partitioning coefficients to foliar nitrogen concentration $[N]_f$ ($mg\,g^{-1}$); Q_{10}, temperature dependence of maintenance respiration; Y, fraction of NPP not consumed in growth respiration; η_j, partitioning coefficient to biomass component j. Subscript j refers to foliage (j = f), branch (j = b), stem (j = s) and roots (j = r).

Net primary production ($kg\,C\,ha^{-1}\,day^{-1}$):

$$NPP = Y(P_n - R_b - R_s - R_r)$$

Maintenance respiration by branch, stem and root ($kg\,C\,ha^{-1}\,d^{-1}$):

$$R_j = r_{jo} \int_0^{24} (Q_{10})^{T/10}\,dt$$

Production of foliage, branch, stem and root tissue (kg dry weight $ha^{-1}\,d^{-1}$):

$$G_j = 2\,\eta_j\,NPP$$

where $\eta_j = a_j\,[N]_f + b_j$

is calculated under the assumption that $1\,kg\,C = 2\,kg$ dry weight [7].

Allocation:
Mechanisms governing assimilate partitioning are among the major unsolved problems in whole-plant physiology and great uncertainty surrounds approaches to modelling biomass allocation by trees. Mechanistic models have been developed but are largely untested. Constant partitioning coefficients are commonly assumed [7]. In BIOMASS, allocation of carbon to leaf, stem, branch and root is determined by partitioning coefficients (η_f, η_s, η_b and η_r, respectively) which vary over the annual cycle and are linear functions of foliar nitrogen concentration (Table 14.3). Conversion of leaf mass to leaf area requires estimates of specific leaf area, including its variability within the canopy.

Litter fall:
Litter fall is subtracted from each component of biomass with rates allowed to vary over an annual cycle. Modelling of foliar retention is particularly important because of feedbacks on both water balance and carbon accumulation. Daily foliar litter fall can be either expressed as a constant fraction of foliage present or calculated from non-linear functions constraining foliage mass to stem biomass.

14.2.3 Water balance

Stomatal conductance:
Stomata control the link between processes of carbon and water uptake. In BIOMASS stomatal conductance (g_s) is specified either by an empirically derived non-linear function of meteorological variables, vapour saturation deficit, photon flux density and soil water content (Table 14.4), or by a relationship linking it to net photosynthesis (**A**), relative humidity (h), and ambient CO_2 concentration (C_a):

$$g_s = g_o + g_1\,(\mathbf{A}h/C_a), \qquad (14.1)$$

where g_o and g_1 are constants [2]. When h is expressed as a percentage, values of the dimensionless parameter g_1 estimated from leaf gas exchange data have ranged between 0.06 and 0.16 for C_3 plants [2,8,11]. The constant g_o has units of conductance, e.g. $g_o = 0.01\,mol\,m^{-2}\,s^{-1}$ [8]. After BIOMASS has calculated an instantaneous rate of canopy net photosynthesis ($\mathbf{A}_c$), canopy conductance (G_c) can be estimated from:

$$G_c = g_o L^* + g_1\,(\mathbf{A}_c h/C_a)$$

where L* is one-sided leaf area index.

Though lacking a sound theoretical base [6], Equation 14.1 has successfully explained much of the variability in several gas exchange data sets [2,6,8,11]. An advantage of the stomatal model (Equation 14.1) is that it contains only two parameters, many fewer than are required by models which link g_s to several meteorological variables. Table 14.4 discusses how photosynthesis and stomatal conductance are related in the BIOMASS model.

Stomatal conductance is assumed to decline under water-stressed conditions which occur when the availability of soil water to plants declines below a threshold, typically 30–40% of the amount available at field capacity. Below the threshold, conductance is assumed to decline linearly with available soil water. Equation 14.1 and equations given in Table 14.4 define how photosynthetic rates decline in response to reduced stomatal conductance.

Stand water balance:
Conductance is obtained for each canopy layer (Table 14.4) and tree transpiration is calculated from the standard Penman–Monteith equation (Table 14.5). Allowance is made for the interception of rainfall by the canopy and the evaporation of intercepted water. The model contains an option to allow for losses through soil evaporation and understorey evapo-transpiration [12].

The water balance model has provision to consider two soil layers. Drainage occurs from the base of a layer only if its volumetric water content exceeds nominal field capacity. This type of drainage model is termed a 'tipping bucket' model. BIOMASS does not consider the role of roots in water uptake. Trees extract water from the upper soil layer until it is depleted to a critical water content, when extraction commences from the lower layer. Rooting depth and parameters of a soil water retention function must be supplied in the user's parameter file.

14.3 APPLICATION OF THE MODEL TO AN EXPERIMENT

14.3.1 The experiment

A major field experiment called the Biology of Forest Growth (BFG), on the response of *Pinus radiata* to irrigation and fertilization, was commenced by CSIRO Division of Forest Research in 1983 at a site near Canberra, Australia (35° 21′S, 148° 56′E, elevation 625 m). Average annual rainfall is 790 mm, but with large year-to-year differences and with frequent summer droughts. Over the period 1973–1986, annual rainfall varied between 359 and 1239 mm. (See Benson *et al.* [3] for a detailed description of the site.) Trees were planted in 1973 at a stocking density of approximately 700 per hectare. Treatments applied in the experiment were irrigation (I), solid fertiliser (F), irrigation and solid fertiliser (IF), and irrigation and liquid fertiliser (IL). The control (C) received no treatment. The irrigation treatment applied to I, IF and IL stands commenced in August (Winter) 1984 with the aim of removing soil moisture as a growth-limiting variable. Water was applied by sprinklers at a rate sufficient to maintain soil moisture at or close to field capacity.

The solid-fertiliser treatment was applied in two doses in September and October (Spring) 1983. The total supplied was 400 kg of nitrogen and 200 kg of phosphorus per hectare. The liquid-fertiliser treatment consisted of regular applications of a complete nutrient solution delivered weekly through the irrigation system at rates designed to provide adequate nutrients for tree growth throughout the season. The IL treatment commenced in August 1984, with annual applications of approximately 300 kg ha^{-1} of nitrogen.

14.3.2 Parameterization of the model

Tables 14.2–14.5 list many parameters, values of which must be supplied by model

Table 14.4. Stomatal conductance and leaf photosynthesis. Equations below define linkages between the mechanistic model of leaf assimilation [4] and an empirical model of stomatal conductance expressed as a function of meteorological variables. Symbols are: c_a and c_i, ambient and intercellular CO_2 concentration (μbar); f_i, functions for dependence of stomatal conductance on Δe, volumetric soil water content Θ, and photosynthetic photon flux density incident on foliage ϕ; g_{smax}, maximum stomatal conductance ($mol\,m^{-2}s^{-1}$); J and J_{max}, rate of electron transport and its light-saturated rate (μmol electrons $m^{-2}s^{-1}$); k_c and k_o, Michaelis constants (μbar); O_2, intercellular oxygen concentration (μbar); r_d, non-photorespiratory rate of respiration in daytime (μmol CO_2 $m^{-2}s^{-1}$); V_{cmax}, maximum RuBP-saturated rate of carboxylation (μmol CO_2 $m^{-2}s^{-1}$); Γ_*, CO_2 compensation point (μbar); and ϕ_s, threshold for the g_s/ϕ relationship (μmol quanta $m^{-2}s^{-1}$).

Stomatal conductance ($mol\,m^{-2}s^{-1}$):

$$g_s = g_{smax}\, f_1(\Delta e)\, f_2(\Theta)\, f_3(\phi)$$

Net photosynthesis of leaves (μmol $CO_2\,m^{-2}s^{-1}$):

$$\mathbf{A} = \min(\beta(1 - \rho_l - \tau_l)\phi, \mathbf{A}_{sat}) - r_d$$

Rate of net photosynthesis by leaves of C_3 plants:

$$\mathbf{A} = (1 - \Gamma_*/c_i)\min(W_j, W_c) - r_d$$

Rubisco-limited rate of carboxylation:

$$W_c = \frac{V_{cmax} c_i}{c_i + k_c(1 + O_2/k_o)}$$

RuBP-regeneration limited rate of carboxylation:

$$W_j = \frac{J/4}{1 + 2\Gamma_*/c_i}$$

Rate of CO_2 supply (assuming negligible boundary layer resistance):

$$\mathbf{A} = g_s(c_a - c_i)/1.6$$

Formula for irradiance-saturated rate $\mathbf{A}_{sat}$ is obtained by setting $\mathbf{A} = \mathbf{A}_{sat}$, $J = J_{max}$ and $g_s = g_{smax}\, f_1(\Delta e)\, f_2$ Θ), and solving the above equations for c_i and $\mathbf{A}_{sat}$, assuming first that photosynthesis is Rubisco-limited and then RuBP-regeneration-limited. The appropriate solution is the one with lower $\mathbf{A}_{sat}$.

A formula for quantum yield ϕ_{abs} is obtained from the limit of low irradiance when $J = 0.385\,\phi$ so that:

$$\phi_{abs} = 0.096\frac{c_i - \Gamma_*}{c_i + 2\Gamma_*}$$

Assuming that assimilation is limited by RuBP-regeneration at low light, and that g_s is proportional to ϕ at low irradiation, with $f_3 = 1$ when $\phi \geq \phi_s$, gives an equation which can be solved for c_i:

$$0.096\,\phi_s\frac{c_i - \Gamma_*}{c_i + 2\Gamma_*} - r_d = \frac{g_{smax}\, f_1(\Delta e)\, f_2(\Theta)\,(c_a - c_i)}{1.6}$$

The slope parameter β is obtained by adjusting ϕ_{abs} for the curvature of the photosynthetic light response [11]:

$$\beta = 0.78\,\phi_{abs}$$

The following parameters are temperature-dependent: Γ_*, r_d, J_{max}, V_{cmax}, k_c and k_o. Parameters J_{max} and V_{cmax} are linear functions of foliar nutrient concentration.

Table 14.5. Equations for tree water use. Symbols as follows: a, albedo; c_p, specific heat capacity of air at constant pressure; f_E, reduction of transpiration when canopy is partially wet; G_b, canopy boundary layer conductance ($mol\,m^{-2}\,s^{-1}$); s, slope of saturation vapour-pressure/temperature curve; γ, psychometric coefficient; ε_v, emissivity of vegetation (= 0.95); λ, latent heat of vapourisation of water; ρ, density of air; σ, Stefan–Boltzmann constant.

Daily canopy transpiration (mm day^{-1}):

$$E_T = f_E \int_0^{\Delta} E(t)\,dt$$

Instantaneous rate of canopy tranpiration (mm $hour^{-1}$):

$$E(t) = E_{eq}\Omega + E_{imp}(1 - \Omega)$$

Equilibrium rate of transpiration:

$$E_{eq} = sK_n/((s + \gamma)\lambda)$$

Imposed rate of transpiration:

$$E_{imp} = \rho\, c_p\, G_c\, \Delta\varepsilon/(\lambda\gamma)$$

Coupling coefficient describing extent to which transpiration is determined by incident radiation *vs.* canopy conductance:

$$\Omega = (s + \gamma)/(s + \gamma + \gamma G_b/G_c)$$

Net radiation (MJ m^{-2} day^{-1}):

$$K_n = (1 - a)K_s - K_{ln}$$

Net long-wave radiation (W m^{-2}):

$$K_{ln} = (0.1 + 0.9 K_s/K_{sc})\,(\varepsilon_a - \varepsilon_v)\,\sigma(\bar{T} + 273)^4$$

Apparent emissivity of the atmosphere:

$$\varepsilon_a = 1 - 0.261 \exp(-7.77 \times 10^{-4}\,\bar{T}^2)$$

Canopy conductance (mol H_2O m^{-2} ground s^{-1}):

$$G_c = \int_0^{L^*} g_s\, dL^*$$

users. For *P. radiata* the parameters were derived from several sources. Several were estimated from experimentation conducted on the site (e.g. J_{max}, V_{cmax}, $[N]_f$, g_{smax}). Some were obtained from laboratory experiments (e.g. ρ_l, τ_l). Others came from the scientific literature or were derived from theory (e.g. k, Y, k_c, k_o, Γ_*, a, ε_v). Parameter estimation for the allocation and litter fall sub-models involved fitting the model to biomass data.

14.3.3 Performance of the model

A simulation of soil moisture to a depth of 2 m for the C plot is presented in Fig. 14.2, together with regular measurements representing means of six neutron-probe readings. In both years drying of the soil profile commenced in late spring with summer droughts of 3–4 months' duration. Recharge occurred in autumn and winter. Similar good fits of the model to soil moisture data have been obtained for other sites [12].

Simulated values for leaf biomass [10] are shown in Fig. 14.3. Measured and simulated annual foliage production are compared in Fig. 14.4. The correspondence between simulated and measured stem and branch biomass over a 3-year period was also good [10]. While Figs. 14.3 and 14.4 indicate that the model fits

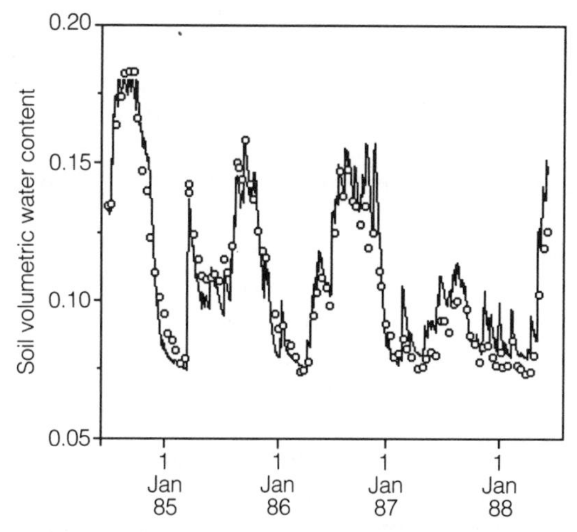

Fig. 14.2. Simulated (solid line) and measured soil water contents (circles) to a depth of 2 m for the control stand of the BFG experiment. The measurements represent means of six neutron probe readings. Redrawn from McMurtrie and Landsberg [10].

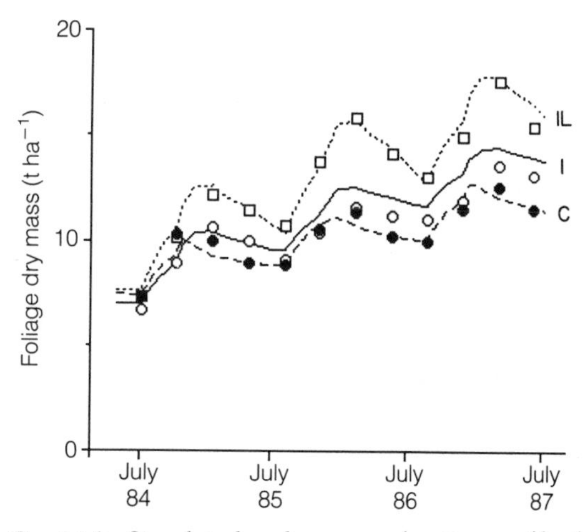

Fig. 14.3. Simulated and measured patterns of leaf biomass development for the period July 1984 to July 1987 for control (C, ●), irrigated (I, ○), and liquid fertilized and irrigated (IL, □) stands. Redrawn from McMurtrie and Landsberg [10].

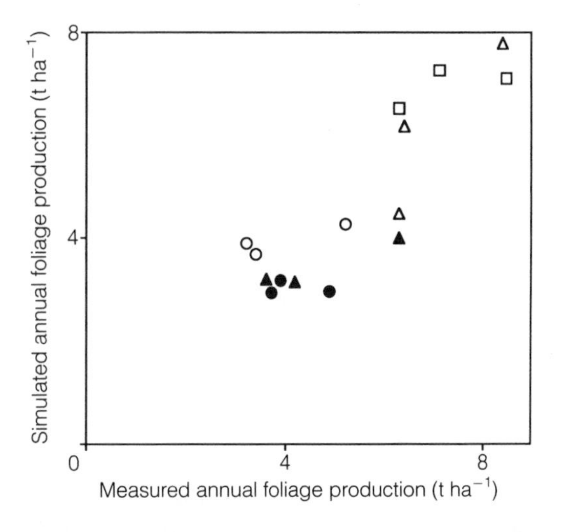

Fig. 14.4. Annual production of foliage dry mass estimated from biomass data and simulated by the BIOMASS model for the period July 1984 to July 1987 for C (●), I (○), F (▲), IF (△) and IL (□) stands.

the data well, and enhance confidence in the model, they do not represent an independent test of the model. Model testing can only be achieved using data completely independent of data used for parameterization. Use of the biomass data to estimate partitioning coefficients and litter fall rates precludes its use for validation purposes.

A simulation of the growth of stands of *P. radiata* over an 8-year period on a fertile site free of water stress is illustrated in Fig. 14.5. The simulation of leaf biomass development shows annual variations associated with the seasonality of foliage production and litter fall. After approximately 6 years of canopy development, the pattern appears to stabilise with annual foliage production balanced by annual litter fall. Both branch and stem biomass accumulate steadily over time, though branches show a decelerating trend.

14.4 DISCUSSION

Models are often advocated as tools to guide experimentation in large-scale field experiments. A simulation model was developed at the outset of the BFG project, and was instrumental in identifying particular measurements to be performed and in initiating some experimentation. However, the model did evolve significantly in the course of the

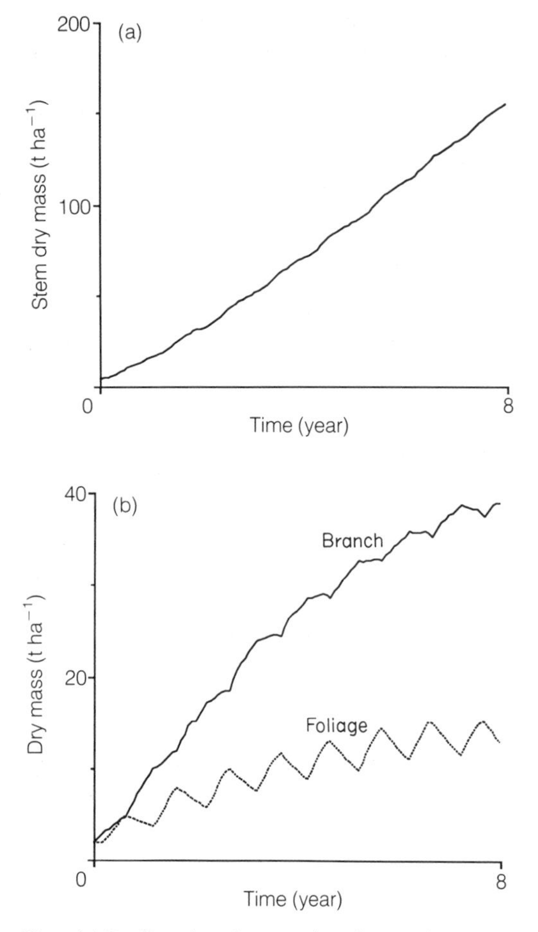

Fig. 14.5. Simulated growth of stands of *Pinus radiata* over an 8-year period on a fertile site with an abundant supply of water: (a) stem biomass; (b) foliage and branch biomass.

experiment and in its current form bears little resemblance to the original model. Models may be regarded as statements of hypotheses, providing formal statements of how it is thought systems work. It is to be expected therefore that the model will evolve as our understanding of ecosystem functioning improves. The evolution occurred as submodels were reformulated either to revise and improve theory or to provide tighter coupling with experimentation underway in the field.

Like all models, BIOMASS has its strengths and weaknesses. Its requirement for standard daily meteorological data can be met readily for many sites. Its representations of water balance and canopy photosynthesis contain sound theory, though both need further testing. The water balance model has been validated against soil moisture data. Validation of canopy CO_2 and water exchange submodels should be conducted using micrometeorological techniques. This would provide a direct test of instantaneous model performance, a test which is more powerful than less direct methods involving comparison with actual biomass data.

Several components of the BIOMASS model are applicable to crops other than trees. The model of water balance is quite general, though care should be taken in applying the model to short, extensive canopies whose transpiration is relatively insensitive to stomatal conductance. Improvements to equations for boundary layer conductance and in particular its dependence on wind speed will be required for vegetation less subject to stomatal control. The mechanistic model of canopy photosynthesis is appropriate for C_3 plants and uses a Blackman photosynthetic light response, which is a reasonable assumption for C_3 but not for C_4 plants. Models of radiation interception and maintenance respiration have wide applicability.

The mechanisms of biomass allocation by trees are poorly understood. This is reflected by the empirical nature of equations describing allocation in models such as BIOMASS. Formulations of allocation to foliage production are particularly important model components because of feedbacks which affect canopy photosynthesis and tree water use. Progress to improve understanding of forest allocation has been hampered by the amount of labour required to acquire and process biomass samples, a problem which is compounded by the long lifetimes of forest crops. In agriculture where generation times are shorter and biomass is more readily estimated, it is easier to develop and independently test empirical models of allocation. Models of carbon allocation by agricultural crops emphasise phenology, the relationship of tissue production to heat sums, partitioning to reproductive organs, and, less frequently, the role of sink strength as a determinant of photosynthetic production. These features are not addressed in BIOMASS. There is a need for specifically tailored experimentation to develop and test more mechanistic models of allocation by trees.

A further shortcoming of the model is its neglect of the dynamics of nutrient cycling. Foliar nutrient concentrations are supplied as model inputs affecting photosynthetic parameters, J_{max} and V_{cmax}, and the partitioning coefficient (Tables 14.3 and 14.4). Interactions with nutrient cycles are not adequately treated by the model. However, steps are currently underway to expand the BIOMASS model to incorporate processes such as soil mineralization, decomposition of soil organic matter, nutrient uptake and its redistribution within vegetation.

ACKNOWLEDGEMENTS

I am grateful to Roberto Inzunza, Ray Leuning, Steve Long and Ying-ping Wang for valuable comments and assistance, and acknowledge the support of the Australian Research Council and the NGAC Dedicated Greenhouse Research Grants Scheme.

REFERENCES

1. Ågren, G.I., R.E. McMurtrie, W.J. Parton, J. Pastor and H.H. Shugart (1991) State-of-the-art of models of production-decomposition linkages in conifer and grassland ecosystems. *Ecological Applications* **1**, 118–138.
2. Ball, J.T., I.E. Woodrow and J.A. Berry (1987) A model predicting stomatal conductance and its contribution to the control of photosynthesis under different environmental conditions. In: *Progress in Photosynthesis Research*, Vol IV (I. Biggins, ed.) Martinus Nijhoff, Dordrecht. pp 221–224.
3. Benson, M.L., J.J. Landsberg and C.J. Borough (1992) The Biology of Forest Growth experiment: an introduction. *Forest Ecology and Management* (in press).
4. Farquhar, G.D. and S. von Caemmerer (1982) Modelling of photosynthetic response to environment. In: *Encyclopedia of Plant Physiology*, New Series Vol. 12B (O.L. Lange, P.S. Nobel, C.B. Osmond and H. Ziegler, eds.). Springer-Verlag, Berlin. pp. 549–587.
5. Goudriaan, J. (1986) A simple and fast, numerical method for the computation of daily totals of crop photosynthesis. *Agricultural and Forest Meteorology* **38**, 249–254.
6. Grantz, D.A. (1990) Plant response to atmospheric humidity. *Plant Cell and Environment* **13**, 667–679.
7. Landsberg, J.J. (1986) *Physiological Ecology of Forest Production*. Academic Press, London.
8. Leuning, R. (1990) Modelling stomatal behaviour and photosynthesis of *Eucalyptus grandis*. *Austr. J. Plant Physiol.* **17**, 159–175.
9. McMurtrie, R.E., M.L. Benson, S. Linder, S.W. Running, T. Talsma, W.J.B. Crane and B.J. Myers (1990) Water/nutrient interactions affecting the productivity of stands of *Pinus radiata*. *Forest Ecology and Management* **30**, 415–423.
10. McMurtrie, R.E. and J.J. Landsberg (1992) Using a simulation model to evaluate the effects of water and nutrients on the growth and carbon partitioning of *Pinus radiata*. *Forest Ecology and Management* (in press).
11. McMurtrie, R.E., R. Leuning, W.A. Thompson and A.M. Wheeler (1992) A model of canopy photosynthesis and water use incorporating a mechanistic formulation of leaf CO_2 exchange. *Forest Ecology and Management* (in press).
12. McMurtrie, R.E., D.A. Rook and F.M. Kelliher (1990) Modelling the yield of Pinus radiata on a site limited by water and nitrogen. *Forest Ecology and Management* **30**, 381–413.
13. Norman, J.M. (1980) Interfacing leaf and canopy light interception models. In: *Predicting Photosynthesis for Ecosystem Models* (J.D. Hesketh and J.W. Jones, eds.). CRC Press, Boca Raton, Florida. pp. 49–67.

15
Carbon partitioning

J.F. Farrar

15.1 INTRODUCTION

Leaves are where most photosynthesis occurs, but seeds and storage roots form the bulk of food sources. It follows that photosynthesis itself is an inadequate measure of food production, and the fate of carbon fixed in photosynthesis needs to be studied carefully. There is every likelihood that the factors controlling partitioning will be different in general or in detail from those that affect photosynthesis, and it is impossible with current knowledge to predict how, for example, an increased supply of photosynthate (which may come from increased partial pressures of atmospheric CO_2) will be partitioned. Indeed, the control of partitioning between organs is scarcely understood, and whole plant and crop models mainly use, as a gesture of despair, a simple partitioning coefficient to describe the proportion of photosynthate passing to the root.

At least part of the reason for our poor understanding is the complexity of the partitioning process. Although descriptions of partitioning abound, there have been relatively few attempts to investigate mechanisms, although the last few years have seen great improvements. We even lack a uniform and sufficiently precise terminology, a sure indication of ignorance. I propose to use the word partitioning generically, to embrace cellular, organ and whole-plant partitioning; and to appropriate other terms where they are needed. This chapter will make the assumption that the study of carbon partitioning is largely the study of carbohydrate transport and metabolism. Indeed a good case can be made for sucrose being the compound central to the growth of plants, so that its production, translocation and utilisation is the focus of much work on carbon partitioning. In some species, stachyose, sorbitol or mannitol are translocated in addition to sucrose.

The literature on carbon partitioning is large, and frequently confused. There are two reasons for the confusion. Firstly, there has been an unwillingness to dissect it into its partial processes; problems have not been analysed sufficiently and complex processes described as if they were far simpler than they are. For example, respiration (which often accounts for 50% of photoassimilate) is often ignored when calculating partitioning into an organ. Secondly, there is a lack of appreciation of the different time-scales over which control of different types may operate. In the short term (seconds, minutes) fine regulation of metabolism, including allosteric effects, feedback inhibition and the effect of fructose 2,6-bisphosphate, are seen; over a number of hours, the size of sugar pools in the cells will alter, with consequences for

Photosynthesis and Production in a Changing Environment: a field and laboratory manual
Edited by D.O. Hall, J.M.O. Scurlock, H.R. Bolhàr-Nordenkampf, R.C. Leegood and S.P. Long
Published in 1993 by Chapman & Hall, London.
ISBN 0 412 42900 4 (HB) and 0 412 42910 1 (PB).

Table 15.1. Typical questions asked in studies of carbon partitioning. Names of specific processes are given in parentheses

(1) Spatial
  (a) Cell
    What are the major compounds (sucrose, starch, alditols, fructan) involved? What is their rate of turnover? (Product partitioning)
    What subcellular compartments (plastids, cytosol, vacuole, apoplast) are involved? (Compartmentation)
    What proportion of carbon is exported? What proportion is respired?
  (b) Organ
    What cell types are involved?
    Is the pathway of sugar movement apoplastic or symplastic?
    What is the route of phloem loading? What is the rate of unloading?
    What proportion of carbon is respired, imported or exported?
    What proportion of the capacity of the organ for C flux is realized *in vivo*?
    What controls the major fluxes (respiration, storage, synthesis of new structure) within sinks?
  (c) Whole-plant (between organs)
    What is the flux from a single source to a single sink? (Acquisition)
    In what proportion is C from a single source divided between sinks? (Allocation)
    What proportion of the capacity of the transport system is realised *in vivo*?
    What controls the acquisition by, and allocation to, a sink?
(2) Temporal
  Where is fine control of metabolism exercised?
  To what extent do stores of carbohydrate buffer carbon fluxes?
  Is coarse control (protein synthesis or degradation) of the process involved?

fluxes through them and for processes that may be controlled by amount of sugar; and over longer periods (hours to days) the amount of metabolic machinery may be altered by coarse control, in response to developmental, external, or internal modulation of gene expression.

The scope of carbon partitioning can be illustrated by asking a series of questions (Table 15.1). It will be appreciated that this is far from an exhaustive list, and that even so it is not possible to deal with methods relevant to all of it in this chapter. It should also be clear that good answers to many of these questions demand quantitative rather than qualitative answers.

15.2 CARBON BALANCE AND DISTRIBUTION OF DRY MASS

The whole plant represents the end result of partitioning, but also of other processes, notably nutrient uptake and respiration. Thus whilst growth analysis (Chapter 3) can give a valuable guide to carbon partitioning, measures such as root:shoot ratio should not be confused with it. The nutrient content of plant material can be substantial, commonly 5–20% of the dry mass, and so can introduce appreciable errors into estimates of carbon partitioning based on dry mass alone. Even more serious is the effect of respiration. A growing organ respires roughly as much carbon as is contributed to its increased mass (about half of the carbon entering it): a storage organ respires somewhat less. It is vital to distinguish carefully between net and gross partitioning, since only the latter can give insights into mechanism. Further, the proportion of total dry mass which is carbon will alter with tissue composition.

A combination of dry mass measurements (destructive) and gas analysis [1], if used with care, can provide adequate descriptions of the carbon balance of plants and their parts. Since gas exchange rate and carbohydrate content, a major component of dry mass, change throughout a diurnal cycle, it is necessary to select sampling times with care and if possible measure gas exchange continuously.

For the carbon balance of an organ, it is necessary to know: rate of C gain (photosynthesis, import); rate of C loss (respiration, leakage, export); change in C content (growth, storage, mobilisation). From such figures it is possible to calculate a number of parameters such as conversion efficiency (growth yield) (Yg = net C gain/(net C gain + C respired less maintenance respiration)) and growth efficiency (GE = net C gain/gross C gain).

15.3 MEASUREMENT OF MAJOR CARBOHYDRATES AND PROTEIN

Material should be sampled at a uniform time into the photoperiod, since large changes in sugar content occur during illumination and darkness. Detailed information on carbohydrate analysis is given in Dey [2].

15.3.1 Soluble sugars

The large pools of soluble sugars are usually measured after killing and extracting the tissue in 90% ethanol at 60°C or, less reliably, after drying at 80°C in a forced-draught oven. Of the methods described here only high performance liquid chromatography (HPLC) will measure polyols (alditols).

Extraction and spectrophotometric assay of soluble sugars:

1. To 10 ml 90% ethanol at 60°C add approx. 100 mg plant tissue containing 2–20 mg soluble carbohydrate. Stopper and heat for one hour at 60°C. Decant the extract into a 25 ml volumetric flask and re-extract the residue. Combine the extracts and make up to 25 ml.
2. Transfer 1 ml of sample containing 20–100 μg sugar to a wide, thick-walled test tube (if necessary take less than 1 ml and make to this volume with water). Add 1 ml 5% phenol (CARE: phenol is caustic) and mix thoroughly.
3. For this step, safety goggles and a laboratory coat MUST be worn. Add, carefully but not slowly, 5 ml analytical grade sulphuric acid. (The heat generated by mixing H_2SO_4 and water speeds the reaction, but can cause spurting from the tube, especially in a warm laboratory). Mix thoroughly by vertical agitation with a glass rod with a broadened end. Allow to cool in air.
4. Read the absorbance at 485 nm against a blank containing water, 5% phenol and H_2SO_4. Calibrate with sucrose at 0–100 $\mu g\, ml^{-1}$. Glass cuvettes should be used.

The phenol–sulphuric acid technique is robust, and gives similar absorbances with a range of standard sugars.

Enzymatic assay of glucose, fructose and sucrose: The reactions shown below result in the formation of NADPH, which may be detected by the increase in absorbance at 340 nm:

$$\text{sucrose} \xrightarrow[\text{(invertase)}]{} \text{glucose} + \text{fructose}$$

$$\text{glucose} + \text{ATP} \xrightarrow[\text{(hexokinase)}]{} \text{Glc-6-P}$$

$$\text{fructose} + \text{ATP} \xrightarrow[\text{(hexokinase)}]{} \text{Fru-6-P}$$

$$\text{Fru-6-P} \xrightarrow[\text{(phosphoglucose isomerase)}]{} \text{Glc-6-P}$$

$$\text{Glc-6-P} + \text{NADP}^+ \xrightarrow[\text{(glucose-6-phosphate dehydrogenase)}]{} \text{6-phosphogluconate} + \text{NADPH} + \text{H}^+$$

This spectrophotometric procedure (see also Chapter 16) is specific and reliable. Ethanol extracts need to be dried down and taken up in water:

1. To a 0.1 ml sample containing 0–20 μg glucose add 1.00 ml triethanolamine buffer (750 mM, pH 7.6 with 10 mM $MgCl_2$), 1.82 ml H_2O, 0.10 ml 11.5 mM NADP and 0.10 ml 80 mM ATP (ATP-Na_2-H_2 in 1 M $NaHCO_3$).
2. After mixing and standing for 3 minutes, start the reaction by adding 20 μl of hexokinase/glucose-6-phosphate dehydrogenase/phosphoglucose isomerase solution (2, 1 and 2 mg ml^{-1}, respectively).
3. Read the absorbance at 340 nm before and after enzyme addition.
4. For sucrose, add 0.10 ml of the sample (0.20 ml containing 0.20 μg of sucrose) to 0.20 ml 320 mM citrate buffer, pH 4.6, and then add 20 μl 10 mg invertase (*Boehringer* or *Sigma* Grade X – high grade) in citrate buffer.
5. After hydrolysis of sucrose at 25°C for 15 minutes, follow the procedure for glucose; sucrose is given by the difference in glucose before and after enzymatic hydrolysis. Alternatively, sucrose hydrolysis may be followed continuously [3].

Chromatography:

In general, thin layer chromatography (TLC) has a rather low loading capacity for work with sugars; gas chromatography is good but involves derivatisation. HPLC is the method of choice, whilst paper chromatography still has much to commend it.

Descending paper chromatography is performed using Whatman 3 MM paper. Between 10 and 100 μg sugar are applied to the origin, with standards in the outer lanes and the chromatograms run in a draught-free environment for 2 days in ethyl acetate : acetic acid : water (14 : 3 : 3), or propan-1-ol : ethyl acetate : water (6 : 1 : 3), the latter being particularly good for separating di- and trisaccharides. Spots are visualised either by dipping through a solution of 1.75 g 4-amino-benzoic acid and 1 ml 88% orthophosphoric acid in 250 ml methanol, drying and heating at 110°C for 6 minutes, or, for alditols as well, the chromatogram is first dipped through acetone (300 ml) to which saturated silver nitrate solution (4 ml) has been added, dried, dipped again through 0.5% w/v NaOH in ethanol and rinsed in concentrated sodium thiosulphate and water (this procedure does not stain sucrose).

High pressure liquid chromatography is performed isocratically with a refractive index detector. Samples, with 20–200 μg of each sugar per 20 μl, are deproteinized and either centrifuged at 14 000 *g* or filtered through 0.2 μm syringe filters to remove particulates. Hexoses and disaccharides may be separated on an amino column (e.g. *Spherisorb* Amino 5 μm) with acetonitrile : water (82 : 18) as mobile phase whilst better separation of di- and trisaccharides is achieved on calcium-based ion-exchange resin columns such as the *Bio-Rad HPX-87C* with water as the mobile phase. Peaks are quantified by height or area and calibrated against a range of concentrations of standard sugars.

15.3.2 Fructan and starch

Fructan and starch occur together in many species, although commonly one dominates or is absent. Fructan occurs in some members of the Gramineae, Compositeae, Liliales, and several other families.

Extraction and assay of fructan:

Fructan is water-soluble and acid-labile. Fructan of low degree of polymerization (d.p.) is also soluble in ethanolic solutions and so soluble sugars are best removed with 95% ethanol, and the fructan so removed assayed separately. The bulk of the fructan then can be extracted into warm water. Alternatively, successive ethanolic (80%; 2 hours at 60°C) and warm water (30°C, 2 hours) fractions can be combined before further analysis. Either a discrete fructan fraction can be assayed for carbohydrate using the phenol–sulphuric acid method (Section 15.3.1) or fructose moieties can be assayed colorimetrically, using a ketose-specific method such as the following:

1. To 2 ml of sample, containing less than 400 μg fructan, add 5 ml of glycerol-copper reagent (200 ml HCl, 8.4 mg $CuSO_4 \cdot 5H_2O$, 260 ml glycerol, 100 ml H_2O) and 1 ml 0.45% resorcinol.
2. Heat to 100°C for 12 minutes. Cool, and read the absorbance at 470 nm.

For separation of fructans by TLC, the combined ethanolic and aqueous extracts are dried with a stream of cold air, redissolved in a small volume (100 μl g^{-1} tissue extracted) of 50% ethanol, and deionized with Amberlite MB-1 monobed resin. Samples (10 μl) are spotted onto silica gel TLC plates (*F1500, Schleicher & Schul, Kassel, Germany*) and the plates developed three times in propan-2-ol:butan-1-ol:water (12:3:4). Following air-drying the fructans are located by spraying with urea–phosphoric acid (5 ml 85% orthophosphoric acid, 3 g urea, 5 ml ethanol, and 100 ml water-saturated butan-1-ol), air-drying and heating at 110°C for 15 minutes.

Extraction and assay of starch:
Starch can be extracted from ground material, following removal of soluble sugars and fructan, using amyloglucosidase (10 units ml^{-1}) in 0.2 M Mes buffer, at pH 4.5 and 40°C for 24 hours. The soluble products may be assayed by the phenol–sulphuric acid method (or by the spectrophotometric method; Section 15.3.1). It is important to run a blank of the enzyme preparation alone as this often contains sugars. This method is likely to remove material other than starch, and so result in overestimates, due largely to contamination of the amyloglucosidase with other enzymes.

A more precise but more laborious method is based on depositing a starch–iodine complex on filters and assaying it after solubilization in H_2SO_4:

1. Homogenise 5 mg dry weight of tissue in 5 ml 32% perchoric acid; keep in $HClO_4$ for 20 minutes at 20°C.
2. Filter through 9.0 cm Whatman GF/A glass fibre disc (previously heated to 400°C for 2 hours).
3. Mix 1 ml of the filtrate (containing <100 μg starch) with 3 ml iodine solution (0.392 g I_2, 0.112 g KI in 15 ml H_2O plus 185 ml 3.2% NaCl); keep at 0–5°C for 30 minutes.
4. Collect the starch–I_2 complex precipitate on 2.5 cm Whatman GF/A glass fibre discs and wash with iodine solution diluted 25% by 32% $HClO_4$, and then with three aliquots of ethanolic NaOH (140 ml 96% ethanol, 10 ml 5 M NaOH, 50 ml H_2O).
5. Dry the filter discs and place them in calibrated test tubes. Add 4 ml 0.75 M H_2SO_4 and heat at 100°C for 30 minutes. Readjust volume to 4 ml.
6. To 1 ml of the eluate in a thick-walled tube add 1 ml 5% phenol and then 5 ml H_2SO_4 as described in Section 15.3.1. Read absorbance at 485 nm.
7. Calculate mass of starch as mass of glucose ×0.9.

15.3.3 Protein and structural material

The residue from previous extractions can be dried and weighed to estimate the insoluble or structural material. Alternatively, protein can be extracted (see also Section 18.7.3):

1. Tissue is extracted by grinding in cold NaCl (100 mM in phosphate buffer, pH 7.5) at 4°C.
2. Centrifuge at 3000 *g* for 15 minutes.
3. To 0.1 ml of the supernatant containing 10–100 μg protein add 5 ml Coomassie blue solution (100 ng Coomassie Brilliant Blue G250 in 50 ml 95% ethanol, plus 100 ml 85% orthophosphoric acid, diluted to 1000 ml with distilled water and filtered immediately before use).

4. Read in a spectrophotometer at 595 nm against a reagent blank within 60 minutes.
5. Prepare a standard curve using bovine serum albumen.

15.4 DIRECT MEASUREMENT OF POOLS AND FLUXES

The export from or import into an organ can be estimated by the difference between the change in its dry mass and gas exchange over a given period. Measurements of photosynthetic and respiratory CO_2 exchange by infra-red gas analysis (Chapter 9), corrected to appropriate units, can be directly related to dry mass; measurements of oxygen exchange must be treated more cautiously. Photosynthetic O_2 evolution is usually measured under high partial pressure of CO_2 (Chapter 10) and so cannot be used for this purpose. Respiratory oxygen uptake can be, as respiration rate is comparatively insensitive to the partial pressure of CO_2, but the ratio of CO_2 evolved to O_2 taken up must be measured, or assumed to be unity. In many growing tissues, respiration rate is proportional to growth rate, and so respiration integrated over a diurnal cycle can be used as a non-destructive indicator of growth.

15.4.1 Measurement of respiration using an oxygen electrode

Respiration of leaf discs can be measured in the leaf-disc oxygen electrode (Chapter 10), with the advantage that the leaf disc is in air. The buffer which is used to maintain high concentrations of CO_2 (often 5%), in order to saturate photosynthesis, is not required for measurements of respiration. The electrode should be darkened, and measurement should continue until the rate of respiration is constant. Respiration can also be measured in leaf pieces suspended in a liquid-phase system.

The theory and practice of liquid-phase polarographic oxygen electrodes have been described in Chapter 10. Working with excised tissues or parts of intact plants requires more care than with organelles, yet this technique still offers the easiest way to test effects of substrates and inhibitors. Two critera must be met: the response of the tissue to oxygen must be linear over a wide range of oxygen concentration, and the rate of respiration must approximate to that measured less intrusively (e.g. leaf respiration determined by infra-red gas analysis). These criteria can be hard to meet, due partly to problems of oxygen diffusion into bulky tissue slices. Tangling of tissue with the magnetic follower may be avoided by placing a supporting steel mesh above the follower. If a custom-made cuvette is used with an electrode probe, efficient temperature control must be provided.

The following protocol works with a range of plant tissue:

1. Excise the material with a sharp blade. Leaves are rapidly dipped 5 times in 1% Tween-20 detergént to remove air pockets from the leaf surface and rinsed in distilled water. A known quantity of material – discs cut from leaves with a steel borer, for example, is taken.
2. Tissues with a high internal volume of air are placed in aerated electrode medium for about 30 minutes to bring their internal gaseous oxygen into equilibrium with that of the medium. (Two 9 mm leaf discs in 1 ml of medium introduce more oxygen into the electrode than is present in the electrode medium.)
3. Tissues are sliced into pieces small enough (1–2 mm for leaves; cutting is not necessary for fibrous roots) to minimise problems with oxygen diffusion into the tissue, and are rapidly introduced into the electrode. For a slice the possible flux J into the tissue can be calculated as $J = 8D (C_o - C_i)/l^2$, where D is the diffusion coefficient of oxygen, C_o and C_i its external and internal concentration and l the thick-

ness of the slice; the possible flux should exceed that measured.

Respiratory oxygen uptake by plant tissues (v_t) can be considered as the sum of three components: $v_t = v_{cyt} + pV_{alt} + v_{res}$, where v_{cyt} is the contribution of the cytochrome path, p the degree of engagement and V_{alt} the capacity of the alternative oxidase, and v_{res} non-mitochondrial residual respiration. With care, these components can be estimated separately using inhibitors [4].

15.4.2 Tissue compartmentation

Sophisticated methods for separating selected organelles from tissues are available, but all that will be considered here is sampling apoplastic fluid, and a crude separation of vacuolar and cytosolic sugars.

Collecting apoplastic sugars:
Approximately 2 g of leaf tissue is weighed accurately, submerged in a beaker of distilled water, and infiltrated under vacuum. Vacuum is applied 3–4 times for 30 s each. Fully infiltrated leaves are uniformly dark, and sink. The leaves are then carefully dried and packed, cut surface down, into ultra-filtration membrane cones (*Amicon CF 50.A*) and centrifuged at 490 *g*, 4°C, for 15–30 minutes. It is necessary to test for contamination with cell contents by assaying for marker enzymes, for example malate dehydrogenase (or alcohol dehydrogenase, where present). This method extracts apoplastic sugars (and other solutes) diluted into the infiltration medium.

Apoplastic volume is measured by introducing a radio-labelled solution of known specific activity that is relatively membrane-impermeable. Leaf discs (max. 6 mm in diameter) are cut and washed briefly in 1 mM $CaCl_2$; one sample consists of five discs placed in 1 ml solution containing: 50 mM MES buffer, pH 6.2; 1 mM $CaCl_2$; 20 mM polythene glycol (PEG) 1000; ^{3}H-PEG-1000 of known activity. Samples are removed and rinsed for 5 s in 1 mM $CaCl_2$ over a 3-hour period; dried discs are decolorized in liquid scintillation vials by successive additions of 200 µl•7% $HClO_4$, 200 µl H_2O_2, and 100 µl 2-methoxyethanol before adding scintillant and counting by liquid scintillation. Apoplastic volume is calculated from when uptake of ^{3}H-PEG is saturating.

Simple separation of cytosolic and vacuolar sugars using DMSO:
This rapid method of separating cytosolic and vacuolar sugars must not be regarded as a substitute for more rigorous methods, and needs careful justification on each tissue used. It is based on the observation that aqueous solutions of dimethyl sulphoxide (DMSO) can render the plasmalemma permeable, but not the tonoplast. The following method works for roots of barley; concentration and time may need modifying for other tissues.

1. Whole roots are excised and lowered very carefully into 15 ml 10% aqueous DMSO in 5 ml Petri dishes held at 20°C, keeping the cut end of the root above the solution.
2. The roots are left in DMSO for 40 minutes, with occasional shaking. Then the non-submerged cut end of the root is discarded and the roots blotted and extracted in ethanol as described above. The DMSO solution is filtered through a 2.1 cm Whatman GF/A glass microfibre filter. Both DMSO solution and root extract are then assayed for sugars as described under Section 15.3.1.

An alternative method of estimating cytosolic and vacuolar sugars, using ^{14}C, is discussed in Section 15.5.7.

15.5 USING RADIOISOTOPES OF CARBON

Many publications estimate partitioning by supplying ^{14}C, using methods that vary from simple to highly sophisticated. As ever, the central issue is to use an experimental protocol designed to address a well-defined question.

15.5.1 Supplying $^{14}CO_2$ to plant material

The most elegant ways of supplying $^{14}CO_2$ fulfil several criteria. They keep the leaf and its carbon metabolism in steady-state by keeping conditions the same as those the plant has been grown in, and by not changing the total CO_2 concentration of the air around the leaf. They maximise $^{14}CO_2$ uptake by maintaining a thin boundary layer around the leaf. Lastly, they are appropriate for the data analysis that will follow. A comprehensive system is described by Geiger [5]. Two simpler systems are shown in Fig. 15.1, one for crude pulse-feeding and the second offering more control. Anything from 10 to 100 µCi (0.34–3.4 MBq) can be supplied, depending on duration of the feed and whether individual metabolites are to be extracted and assayed. Since $^{14}CO_3^{2-}$ in solution is in equilibrium with $H^{14}CO_3^-$ and $^{14}CO_2$, solutions lose activity when exposed to the air. This is minimised by making them up in Tris buffer at pH 7–8, and by dividing the ^{14}C stock into aliquots kept frozen until needed, rather than repeatedly thawing and refreezing the stock. Duration of the feed is dictated by its purpose; we use anything from 2 minutes (to label metabolites which turn over rapidly) to 72 hours (to ensure all pools in the leaf are at isotopic equilibrium).

15.5.2 Measurement of ^{14}C in plant material

If the total amount of ^{14}C in a tissue is required, the tissue is dried at 70°C in a forced-draught oven, ground in a mill and a subsample burnt in a sample oxidiser; the $^{14}CO_2$ evolved is trapped in scintillant:phenylethylamine:95% ethanol (1.7:1:1) and counted by liquid scintillation spectrometry. For many comparative purposes, tissues can be cleared and counted directly by liquid scintillation. Leaf discs are placed in scintillation vials and decolorised by sequential addition of 200 µl 7% $HClO_4$, 200 µl 30% H_2O_2 and 100 µl 2-methoxyethanol; after 3 hours 10 ml *Aquasol* (*New England Nuclear*) is added. Many fibrous roots clear if left overnight in 10 ml *Aquasol*, and the ^{14}C in them can be counted with good efficiency. Sequential extraction of soluble carbohydrates and starch (Section 15.3.2) can of course precede these assays.

15.5.3 Continuous monitoring of ^{14}C within intact plants

In spite of the low energy of its radiation (stopped by 0.3 mm of H_2O) it is possible to monitor ^{14}C within intact plants using end-window Geiger–Muller tubes. The tubes should have a large effective area and thin end-windows for sensitivity, and count rates should be as high as is possible without encountering reduced counting efficiency due to the dead-time of the counter; 500 c.p.s. is a suitable target. For most purposes it is best to use a ratemeter with the capability of long time constants or integrating times.

Efflux of ^{14}C from leaves in the phloem can readily be monitored by pulse-labelling a leaf with 50 µCi $^{14}CO_2$ for 15 minutes. The leaf is then held over the end-window of a Geiger–Muller tube, preferably between two frames supporting nylon thread to hold the leaf. The leaf can be left in place for up to 24 hours. Using a stirred leaf chamber containing two Geiger–Muller tubes, one to measure the $^{14}CO_2$ content of the gas stream, it is possible to supply $^{14}CO_2$ to a leaf already held over a Geiger–Muller tube. Information can be obtained on the rate of translocation of $^{14}CO_2$ relative to its rate of fixation, by subtracting integrated translocation from integrated fixation after adjusting for the relative sensitivity of the two counters, and the counts that are due to gaseous $^{14}CO_2$ recorded by the detector to which the leaf is attached. The latter is achieved by darkening the leaf chamber with the leaf in place, starting the flow of $^{14}CO_2$, and comparing counts on the two detectors, before starting photosynthesis by illuminating the chamber. An additional

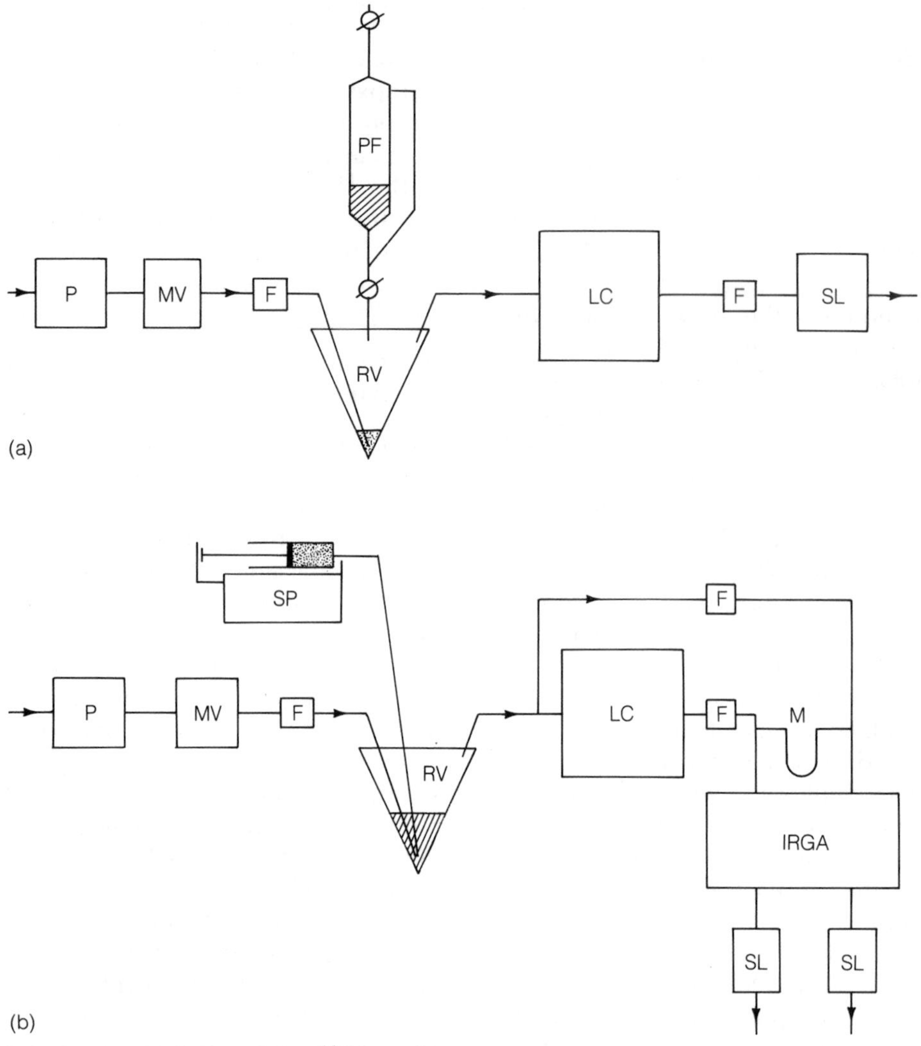

Fig. 15.1. Apparatus for supplying $^{14}CO_2$ to leaves.
(a) Simple system for pulse-labelling (not at constant specific activity). $^{14}CO_2$ is liberated by adding 5 ml lactic acid from a pressure-equalising funnel (PF) to $Na_2{}^{14}CO_3$ in solution in a reaction vessel (RV). A stream of ambient air, monitored and regulated by flowmeters with needle valves (F) enters the reaction vessel from a pump (P) via a 1–5 litre mixing vessel (MV). It sweeps $^{14}CO_2$ through the leaf chamber (LC) and into a tower containing soda lime granules (SL). The leaf chamber should offer a small boundary layer resistance around the leaf, achieved either by stirring the gas in the chamber or by having a cross-sectional area small in relation to the flow rate.
(b) System for feeding $^{14}CO_2$ at constant specific activity and measuring photosynthetic rate. $Na_2{}^{14}CO_3$ in solution is driven by a syringe (perfusor) pump (SP) at a constant rate into 5–10 ml lactic acid in a reaction vessel (RV). The lactic acid is purged by a stream of air (which can be scrubbed of CO_2 by passage through soda lime) from a pump (P), 1–5 litre mixing vessel (MV) and flowmeter with needle valve (F). The gas stream leaving the reaction vessel is split. A small proportion is taken, via a flowmeter (F) to the reference channel of a differential CO_2 – infra red gas analyser (IRGA); the rest goes through a leaf chamber (LC) and flowmeter to the sample channel of the IRGA. The inlet channels of the IRGA are connected by a manometer (M) to facilitate pressure equalisation, and both outlet channels pass through towers of soda lime (SL). Note that a syringe pump capable of delivering 0.5–100 ml h^{-1} (e.g. *Braun Perfusor IV*) is suitable; the leaf chamber can incorporate one or more Geiger-Muller tubes to monitor the ^{14}C content of the air and/or leaf; and the $Na_2{}^{14}CO_3$ in the syringe can contain carrier carbonate at a concentration calculated to give 340 μl CO_2 l^{-1} in the gas phase.

advantage of this approach is that it minimises handling of the leaf; we find that better efflux is obtained when the leaf has been set up in the feeding chamber, and left undisturbed for several hours before supplying $^{14}CO_2$.

Use of efflux and efflux curves yields precisely quantified date on translocation which can be analysed in several ways (Section 15.5.4). Without analysis, it enables comparison of treatments; with analysis, it is possible to obtain information on how carbon in compartmented within the leaf.

It is also possible to measure import into sink regions using Geiger–Muller tubes. The favoured material is young leaves, but with care sections of fibrous root systems sandwiched between moist tissue and 10 μm polyester film can be monitored successfully.

Speed of phloem transport can be measured by briefly pulse-labelling a leaf with $^{14}CO_2$ and monitoring continuously with Geiger–Muller tubes at two points, 10–20 cm apart, downstream of the fed region. Alternatively, two closely similar leaves can be pulse-labelled; at the end of the labelling period, one leaf and the downstream region of the plant is sampled, and the other leaf is sampled 10–15 minutes later. Sampling involves rapidly cutting the plant, from fed leaf downwards, into 1 cm sections which can then be decolorised and counted by liquid scintillation (Section 15.5.2). A semi-logarithmic plot of the date shows a shift in the tracer front with time from which speed can be calculated.

15.5.4 Analysis of data

As ever, the base against which data or radioactive content is expressed must reflect the question being asked in the experiment and must itself not vary with the treatment imposed. In general it is better not to express data as per cent distribution, especially for each of a number of harvests, as the respiratory loss of $^{14}CO_2$ means that a constant ^{14}C content can appear as an increasing percentage with time. In general, expression of counts per organ, per unit dry weight or (in leaves) per unit area, is preferred.

Compartmental analysis is frequently used to analyse isotope redistribution between compartments, and to gain information on sizes of pools, and fluxes, that cannot be measured directly. Assuming a pool of mass Q exchanging isotope with its environment in steady state has a radioisotope content of q at t = 0, that the pool is well mixed, the radioisotope is a perfect tracer and that the pool shows first-order kinetics, then:

$$\frac{dq}{dt} = -kq$$

where k is a rate constant (with $t_{1/2} = 0.693/k$). This yields $q = q_o e^{-kt}$, where a one-compartment system is then described by a single exponential and the flux ϕ through it is given by $\phi = kQ$; q_o is the value of q at zero time.

A two-compartment system is described by the sum of two exponential terms, and the equations describing it depend on the connections between the pools, and between the pools and the environment. In theory, most of the properties of a compartmental system can be obtained from an efflux curve plus the flux entering the system. Since a compartment can be a pool of a single metabolite, a subcellular compartment or a component of a tissue, the use of compartmental analysis seems very attractive. In practice, it is far from simple. Systems of more than two compartments are too complex to analyse with confidence unless a great deal is known about them; the assumptions underlying the analysis are rarely met; and it is very difficult to obtain data of sufficient quality. The use of compartmental analysis has been critically reviewed [6].

An alternative to compartmental analysis, which has been used for analysis of data from experiments with ^{11}C (Section 15.5.6), is the recursive instrumental variable method [7]. This procedure essentially describes data in an assumption-free way and enables parameters describing isotope movement to be determined objectively. The parameters are not based on mechanistic ideas about how the system might work.

15.5.5 Autoradiography

Leaves and roots, as well as whole plants, can readily be autoradiographed to show the distribution of ^{14}C. Freshly radiolabelled material is rapidly arranged on blotting paper, and sandwiched between blotting paper and two wire mesh plates fastened with bulldog clips; this process should take less than 2 minutes. The clamped material is then freeze-dried at −75°C for 24 hours. The dried material is then arranged on brass plates which are covered with foam rubber and cut to be the same width as photographic film (35 or 61 mm). The leaves and plates are covered with a layer of 10 μm thick polyester film and, in total darkness, a length of black and white film (e.g. *Kodak Tri-X*) is placed over them, emulsion side towards the leaf. The film is sandwiched between two brass plate which are compressed very tightly together and left for between 12 hours and 8 days before developing the film strip to high contrast (gamma = 1.0). Radioactive areas appear black on the film and white on a print from it.

When interpreting autoradiographs it should be remembered that density of the image on film is proportional to the logarithm of the exposure, and that above a threshold value increased exposure is not reflected in increased density of image.

15.5.6 Use of the short-lived isotope ^{11}C

^{11}C is a short-lived isotope ($t_{1/2}$ = 20.4 minutes) which emits positrons; these have a path of about 4 m in air, annihilating to emit two gamma rays. ^{11}C has several advantages for studies of carbon movement around plants. The high energy of the γ-rays means that self-absorption by the plant is small and large regions of a plant can be monitored uniformly by a remote scintillation detector. The short half-life means a high rate of positron emission, so even small amounts of ^{11}C yield an appreciable count rate. This also means that the same plant can be fed twice or more, perhaps the first time before and the second after a treatment, using the plant as its own control. Furthermore, it is relatively easy to produce time-courses of the effects of treatment. The disadvantage of ^{11}C is that it is manufactured in a cyclotron or linear accelerator, and experiments must be done very close to the site of manufacture. More detail on the use of ^{11}C, and the treatment of data that result from ^{11}C experiments can be found in Minchin [7].

15.5.7 Applying and analysing ^{14}C-labelled metabolites

Work with metabolites is straightforward. Care is needed in extracting metabolites from tissues to ensure that the tissue is killed rapidly and effectively (Chapter 16), and if ^{14}C in a specific metabolite is to be measured it is necessary to separate it, usually chromatographically, from other compounds.

Uptake of ^{14}C-labelled sugars:
Small pieces of tissue (root tips, leaf discs up to 6 mm in diameter) can be used for studies of rate of sugar uptake. They should be cut from the plant and used as soon as possible as their endogenous substrate will only sustain maximal respiration for a short period. A typical protocol for glucose uptake is given below:

1. Six tissue pieces per replicate are preincubated in small petri dishes containing 5.0 ml of 50 mM MES, pH 6.0, 0.1 mM $CaCl_2$ and the chosen concentration of glucose, at 20°C and defined photon flux density.
2. Transfer tissue to a solution identical except for the inclusion of 1 μCi (34 kBq) ^{14}C-glucose, and agitate for 30 minutes.

3. Rinse tissue in two changes of distilled water at 20°C for 10 minutes each to remove label from the free space.
4. Count radioactivity in cleared or digested material by liquid scintillation spectrometry (Section 15.5.2).

Autoradiography of the tissue (Section 15.5.5) can be used to check penetration and distribution of isotope.

Analysis of ^{14}C-labelled metabolites:
For metabolites that are turned over rapidly, the tissue should be freeze-clamped (Chapter 16) and processed at low temperature. Only relatively stable and more abundant compounds are considered here. Mono- and disaccharides can be extracted as described in Section 15.3.1 and must then be separated chromatographically. The appropriate region of a TLC plate is scraped off and the particles suspended for liquid scintillation counting in a thixotropic gel containing scintillant (toluene:Triton X-100:PPO:POPOP, 667:333:5:0.1, mixed 3:2 with water). Separation by paper chromatography is followed by either eluting the desired spot and counting by liquid scintillation, or counting the paper directly with a Geiger–Muller tube. Automatic scanners are available to quantify radioisotopes on paper and TLC, and spark chambers to give an image of isotope distribution. HPLC can be used, delivering the mobile phase from the detector to a fraction collector. A programmable fraction collector enables only those peaks of interest to be collected. ^{14}C in fructan and starch can be measured after separation and hydrolysis (Section 15.3.2).

It is possible to estimate the distribution of sugars between cytosol and vacuole using ^{14}C. A plant is fed with $^{14}CO_2$ for a prolonged period (24–72 hours) at constant specific activity (Section 15.5.1) until all pools are at uniform specific activity. Then the tissue of interest (roots, leaf discs) is rapidly prepared and passed through a series of solutions, buffered and at a controlled temperature, and containing sucrose at a concentration approximating to that in the apoplast of the tissue concerned (typically between 1 and 10 mM). The ^{14}C is washed out into the medium, by exchange with the external sucrose. The washout should be continued for 6–10 hours, with solutions changed more frequently initially than later; the ^{14}C content of the tissue at the end of the washout is measured. A three-phase exponential decay of ^{14}C remaining in the tissue, with half-times of about 2 minutes, 20 minutes and 20 hours, is found, representing free space, cytosol and vacuoles, respectively, and the proportion of ^{14}C in each at the beginning of washout is a good measure of sugar compartmentation [8].

15.5.8 Short-term partitioning in leaves using a leaf-disc electrode

Changes in partitioning between starch and sucrose (or other products) can be measured by short-term feeding of $^{14}CO_2$ in a leaf-disc oxygen electrode (Chapter 10), followed by separation of the neutral, acidic and basic fractions (sugars, organic acids and amino acids, respectively).

$H^{14}CO_3^-$ (about 0.1 μCi/μmol) is added to the bicarbonate buffer used in the leaf-disc electrode. Leaf discs or pieces which have reached a steady rate of photosynthesis are then illuminated in the presence of the $^{14}CO_2$ for 20–30 minutes, the chamber opened and the leaf disc quickly removed into liquid nitrogen. It can then be extracted in $HClO_4$ (Chapter 16). This entire procedure should be carried out in a fume cupboard.

Ion exchange analysis:
Glass or plastic tubing (e.g. Pasteur pipettes, plastic syringes or, for very small volumes, plastic pipette tips) can be used for ion exchange columns. Use a wad of glass wool to support the resin and a piece of clamped plastic tubing to control the flow.

Cation exchange: Dowex-50, a strongly acidic cation exchange resin (100–200 mesh, 8%

cross-linked, in the H^+ form) is used. The performance of this resin is greatly improved by acid and alkali washes. The resin is suspended in 2 vol. 4 M HCl in a boiling water bath for 1 hour. The resin is then washed with distilled H_2O until the pH of the washings equals pH 7, and resuspended in 2 vol. 10% NaOH in a boiling water bath for 1 hour. The resin is again washed with distilled H_2O to pH 7, and then converted back to the H^+ form by stirring with 2 M HCl (at least three changes, each of 2 vol.). The resin is suspended in several volumes H_2O. The water is decanted just before the resin has completely settled so that the finest particles of resin (the fines) are discarded (this improves the flow speed of column). This is repeated twice. Finally the resin is washed with distilled H_2O until the washings reach pH 7.

Anion exchange: Dowex-1, a strongly basic anion exchange resin (100–200 mesh, 8% cross-linked, in the Cl^- form) is used. The resin is suspended in several volumes H_2O and the fines removed. This resin is used in the acetate (not the chloride) form, so the resin is suspended in 1 M sodium acetate (at least three changes, each of 2 vol.) and then in 0.1 M acetic acid (changed until pH equals that of the acetic acid).

If using $HClO_4$ extracts, the leaf extracts are acidified and left to stand for 15 minutes before being added to the Dowex-50 columns. Acidification is needed both because amino acids bind to Dowex-50 at pH 3–4 and because, if the extract contains residual K_2CO_3, the column is disrupted by bubbles of CO_2. The acidified extract is added to the top of a column of Dowex-50 with a Pasteur pipette and the column washed through with excess water. The eluate contains the acidic and neutral fraction of the extract. The contents of the flask are reduced to dryness in a rotary evaporator at 30°C (reduce the pressure gradually so that the mixture does not boil over). At this temperature there is no significant hydrolysis of sucrose. The Dowex-50 column is then eluted by addition of 1 M NH_4OH.

The combined acidic and neutral fractions are taken to dryness, redissolved in about 1 ml water and applied to the top of a column of Dowex-1 with a Pasteur pipette. An excess of water is then used to wash the column, yielding the neutral fraction. The Dowex-1 column is eluted by addition of 75 ml 0.6 M HCl.

The fractions are taken to dryness in a rotary evaporator and the acidic, neutral and basic fractions redissolved in water (about 1 ml). Small aliquots of the basic, neutral and acidic fraction are counted, and the remainder frozen for further analysis (e.g. to check the sugar composition of the neutral fraction by paper chromatography).

The pellet containing insoluble material (principally starch) can be counted separately. Its composition should also be checked by digestion of the starch with amyloglucosidase (Section 15.3.2).

15.6 AN EXAMPLE: CARBON FLUX THROUGH BARLEY LEAF BLADES

Carbon, photosynthetically fixed in barley leaf blades, is mainly used for the synthesis of sucrose; relatively little starch is formed. An appreciable proportion of the sucrose enters vacuoles, so that sucrose exported from the leaf blade includes that just synthesized and that previously stored in vacuoles.

Figure 15.2 shows the content of ^{14}C in a leaf blade of barley, monitored continuously with a Geiger–Muller tube, during and after intermittent supply of $^{14}CO_2$ at varying rates. A second Geiger–Muller tube monitored the air in the leaf chamber and enables the count for the leaf detector to be corrected for $^{14}CO_2$ in the air (Section 15.5.3). Note how sensitive the ^{14}C content of the leaf is to the supply of $^{14}CO_2$. When $^{14}CO_2$ supply ceased, ^{14}C content of the leaf fell due to translocation

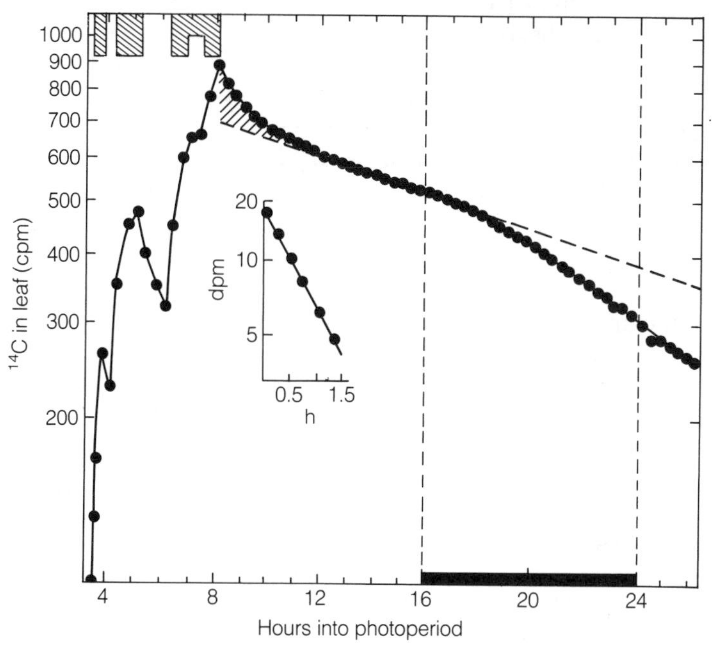

Fig. 15.2. The ^{14}C content of a mature barley leaf blade, monitored continuously by a Geiger– Muller tube, during and after feeding with $^{14}CO_2$ (the duration and relative intensity of $^{14}CO_2$ presentation is shown by the hatched region on the upper x-axis). Note the semi-logarithmic plot. The translocatory efflux of ^{14}C (8 h into the photoperiod and later) has three distinct components (see text). The inset is the shaded region of the curve replotted to show the loss of activity, i.e. translocated ^{14}C. The bar on the lower x-axis shows the period of darkness.

Table 15.2. Characteristics of sucrose in barley mesophyll vacuoles at two different times during the 16 h photoperiod

	0.5 h	15.5 h
Proportion of sucrose in the leaf which is vacuolar (%)	81	82
Half-time of vacuolar sucrose (h)	25	12
Concentration within vacuole (mM)	45	124
Efflux of sucrose across tonoplast ($mg\,m^{-2}$ leaf h^{-1})	40	233
%C fixed in photosynthesis passing through vacuole	57 (40% of total photosynthesis)	3
Half-time of pool of non-vacuolar sucrose (h)	0.85	0.32

(respiratory loss of $^{14}CO_2$ was below the sensitivity of the detector monitoring the air) and showed a two-phase exponential decay. The first, rapid phase, with a half-time of about 0.75 hours, is due to loss of recently-synthesized sucrose; the second, slow phase is loss of sucrose stored in vacuoles, and demonstrates continual turnover of the vacuolar sucrose pool even in the light. Following onset of the dark period, the rate of loss of ^{14}C from the leaf increases due to mobilization of ^{14}C-starch synthesized during $^{14}CO_2$ feeding.

Analysis of efflux data of this type, and washout of ^{14}C-labelled leaf discs, has led to the conclusions given in Table 15.2. Most of the sucrose in a barley leaf is within vacuoles; this is turned over continuously, and the rate of turnover varies during the photoperiod. The non-vacuolar sucrose is readily available for export in the phloem, and this pool too has a turnover rate that varies during the day, but is never less than $t_{1/2} = 0.85$ hours, reflecting the high throughput from photosynthesis to translocation and storage.

ACKNOWLEDGEMENTS

I would like to thank J.H.H. Williams, I.J. Tetlow and M.R. Thorpe for advice and help.

REFERENCES

1. McCree, K.J. (1986) Measuring the whole-plant daily carbon balance. *Photosynthetica* **20**, 82–93.
2. Dey, P.M. (ed.) (1990) *Methods in Plant Biochemistry. Vol. 2, Carbohydrates.* Academic Press, London.
3. Jones, M.G.K., W.H. Outlaw and O.H. Lowry (1977) Procedure for assay of sucrose in the range 10^{-1}–10^{-14} moles. *Plant Physiol.* **60**, 379–383.
4. Moller, I.M., A. Berczi, L.H.W. van der Plas and H. Lambers (1988) Measurement of the activity and capacity of the alternative pathway in intact tissues: identification of problems and possible solutions. *Physiologia Plantarum* **72**, 642–9.
5. Geiger, D.R. (1980) Measurement of translocation. *Methods in Enzymology* **69**, 33–54.
6. Zierler, K. (1981) A critique of compartmental analysis. *Ann. Rev. Biophysics and Bioengineering* **10**, 531–62.
7. Minchin, P.E.H. (1989) Systems indentification in plant physiology: an application in the study of carbon movement within whole plants. In: *Concise Encyclopaedia of Environmental Systems* (P. Young, ed.). Pergamon, Oxford.
8. Farrar, S.C. and J.F. Farrar (1986) Compartmentation and fluxes of sucrose in intact leaf blades of barley. *New Phytologist* **103**, 645–57.

16
Carbon metabolism

R.C. Leegood

16.1 INTRODUCTION

Plants face constant change in their environment. Part of their success in dealing with such change lies with the ability of carbon metabolism to be responsive to changes in the relationship between the supply of CO_2 or the products of electron transport and the demand for assimilated carbon. Metabolic adjustment in the leaf must occur both in the short-term (e.g. to fluctuations in temperature or light, such as sunflecks), and during longer-term changes in environmental conditions (e.g. acclimation to temperature or to sun and shade). At a time of increasing atmospheric CO_2 and rapid progress in the genetic manipulation of crop plants, it is becoming more important than ever for the plant biochemist to understand the impact of environmental and other changes on the regulation of the metabolism of carbon, nitrogen, etc. in plants.

The measurement of photosynthetic fluxes, metabolites and enzymes in an intact leaf provides a valuable complement to studies of the regulation of isolated enzymes. The additional use of non-destructive and non-intrusive probes, such as analysis of chlorophyll fluorescence by the use of light-doubling (particularly to estimate the efficiency of electron transport *in vivo*; Chapter 12), together with the measurement of $P_{700}{}^+$, can also yield information about the regulation of metabolism *in vivo*; this is because electron transport is subject to regulation which enables it to respond to variations in the metabolic demands of the leaf. This chapter describes how to make some of these metabolic measurements and the reasons for so doing. It should also be noted that most of these procedures can be applied to the study of protoplasts, cells or chloroplasts (Chapter 17).

16.2 THE REGULATION OF CARBON METABOLISM IN LEAVES

Photosynthetic carbon metabolism is only able to respond rapidly to external factors because it is a highly integrated process. The Calvin cycle (or Reductive Pentose Phosphate pathway) lies at the interface between electron transport and product synthesis (Fig. 16.1). On the one hand there are changing inputs from electron transport (ATP and NADPH) and the regulation of metabolism by electron transport (light-activation of enzymes) while, on the other hand, the processes of photorespiration, respiration and the synthesis of products such as starch and sucrose are carefully adjusted to meet the rate at which the Calvin cycle can supply carbon. In addition, C_4 plants ensure a satu-

Photosynthesis and Production in a Changing Environment: a field and laboratory manual
Edited by D.O. Hall, J.M.O. Scurlock, H.R. Bolhàr-Nordenkampf, R.C. Leegood and S.P. Long
Published in 1993 by Chapman & Hall, London.
ISBN 0 412 42900 4 (HB) and 0 412 42910 1 (PB).

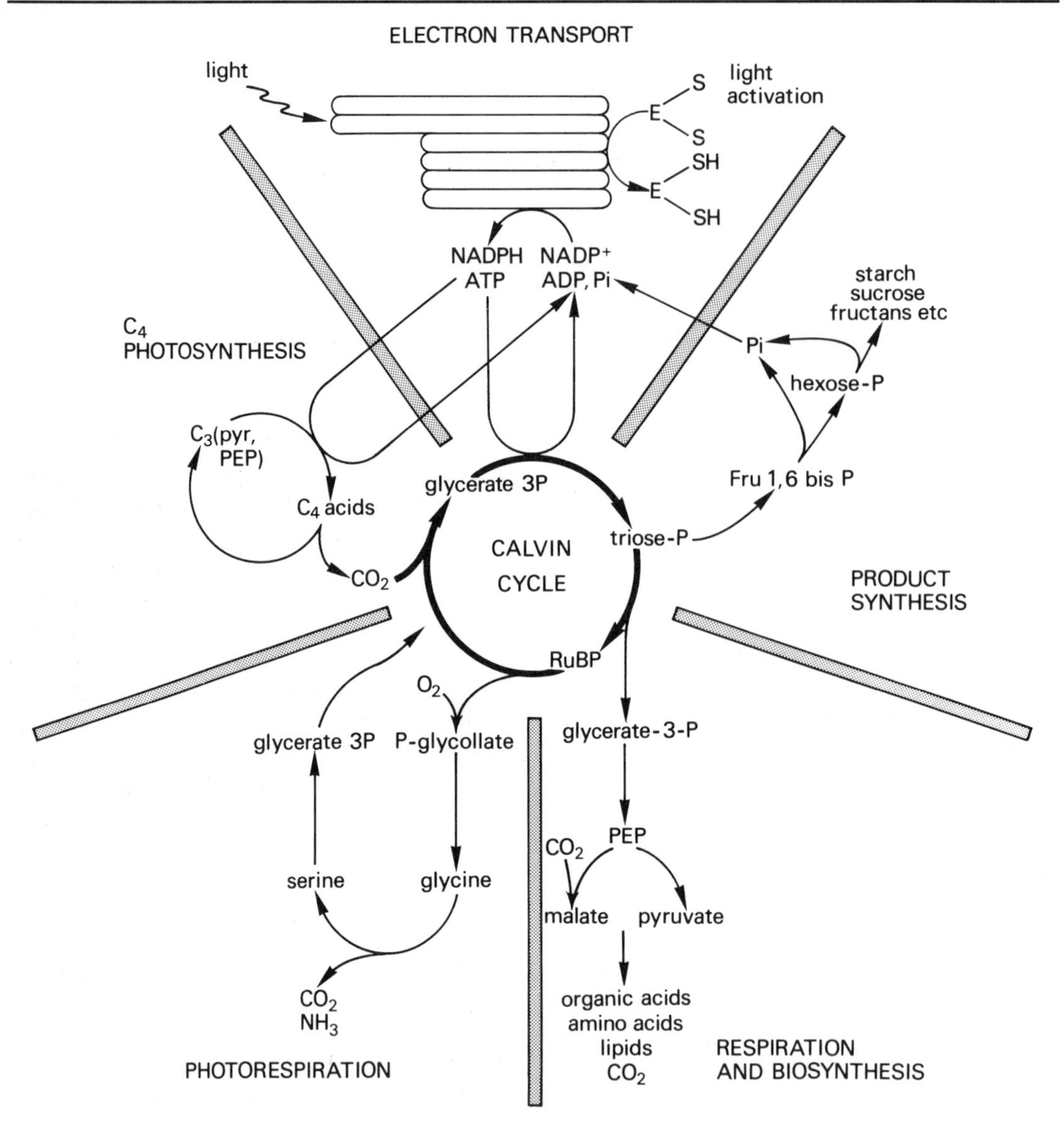

Fig. 16.1. An integrated picture of photosynthetic carbon metabolism in leaves. *Electron transport* in the thylakoids provides ATP and NADPH to power the Calvin cycle and the CO_2-concentrating mechanism of *C_4 photosynthesis*. It also provides reductant, via thioredoxin, for the light-activation of enzymes. Triose-P generated in the Calvin cycle can be used for *product synthesis* (starch in the chloroplast or sucrose in the cytosol): this releases inorganic phosphate (P_i) which is then returned to the chloroplast to be used in photophosphorylation. Carbon is also utilised in *biosynthesis* to make organic acids, amino acids, lipids etc., either in the chloroplast or in the cytosol. Carbon exported to the cytosol can enter glycolysis and be respired by the mitochondria or be converted, via PEP carboxylase, to malate and aspartate. RuBP can also be oxygen-ated, generating P-glycollate. This carbon leaves the chloroplast as glycollate and the process of *photorespiration* regenerates glycerate-3-P which can re-enter the Calvin cycle. However, photorespiratory glycine decarboxylation (which occurs in the mitochondria) results in a loss of ammonia and one-quarter of the carbon in glycollate (Chapter 20). Note that this diagram is highly simplified and contains no indication of compartmentation.

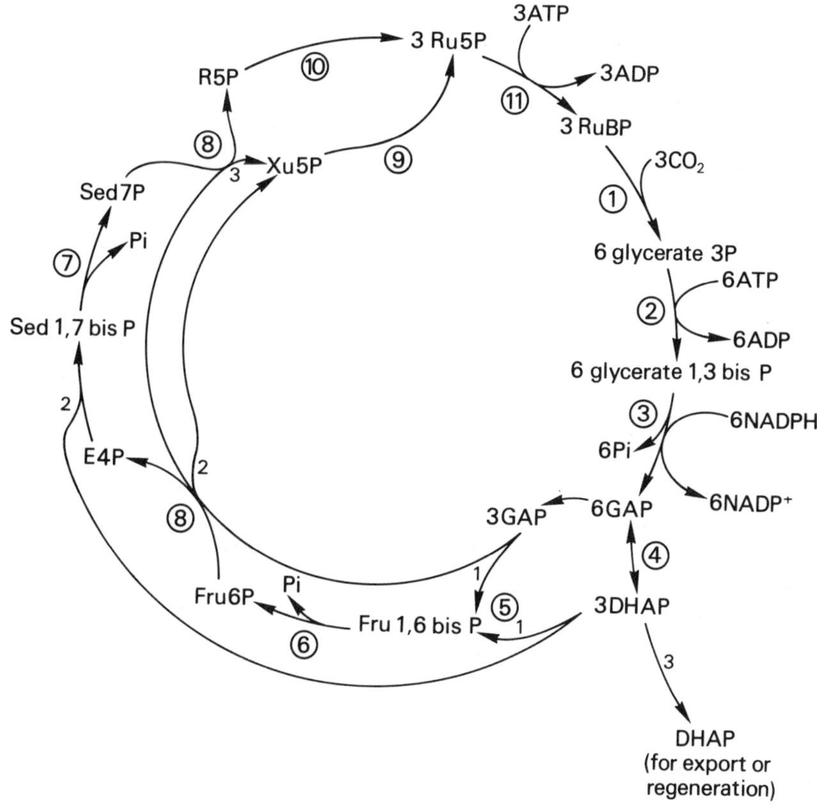

Fig. 16.2. The Calvin cycle comprises 13 reactions catalysed by 11 enzymes. It can be divided into three phases. The first phase is carboxylation catalysed by RuBP carboxylase-oxygenase (Rubisco) (1). The second phase is the reductive phase by which glycerate-3-P is converted to triose-P at the expense of ATP and NADPH, catalysed by glycerate-3-P kinase (2) and $NADP^+$-dependent glyceraldehyde-P dehydrogenase (3). The third phase is regeneration of the CO_2 acceptor in the sugar phosphate shuffle, in which five C_3 units are rearranged to form three C_5 units (Xu5P and ribose-5-P) by the actions of triose-P isomerase (4), aldolase (5), Fru-1,6-bisPase (6), Sed-1,7–bisPase (7) and transketolase (8). Xu5P and ribose-5-P are then converted to ribulose-5-P by ribulose-5-P 3-epimerase (9) and ribulose-5-P isomerase (10). Finally, ribulose-5-P is converted to RuBP by phosphoribulokinase (11) at the expense of ATP. The destination of each of the three molecules of GAP and DHAP is shown by the small numbers (1–3).

rating supply of CO_2 to the enzyme ribulose-1,5-bisphosphate carboxylase-oxygenase (Rubisco), and hence the suppression of photorespiration, by the possession of an ATP-dependent CO_2 pump, which is closely co-ordinated with the Calvin cycle.

Although some bacteria possess alternative mechanisms for the net fixation of CO_2, the vast majority of photosynthetic organisms (all plants, algae and cyanobacteria) fix carbon via the Calvin cycle. The Calvin cycle comprises 13 reactions catalysed by 11 enzymes (Fig. 16.2). It can be divided into three phases. The first phase is carboxylation catalysed by Rubisco. The second phase is the reductive phase by which glycerate-3-P is converted to triose-P at the expense of ATP and NADPH. The third phase is regeneration of the CO_2 acceptor in the sugar phosphate shuffle, in which five C_3 units are rearranged to form three C_5 units. Finally, ribulose-5-P is converted to ribulose-1,5-bisphosphate (RuBP) at the expense of ATP. Glyceraldehyde-P dehydrogenase, the bisphosphatases and synthesis of carbohydrate, etc., recycle the inorganic phosphate (P_i) required for continued ATP synthesis. The cycle acts as an autocatalytic 'breeder' reaction and, for every three turns of the cycle, one molecule of triose-P is generated from three molecules of CO_2. Triose-P may be utilised to regenerate the CO_2 acceptor, RuBP, or in the synthesis of starch or sucrose.

16.3 MEASUREMENT OF FLUXES IN LEAVES

The measurement of CO_2 uptake by leaves and of O_2 evolution by leaf pieces and by protoplasts and chloroplasts has been described in Chapters 9, 10 and 17. It is also possible to use $^{14}CO_2$ both to measure net photosynthesis and to measure partitioning of assimilated carbon into starch, sucrose and other products (Chapter 15). Measurement of photorespiratory fluxes is exceedingly difficult and liable to a variety of interpretations ([1], Chapter 9).

16.4 FREEZE-CLAMPING LEAF TISSUE

Most metabolites involved in photosynthesis turn over in a few seconds or less. This means that when measuring any metabolite, it is important to stop leaf metabolism immediately under the conditions to which the leaf is exposed. It is also necessary to prevent the subsequent action of endogenous enzymes (especially phosphatases) on metabolites. Similarly, measurement of the activation state of an enzyme requires that the activation state of the enzyme be maintained during extraction and assay, because modulation of enzyme activity following exposure of leaves or chloroplasts to light can be very rapid.

The best method of stopping metabolism is to freeze the leaf rapidly to the temperature of liquid N_2. This is done by freeze-clamping, which causes both rapid cooling and effective disruption of the tissue. The activation states of enzymes are maintained in tissue frozen in liquid N_2 for several days or weeks. The amount of leaf material required is quite small, as 30 separate measurements of enzymes and metabolites can be made routinely on the same $10\,cm^2$ leaf disc. In principle, many more could be made with determination of amino acids, etc.

The measurement of fluxes by gas exchange and by rapid stopping of leaf metabolism might at first sight seem irreconcilable. However, various devices have been designed, although none are yet available commercially. These also require specially designed gas-exchange chambers which have parafilm or kitchen-film (e.g. *Clingfilm*) windows. Ideally the chamber design should permit the use either of attached leaves or of leaf pieces or discs.

(i) The simplest method is to use freeze-clamping tongs, originally introduced by Wollenberger, which comprise a pair of aluminium discs fixed to a pair of long-handled tongs (Fig. 16.3). The tongs are placed in liquid N_2 until the temperature of the aluminium discs is as low as that of the nitrogen. These are cheap to construct and give perfectly acceptable results for non-photosynthesizing tissues, and can also be used to freeze-clamp in the field. However, caution must be exercised with illuminated samples, since the leaf is momentarily darkened when it is freeze-clamped (the time taken will obviously depend upon the skill of the operator). Freeze-clamping tongs can also be used in conjunction with a simple chamber, although this may not be ideal for accurate measurements of gas exchange.

(ii) An automated system for freeze-clamping is preferable. Again, this essentially comprises two copper rods, which can be removed and cooled to the temperature of liquid N_2. The leaf chamber is held so that the plane of the leaf is held mid-way between the

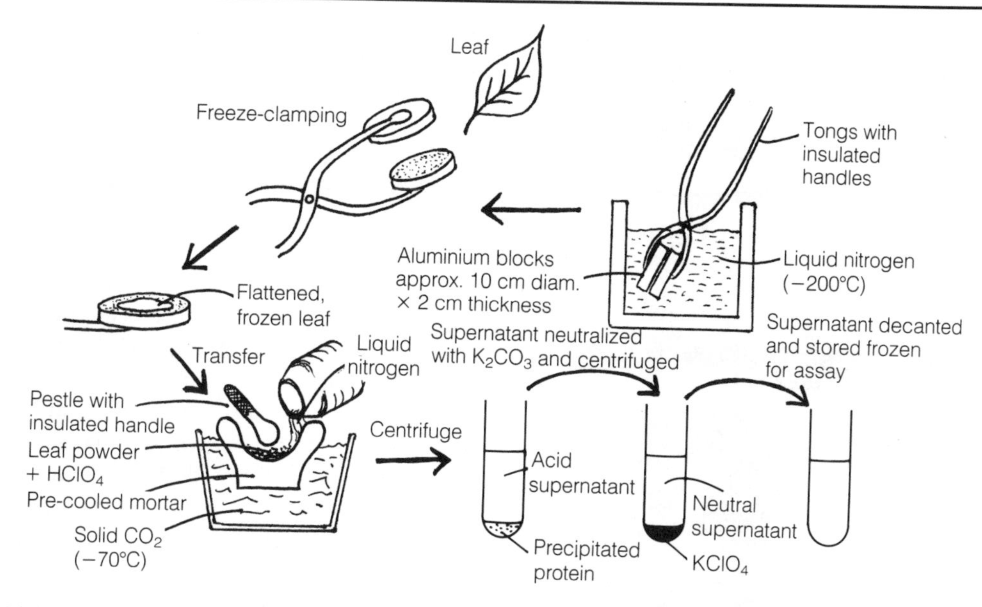

Fig. 16.3. Typical procedure for preparation of extracts from plant tissues by 'freeze-clamping'. The tissue is rapidly frozen by pressure between two relatively massive aluminium blocks pre-cooled to −200°C. The frozen material is then pulverised using a pestle and mortar. The low temperature can be maintained by direct addition of liquid nitrogen to the sample. The powder is dispersed rapidly in ice-cold $HClO_4$ by hand or by mechanical homogenisation, and the extract is finally neutralised as shown.

two copper rods. The head of the lower rod is milled so that it has a rim 0.3–0.4 mm higher than the rest. This ensures that the leaf material is not squeezed out when the heads meet. A detailed diagram of a spring-loaded apparatus can be found in Badger *et al.* [2]. The time between the rods interrupting the light beam and cooling of the leaf to below 0°C is 0.05–0.1 s. A solenoid-driven system is employed by the author's own research group in Sheffield (Fig. 16.4), while the chamber is an adaptation of a design marketed by ADC (*Analytical Development Co. Ltd., Hoddesdon, UK*; see also Chapter 9).

It may also be necessary to measure metabolites in leaf pieces which are in a leaf-disc oxygen electrode (Chapter 10). In this case freeze-clamping cannot be employed. The only way in which a rapid cessation of metabolism can be achieved is to illuminate the area around the leaf disc electrode to the same photon flux density (PFD) as occurs in the chamber, and to direct a stream of 5% CO_2 around the chamber (this is easiest to do if the chamber is enclosed within an open box). The chamber can then be opened and the leaf piece rapidly transferred, using forceps, to liquid N_2.

16.5 MEASUREMEMT OF METABOLITES

16.5.1 Preparation of leaf extracts for metabolite assays

Once freeze-clamped, leaf samples can temporarily be stored in envelopes of aluminium foil in liquid N_2. When samples are frozen in liquid N_2, enzymes are not inactivated when the samples are subsequently thawed. Extraction must be carried out by grinding the frozen leaf sample, using a small pestle and mortar, with a small quantity of

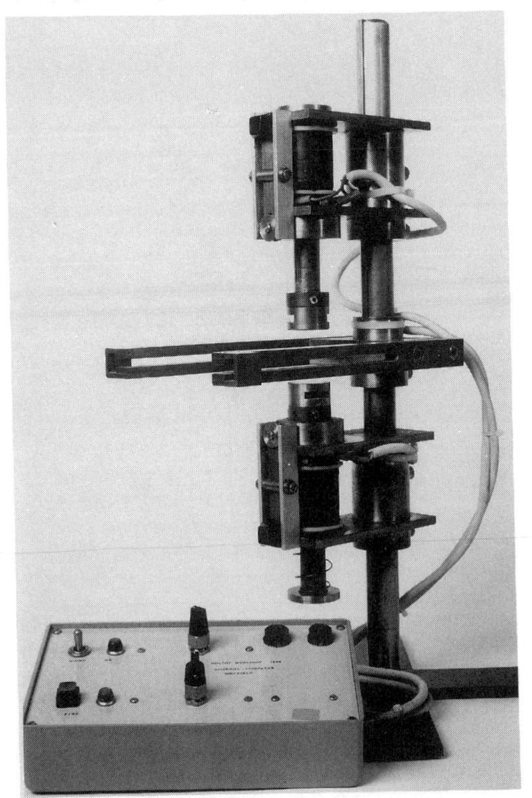

Fig. 16.4. A solenoid-driven freeze-clamping device. Detachable copper rods are attached to solenoids. When fired, the copper rods, which have previously been cooled in liquid N_2, meet and rapidly freeze the leaf. The leaf chamber (not shown) has windows of plastic film and is illuminated using fibre optics. It is mounted on the horizontal rail shown in the middle of the photograph. This allows the leaf chamber to be moved into position for freeze-clamping.

$HClO_4$ or $CHCl_3/CH_3OH$, both of which inactivate enzymes in the tissue and precipitate the protein.

16.5.2 Extraction in perchloric acid

Frozen pellets of 1 M $HClO_4$ are prepared by pipetting the required amount of $HClO_4$ (1.0 ml for a $10\,cm^2$ leaf disc) into parafilm 'boats'. These can be made by slightly stretching parafilm over the end of the pestle. The boat is immersed in liquid N_2, and the pellet of perchloric acid separates from the parafilm. If necessary, the pellets can be stored in liquid N_2. Pre-cool the mortar and pestle by filling once with liquid N_2, but do not over-cool, otherwise it will take too long to warm up to room temperature and the volume of the extract will be increased considerably by condensation.

The leaf material is added to the mortar and, when all the liquid N_2 has evaporated, it is ground with the pellet of $HClO_4$ to a pale-green powder. The mixture should take about 20 minutes to thaw. Grind it at regular intervals to ensure thorough mixing. When the mixture has thawed (this occurs below 0°C), transfer it to a plastic centrifuge tube (using an automatic pipette with the narrow end of the tip cut off at an angle), and wash out the pestle and mortar with 1.0 ml cold 0.1 M $HClO_4$. Centrifuge the extract at 2000 *g* for 2 minutes to remove the protein, starch, phaeophytin, etc. Decant the supernatant into a plastic tube. The pellet can be reserved for the determination of phaeophytin (Section 16.7). Neutralise the ice-cold supernatant carefully, stirring constantly, with 5 M K_2CO_3 (neutralization will require approximately one tenth of the volume of $HClO_4$ added). The final pH should be between 6 and 7 (using pH paper, but note that most plant extracts turn a bright yellow colour under alkaline conditions and, therefore, already contain a pH indicator). $KClO_4$ is precipitated and is removed by centrifugation at 2000 *g* for 120 s. The colder the extract, the more $KClO_4$ is precipitated. (Precipitation of the $KClO_4$ is enhanced by using K_2CO_3 (since CO_2 is released) but can be further enhanced by adding KCl. 2 M KOH/0.4 M KCl may also be employed for neutralisation, but do not use NaOH because $NaClO_4$ is very soluble. A buffer such as imidazole or triethanolamine (0.4 M) may also be included in the mixture used to neutralise the sample.)

At this point the volume of the extract should be measured, either by measuring aliquots using an automatic pipette, or by marking the side of the (plastic) tube at the meniscus with a razor blade. The extract should be clarified, by centrifugation, before using it to measure metabolites.

16.5.3 Extraction in chloroform/methanol

An alternative to extraction in $HClO_4$ is the use of a chloroform/methanol mixture [3]. This is necessary for the measurement of acid-labile compounds such as fructose 2,6-bisphosphate [3] and is useful for the extraction of sucrose, for the same reason. However, in principle, extraction may be done at any pH value. Care should be taken to check the final pH of the extract if these labile compounds are to be measured, as the mixture often becomes considerably more acid during extraction. Frozen leaf material (0.2 g) is mixed in a pestle and mortar (pre-cooled with liquid N_2) with 4.2 ml extraction medium (1.2 ml $CHCl_3$, 2.4 ml CH_3OH, 0.6 ml buffer containing 50 mM NaF, 10 mM EGTA, 50 mM Hepes buffer (pH 8.5). NaF is included to prevent the possible action of phosphatases.) The paste is then held at 4°C until it thaws. After transfer to a glass tube, 4 ml H_2O is added, the mixture shaken vigorously, and centrifuged at 2000 *g* for 120 s. Two phases are formed, a lower (chloroform) phase containing the chlorophyll etc. (for measurement of chlorophyll, see below) and an upper (methanol/water) phase containing metabolites, sugars etc. (This distribution should be checked through the recovery procedures, as metabolites may not always partition in the expected fashion in different plant material. If necessary, the lower phase can be re-extracted with water.) Starch, cell wall material, protein etc. lies in between the two phases. The supernatant is removed and taken to dryness at 35°C, then redissolved in 1 ml water and stored in liquid N_2, or in a freezer at −80°C, or it can be lyophilised.

16.5.4 Metabolite assays

The abundance of secondary products such as phenolics and tannins in leaves which interfere with metabolite assays means that it is extremely important to establish whether the compounds of interest can be measured accurately. Measurement of metabolites in any plant tissue should be accompanied by a rigorous check that they have not been degraded during the killing and extraction procedure. The best method is to prepare and extract duplicate samples of tissue, adding a measured amount of the metabolite to a sample in a pestle and mortar. The amount of the compound added should be comparable to that present in the tissue. Both samples are then extracted and measured, and the recovery of the metabolite estimated. This should normally be within 10% of the expected value.

The extraction procedures described above are suitable for the extraction of starch, sugars, phosphorylated metabolites, and amino and organic acids. There are, of course, numerous methods for the analysis of these compounds, such as high-pressure liquid chromatography (HPLC) and gas-liquid chromatography (GLC). Many of these compounds can be assayed enzymically in pyridine-nucleotide-coupled assays using a dual-wavelength spectrophotometer (340–400 nm). However, in the case of C_4 plants, metabolite contents can be very high and, if care is taken, can be measured in a conventional double-beam spectrophotometer. Sugars (and starch) are also usually present in such large amounts that they can conveniently be measured in either a single-beam or a double-beam spectrophotometer.

The most important aim in sample preparation for sensitive spectrophotometric assay is to reduce the background absorbance and hence to reduce the noise level. This is why dual-wavelength measurements result in such dramatic improvements in sensitivity (it is quite possible to measure less than 1 nmol of a metabolite, i.e. an absorbance change as low as 0.001, in extracts of leaf tissues such as young maize or spinach). However, nearly all plant tissues contain substances which, although colourless at acid pH values, are visibly bright yellow at pH values above about pH 7. These substances absorb strongly in the ultra-violet and therefore increase noise. Following neutralisation of $HClO_4$ extracts or preparation of $CHCl_3$/CH_3OH extracts, sufficient charcoal is added to remove most of these yellow substances and to render the extract completely colourless (this cannot be done where adenylates such as ATP or other aromatic compounds are to be measured, since they are absorbed by the charcoal).

First add about 10 mg charcoal from a 100 mg ml^{-1} suspension of activated charcoal in water (ensure that the suspension is free of fine particles). Considerably more charcoal may be required to decolorise the extract completely, but check that the addition of very large amounts does not result in losses of metabolites. The extract is centrifuged at 2000 *g* for 2 minutes. At this point the volume of $HClO_4$ extracts should be measured. Measured aliquots of the supernatant can be stored in liquid N_2 or at −80°C, or can be lyophilised. In tissues which are particularly rich in phenolics, etc., it may be found that charcoal treatment is insufficient to remove all interfering substances from enzyme-linked assays. Treatment of the extract directly, or addition to the reaction cuvette of BSA (5 to 10 mg ml^{-1}) or dithiothreitol (10 mM), or even the addition of more coupling enzymes, can often overcome problems due to inactivation of coupling enzymes or lack of a stable endpoint (drift). Note that such interference can also be particularly severe in fluorimetric or luminometric assays.

The following spectrophotometric procedures are largely based on those described by Lowry and Passonneau [4], with some modifications. The first of these (the assay for hexose-P) can be regarded as a specimen assay, and describes in detail the procedures which should be followed.

(a) Hexose phosphates:

When developing any metabolite assays, or when employing new plant material, it is usually best to try this assay first. If it does not work satisfactorily, then it is unlikely that any of the other assays will work. However, it generally works well (a) because hexose phosphates are relatively abundant in leaves (100–300 nmol mg^{-1} chlorophyll), (b) because $NADP^+$- or NAD^+-linked assays are less noisy (because the background absorbance is less) and (c) because the reaction usually reaches a stable end-point within 1–2 minutes.

The assay for glucose-6-P and fructose-6-P is based on the following reactions:

$$\text{Fru-6-P} \xrightarrow[\text{(phosphoglucose isomerase)}]{} \text{Glc-6-P}$$

$$\text{Glc-6-P} + \text{NADP}^+ \xrightarrow[\text{(glucose-6-P dehydrogenase)}]{} \text{6-phosphogluconate} + \text{NADPH} + \text{H}^+$$

The assay buffer contains 100 mM Tris-HCl (pH 8.1). Ensure that the buffer is absolutely free of particulate matter, by passing it through a membrane filter if necessary. $NADP^+$ (0.5 mM) is added to the cuvette and mixed with the sample. It is necessary to restrict the amount of extract added because the perchloric anion and other components of the extract will interfere with the assays if present at high concentrations. In practice a total volume of between 0.6 and 1 ml is used in the cuvette, to which is added 50 to 100 μl extract. This amount of extract is usually equivalent to between 5 and 10 μg of chlorophyll. It may be necessary to add more, but this increases noise and slows the assays. Although it is sufficient to conduct most of these assays at room temperature, make sure that the buffers which are stored cold have reached room temperature, otherwise air bubbles will be released which will cause an apparent increase in absorbance. Ensure that the cuvette is held firmly within the sample holder, using clips if necessary. Zero the spectrophotometer, stir gently several times, and check that the reading returns to zero. This should be repeated several times. Use a Sarstedt stirring rod (which is shaped like a paddle) to stir by rotating it gently. Do not stir so vigorously that you move the cuvette or create bubbles.

Most coupling enzymes used for these assays are sold as suspensions in ammonium sulphate. First, the protein must be removed from the $(NH_4)_2SO_4$ by centrifugation (a small Eppendorf tube is best). The $(NH_4)_2SO_4$ solution can then be removed almost totally using a fine-needled syringe. The enzyme should be redissolved in a minimum volume of buffer so that enzyme additions can be made in a volume of 1–2 µl (this minimises absorbance changes due to dilution). Some coupling enzymes are sold as solutions in glycerol. In this case the enzyme can be added direct, but always transfer the enzyme solution/suspension to another container so that there is no danger of cross-contaminating the stocks of coupling enzymes.

Add 0.06 unit ml^{-1} glucose-6-P dehydrogenase by placing a 1–2 µl drop of the enzyme solution on the blade of a Sarstedt stirring rod. Stir it gently into the cuvette and follow the increase in absorbance. After the reaction is complete, add 0.35 unit ml^{-1} phosphoglucose isomerase. You should not encounter drift (see below) in an assay such as this. Usually the amount of fructose-6-P will be substantially less (up to five times less) than the amount of glucose-6-P (the ratio of glucose-6-P to fructose-6-P at equilibrium is usually between 1 and 2 in the chloroplast and 4–5 in the cytosol; whole leaf ratios are normally around 4). If there is more fructose-6-P than glucose-6-P, then something *may* have gone wrong! Glucose-1-P can also be measured in the same assay by adding phosphoglucomutase, though it is present in amounts which are 10-fold less than glucose-6-P.

The amount of metabolite is calculated using the following extinction coefficient:

$$E^{1\,cm}_{1\,mM}\ \text{NAD(P)H} = 6.22\ (\text{at } 340\,\text{nm})$$

Thus conversion of 1 nmol of hexose-P to 6-phosphogluconate will result in an absorbance change of 0.00622 in a cuvette of 1 cm path length, in a volume of 1 ml.

(b) Glycerate-3-P and ribulose-1,5-bisphosphate:

$$\text{RuBP} + CO_2 \xrightarrow[\text{(Rubisco)}]{} \text{glycerate-3-P}$$

$$\text{glycerate-3-P} + \text{ATP} \xrightarrow[\text{(glycerate-3-P kinase)}]{} \text{glycerate-1,3-bisP} + \text{ADP}$$

$$\text{glycerate-1,3-bisP} + \text{NADH} + H^+ \xrightarrow[\text{(glyceraldehyde-P dehydrogenase)}]{} \text{glyceraldehyde-3-P} + NAD^+ + P_i$$

A frequent problem with this assay, and a number of other NADH-linked assays in plant extracts is that of drift, i.e. the reaction does not apparently reach a clear end-point. In the case of the assay for glycerate-3-P, much of this drift can be overcome by pre-incubation of the extract, in the assay cuvette, with ATP. However, if the drift persists, estimate the absorbance change by drawing a tangent to the curve. It is also important not to add more than a 10–20% excess of NADH in all the NADH-linked assays, as this causes an unnecessary increase in background absorbance and hence more noise.

Preincubation step: 50 mM imidazole–HCl (pH 7.1), 2 mM mercaptoethanol, 1 mM $MgCl_2$, 20 mM NaCl, 1 mM ATP, 50 µl extract.

Assay of glycerate-3-P: add 150 µM NADH immediately before the assay. Then 0.36 unit ml^{-1} glycerate-3-P kinase and 0.18 unit ml^{-1} glyceraldehyde-P dehydrogenase are added together to start the reaction. The assay takes between 5 and 15 minutes.

Assay of RuBP: assay conditions are modified for the combined assay of RuBP and glycerate-3-P. The assay pH is changed to pH 8.0 and 10 mM $NaHCO_3$ is included. The more alkaline pH is necessary to maintain adequate activity of Rubisco, but it slows the assay for glycerate-3-P. The pre-incubation step contains 50 mM Tricine-KOH (pH 8), 2 mM mercaptoethanol, 10 mM $MgCl_2$, 10 mM $NaHCO_3$, 1 mM ATP, 50 µl extract. The assay then proceeds as for the assay of glycerate-3-P, with the addition of 150 µM

NADH immediately before the assay and 0.36 unit ml^{-1} glycerate-3-P kinase and 0.18 unit ml^{-1} glyceraldehyde-P dehydrogenase added together to determine glycerate-3-P. Then 0.05 unit ml^{-1} Rubisco is added to determine RuBP (the commercially-available enzyme is just satisfactory, but it is best to purify Rubisco; see Keys and Parry [10]). The Rubisco should be activated with CO_2 and Mg^{2+} prior to use as described in Section 16.6.3.

(c) Triose-P and fructose-1,6-bisphosphate:

$$\text{Fru-1,6-bisP} \xrightarrow[\text{(aldolase)}]{} \text{dihydroxyacetone-P} + \text{glyceraldehyde-3-P}$$

$$\text{dihydroxyacetone-P} \xrightarrow[\text{(triose-P isomerase)}]{} \text{glyceraldehyde-3-P}$$

$$\text{glyceraldehyde-3-P} + \text{NAD}^+ \xrightarrow[\text{(glyceraldehyde-P dehydrogenase)}]{\text{arsenate}} \text{glycerate-3-P} + \text{NADH} + \text{H}^+$$

Arsenate takes the place of P_i in the reaction catalysed by glyceraldehyde-P dehydrogenase and renders the reaction irreversible. The assay mixture contains 50 mM imidazole–HCl (pH 7.5), 1 mM arsenate, 1 mM EDTA, 2 mM 2-mercaptoethanol, 1 mM NAD^+. Add 1.8 units ml^{-1} triose-P isomerase and 1.8 unit ml^{-1} glyceraldehyde-P dehydrogenase to determine total triose-P. Since, at equilibrium (which probably obtains *in vivo*), the amount of glyceraldehyde-P is only 1/22 of the amount of dihydroxyacetone-P, it is not usually worth attempting to measure glyceraldehyde-P independently. Fru-1,6-bisP is determined by adding 0.09 unit ml^{-1} aldolase. Note that 1 mol of Fru-1,6-bisP results in the generation of 2 mol NADH.

It is important to note that the measurement of Fru-1,6-bisP includes half of the measurement of sedoheptulose-1,7-bisP because of the dual role of aldolase:

$$\text{sedoheptulose-1,7-bisP} \xrightarrow[\text{(aldolase)}]{} \text{DHAP} + \text{erythrose 4-P}$$

In practice both these metabolites can be measured if the order of additions is changed, i.e. add glyceraldehyde-P dehydrogenase followed by aldolase to estimate the amount of glyceraldehyde-P deriving from Fru-1,6-bisP, followed by triose-P isomerase to determine the amount of dihydroxyacetone-P deriving from both Fru-1,6-bisP and Sed-1,7-bisP. Since Fru-1,6-bisP generates equimolar amounts of glyceraldehyde-P and dihydroxyacetone-P, the amount of dihydroxyacetone-P deriving from Sed-1,7-bisP can be calculated.

(d) ATP:

$$\text{ATP} + \text{glucose} \xrightarrow[\text{(hexokinase)}]{} \text{ADP} + \text{glucose-6-P}$$

$$\text{glucose-6-P} + \text{NADP}^+ \xrightarrow[\text{(glucose-6-P dehydrogenase)}]{} \text{6-P-gluconate} + \text{NADPH} + \text{H}^+$$

The assay mixture contains 50 mM Tris-HCl (pH 8.1), 1 mM $MgCl_2$, 0.5 mM DTT, 1 mM glucose and 0.5 mM $NADP^+$. Add 0.07 unit ml^{-1} glucose-6-P dehydrogenase to determine the amount of glucose-6-P present in the extract, followed by 0.28 unit ml^{-1} hexokinase to determine ATP.

Do not treat extracts with charcoal if adenylates are to be measured. ATP and other adenylates may also be determined using the luciferase assay.

(e) ADP, AMP, pyruvate, PEP:

$$\text{AMP} + \text{ATP} \xrightarrow[\text{(adenylate kinase)}]{} \text{2ADP}$$

$$\text{PEP} + \text{ADP} \xrightarrow[\text{(pyruvate kinase)}]{} \text{pyruvate} + \text{ATP}$$

$$\text{pyruvate} + \text{NADH} + \text{H}^+ \xrightarrow[\text{(lactate dehydrogenase)}]{} \text{lactate} + \text{NAD}^+$$

For the assay of ADP and AMP, the reaction mixture contains 50 mM imidazole–HCl (pH 7.0), 2 mM $MgCl_2$, 75 mM KCl, 100 μM ATP, 150 μM NADH, 0.3 mM PEP and 0.4 unit ml^{-1} lactate dehydrogenase. Pyruvate kinase (0.3 unit ml^{-1}) is added to determine ADP, followed by 0.36 unit ml^{-1} myokinase to determine AMP.

For the assay of pyruvate and PEP, the reaction mixture contains 30 mM K_2HPO_4, 20 mM Na_2HPO_4 (these buffer to pH 7.0), 2 mM $MgCl_2$, 150 µM NADH and 0.2 mM ADP. Pyruvate is determined by adding 0.2 unit ml^{-1} lactate dehydrogenase, followed by 0.15 unit ml^{-1} pyruvate kinase to determine PEP. The reaction time is 1–2 minutes for pyruvate and 2–5 minutes for PEP.

16.6 MEASUREMENT OF ENZYME ACTIVITY

16.6.1 Regulation of photosynthetic enzymes

The activities of chloroplastic enzymes are under several forms of control. Various forms of activation, including activation by light and metabolite modulation allow rapid short-term modulation of enzyme activity, while protein synthesis allows long-term modulation of the amounts of enzymes.

(a) Reductive activation:
In photosynthetic organisms the catalytic activity of Calvin cycle enzymes and others is linked to the availability of photosynthetically-generated reductants. Five enzymes of the Calvin cycle (RuBP carboxylase, Fru-1,6-bisPase, Sed-1,7-bisPase, Ru5P kinase and NADP-GAP dehydrogenase) show light-induced transitions to active forms. The last four of these undergo reductive activation by thioredoxin, as do a number of other enzymes in the chloroplast, such as NADP-malate dehydrogenase. This type of activation involves the photosynthetic reduction of ferredoxin, which reduces thioredoxin in a reaction catalysed by ferredoxin–thioredoxin reductase (FTR) and the reduced thioredoxin, in turn, reduces the enzyme:

$$\text{ferredoxin}_{red} + \text{thioredoxin}_{ox} \xrightarrow[\text{(ferredoxin–thioredoxin reductase)}]{} \text{thioredoxin}_{red} + \text{ferredoxin}_{ox}$$

$$\text{thioredoxin}_{red} + \text{enzyme}_{ox} \rightarrow \text{thioredoxin}_{ox} + \text{enzyme}_{red}$$

(b) Regulation of Rubisco:
Although Rubisco is certainly the most commonly measured enzyme of carbon metabolism, it is nevertheless one of the most complicated. Rubisco is a bifunctional enzyme, catalysing the incorporation of both O_2 and CO_2:

$$\text{RuBP} + H_2O + CO_2 \xrightarrow[\text{(RuBP carboxylase)}]{} 2(\text{glycerate-3-P}) + 2H^+$$

$$\text{RuBP} + O_2 \xrightarrow[\text{(RuBP oxygenase)}]{} \text{glycerate-3-P} + \text{glycollate-2-P} + 2H^+$$

It is also subject to two forms of regulation.

(i) The enzyme is converted from an inactive to an active, carbamylated, form by reaction with CO_2, then Mg^{2+}:

$$\text{E (inactive)} + CO_2 \rightarrow \text{E.CO}_2 + Mg^{2+} \rightarrow \text{E.CO}_2\text{.Mg}^{2+} \text{ (active)}$$

These reactions occur on a lysine residue at an activator site separate from the catalytic site. The enzyme must therefore be pre-incubated with Mg^{2+} and high concentrations of CO_2 to obtain maximum activity. This activation is promoted by a number of phosphorylated metabolites. However, at the very low concentrations of CO_2 which occur *in vivo*, it is impossible to activate the enzyme adequately *in vitro*. *In vivo* this activation is accomplished by an ATP-dependent enzyme, Rubisco activase [5].

(ii) In some plants it proves impossible to activate the enzyme fully by preincubation with CO_2 and Mg^{2+}. This is particularly the case at night or in low light because many plants, particularly legumes, possess a natural tight-binding competitive inhibitor of the enzyme, carboxyarabinitol-1-P [5] which is synthesised at night or in low light.

(c) Regulation by phosphorylation:
In leaves of C_3 and C_4 plants, the activity of sucrose phosphate synthase is regulated indirectly by light through interconversion of distinct kinetic forms by protein phosphorylation. The enzyme extracted from

illuminated leaves is markedly less sensitive to inhibition by P_i than is the enzyme extracted from darkened leaves [6], although this varies considerably between different species [7]. In C_4 and CAM plants the activities of PEP carboxylase and pyruvate-P_i dikinase are regulated by phosphorylation. In the case of PEP carboxylase, phosphorylation of serine residues results in reduced sensitivity of the enzyme to inhibition by malate.

16.6.2 Measurement of enzyme activity in leaf or chloroplast extracts

Measurement of the activity of enzymes may be attempted for a variety of reasons. We may wish to know the maximum catalytic activity in order to assess whether the total amount of an enzyme is changing (coarse control). On the other hand, we may wish to know how the activity of an enzyme is regulated in the short-term (fine control) by measuring changes in its activation state. Apart from influencing the direction of metabolism, changes in environmental conditions are likely to alter both the amounts of enzymes and their activation states. Although it is not possible to indicate that changes in any particular enzyme are likely to be more important in any particular circumstance, a number of assays for key enzymes are included in the following section. Methods for the assay of a wide range of enzymes in plants may be found in compendia such as Lea [8].

Measurement of the maximum catalytic activity of an enzyme requires optimisation of the pH and all of the components of the extraction and assay media for each tissue studied. Evidence should also be sought that all of the enzyme has been recovered from the tissue. This can be done by checking for complete cell breakage, and by recovery experiments using comparable amounts of the commercially-available enzyme or by mixing known quantities of the tissue under study with known quantities of other tissues (e.g. spinach leaf) in which the lability of the particular enzyme is believed to be low. If the maximum activity of thiol-regulated enzymes is to be measured, then extraction and preincubation with DTT is necessary.

Many plant tissues, particularly older leaf tissues, roots etc., are not easily amenable to the assay of enzymes, but attempts can be made to remove interfering substances. Enzyme inactivation by phenolics, etc. may sometimes be countered by the addition of compounds such as polyvinylpyrrolidone or polyethylene glycol (which complex phenols), thiols or mercapto compounds (e.g. dithiothreitol, 2-mercaptoethanol), BSA, and use of reducing compounds (e.g. isoascorbate) or anaerobic media (see also Chapter 18). Addition of chelators (EDTA or EGTA), stabilisers, such as glycerol, and protease inhibitors, such as PMSF (phenylmethylsulphonyl fluoride; but note that this is unstable above pH 7), may also be useful. Affinity columns have also been used successfully to remove interfering substances, e.g. for measurement of the activity of sucrose-P synthase in difficult tissues such as cell cultures and in leaves of plants such as grapevine (see under sucrose-P synthase).

Rapid extraction of leaf tissue can be achieved in less than 10 s using a small pestle and mortar or a glass-in-glass homogeniser. If necessary, removal of solids can be accomplished by a short centrifugation (10 s) in an Eppendorf centrifuge, or by rapid passage through a fine nylon mesh (e.g. 20 μm). A wide range of extraction media may be employed, but the following features are often desirable for the measurement of enzymes deriving from the chloroplasts: (i) pH between 7.5 and 8.5; (ii) a relatively high concentration of Mg^{2+} (10 mM or more); (iii) inclusion of thiol-protecting agents such as 2-mercaptoethanol or DTT which protect the enzymes but which do not themselves lead to appreciable reductive activation; (iv) inclusion of a detergent such as Triton in the extraction medium is useful because it solubilises the thylakoid membrane, thereby preventing binding of enzymes to the membrane (which is promoted by Mg^{2+}), and it facilitates chlorophyll estimation. For young leaves of cereals etc., a suitable medium contains: 1 mM EDTA, 10 mM $MgCl_2$, 15 mM 2-mercaptoethanol, 0.05% Triton X-100, 100 mM Tris-HCl (pH 7.8).

In order to detect changes in activation state, it is usually necessary to assay the enzyme under sub-optimal conditions. For example, the V_{max} of Fru-1,6-bisPase is unaffected by light activation and the V_{max} of sucrose-P synthase in many tissues is unaffected by activation *in vivo*, so that when measured under optimal conditions no difference in the activity is detected between illuminated and darkened samples.

16.6.3 Assay of enzymes of the Calvin cycle

All the light-activated enzymes of the Calvin cycle exhibit hysteretic behaviour (i.e. catalysis during the activation phase may be slower than the subsequent rate of catalysis). Two aspects of an enzyme's behaviour in relation to its substrate are therefore open to investigation. A rapid assay (generally lasting less than a minute) will reveal the instantaneous effect of changed conditions of substrate upon the rate of catalysis. However, prolonged incubation of an enzyme under assay conditions which differ from those of its previous environment will often lead to slow, but reversible, changes in activation state. Rapid assays of hysteretic enzymes have the advantage that they will give an estimate of the activation state of the enzyme under *in vivo* conditions rather than under the conditions of the assay. They also allow accurate determination of enzyme activities at low concentrations of substrate and are thus useful for the determination of kinetic constants. Care must be taken to ensure linearity during the course of the assay. These types of assays occur in two stages since they involve measurement of the reaction products in a separate assay from measurements during the catalytic step. Two-stage assays have the advantage that large numbers of samples may be handled in parallel, whereas direct spectrophotometric assays have the advantage of speed and simplicity.

(a) Fructose-1,6-bisphosphatase:
The chloroplastic Fru-1,6-bisPase differs from the cytosolic enzyme (and mammalian and yeast enzymes) in not being inhibited by AMP and in being light-activated. The native oxidised enzyme contains six disulphide bridges, two of which are cleaved upon activation by thioredoxin. A complex interplay of electron transfer, ligand binding and the stromal pH determines the activation state of the enzyme. In its oxidised form the enzyme has a pH optimum of pH 8.8, whereas the reduced form is optimally active at pH 7.5–8.5. The activity of the enzyme in illuminated leaves is predominantly stromal, rather than cytosolic. Extraction is best done in the presence of 1 mM Fru-1,6-bisP as this stabilises the enzyme against oxidative inactivation. Desalting and storage on ice causes a loss of activity.

The enzyme is most conveniently assayed by a continuous spectrophotometric method:

$$\text{Fru-1,6-bisP} \xrightarrow[\text{(fructose-1,6-bisPase)}]{} \text{Fru-6-P} + P_i$$

$$\text{Fru-6-P} \xrightarrow[\text{(phosphoglucose isomerase)}]{} \text{Glc-6-P}$$

$$\text{Glc-6-P} + NADP^+ \xrightarrow[\text{(glucose-6-P dehydrogenase)}]{} \text{6-phosphogluconate} + NADPH + H^+$$

The assay contains 1 mM EDTA, 10 mM $MgCl_2$, 0.05% Triton X-100, 100 mM Tris–HCl (pH 8.2), to which is added 0.4 mM $NADP^+$, 0.1 to 1 mM Fru-1,6-bisP, 4 units ml^{-1} phosphoglucose isomerase and 2 units ml^{-1} glucose-6-P dehydrogenase. Initiate the reaction (total volume 1 ml) with extract containing 1–5 μg chlorophyll. Assays are less linear at low pH (<pH 7.8), and at low concentrations of Mg^{2+} and Fru-1,6-bisP. Different concentrations of Fru-1,6-bisP will be appropriate for different plants. Excessive Fru-1,6-bisP (>1 mM) or too high a pH (greater than pH 8.2) will tend to minimise apparent light-activation. The maximum activity of the enzyme can be determined by preincubating crude extracts with 10 mM dithiothreitol under N_2.

An alternative to the spectrophotometric assay is the assay of P_i release. The reaction mixture is the same as for the continuous spectrophotometric assay (omitting coupling enzymes), but in a total volume of 500 µl. Terminate the reaction after 30 s with 200 µl 1 M $HClO_4$. Centrifuge the acidified sample and use the supernatant for the determination of P_i. Malachite green is employed to increase the sensitivity of the standard acid molybdate assay for P_i. Care must be taken to ensure that all glassware, etc. is scrupulously clean. The assay includes 500 µl supernatant, 1 ml 1 M HCl and 1 ml molybdate reagent. (The molybdate reagent is prepared by mixing 1 vol. 4.2% $(NH_4)_6Mo_7O_{24}$ in 5 M HCl with 3 vol. 0.2% (w/v) malachite green. The solution is allowed to stand for 30 minutes and then filtered.) After 15 minutes add 100 µl 1.5% Tween-20 detergent. Read the absorbance at 660 nm and estimate P_i by reference to a standard curve of mixtures containing between 0.1 and 2 µg P_i.

The amount of fructose-6-P released may also be determined (Section 16.5.4).

(b) Ribulose-1,5-bisphosphate carboxylase-oxygenase (Rubisco):

Rubisco has been termed the world's most abundant protein. It is by far the most abundant protein in leaves of C_3 plants, constituting up to half the total leaf protein. For this reason it plays a crucial role in the nitrogen economy of plants. It comprises eight large and eight small subunits (L_8S_8). The large subunits, encoded by the chloroplast genome, are made on chloroplastic ribosomes, while the small subunits, encoded by the nucleus, are made in the cytoplasm and imported into the chloroplasts.

Rubisco is a notoriously difficult enzyme to assay. Besides the complexity of the factors which regulate its activity, Rubisco is a bifunctional enzyme, catalysing the incorporation of both O_2 and CO_2, with oxygenation acting as a competitive inhibitor of carboxylation and thereby providing the glycollate for the photorespiratory pathway. Close study of Lorimer *et al.* [9] and Keys and Parry [10] is advised before embarking on measurements of Rubisco activity, and particular care should be taken in the design of experiments to measure the enzyme in leaf extracts. An added complication is that the enzyme appears to be cold-inactivated in a number of plant extracts, such as wheat, particularly in the absence of CO_2.

When the enzyme is extracted from the tissue its activation state will start to change. In the absence of added CO_2 and Mg^{2+} the enzyme will be quite rapidly inactivated as the active carbamate (formed by CO_2 addition) breaks down. However, in the presence of CO_2 and Mg^{2+} the enzyme will be slowly activated (unless it is already fully activated *in vivo*). The initial activity must therefore be estimated quickly after rapid extraction, with the extract kept as cold as possible, so as to minimise changes in the carbamylation state.

The total activity should also be measured in the same extract. This requires preincubation of the leaf extract (or the purified enzyme) with 100 mM Tris–HCl (pH 8), 20 mM $MgCl_2$, 10 mM $NaHCO_3$ and 1 mM dithiothreitol at 25°C for 10–20 minutes. Alternatively, the extract can be applied to a gel column of Sephadex G-25 equilibrated with the same buffer, and eluted in a minimal volume (this procedure removes phenolics, etc.).

It is not possible fully to activate the enzyme when the natural tight-binding inhibitor, carboxyarabanitol-1-P, is present. This inhibitor is synthesised at night and degraded during the day, so that leaf samples taken during the middle of the day are likely to provide the best estimates of total activity. In such circumstances it is best to use alternative means to quantify the total Rubisco protein, either using gel electrophoresis [10] or antibodies to Rubisco.

Rubisco may be assayed by both radiometric and spectrophotometric assays:

$$RuBP + H_2O + {}^{14}CO_2 \xrightarrow[\text{(Rubisco)}]{} 2({}^{14}C\text{-glycerate-3-P}) + 2H^+$$

For the radiometric assay, an aliquot of the activated enzyme is added to the reaction mixture which contains 100 mM Tris–HCl (pH 8.2) (buffer made CO_2-free), 5 mM dithiothreitol, 20 mM $MgCl_2$, 20 mM $NaH^{14}CO_3$ (specific activity *ca.* 1 TBq mol^{-1}) and 0.4 mM RuBP. The reaction is allowed to proceed for 1 minute before being stopped by the addition of 2 M HCl (100 µl added to 0.5 ml), and the acid-stable radioactivity determined.

Before use the solutions should be degassed. Distilled water is acidified with a drop of concentrated HCl and degassed by boiling it vigorously for about 20 minutes. The container should then be capped with a sodalime tube and allowed to cool. The buffer is dissolved and adjusted to about pH 4 with HCl. The solution is then purged with N_2 for about 10 minutes. The pH is then adjusted with KOH prepared in CO_2-free water from pellets which have been rinsed with CO_2-free water. The solution is again purged with N_2 and sealed.

The spectrophotometric assay for Rubisco [12] is linked to phosphoglycerate kinase and glyceraldehyde-P dehydrogenase. Creatine phosphokinase is also added to regenerate ATP. Although the spectrophotometric assay is more suitable for the determination of total activity than of initial activity [11], if care is taken the initial activity can be measured [23]:

$$RuBP + H_2O + CO_2 \xrightarrow[\text{(Rubisco)}]{} 2(\text{glycerate-3-P}) + 2H^+$$

$$\text{glycerate-3-P} + ATP + NADH \xrightarrow[\text{(glycerate-3-P kinase/glyceraldehyde-P dehydrogenase)}]{} \rightarrow \text{glyceraldehyde-P} + ADP + NAD+ + P_i$$

$$ADP + \text{creatine-P} \xrightarrow[\text{(creatine phosphokinase)}]{} ATP + \text{creatine}$$

The assay mixture contains 50 mM Hepes (pH 7.8), 10 mM KCl, 1 mM EDTA, 15 mM $MgCl_2$, 5 mM dithiothreitol, 5 mM ATP, 1 mM NADH, 10 mM $NaHCO_3$, 5 units ml^{-1} glyceraldehyde-P dehydrogenase, 8 units ml^{-1} glycerate-3-P kinase, 5 mM creatine-P, 2 units ml^{-1} creatine phosphokinase, 0.5 mM RuBP and chloroplast or leaf extract equivalent to 5 µg chlorophyll ml^{-1}. Ribose-5-P (0.5 mM) may be substituted for RuBP in crude extracts as there is usually enough endogenous ribose-5-P isomerase and ribulose-5-P kinase to convert ribose-5-P to RuBP. However, this must be checked. Induction lags in this assay may be largely eliminated by the addition of an excess of coupling enzymes.

16.6.4 Assay of other enzymes involved in photosynthesis

(a) $NADP^+$-malate dehydrogenase:
$NADP^+$-dependent malate dehydrogenase occurs in the chloroplasts of both C_3 and C_4 plants, though its activity is much higher in C_4 plants. Like Fru-1,6-bisPase, its activity is modulated by the ferredoxin-thioredoxin system, which reduces cysteine residues on the enzyme. Maximum activity must therefore be measured after preincubation with dithiothreitol (Section 16.6.3). It is a good indicator of the redox state of the acceptor side of Photosystem I and provides a useful complement to measurement of the redox states of Q_A (by chlorophyll fluorescence, Chapter 12) and of P_{700}.

$$\text{oxaloacetate} + NADPH + H^+ \xrightarrow[\text{(NADP-malate dehydrogenase)}]{} \text{malate} + NADP^+$$

The assay mixture contains 1 mM EDTA, 15 mM 2-mercaptoethanol, 0.05% (v/v) Triton X-100, 100 mM Tris–HCl (pH 7.8), 0.5 mM oxaloacetate (made up freshly and neutralised with dilute KOH) and 0.2 mM NADPH. Rates are linear for between 2 and 4 minutes.

(b) Phosphoenolpyruvate carboxylase:
PEP carboxylase occurs in C_3 plants, as well as being involved in photosynthesis in C_4 and CAM plants. It catalyses the carboxylation of PEP to the C_4 acid, oxaloacetate.

$$\text{PEP} + \text{HCO}_3^- \xrightarrow[\text{(PEP carboxylase)}]{} \text{oxaloacetate} + \text{P}_i$$

$$\text{oxaloacetate} + \text{NADH} + \text{H}^+ \xrightarrow[\text{(malate dehydrogenase)}]{} \text{malate} + \text{NAD}^+$$

In C_4 and in CAM plants its activity is regulated by phosphorylation, such that the enzyme from illuminated leaves of C_4 plants, or darkened photosynthetic tissue of CAM plants, is less sensitive to inhibition by malate. Whether the same is true of C_3 plants has not yet been shown. The activity can be measured either radiometrically or spectrophotometrically.

Leaf extracts should be made at about pH 7 in the presence of BSA [13]. A loss of activity occurs with prolonged storage of extracts on ice, rather than at room temperature. A suitable extraction medium for maize leaf contains 50 mM Mops-KOH (pH 7.2), 10 mM $MgCl_2$, 1 mM dithiothreitol, 10% (v/v) glycerol, 2% (w/v) BSA. For the continuous spectrophotometric assay, the reaction mixture contains 100 mM Tris-HCl (pH 7.5), 10 mM $MgCl_2$, 0.4 mM NADH, 10 mM $NaHCO_3$, 0.5 mM phosphoenolpyruvate, 2.7 units ml^{-1} malate dehydrogenase and 50 µl extract. Glucose-6-P (5 mM; a potent activator of PEP carboxylase) is often added if maximum activities are being measured.

Note that oxaloacetate is unstable even during the course of the assay. Errors may occur due to the non-enzymatic decarboxylation of oxaloacetate to pyruvate. However, these can generally be overcome by adding a large excess of malate dehydrogenase (most tissues already contain a substantial amount of malate dehydrogenase). Addition of lactate dehydrogenase to estimate pyruvate production is liable to error because of the presence of PEP phosphatase [8] in crude extracts.

For the radiometric assay, substitute 10 mM $NaH^{14}CO_3$ (1.3 TBq mol^{-1}) in the above assay. After 30 s, reactions are stopped with 500 µl 10 M HCOOH, the mixtures are taken to dryness under heating lamps or a stream of air, a further 100 µl of HCOOH is added and the mixtures are dried again to measure acid-stable radioactivity.

(c) Sucrose phosphate synthase:
Sucrose-P synthase catalyses the formation of sucrose phosphate in the cytosol:

$$\text{UDP-glucose} + \text{fructose-6-P} \xrightarrow[\text{(sucrose-P synthase)}]{} \text{sucrose-P} + \text{UDP}$$

The enzyme extracted from illuminated leaves of certain plants (e.g. spinach, barley) is markedly less sensitive to inhibition by P_i than is the enzyme extracted from darkened leaves [7]. If sucrose-P synthase activity is measured in the 'unselective' assay, with high hexose-P and no P_i, then the two kinetic forms are indistinguishable and the activity measured can be considered to be the maximum catalytic activity of the enzyme. In the presence of lesser amounts of hexose-P, the two kinetic forms of sucrose-P synthase can be distinguished on the basis of their sensitivity to inhibition by 5 mM P_i (i.e. the activated enzyme is less sensitive to inhibition by P_i). The activation state can then be expressed as the ratio of the activity of sucrose-P synthase in the presence of P_i to the activity in the absence of P_i (Fig. 16.5).

Leaf material (about 5 cm^2) is extracted in 1 ml of a medium containing 50 mM Hepes (pH 7.4), 5 mM $MgCl_2$, 1 mM EDTA, 1 mM EGTA, 10% glycerol, 0.1% Triton X-100. In difficult tissues (e.g. grape leaves and cell suspension cultures) it may be necessary to use an affinity column to remove interfering substances from the extract [14]. For example, with sucrose-P synthase, an ω-aminohexyl-agarose column (*Sigma Chemicals*, 0.2 ml volume, held in the tip of a micropipette by a piece of glass wool) is pre-equilibrated with 1 ml 0.1 M Tris-HCl (pH 8.5), 0.5 M NaCl, followed by 1 ml 0.1 M potassium acetate (pH 4.5), 0.5 M NaCl (to convert the

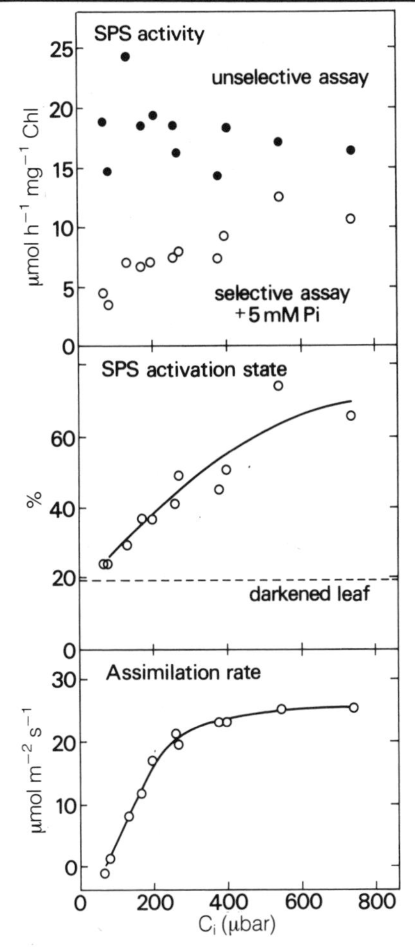

Fig. 16.5. Relationship between the assimilation rate (lower panel) and the activation state of sucrose-P synthase (SPS) in relation to the intercellular concentration of CO_2 (c_i) in spinach leaves. The top panel shows the activity of SPS measured in the unselective assay (under which conditions the activity remains essentially constant, although it varies between different leaves) and in the selective assay including 5 mM P_i (under which conditions the change in activation can be detected; see Section 16.6.4). The middle panel shows the activation state (defined as the activity under limiting conditions (measured in the presence of 5 mM P_i) as a percentage of the activity under non-limiting conditions (measured in the absence of P_i). The dashed line shows the activation state of SPS in the dark. Each point represents a different leaf sample. Leaves were illuminated at a PFD of 1400 μmol m^{-1} s^{-1} and a leaf temperature of 25°C [24].

resin to the acetate form), followed by 1 ml of extraction buffer containing 50 mM Hepes (pH 7.4), 5 mM $MgCl_2$, 1 mM EDTA, 1 mM EGTA, 10% glycerol, 0.1% Triton X-100, 1 mM dithiothreitol, 2 mM benzamidine, 2 mM 6-amino-*n*-hexanoic acid 0.5 mM phenylmethylsulphonyl fluoride (the last two are protease inhibitors). Plant material is extracted in 10 volumes of extraction buffer, centrifuged and 200 μl applied to the column. Activity is eluted from the column by six successive washes of 400 μl extraction buffer containing 0, 50, 100, 200, 350 and 500 mM NaCl. The 100, 200 and 500 mM fractions are pooled and assayed for activity (the remaining fractions should also be checked for enzyme activity). This method can, of course, be modified for other enzymes.

Enzyme activity is measured by the release of uridine diphosphate (UDP) from UDP-glucose. Care must be taken that enzymes which degrade UDP do not interfere with the assay [15]. Accordingly, it is safest to run a parallel control sample containing, for example, 1 nmol UDP. The assay medium contains 50 mM Hepes buffer (pH 7.4), 4 mM $MgCl_2$, 1 mM EDTA, and extract equivalent to about 5 μg chlorophyll, but it is varied in respect of the following additional constituents:

Unselective assay:	4 mM fructose-6-P; 20 mM glucose-6-P, 3 mM UDP-glucose
Selective assay:	2 mM fructose-6-P, 10 mM glucose-6-P, 3 mM UDP-glucose ± 5 mM P_i

Glucose-6-P is included to avoid depletion of fructose-6-P by the action of phosphoglucose isomerase.

Incubate the reaction mixtures for 15 minutes at 25°C. Stop the reaction by boiling the mixtures and centrifuge them to remove precipitated protein. Controls should be included which lack hexose-P. Assay for UDP using the following reaction mixture (cf. assay for ADP described in the section on metabolite assays, but note that UDP reacts more slowly than does ADP):

$$\text{PEP} + \text{UDP} \xrightarrow[\text{(pyruvate kinase)}]{} \text{pyruvate} + \text{UTP}$$

$$\text{pyruvate} + \text{NADH} + \text{H}^+ \xrightarrow[\text{(lactate dehydrogenase)}]{} \text{lactate} + \text{NAD}^+$$

The assay contains 5 mM KCl, 50 mM Hepes (pH 7.0), 5 mM $MgCl_2$, 0.8 mM phosphoenolpyruvate, 0.3 mM NADH, 50 µl supernatant, lactate dehydrogenase (20 units ml^{-1}) and pyruvate kinase (6 units ml^{-1}).

16.7 MEASUREMENT OF CHLOROPHYLL AND PHAEOPHYTIN

For chlorophyll estimation in leaves, the simplest method is to freeze the leaf in liquid N_2, grind it to a powder, extract with 80% acetone in the dark and then centrifuge. The absorbance is measured at 652 nm and the chlorophyll concentration (in $mg\,l^{-1}$) is given by the expression: $(A_{652}) \times \frac{1000}{34.5}$. For the accurate determination of chlorophyll *a* and *b*, the extinction coefficients of Graan and Ort [16] can be used, based on absorbance readings made in 80% acetone at 664 and 647 nm.

$$\text{Chlorophyll } a\ (\mu\text{M}) = 13.19\ A_{664} - 2.57\ A_{647}$$
$$\text{Chlorophyll } b\ (\mu\text{M}) = 22.10\ A_{647} - 5.26\ A_{664}$$
$$\text{Total chlorophyll } (\mu\text{M}) = 7.93\ A_{664} + 19.53\ A_{647}$$

Particular care must be taken when measuring chlorophyll that (a) the extract is not exposed to light and (b) that the extract is not acid. Both these conditions can result in chlorophyll destruction.

Although extraction in acidic conditions results in chlorophyll destruction, the product, phaeophytin, can be measured [17]. The pellet resulting from extraction in $HClO_4$ should be homogenised in water and acetone added to a final concentration of 80% (v/v). Samples may require overnight extraction. Absorbance readings are taken at 536, 655 and 666 nm. In 80% acetone, the content of phaeophytin (in $\mu g\,ml^{-1}$) is given by one of two expressions. Either

$$\text{Phaeophytin} = (77.58\ A_{536} - 0.33\ A_{666})$$

or

$$\text{Phaeophytin} = (6.75\ A_{666} + 26.03\ A_{655}).$$

It is useful to measure at all three wavelengths and to check the phaeophytin content using both equations.

16.8 DATA INTERPRETATION

Studies of changes in metabolites, enzyme activities and carbon fluxes in relation to environmental variables such as light, CO_2 and temperature (or even fungal and viral infection) can yield information about how photosynthetic metabolism in a leaf is limited and regulated. A considerable amount of information can be derived from gas-exchange data alone. For example, studies *in vivo* have shown that it is possible to select conditions of gas phase, temperature, illumination, etc. so as to impose limitations on photosynthesis by electron transport (low light, high CO_2), by carboxylation (low CO_2, high light) or by the capacity for sucrose synthesis (low temperature, high light and high CO_2). Thus if a reduction in PFD leads to a decrease in the initial slope of the assimilation rate (**A**) vs. intercellular CO_2 (c_i) curve (the carboxylation efficiency) then it will have done so via a reduction in the activity of Rubisco. If, on the other hand, a selective restriction of sucrose synthesis occurs (e.g. by lowering the temperature), this will be evident as a diminished sensitivity of photosynthesis to O_2 and CO_2 in high light and high CO_2 (i.e. a flattening of the **A**/c_i curve at higher concentrations of CO_2) because the rate of sucrose synthesis is not fast enough to support high rates of carbon assimilation [1,19]. This will be accompanied by enhanced oscillations and a stimulation of CO_2 assimilation rate by feeding P_i under the same conditions (Chapter 10).

Metabolite and enzyme data can be used in a variety of ways in the identification of regulatory reactions. The following examples give only an indication of their use. Further information can be found in [18–20,22] and references therein.

(a) Identification of those enzymes which have maximum catalytic activities of the same order of magnitude as the fluxes which they catalyse. In Fig. 16.5, for example, we can see that the *in vitro* activity of sucrose-P synthase is about 18 μmol h^{-1} mg^{-1} chlorophyll. This would allow a rate of carbon fixation leading solely to sucrose synthesis of 216 (= 18 × 12) μmol CO_2 h^{-1} mg^{-1} chlorophyll. The rate of photosynthesis is 25 μmol m^{-2} s^{-1}, which is equivalent to 318 μmol h^{-1} mg^{-1} chlorophyll, while the flux into sucrose under comparable conditions would be about half the rate of CO_2 fixation (i.e. about 150 μmol h^{-1} mg^{-1} chlorophyll). In other words, the capacity of sucrose-P synthase is of the same order as the rate of sucrose synthesis in leaves and small changes in the activity of the enzyme could reasonably be expected to influence the flux.

(b) Measurement of changes in metabolite pools and fluxes. The observation that the substrate changes in a direction opposite to that of the flux identifies a regulatory reaction (although this does not necessarily mean that the enzyme is regulatory if other substrates or cofactors are involved). Fig. 16.6 illustrates the behaviour of the assimilation rate (**A**) and metabolite pools in leaves of *Amaranthus edulis* with changes in temperature. In leaves of C_4 plants, the exceptionally large leaf pools of photosynthetic intermediates such as glycerate-3-P, triose-P and pyruvate reflect the demands of intercellular transport between mesophyll and bundle-sheath (Chapter 7). In general it would be expected that, as the photosynthetic flux falls, so metabolite gradients of C_4 acids (such as aspartate), glycerate-3-P, triose-P and pyruvate between the bundle-sheath and mesophyll would decline. This is the reason

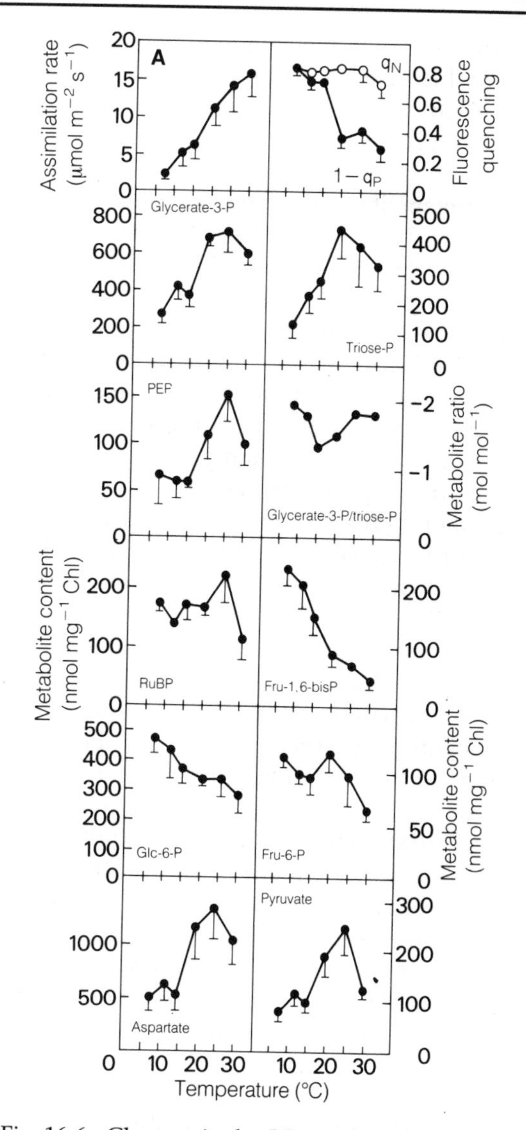

Fig. 16.6. Changes in the CO_2 assimilation rate (**A**), metabolite pools and the photochemical (1-q_Q) and non-photochemical (q_{NP}) components of chlorophyll fluorescence quenching in leaves of *Amaranthus edulis* at different temperatures. The photon flux was 1050 μmol quanta m^{-1} s^{-1} and the concentration of CO_2 was 350 μbar. Data are means of four replicates, with Standard Error bars shown. (M.D. Adcock and R.C. Leegood, unpublished results).

for the decline in metabolite pools (including PEP, which is in equilibrium with glycerate-3-P) with a decrease in temperature below 25°C. Intermediates of the Calvin cycle or of product synthesis behave differently from the metabolites involved in intercellular transport. The amount of RuBP remains more or less constant as the flux falls at low temperature but, as in the leaves of C_3 plants [19], the amounts of Glc6P, Fru6P and Fru1,6bisP rise as the temperature is decreased. These data indicate the development of a restriction on the consumption of these metabolites at lower temperatures which results from specific temperature-dependent regulation of the enzymes of the Calvin cycle (in the case of RuBP and Fru-1,6-bisP) and of starch or sucrose synthesis (in the case of hexose-P).

(c) Measurement of changes in metabolite ratios can also provide information. First, the mass-action ratio, from contents of metabolites measured *in vivo*, can be compared with the equilibrium constant. This method can be used to identify reactions which are strongly removed from equilibrium *in vivo* and which enzymes are therefore likely to be regulatory (although this does not mean that enzymes which catalyse reactions which are close to equilibrium are not regulatory). Second, metabolite ratios can be employed to estimate other intracellular metabolite concentrations or ratios if it is assumed that a particular enzyme catalyses a reaction which is close to equilibrium. Thus:

$$\frac{[\text{glycerate-3-P}]}{[\text{triose-P}]} \times \frac{[\text{ATP}][\text{H}^+][\text{NADPH}]}{[\text{ADP}][\text{P}_i][\text{NADP}]} = \text{K}$$

Electron transport provides ATP and NADPH for the reduction of glycerate-3-P to triose-P, and the ratio between the amounts of triose-P and glycerate-3-P can be used as an indication of the assimilatory power, i.e. the combined redox and phosphorylation potential [18].

In Fig. 16.6, the glycerate-3-P/triose-P ratio declined with decreasing temperature to 15°C and then showed a rise to 8°C. At low temperatures a restriction on the generation of ATP and NADPH by electron transport therefore appears. This is because a decrease in the utilisation of the products of electron transport by metabolism would be expected to lead to a fall in the glycerate-3-P/triose-P ratio, not the rise which is observed. Figure 16.6 also shows two chlorophyll fluorescence parameters (Chapter 12). The redox state of Q_A (shown as 1-q_P) and the non-photochemical (q_{NP}) component of chlorophyll fluorescence quenching increased as the temperature decreased. These changes in fluorescence quenching are consistent with the occurrence of photosynthetic control and the notion that a high thylakoid ΔpH at low temperature restricts electron transport and hence the generation of ATP and NADPH.

It should be noted that most of these approaches are qualitative, in that points of control can be identified, but not quantified. It then becomes important to ask how much an enzyme controls the flux, i.e., what fractional change in flux is consequent upon a fractional change in enzyme activity. This type of analysis of the control of metabolism, using the approach of Kacser and Burns [21] is increasingly finding application in plants. This follows the selection of a number of mutants in carbon metabolism and advances in genetic manipulation which allow the generation of plants with varying amounts of specific enzymes (e.g. [20,22]).

REFERENCES

1. Sharkey, T.D. (1988) Estimating the rate of photorespiration in leaves. *Physiologia Plantarum* **73**, 147–152.
2. Badger, M.R., T.D. Sharkey and S. von Caemmerer (1984) The relationship between steady-state gas exchange of bean leaves and the levels of carbon-reduction-cycle intermediates. *Planta* **160**, 305–313.
3. Stitt, M. (1990) Fructose 2,6-bisphosphate. In: *Methods in Plant Biochemistry* (Lea, P.J., ed.). Academic Press, New York. pp. 87–92.
4. Lowry, O.H. and J.V. Passonneau (1972) *A*

flexible system of enzymatic analysis. Academic Press, New York.

5. Salvucci, M.E. (1989) Regulation of Rubisco activity *in vivo. Physiologia Plantarum* **77**, 164–171.
6. Stitt, M., I. Wilke, R. Feil, and H.W. Heldt (1988) Coarse control of sucrose-phosphate synthase in leaves: Alterations of the kinetic properties in response to the rate of photosynthesis and the accumulation of sucrose. *Planta* **174**, 217–230.
7. Huber, S.C., T.H. Nielsen, J.L. Huber and D.M. Pharr (1989) Variation among species in light activation of sucrose-phosphate synthase. *Plant Cell Physiol.* **30**, 277–285.
8. Lea, P.J. (ed.) (1990) *Methods in Plant Biochemistry.* Academic Press, New York.
9. Lorimer, G.H., M.R. Badger and T.J. Andrews (1977) D-ribulose-1,5-bisphosphate carboxylase-oxygenase. Improved methods for the activation and assay of catalytic activities. *Anal. Biochem.* **78**, 66–75.
10. Keys, A.J. and M. Parry (1990) Ribulose bisphosphate carboxylase/oxygenase and carbonic anhydrase. In: *Methods in Plant Biochemistry* (Lea, P.J., ed.). Academic Press, New York. pp. 1–14.
11. Ward, D.A. and A.J. Keys (1989) A comparison between the coupled spectrophotometric and uncoupled radiometric assays for RuBP carboxylase. *Photosynth. Research* **22**, 167–171.
12. Lilley, R.McC. and D.A. Walker (1974) An improved spectrophotometric assay for ribulose bisphosphate carboxylase. *Biochim. Biohys. Acta* **358**, 226–229.
13. Ashton, A.R., J.N. Burnell, R.T. Furbank, C.L.D. Jenkins and M.D. Hatch (1990) Enzymes of C_4 photosynthesis. In: *Methods in Plant Biochemistry* (Lea, P.J., ed.). Academic Press, New York. pp. 39–72.
14. Dancer, J., W.-D. Hatzfeld and M. Stitt (1990) Cytosolic cycles regulate the turnover of sucrose in heterotrophic cell-suspension cultures of *Chenopodium rubrum* L. *Planta* **182**, 223–231.
15. Crafts-Brandner, S.J. and M.E. Salvucci (1989) Species and environmental variations in the effect of inorganic phosphate on sucrose-phosphate synthase activity. Reliability of assays based upon UDP formation. *Plant Physiol.* **91**, 469–472.
16. Graan, T. and D.R. Ort (1984) Quantitation of the rapid electron donors to P700, the functional plastoquinone pool, and the ratio of the photosystems in spinach chloroplasts. *J. Biol. Chem.* **259**, 14003–14010.
17. Vernon, L.P. (1960) Spectrophotometric determination of chlorophylls and phaeophytins in plant extracts. *Anal. Chem.* **32**, 1144–1150.
18. Heber, U., S. Neimanis, K.-J. Dietz and J. Viil (1986) Assimilatory power as a driving force in photosynthesis. *Biochim. Biophys. Acta* **852**, 144–155.
19. Labate, C.A., M.D. Adcock and R.C. Leegood, (1990) Effects of temperature on the regulation of photosynthetic carbon assimilation in leaves of maize and barley. *Planta* **181**, 547–554.
20. Stitt, M., A. von Schaewen and L. Willmitzer (1991) 'Sink' regulation of photosynthetic metabolism in transgenic tobacco plants expressing yeast invertase in their cell wall involves a decrease of the Calvin cycle and an increase of glycolytic enzymes. *Planta* **183**, 40–50.
21. Kacser, H. and J.A. Burns (1973) The control of flux. *SEB Symposia* **27**, 65–104.
22. Neuhaus, H.E., A.L. Kruckeberg, R. Feil and M. Stitt (1989) Reduced-activity mutants of phosphoglucose isomerase in the cytosol and chloroplast of *Clarkia xantiana*. II. Study of the mechanisms which regulate photosynthate partitioning. *Planta* **178**, 110–122.
23. Sharkey, T.D., L.V. Savitch and N.D. Butz (1991) Photometric method for routine determination of K_{cat} and carbamylation of Rubisco. *Photosynth. Research* **28**, 41–48.
24. Battistelli, A., M.D. Adcock and R.C. Leegood (1991) The relationship between the activation state of sucrose-phosphate synthase and the rate of CO_2 assimilation in spinach leaves. *Planta* **183**, 620–622.

17

Chloroplasts and protoplasts

R.C. Leegood and D.A. Walker

17.1 INTRODUCTION

Much of our present knowledge concerning the nature and regulation of processes such as electron transport, photophosphorylation and CO_2 assimilation derives from experiments with isolated chloroplasts and thylakoid membranes. Sub-cellular systems, if isolated with due care, continue to catalyse their component reactions at rates which equal those observed *in vivo*. It is widely accepted that fully functional organelles provide a more realistic basis for such studies than those which have been unintentionally stripped of their limiting envelopes during isolation.

The isolation of chloroplasts which retain their outer double membrane (or *envelope*) is clearly a prerequisite for the maintenance of any process whose activity is located in the chloroplast stroma, such as assimilation of CO_2, metabolism of nitrite, ammonia and amino acids, reduction of sulphur compounds, lipid and protein synthesis or the uptake of small molecules and proteins across the chloroplast envelope. The chloroplast envelope also prevents loss of, or damage to, stromal components by the action of proteases, phenolics, etc. Thus thylakoids obtained from intact chloroplasts (isolated previously) show higher P/O ratios and better photosynthetic control (Chapter 18) than those isolated directly from leaves.

Similarly, thylakoid membranes isolated from chloroplasts show higher rates of electron transport and are better coupled than thylakoids isolated directly from leaf tissue. Both thylakoids and chloroplasts are more accessible than the intact leaf because of the limited permeability offered to many compounds by the plasmalemma of the cell (and for thylakoids by the chloroplast envelope). However, until relatively recently, intact chloroplasts which fix CO_2 at high rates have only been separated routinely from spinach and peas. Other species have usually been found unsuitable for this purpose because of the presence of phenolics in their leaves or because they possess a high proportion of tissue with thickened cell walls.

The introduction of techniques which allow the isolation of protoplasts has greatly extended the range of plants from which intact chloroplasts may be isolated. The technique has been used successfully with C_3 grasses such as wheat and barley and with some C_3 dicotyledons such as sunflower, tobacco, peas and spinach, as well as many C_4 species. In certain respects this method of preparing chloroplasts may be considered superior to direct mechanical disruption. The mechanical force required to rupture the cell plasma membrane is much less than that required to break cell walls, so that disrup-

Photosynthesis and Production in a Changing Environment: a field and laboratory manual
Edited by D.O. Hall, J.M.O. Scurlock, H.R. Bolhàr-Nordenkampf, R.C. Leegood and S.P. Long
Published in 1993 by Chapman & Hall, London.
ISBN 0 412 42900 4 (HB) and 0 412 42910 1 (PB).

tion of the plasma membrane alone greatly reduces damage to organelles. Chloroplasts isolated from protoplasts thus show very high intactness. Secondly, interfering substances present in the cell walls or vascular tissues are removed during preparation of the protoplasts. These would be released during the mechanical maceration of the tissue, and their effects can only be minimised by rapid centrifugation of the organelles. For example, with sunflower, a species with phenolics present in the leaves, mechanical procedures yield chloroplasts which assimilate CO_2 at rather low rates despite the inclusion of a great many protective agents in the isolation medium. By contrast, protoplasts of sunflower yield chloroplasts of high activity. The principal disadvantages of protoplasts are the time-consuming preparation, the relatively low yield by comparison with mechanical techniques and the expense of the cell-wall degrading enzymes.

The isolation of organelles with a high degree of integrity is also of particular value in studies of the intracellular location of enzymes. Thus protoplasts have been a valuable tool in the study of the inter- and intracellular compartmentation of photosynthesis in C_3, C_4 and CAM plants. Chloroplasts can also be isolated from illuminated protoplasts [1], allowing studies of the intracellular distribution of metabolites during photosynthesis.

There can be little doubt that time and persistence will lead to an extension of the number of species from which protoplasts and chloroplasts may be isolated by modification of existing methods, although it is not recommended that experience in either method of preparation be gained with anything other than established species.

17.1.1 Plant material

Whichever method of preparation is chosen, it is impossible to isolate protoplasts or chloroplasts with high rates of photosynthesis from poor or indifferent material. Even spinach and peas will only yield 'good' chloroplasts if the parent tissue is itself in 'good' condition. Thus adequately controlled growth facilities are essential to allow reasonably reproducible results throughout the year. Spinach is a particularly demanding plant in this respect, requiring a short day length, together with a high light intensity, high humidity and a moderate temperature. Good results have been obtained with plants grown in water culture. As plants are grown for 6–8 weeks, a reasonably large stock is needed. Peas, usually of a dwarf variety, can be grown more simply in a medium like vermiculite, until about 10–12 days old. They are best grown under low light intensities in order to reduce starch formation. Whole young shoots may be employed in mechanical isolation procedures and young expanded leaves for protoplast preparations. Wheat, barley and maize, grown similarly for 1–2 weeks in vermiculite watered with a nutrient solution, are ideal material for protoplast isolation.

17.2 MESOPHYLL PROTOPLASTS FROM C_3, C_4 AND CAM PLANTS

17.2.1 Preparation of leaf tissue and the digestion procedure

The method used for preparing leaf tissue for digestion depends upon the species used. With monocots, such as wheat, barley and maize, leaf segments can be cut using a razor blade (transverse segments of about 0.5–1.0 mm). In some plants, the epidermis can often be removed from the lower surface of the leaf (e.g. tobacco, peas, spinach). The epidermis may also be stripped (from the apex down) from the turgid leaves of some varieties of barley. The surface of the leaf can be rubbed with carborundum or brushed gently (e.g. with a tooth-brush) in order to break through the epidermal tissue, a method particularly successful with sunflower. In cases where air might enter the leaf during cutting, this can be counteracted either by vacuum infiltration or else by cutting the tissue into thin segments under 0.5 M sorbitol. This latter method works well with spinach.

The enzymes used to digest cell walls are available commercially (Section 17.8), and have been found to be useful with a large number of species including most monocots. For species resistant to digestion, either some variation in growth conditions or alternative sources of digestive enzymes may prove useful. For example, Rohament P pectinase is particularly effective with sunflower leaves.

One of the main disadvantages of enzymic digestion is the cost of the cellulase and pectinase, although both enzymes can be recycled. After use the incubation medium may be frozen, and when sufficient has accumulated the enzymes can be purified by filtration and precipitated by $(NH_4)_2SO_4$ and freeze-dried for storage. Pectolyase Y23, which is an extremely effective pectinase, can be purified very simply and inexpensively from a culture of *Aspergillus japonicus* grown on wheat bran [2].

The following digestion medium is suitable for young leaves of wheat (*Triticum aestivum*) and barley (*Hordeum vulgare*): 2% (w/v) cellulase (Onozuka R10 or RS), 0.3% (w/v) pectinase (Macerozyme R10), 0.5 M sorbitol, 5 mM Mes. Adjust to pH 5.5 with dilute HCl.

The enzyme solution can be made up and left overnight at 4°C, if desired. Usually digestion is done at pH 5.0–5.5, a compromise between the pH optimum for the enzymes (which is more acid) and avoidance of damage to the tissue. The following are some of the variations in digestion media which are commonly employed, but optimum concentrations can readily be determined in a series of trial incubations.

Spinach (*Spinacia oleracea*): 3% Onozuka R10, 0.5% Macerozyme R10
C_4 spp.: 2% Onozuka R10, 0.2% Macerozyme R10
Sunflower (*Helianthus annuus*): 2% Onozuka R10, 0.5% Rohament P.
Pea (*Pisum sativum*): 2.5% Onozuka R10, 0.5% Macerozyme R10, 0.025% Rohament P.
Bovine serum albumin (0.05%; defatted) can also be added as a wetting agent.

Cut up leaf tissue with a new single-edged razor blade into segments approximately 0.7 mm wide for C_3 and 0.5 mm for C_4. (Note: it is important to cut small segments of C_4 leaves for bundle-sheath strands to be released during the digestion period.) Take care not to bruise the tissue and change the blade frequently. Put the tissue (about 3 g) with 50 ml of enzyme medium in a large crystallising dish (about 19 cm diameter) or into one or more large petri dishes, but do not crowd the tissue in the dish. Incubate at 25–30°C for 3–4 hours under low light; a 150 W light bulb suspended about 25 cm above the dish is sufficient. For maximum photosynthetic activity of maize mesophyll protoplasts, or when Pectolyase Y23 is employed, a shorter incubation period of about 2 hours suffices. Gently remove the enzyme medium from the dish using a pipette.

17.2.2 Isolation and purification

Following incubation of the leaf tissue with digestive enzymes, the isolation medium can be carefully removed (using a 25 ml pipette with its end covered by a piece of mesh) and either discarded or kept for recycling enzymes (see above). Sometimes appreciable numbers of protoplasts may be released into the incubation medium, in which case it can be centrifuged and the protoplasts retained. Successive washing with an osmoticum (e.g. in 0.5 M sorbitol, 5 mM Mes (pH 6.0), 1 mM $CaCl_2$) releases a mixture of protoplasts, chloroplasts and vascular tissue. Undigested material, including undigested tissue, epidermal tissue and vascular strands, can be removed by filtration through nylon sieves. Bundle-sheath strands from C_4 plants are resistant to digestion and can be collected on

80 µm nylon mesh. Low speed centrifugation of the extract, at 100 *g* for 5 minutes, gives a pellet containing protoplasts and chloroplasts. Any bench centrifuge will suffice, but it should preferably have a swing-out rotor. Determine a speed adequate to produce a compact pellet, yet not so compact as to be difficult to resuspend by gentle shaking.

Mesophyll protoplasts will float when resuspended in a solution of sufficiently high density. For example, if mesophyll protoplasts of wheat, barley or sunflower are suspended in a solution containing 0.5 M sucrose and centrifuged at low speeds (250 *g* for 5 minutes), they float to the top of the medium. Layering a lower density osmoticum (e.g. sorbitol) on top of the sucrose medium prior to centrifugation results in protoplasts partitioning at the interphase. They are then easily collected with a Pasteur pipette (Fig. 17.1). The yield of protoplasts should be at least 10% on a chlorophyll basis.

If protoplasts do not float readily the concentration of the whole gradient can be increased (e.g. to 0.6 M sucrose and sorbitol) [3].

(a) C_3 Tissue:
Make up the following media and use them chilled throughout the protoplast purification:

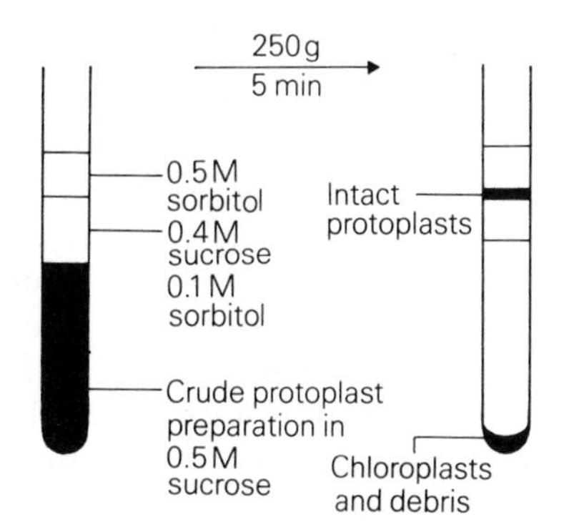

Fig. 17.1. Scheme for the purification of protoplasts by flotation in a step gradient of sucrose and sorbitol.

(I) 0.5 M sorbitol, 1 mM $CaCl_2$, 5 mM Mes (pH 6.0)
(II) 0.4 M sucrose, 0.1 M sorbitol, 1 mM $CaCl_2$, 5 mM Mes (pH 6.0)
(III) 0.5 M sucrose, 1 mM $CaCl_2$, 5 mM Mes (pH 6.0)

Gently wash the segments in 20 ml medium I. Filter the extract through a 500 µm nylon mesh (e.g. a tea-strainer), which retains larger pieces of tissue and 200 µm nylon mesh, which retains vascular strands. Wash the segments twice more and centrifuge the combined extract at 100 *g* for 5 minutes (this procedure should yield a pellet which is loosely compacted). Discard the supernatant (by rapid inversion of the tubes) and gently resuspend the pellet in a few drops of medium III. Add a further 5 ml medium III, then layer 2 ml of medium II on top (this washes the protoplasts), followed by a layer of 1 ml of medium I. Centrifuge at 250 *g* for 5 minutes and collect the pure protoplast fraction at the interphase with a Pasteur pipette. Usually the pellet consists of debris (chloroplasts and broken cells) but it is worth checking by light microscopy that no protoplasts are discarded. If many protoplasts are present in the pellet, some Percoll may be added to the sucrose medium and the purification repeated.

(b) C_4 Tissue:
The incubation and purification procedure is similar to that of C_3 protoplasts with the following modifications. After washing the segments, filter the extract through 500 µm and 80 µm nylon mesh. The bundle-sheath strands will collect on the 80 µm mesh. The bundle-sheath strands may be resuspended in sorbitol medium I. With maize, the media have been optimized for the preparation of mesophyll protoplasts by Day *et al.* [3] to include 0.5 M sorbitol, 0.2 mM $CaCl_2$, 0.2 mM KH_2PO_4, 1 mM $MgCl_2$, 0.2% BSA, 10 mM Mes (pH 5.5), 2% cellulase 3S and 0.2% pectinase in the digestion medium, and for the purification: 0.6 M sorbitol, 0.2 mM $CaCl_2$, 0.2 mM KH_2PO_4, 1 mM $MgCl_2$,

0.2% BSA, 5 mM Hepes (pH 7.8) (replacing medium I) and 0.6 M sucrose, 0.2 mM $CaCl_2$, 0.2 mM KH_2PO_4, 1 mM $MgCl_2$, 0.2% BSA, 5 mM Hepes (pH 7.8) (replacing medium III). The mesophyll protoplasts are purified from the filtrate by a method similar to that used for C_3 protoplasts except that a layer of 1 ml of sorbitol medium is put directly on top of the sucrose layer, omitting medium II.

In the case of C_4 plants (e.g. maize) strands of bundle-sheath cells are usually isolated together with mesophyll protoplasts. Protoplasts cannot usually be isolated in appreciable amounts from bundle-sheath cells because of the nature of the cell wall. However, Edwards *et al.* [4] have isolated bundle-sheath protoplasts from *Panicum milaceum* and Moore *et al.* [5] have had success with C_4 dicots such as *Atriplex spongiosa* and *Flaveria trinervia*. Boag and Jenkins [6] have also isolated extremely small quantities of bundle-sheath protoplasts from maize.

In some species such as maize, the photosynthetic capacity of the bundle-sheath strands is readily lost with prolonged enzymic digestion. For this reason, a rapid procedure combining enzymic digestion and mechanical separation yields the most active preparations [11].

(c) CAM Tissue:
Attempts to isolate cells should include only pectinase in the digestion medium. Attempts to isolate protoplasts should have pectinase and cellulase in the digestion medium.

For cells, wash segments after digestion with a medium containing 0.5 M sorbitol, 1 mM $CaCl_2$, 50 mM Tricine–KOH (pH 7.8) (sorbitol medium). Filter through 1 mm and 500 μm nylon mesh. Centrifuge cells at 100 *g* for 3 minuters and resuspend in the sorbitol medium followed by storage on ice. In the case of protoplasts, overlay the extract onto a cushion of 0.5 M sucrose, 1 mM $CaCl_2$, 50 mM Tricine–KOH pH 7.8 and 20% (w/v) dextran T_{20}. Alternatively, because protoplasts from CAM species are usually larger and more dense than those from C_3 and C_4 species, they can be purified by allowing the intact protoplasts to settle out. The supernatant may then be discarded.

17.2.3 Storage of protoplasts

Protoplasts for immediate use can be resuspended in sorbitol medium and stored on ice. Mesophyll protoplasts from C_3 plants are most stable if stored at a relatively low pH and with some divalent cations (e.g. 0.4 M sorbitol, 1 mM $CaCl_2$ and 20 mM Mes, pH 6.0). Studies with barley, wheat and tobacco indicate that sorbitol is an adequate osmoticum for at least 8–10 hours if the protoplasts are stored on ice. If stored at 25°C, the protoplasts lose their photosynthetic capacity when sorbitol medium is used. However, the protoplasts are photosynthetically stable for at least 10–20 hours at 25°C with sucrose as the osmoticum.

Protoplast intactness can be assessed by microscopic examination (a typical leaf protoplast of wheat contains about 200 chloroplasts). A quantitative estimate of protoplast intactness may also be made using an oxygen electrode. This involves measurement of the activity of glycollate oxidase, an enzyme which is released from damaged protoplasts, before and after the addition of a detergent which disrupts the cell membrane [10].

17.2.4 Isolation of plant cells

Apart from bundle-sheath cells, intact cells can be isolated from the leaves of an appreciable number of plant species by mechanical means or by using pectinase in the absence of cellulase (see Jensen [7] and references therein). Their great advantage over protoplasts is the ease of preparation, although because they possess tough cell walls it is extremely difficult to isolate organelles or enzymes from them. Isolated cells are also a cheaper alternative to protoplasts for the screening of herbicides, etc. Where mechanical procedures are effective in the isolation of cells, the cells of such leaves are clearly remarkably tough yet relatively loosely attached to one another. Leaves of many species (e.g. maize) do not yield any intact mesophyll cells after maceration in a blender. Examples of plants which give good yields of intact cells after mechanical separation are *Xanthium strumarium* (cocklebur), *Digitaria sanguinalis* (crab-grass), a particularly useful C_4 species because the cells are permeable to photosynthetic substrates [8], and *Asparagus* spp. (cells can be scraped from the cladophylls using a plastic card without any need for blending; an old credit card is suitable). Leaves of other plants such as spinach and *Glycine max* (soybean) are suitable for the preparation of cells by use of pectinase alone in the enzymic digestion medium, although a high concentration of pectinase may be required. Usually maceration is followed by filtration through nylon mesh or muslin, then centrifugation and storage in aerated buffer.

For the preparation of cells from *X. strumarium* [9], derib the leaves and grind $2\,cm^3$ pieces for about 3 s in 100 mM Hepes (pH 7.0) + $3\,mg\,ml^{-1}$ PVP-40 in a Polytron or a blender. Filter through 10 layers of muslin and centrifuge at low speed for 1 minute. Resuspend the cells and wash them twice. Isolated cells should be stored in the same medium bubbled with air. Cells can be assayed for photosynthesis in the same medium supplemented with bicarbonate. Cell viability is easily assessed by the ability of living cells to exclude the dye *Evans' Blue*.

17.3 PHOTOSYNTHESIS BY ISOLATED PROTOPLASTS

17.3.1 C_3 protoplasts

When using the oxygen electrode, care should be taken to set the stirrer at the lowest setting consistent with an adequate response of the electrode to the oxygen evolved. In this way breakage of the fragile protoplasts is minimised. A suitable concentration of protoplasts is $50\,\mu g\,Chl\,ml^{-1}$; they can be illuminated with white light. A wide range of media may be used for the assay of CO_2-dependent O_2 evolution. The pH optimum is broad as long as the concentration of bicarbonate is kept low at acid pH values (< pH 7.0). $CaCl_2$ is generally included, not only to prevent clumping but also to inhibit photosynthesis by chloroplasts, which are inevitably present in the preparation as a result of protoplast rupture. A suitable medium for wheat protoplasts is 0.4 M sorbitol, 10 mM $NaHCO_3$, 5 mM $CaCl_2$, 50 mM Tricine–KOH (pH 7.6).

17.3.2 C_4 mesophyll protoplasts

Mesophyll protoplasts from C_4 plants have a rather limited applicability for studies of photosynthetic carbon fixation since they do not fix CO_2 and the plasmalemma is impermeable to natural substrates (pyruvate, phosphoenolpyruvate and oxaloacetate) [4], although they are useful for studies of enzyme localisation. The alternative is to make intact mesophyll cells from plants such as *Digitaria sanguinalis*. Such cells retain permeability to the natural substrates.

17.3.3 CAM cells and protoplasts

$^{14}CO_2$ fixation and light-dependent O_2 evolution with and without bicarbonate may be measured in an assay medium containing 0.33 M sorbitol, 1 mM $MgCl_2$, 50 mM Tricine, pH 8.0.

17.4 CHLOROPLAST ISOLATION FROM PROTOPLASTS

The mesophyll protoplasts of C_3 and C_4 plants have an average diameter of 30–40 µm whilst those of CAM plants are at least 60–100 µm in diameter. A quick effective procedure for isolating chloroplasts is to pass the protoplasts several times though a 20 µm nylon mesh (e.g. Nybolt P20). This breaks all of the protoplasts but leaves chloroplasts and other organelles largely intact. Breakage can be achieved by fitting nylon mesh, using a ring cut from the plastic tip of a automatic pipette, to the end of a 1.0 ml disposable syringe which has had the tip excised to provide a pore of about 3 mm diameter. An aliquot of the protoplasts will be ruptured when taken up and ejected two or three times from the syringe. Intact chloroplasts (which have a diameter of 3–5 μm) will be released.

For storage (Section 17.2.3), for measurement of photosynthesis and for chloroplast preparation, the protoplasts should be diluted with sorbitol medium (about 10-fold) and centrifuged at 100 *g* for 5 minutes. The protoplast extract can be centrifuged at low speed, e.g. 250 *g* for 1 minute, to obtain a chloroplast pellet largely free of the cytosol fraction. This is sufficient for many studies on photosynthesis with chloroplasts from various species. For studies on the intracellular localisation of enzymes, the total protoplast extract can be fractionated by standard procedures on a Percoll density gradient.

17.4.1 C_3 chloroplasts

Take an aliquot of protoplasts in sorbitol medium and centrifuge at 100 *g* for 5 minutes. Resuspend the protoplasts in the isolation, resuspension and assay medium containing 0.4 M sorbitol, 10 mM EDTA, 25 mM Tricine–KOH (pH 8.0) at a concentration of about 200 μg Chl ml^{-1}. Break the protoplasts and centrifuge at 250 *g* for 1 minutes. Gently resuspend the chloroplast pellet and store on ice. Inclusion of $NaHCO_3$ (5–10 mM) in the breakage and resuspension medium may prove beneficial [12].

17.4.2 C_4 chloroplasts

Chloroplasts from mesophyll protoplasts can be made in the same way as C_3 chloroplasts. Again, the presence of relatively high concentrifuge at 250 *g* for 1 minute. Gently resuspend the chloroplast pellet and store on

17.4.3 CAM chloroplasts

For breakage of the protoplasts a 30 μm mesh rather than a 20 μm mesh should be used. For the breaking medium use 150 mM sorbitol, 1% (w/v) PVP (soluble), 5 mM EDTA, 5 mM $Na_4P_2O_7$, 250 mM Tricine–KOH (pH 8.2). A higher buffer concentration is used than with C_3 or C_4 protoplasts owing to the high acidity of CAM tissue. Centrifuge the protoplast extract at 100–200 *g* for 2 minutes. Resuspend the chloroplast pellet in 0.33 M sorbitol, 5 mM $Na_4P_2O_7$, 50 mM Tricine–KOH (pH 7.8) [17].

17.5 MECHANICAL SEPARATION OF INTACT CHLOROPLASTS

At all steps of the isolation, glassware and solutions should be kept chilled and the entire procedure conducted as rapidly as possible. The glassware must be clean and quite free from traces of detergent.

17.5.1 Isolation of chloroplasts from spinach

(a) Grinding medium:
Any one of a number of sugars or sugar alcohols may be used as osmotica and any one of a number of buffers to maintain the pH. Some advantage is derived from the use of a slightly acid pH value and small quantities of $MgCl_2$ and EDTA are often included. However 'good' chloroplasts can be prepared in the absence of Mg^{2+} (e.g. in 0.33 M sorbitol + 10 mM pyrophosphate at pH 6.5), but the presence of Mg^{2+} together with EDTA appears to be beneficial even though the reasons remain obscure.

The following grinding medium continues to be the most useful for many purposes: 0.33 M sorbitol, 10 mM $Na_4P_2O_7$, 5 mM

$MgCl_2$, 2 mM Na isoascorbate, adjusted to pH 6.5 with HCl.

The sodium ascorbate (or isoascorbate) should be added immediately prior to pH adjustment. The pyrophosphate should be freshly prepared each day. The $MgCl_2$ is best added as an appropriate volume of a 1.0 M stock solution.

(b) Harvesting:
Most leaves yield better chloroplasts if freshly harvested, but spinach has been known to give good rates after 4 weeks of cold storage. Conversely, pea shoots deteriorate very rapidly and should always be used immediately. If leaves are brightly illuminated for 20–30 minutes prior to grinding, chloroplast yield is increased and induction shortened.

(c) Grinding:
This is a compromise between opening as many cells and as few chloroplasts as possible; 3–5 s is appropriate for a conventional Waring blender (at full speed) and 3 s or less for the Polytron blender (*Kinematica, Switzerland; Brinkmann, USA*). The latter gives a higher yield. Freshly prepared media are best used as semi-frozen slush (solutions may be stirred in a bath of chilled alcohol at about −15°C until a suitable consistency is achieved).

(d) Filtration:
The brei is usually squeezed through two layers of muslin (to remove coarse debris) and filtered through a sandwich of cotton wool between eight layers of muslin.

(e) Centrifugation:
Many variants may be used but it is important to separate the chloroplasts quickly from the supernatant. For example, if a brei of 200–300 ml is divided between four unscratched glass tubes of about 100 ml capacity in a swing-out head, adequate precipitation may be achieved by accelerating to approximately 600 *g* and returning to rest in 90 s.

(f) Resuspension:
The supernatant is decanted and the pellet surface washed (e.g. with grinding or resuspending medium which is then discarded). This removes the upper layer of the pellet in which broken chloroplasts are more abundant. Small quantities of resuspending medium (0.5 ml per tube) are added and the chloroplasts resuspended by shaking the tubes or by using a small paintbrush or glass rod wrapped with cotton wool.

If a white ring can be seen lying under the chloroplast pellet, then it is either starch or calcium oxalate. If large amounts of starch are present (this can be confirmed by staining with iodine) then the intactness is likely to be low and steps should be taken to reduce the starch content of the leaves, either by harvesting at the beginning of the photoperiod, or by reducing the light intensity or daytime temperature.

(g) Resuspension and Assay Medium:
This naturally depends on the nature of the investigation but the following, based on that of Jensen and Bassham, is useful for many purposes and (minus bicarbonate) is often used as a resuspending medium for chloroplasts of spinach and peas: 0.33 M sorbitol, 2 mM EDTA, 1 mM $MgCl_2$, 1 mM $MnCl_2$, 50 mM Hepes, 10 mM $NaHCO_3$, 5 mM $Na_4P_2O_7$ (adjusted to pH 7.6 with KOH). Note that phosphate must be included in the assay medium (see below).

17.5.2 Variations in the procedure for spinach

(a) Peas:
The grinding medium used for spinach is less useful for peas because pyrophosphate (PP_i) inhibits photosynthesis in pea chloroplasts unless added together with ADP [13]. The solution first used for the isolation of active pea chloroplasts is still as good as most for this purpose, despite the high concentration of inorganic phosphate (P_i) employed,

although a more inert buffer can be used at the outset or for washing. This comprises 0.33 M glucose, 50 mM Na_2HPO_4, 50 mM KH_2PO_4, 5 mM $MgCl_2$, 0.1% NaCl, 0.2% Na isoascorbate, 0.1% BSA, all adjusted to pH 6.5 with HCl.

(b) Other plants (including C_4 plants):
The range of plants from which chloroplasts may be isolated mechanically is quite limited. Reasonably good chloroplasts (i.e. those which fix CO_2 at high rates) have been obtained from the flag leaves of wheat by using media containing EDTA. Isolation of good chloroplasts from leaves of various C_4 plants (*Zea mays, Sorghum bicolor, Panicum miliaceum, Atriplex spongiosa*) is possible [15]. The grinding medium employed contains: 0.35 M sorbitol, 25 mM Hepes–KOH (pH 7.8), 1 mM $MgCl_2$, 5 mM EDTA, 0.2% (w/v) BSA, with 2 mM sodium isoascorbate and 5 mM dithiothreitol added just before use; the procedure is essentially that described for spinach. An improvement in intactness can be effected by centrifugation through a cushion of Percoll (Section 17.7).

With C_4 plants and the flag leaf of wheat, the resuspension and assay medium should include EDTA (e.g. 0.3 M sorbitol, 25 mM Hepes–KOH (pH 8.0), 5 mM EDTA, 0.1% (w/v) BSA for maize [15]), although resuspension and assay media containing EDTA are not suitable for the assay of intactness (Section 17.6.1), in which case extra Mg^{2+} must be added.

17.5.3 Improvement in chloroplast intactness

Media low in cations have been used in the past because they can result in a considerable improvement in intactness (but only if the intactness is not already high). However, the combination of Mes and Tris which is often employed sacrifices chloroplast stability with respect to CO_2 fixation. However, Cerovic and Plesnicar [16] have found that use of a low cation grinding and washing medium is particularly effective with peas and spinach. This comprises 340 mM sorbitol, 0.4 mM KCl, 0.04 mM EDTA and 2 mM Hepes–KOH (pH 7.8). The resuspension medium for peas and spinach contains 330 mM sorbitol, 10 mM KCl, 1 mM EDTA, 1 mM $MgCl_2$, 1 mM $MnCl_2$, 10 mM KH_2PO_4, 1% BSA and 50 mM Hepes–KOH (pH 7.9).

For many years, sucrose gradients or cushions have been used to further purify organelles such as chloroplasts. However, the great disadvantage of such methods lay in the high osmotic potential of the sucrose solutions, which dehydrated the organelles and left them incapable of carbon fixation. The introduction of silica sols such as *Percoll*, which have high densities and low osmotic potentials, provides an easy way of improving chloroplast intactness, often with a commensurate increase in the rate of CO_2 fixation. For a simple purification through a cushion of Percoll, 5 ml of 30–40% (v/v) Percoll in resuspension medium is put into a centrifuge tube and the chloroplast suspension is gently laid on top (the precise concentration of Percoll required should be determined by experiment). The tube is then centrifuged for 1–2 minutes at 2000 *g*, the supernatant is discarded and the pellet resuspended in assay medium. This procedure yields largely intact (>95%) chloroplasts.

17.6 PHOTOSYNTHESIS BY ISOLATED CHLOROPLASTS

17.6.1 Criteria of intactness

Although lack of damage to chloroplasts during isolation procedures cannot be guaranteed at present, envelope integrity can be assessed easily and quickly. Unless illuminated in mixtures containing additional ferredoxin, $NADP^+$, ADP, Mg^{2+}, etc., envelope-free chloroplasts will not show CO_2-dependent or glycerate-3-P-dependent O_2 evolution. The best test of photosynthetic function is therefore to illuminate in near-saturating red light in an appropriate assay medium. At 20°C, wheat or spinach chloroplasts which exhibit rates of CO_2-dependent O_2 evolution of less than 50 μmol h^{-1} mg^{-1} chlorophyll may be regarded as poor; rates of 50–80 are reasonable, 80–150 are good, and rates above 150 are exceptional.

Chloroplasts with intact envelopes ('Class A chloroplasts') will not carry out many functions at fast rates because of the permeability barrier afforded by the inner envelope. They will not, for example:

1. Reduce exogenous oxidants such as ferricyanide or $NADP^+$.
2. Rapidly phosphorylate exogenous ADP.
3. Support O_2 uptake with non-permeating Mehler reagents such as ferredoxin.
4. Fix CO_2 in the dark with ribulose bisphosphate or ribose 5-P + ATP as substrates.
5. Hydrolyse exogenous inorganic pyrophosphate.

All of the above functions could be used as the basis of 'intactness' assays. Possibly the simplest and certainly that which has gained most favour, is the use of ferricyanide as a Hill oxidant [14]. The reduction of this compound may be followed in the oxygen electrode provided that CO_2-dependent O_2 evolution is inhibited by glyceraldehyde. The assay probably overestimates intactness (there is evidence which suggests that rupture and loss of stromal protein may be followed by resealing) but it provides a useful basis for comparison. The response to the uncoupler is also useful. Well-coupled chloroplasts may show as much as a 14-fold increase in rate following the addition of NH_4Cl. Chloroplasts which show a four-fold response, or less, are unlikely to support carbon assimilation at high rates.

17.6.2 Intactness assay

The following chloroplast resuspension and assay medium [13]: 0.33 M sorbitol, 1 mM $MgCl_2$, 1 mM $MnCl_2$, 2 mM EDTA, 50 mM Hepes–KOH, pH 7.6, is optimal for the intactness assay of all chloroplasts, regardless of the requirements for carbon assimilation. For osmotic shock of the chloroplasts, add 0.1 ml of chloroplast suspension to 0.9 ml of water and stir for one minute. Then add 1 ml of double strength assay medium (0.66 M sorbitol, 2 mM $MgCl_2$, 2 mM $MnCl_2$, 4 mM EDTA, 100 mM Hepes–KOH, pH 7.6). For assay of the intact preparation, add 0.1 ml chloroplasts to 1.9 ml of assay medium. Chlorophyll should not exceed 100 μg ml^{-1}.

Add 10 mM D,L-glyceraldehyde to the assay (this inhibits the Calvin cycle and its associated oxygen evolution) and 3 mM potassium ferricyanide. Two or three minutes after illumination add 5 mM NH_4Cl. Measure the rate of light-dependent oxygen evolution (μmol O_2 evolved h^{-1} mg^{-1} Chl) after the addition of NH_4Cl with shocked (A) and unshocked (B) chloroplasts (Fig. 17.2). The percentage intactness of the original preparation is calculated as follows:

$$\% \text{ intactness} = \frac{A - B}{A}(100) = 100 - (100B/A)$$

17.7 CARBON ASSIMILATION BY C_3 CHLOROPLASTS

The number of different media used for the assay of chloroplasts has increased considerably with the introduction of protoplasts. Their common features are the inclusion of an osmoticum (usually 0.33 M sorbitol); Hepes or Tricine at pH 7.6–8.4; the presence of PP_i or EDTA or a combination of both (the EDTA being particularly useful with chloroplasts from protoplasts or species other than spinach and peas); and the inclusion of a saturating concentration of $NaHCO_3$ (10 mM) and orthophosphate.

Chloroplasts require P_i for carbon assimilation because their major product is triose phosphate rather than free carbohydrate. Chloroplasts isolated from different tissues

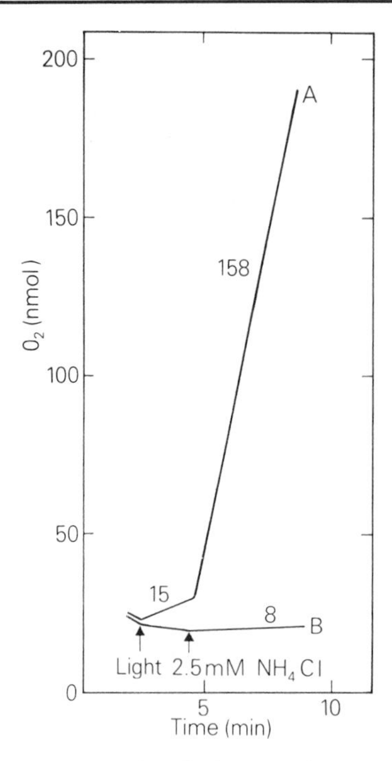

Fig. 17.2. Example of the test for intactness of chloroplasts using ferricyanide as a Hill oxidant. Shocked (A) and unshocked chloroplasts (B), isolated from protoplasts of wheat. The calculated percentage intactness is 95%. Numbers along the traces indicate rates of oxygen evolution (μmol O_2 $h^{-1}mg^{-1}$ Chl).

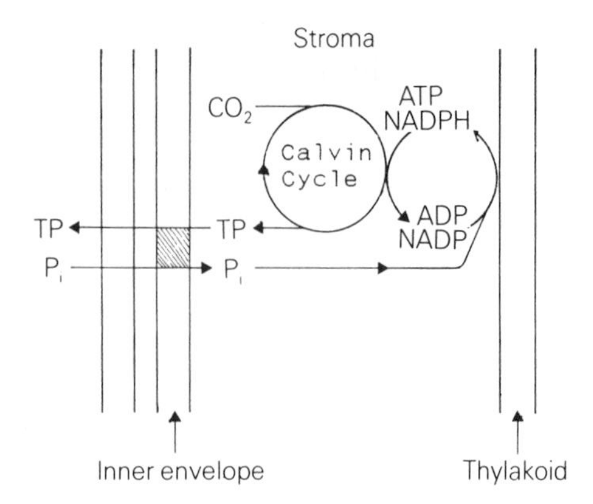

Fig. 17.3. The phosphate translocator, situated in the inner membrane of the chloroplast envelope, exchanges cytosolic phosphate for triose phosphate (TP) (comprising glyceraldehyde phosphate + dihydroxyacetone phosphate) and, to a lesser extent, for glycerate-3-P. Synthesis of sucrose etc. in the cytosol recycles P_i so that, in the steady state, there is a constant uptake of P_i and export of TP by the chloroplast.

differ markedly in their P_i optima. Whilst chloroplasts from spinach display maximum rates at 0.25–0.5 mM P_i and frequently show tolerance of high concentrations of P_i (5–10 mM), chloroplasts from wheat have an optimum around 0.2 mM P_i. In all experiments involving CO_2 fixation or CO_2-dependent O_2 evolution, the initial lag and the final rate will depend upon the balance between the levels of sugar phosphates within the chloroplast and the exogenous concentration of P_i. This dependency is a consequence of the operation of the phosphate translocator. The translocator (Fig. 17.3) brings about a rapid obligatory exchange between P_i and triose phosphate or glycerate-3-P (and to a lesser extent with some other sugar phosphates). Thus too much P_i is inhibitory to photosynthesis (e.g. Fig. 17.4b), especially during induction, because triose phosphate is forcibly exported from the chloroplast. In darkness, the level of sugar phosphates within the chloroplast is low, but following illumination a proportion of the triose phosphate and glycerate-3-P is retained within the chloroplast, and this results in an increase in the total amount of the CO_2 acceptor (ribulose bisphosphate). This allows a higher rate of photosynthesis, thus the generation of even more triose phosphate, and so on. The autocatalytic increase in total sugar phosphate reaches a ceiling imposed by other factors such as enzyme activity, etc., and is largely responsible for the induction period or lag in photosynthesis in isolated chloroplasts.

For these reasons, glycerate-3-P or triose phosphate added to the medium will overcome the initial lag (e.g. Fig. 17.4a). It is useful to examine glycerate-3-P-dependent O_2 evolution, especially with chloroplasts of suspect activity. If high rates of glycerate-

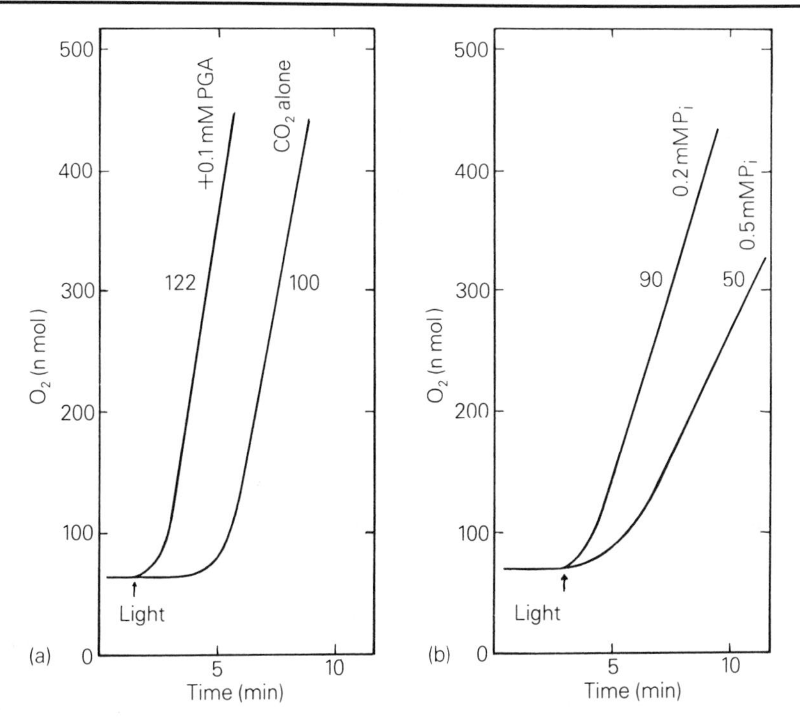

Fig. 17.4. (a) The shortening of the lag in CO_2-dependent oxygen evolution by glycerate-3-P (PGA) and (b) its extension by super-optimal P_i in wheat chloroplasts. Numbers along the traces indicate rates of oxygen evolution (μmol O_2 h^{-1} mg^{-1} Chl).

3-P-dependent O_2 evolution can be achieved, there is every reason to hope for good CO_2-dependent O_2 evolution following manipulation of P_i concentration, media, etc. 1 mM glycerate-3-P usually gives rates between 60 and 100% of the rate of CO_2-dependent O_2 evolution.

The most useful medium for mechanically-derived chloroplasts from spinach and peas is based upon that of Jensen and Bassham (Section 17.5) The role of pyrophosphate is complex, but several points are clear. Firstly, orthophosphate is less inhibitory to chloroplasts in the presence of PP_i because the latter is a competitive inhibitor of the phosphate translocator. Secondly, PP_i does not seem to cross the spinach chloroplast envelope, but in the presence of external Mg^{2+} and pyrophosphatase (released from damaged chloroplasts) is slowly hydrolysed to P_i. It can thus act as an optimal source of P_i. With spinach chloroplasts the optimal concentration of P_i is about 0.25 mM if PP_i is not employed but the optimum is so sharp that it is very difficult to achieve. Under such conditions the concentration of P_i will change rapidly during illumination unless the chlorophyll concentration is kept very low. Work with chloroplasts from young pea leaves and young wheat leaves has shown that ADP and ATP within the chloroplasts exchange with external PP_i. For this reason PP_i is inhibitory if used alone with pea or wheat chloroplasts [12,13] but it stimulates photosynthesis in the presence of small quantities (0.2–0.4 mM) of ADP or ATP.

The most significant respects in which media for chloroplasts of wheat, etc., differ from that based on Jensen and Bassham are the omission of PP_i (for the reasons outlined above) and the inclusion of relatively high concentrations (5–10 mM) of EDTA (Section 17.5). The precise function of EDTA remains unclear, but it seems that its main function may be to provide a high concentration of negative charge in the medium, and it can be replaced with varying degrees of efficiency by pyrophosphate, citrate, salts such as KCl, or choline chloride. It should also be noted that wheat chloroplasts tend to have a more alkaline pH optimum for photosynthesis in such media (pH $\geqslant$ 8.0) than do spinach chloroplasts.

17.7.1 Measurement of chloroplast photosynthesis

Light-dependent O_2 evolution and $^{14}CO_2$ fixation may be measured using the above media. Chloroplasts are normally used at a concentration of 50–100 μg Chl ml^{-1} in a volume of 1–2 ml. Above this concentration the light may not be saturating. A 150 W quartz-iodine lamp (in a slide projector) together with a round-bottomed flask of water (which focuses the light and absorbs heat), a red filter (e.g. ICI red perspex 400) and a *Calflex C* heat filter give a quantum flux of about 1500 μmol m^{-2} s^{-1} (for details of the oxygen electrode apparatus see Chapter 10). The temperature employed for C_3 chloroplasts is normally 20°C although it may be increased (up to 35°C) for material from C_4 plants.

It is usually beneficial to add catalase to the assay medium (100 units ml^{-1}). This prevents H_2O_2 accumulation (from pseudocyclic electron flow) which inhibits chloroplast photosynthesis.

(a) C_3 chloroplasts:
To determine the phosphate optimum, do separate assays with varying concentrations of P_i (e.g. 0, 0.1, 0.3, 0.5, 1.0, 5.0 and 10.0 mM). Other assays which can be attempted are:

1. 10 mM $NaHCO_3$ 0.3 mM P_i (or optimal P_i);
2. 10 mM $NaHCO_3$, 0.3 mM P_i, 10 mM D,L-glyceraldehyde;
3. 1 mM glycerate-3-P, 10 mM $NaHCO_3$;
4. 5 mM oxaloacetate;
5. 1 mM ribose 5-phosphate, 10 mM $NaHCO_3$;
6. 1 mM fructose 6-phosphate, 10 mM $NaHCO_3$.

Note that the chloroplast is not solely concerned with carbon fixation, and that it will reduce compounds involved in nitrogen metabolism (Chapters 19, 20) and will metabolise many other compounds. As an example, oxaloacetate enters the chloroplast on the dicarboxylate translocator and is reduced to malate by NADP-malate dehydrogenase. This can be seen by adding 5 mM oxaloacetate (freshly made solution, pH 7) to a suspension of illuminated chloroplasts in the absence of bicarbonate. After 3 or 4 minutes, add 5 mM NH_4Cl to uncouple the chloroplasts.

(b) C_4 mesophyll chloroplasts:
Assay the chloroplasts for oxygen evolution in a medium [3] containing 0.33 M sorbitol, 5 mM EDTA, 25 mM Tricine (pH 8.0), 0.3 mM P_i, and 1100 units ml^{-1} catalase using an assay volume of 1–2 ml, with about 50 μg Chl ml^{-1}. Compare the following substrates:

1. 2 mM glycerate-3-P;
2. 0.5 mM oxaloacetate + 5 mM pyruvate after 3 minutes of illumination;
3. 0.5 mM oxaloacetate + 5 mM NH_4Cl after 3 minutes of illumination;
4. 10 mM $NaHCO_3$;
5. 5 mM pyruvate.

17.8 MATERIALS

Cellulase from *Trichoderma viride*:
Available as Onozuka RS or R10 (Yakult Biochemical Co. Ltd., 8–12 Shingikancho,

Nishinomiya, Japan) or as Cellulysin (Calbiochem, Ltd., C.P. Laboratories Ltd., P.O. Box 22, Bishops Stortford, Herts CM23 3AL, UK or P.O. Box 12087, San Diego, CA 92112, USA). Onozuka RS is about twice the price of R10, but has a higher specific activity and dissolves cell walls from a wider range of material [5].

Pectinase from *Rhizopus sp.*:
Macerozyme R10 (Yakult Biochemical Co. Ltd., (see above)) or Macerase (Calbiochem Ltd. (see above)).

Pectinase from *Aspergillus sp.*:
As Rohament P (Rohm GmbH Chemische Fabrik, D-1600 Darmstadt, Kirschenalle, Postfach 4242, Germany) or as Pectolyase Y-23 (Seishin Pharmaceutical Co. Ltd., 9-50-1, Nagareyama, Nagareyama-shi, Chiba-ken, Japan).

Nylon mesh:
(i) Henry Simon Ltd., P.O. Box 31, Stockport, Cheshire, SK3 ORT, UK; (ii) Northern Mesh, Bleasdale Street, Royton, Oldham OL2 6PZ, UK; (iii) Tetko Inc., Precision Woven Screening Media, 420, Saw Mill River Road, Elmsford, NY 10523, USA.
Nybolt P20 is obtained from Schweizerische Seidengazefabrik, Zürich, Switzerland.

REFERENCES AND FURTHER READING

1. Robinson, S.P. and D.A. Walker (1979) Rapid separation of the chloroplast and cytoplasmic fractions from intact leaf protoplasts. *Arch. Biochem. Biophys.* **196**, 319–323.
2. Nagata, T. and S. Ishii (1979) A rapid method for isolation of mesophyll protoplasts. *Canadian J. Bot.* **57**, 1820–1823.
3. Day, D.A., C.L.D. Jenkins and M.D. Hatch (1981) Isolation and properties of functional mesophyll protoplasts and chloroplasts from *Zea mays*. *Australian J. Plant Physiol.* **8**, 21.
4. Edwards, G.E., R.McC. Lilley and M.D. Hatch (1979) Isolation of intact and functional chloroplasts from mesophyll and bundles sheath protoplasts of the C_4 plant *Panicum miliaceum*. *Plant Physiol.* **63**, 821.
5. Moore, B.D., M.S.B. Ku, and G.E. Edwards (1984) Isolation of leaf bundle sheath protoplasts from C_4 dicot species and intracellular localization of selected enzymes. *Plant Sci. Lett.* **35**, 127–138.
6. Boag, S. and C.L.D. Jenkins (1985) CO_2 assimilation and malate decarboxylation by isolated bundle sheath chloroplasts from *Zea mays*. *Plant Physiol.* **79**, 165–170.
7. Jensen, R.G. (1979) The isolation of intact leaf cells, protoplasts and chloroplasts. In: *Encyclopedia of Plant Physiology* Photosynthesis II: Photosynthetic carbon metabolism and related processes (M. Gibbs and E. Latzko, eds.) Springer-Verlag, Berlin. pp. 31–40.
8. Huber, S.C. and G.E. Edwards (1975) C_4 photosynthesis. Light dependent CO_2 fixation by mesophyll cells, protoplasts, and protoplast extracts of *Digitaria sanguinalis*. *Plant Physiol.* **55**, 835–844.
9. Sharkey, T.D. and K. Raschke (1980) Effects of phaseic acid and dihydrophaseic acid on stomata and photosynthetic apparatus. *Plant Physiol.* **65**, 291–297.
10. Nishimura, M., I. Hara-Nishimura and S.P. Robinson (1984) Isolation of metabolically competent protoplasts from grapevine leaves. *Plant Sci. Lett.* **37**, 171–175.
11. Chapman, K.S.R., J.A. Berry and M.D. Hatch (1980) Photosynthetic metabolism in bundle sheath cells of the C_4 species *Zea mays*: Sources of ATP and NADPH and the contribution of Photosystem II. *Arch. Biochem. Biophys.* **202**, 330–341.
12. Edwards, G.E., S.P. Robinson, N.J.C. Tyler and D.A. Walker (1978) Photosynthesis by isolated protoplasts, protoplast extracts, and chloroplasts of wheat. Influence of orthophosphate, pyrophosphate, and adenylates. *Plant Physiol.* **62**, 313–317.
13. Robinson, S.P. and J.T. Wiskich (1977) Pyrophosphate inhibition of carbon dioxide fixation in isolated pea chloroplasts by uptake in exchange for endogenous adenine nucleotides. *Plant Physiol.* **59**, 422–427.
14. Lilley, R. McC., M.P. Fitzgerald, K.G. Rienits, and D.A. Walker (1975) Criteria of intactness and the photosynthetic activity of spinach chloroplast preparations. *New Phytol.* **75**, 1–10.
15. Jenkins, C.L.D. and V.J. Russ (1984) Large scale, rapid preparation of functional mesophyll chloroplasts from *Zea mays* and other C_4 species. *Plant Science Lett.* **35**, 19–24.
16. Cerovic, Z.G. and M. Plesnicar (1984) An

improved procedure for the isolation of intact chloroplasts of high photosynthetic capacity. *Biochem. J.* **223**, 543–545.

17. Spalding, M.H. and G.E. Edwards (1980) Photosynthesis in isolated chloroplasts of the Crassulacean acid metabolism plant *Sedum praealtum*. *Plant Physiol.* **65**, 1044–1048.
18. Robinson, S.P., Z.G. Cerovic and D.A. Walker (1986) Isolation of intact chloroplasts – general principles and criteria of integrity. *Methods Enzymol.* **148**, 145–157.

18

Thylakoid components and processes

G. Hind

18.1 INTRODUCTION

Bioproductivity is powered by solar energy conversion within the pigment–protein complexes of the thylakoid membrane. If these complexes are somehow modified, changes in cell physiology, growth and biomass yield are inevitable. It must be emphasised that effects far removed in time and space from the photochemical reactions are frequently traced to damage or a defect at the thylakoid level. Normal functioning of thylakoid processes should thus be established as a first step in seeking the root cause of a change in plant growth perceived in a field setting. Pathological conditions brought on by infection, stress or herbicide, have the potential to promote photobleaching of pigments, so techniques for monitoring pigment content and composition are discussed here, along with procedures for the analysis of individual pigment–protein complexes.

In the previous chapter, Section 17.1 outlines the importance of basic research with isolated chloroplasts and thylakoids towards understanding regulatory processes in photosynthesis. Plants such as pea and spinach are popular for basic studies since direct isolation of active thylakoids from the leaves is possible. However, these are exceptional cases, and Chapter 17 should be consulted for the route towards thylakoid preparation from protoplasts and intact chloroplasts.

The function of the photosynthetic electron transfer pathway is to convert light into chemical energy, which is stored as reductant in the form of NADPH and as phosphorylation potential in the form of ATP. The sequential operation of two photosystems is required to remove electrons from water and accomplish the reduction of $NADP^+$ to NADPH. The associated evolution of oxygen provides a convenient measure of the electron transport rate. In intact chloroplasts, electron transport is normally limited by the rate at which CO_2 fixation regenerates $NADP^+$ and ADP; thus, this CO_2-dependent O_2 evolution rate is also a measure of the rate of carbon assimilation.

The link between electron transport and phosphorylation is a gradient of pH across the thylakoid membrane. This is a sealed vesicle or 'bag' of lipid enclosing an aqueous phase which is chemically and electrically isolated from the stroma (in an intact chloroplast) or the suspension medium. The pH difference arises when H^+ ions (*protons*) are deposited in the inner phase of the membrane by (i) the oxidation of water, and (ii) the oxidation of reduced plastoquinone by Photosystem I (PS I) and the cytochrome *b/f* complex. Plastoquinone is, in turn, reduced

Photosynthesis and Production in a Changing Environment: a field and laboratory manual
Edited by D.O. Hall, J.M.O. Scurlock, H.R. Bolhàr-Nordenkampf, R.C. Leegood and S.P. Long
Published in 1993 by Chapman & Hall, London.
ISBN 0 412 42900 4 (HB) and 0 412 42910 1 (PB).

by Photosystem II (PS II). Plastoquinone is a small, lipid-soluble molecule that can move laterally in the membrane to serve its function as a mediator of electron flow between the much larger, and relatively immobile, PS II and cytochrome complexes.

Protons move *out* of the thylakoid via the coupling factor which consists of a proton channel (CF_0) spanning the membrane and a reversible ATP synthase (CF_1) that protrudes from the outer face of the thylakoid into the stroma. The ATP synthase activity is optimal around pH 8.2, where the condensation of ADP and orthophosphate (P_i) requires removal of a hydroxyl ion (OH^-). Somehow, CF_1 uses the proton gradient to drive the following process:

$$ADP^{3-} + P_i^{2-} = ATP^{4-} + OH^- \quad (pH > 8, + Mg^{2+})$$

When ADP is not available to CF_1, the proton gradient cannot be used and its back-pressure inhibits electron transport.

The inward proton movement driven by electron transport is linked to efflux of Mg^{2+} ions from the thylakoid. In the light, the stroma experiences a shift to more alkaline pH and an increase in Mg^{2+} activity, which in turn bring about activation of key enzymes in the carbon-reduction pathway. Through these and other processes, the soluble enzymes for CO_2 fixation are regulated by electron transport in the thylakoid membrane.

When the chloroplast outer envelope is ruptured by osmotic shock, the stroma contents are lost and CO_2 fixation cannot occur (unless the stroma is experimentally reconstituted). Most early studies of electron transport and photophosphorylation were made with such preparations of broken chloroplasts. Removal of the permeability barrier imposed by the chloroplast envelope makes possible the use of a wide range of unnatural terminal oxidants and reductants, and the addition of substrate amounts of ADP and phosphate; it is consequently possible to study partial reactions of photophosphorylation and the activity of isolated segments of the electron transport chain. The study of partial reactions in broken chloroplasts can yield valuable information regarding the following:

(i) The fundamental mechanisms of proton and electron transport.
(ii) The site of action of herbicides and injurious treatments.
(iii) The site of lesions resulting from mutation. In higher plants, these are usually lethal to seedlings, but may be studied readily in mutants of algae or *Lemna* growing under heterotrophic conditions.
(iv) The relative activities of the photosystems in a specialized tissue (e.g. bundle-sheath cells of the leaf, heterocysts of cyanobacteria, guard cells of the leaf epidermis) or chloroplast thylakoid membrane fractions.
(v) The processes of de-etiolation (as plants 'green-up') and normal chloroplast development.

In (iii)–(v), conclusions may be checked by studies of thylakoid membrane complexes and polypeptides, using the techniques of native and SDS gel electrophoresis (Section 18.5).

The methods described below for monitoring partial reactions require only simple and inexpensive equipment and reagents.

18.2 MEASUREMENT OF PROTON FLUX, PHOTOPHOSPHORYLATION AND ELECTRON TRANSPORT USING BROKEN CHLOROPLASTS AND A pH ELECTRODE

The coupling status of the thylakoid membrane can profoundly influence the rate of electron transport reactions, as discussed in Chapter 17 for the reduction of ferricyanide by osmotically shocked chloroplasts (see Fig. 17.2). As a precautionary measure, inhibitors of electron transport are studied in fully uncoupled membranes, lest they also possess some uncoupling activity. Uncouplers act

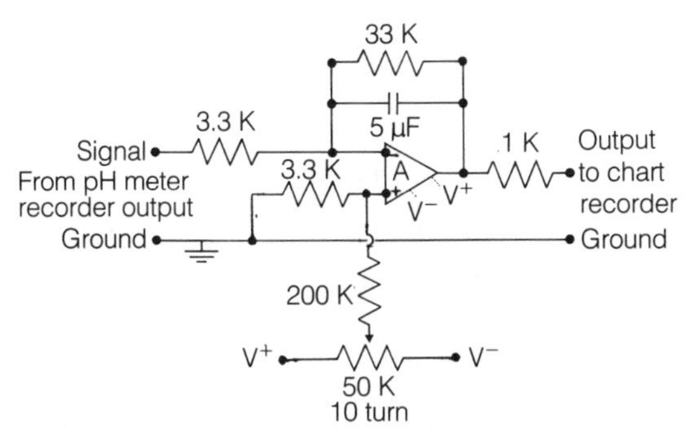

Fig. 18.1. pH amplifier and offset circuit. A is any cheap operational amplifier; +V and −V are ±9–15 V D.C. from a solid-state power supply or from two MN1604 (6LR61) alkaline batteries connected in series and centre-tapped to ground.

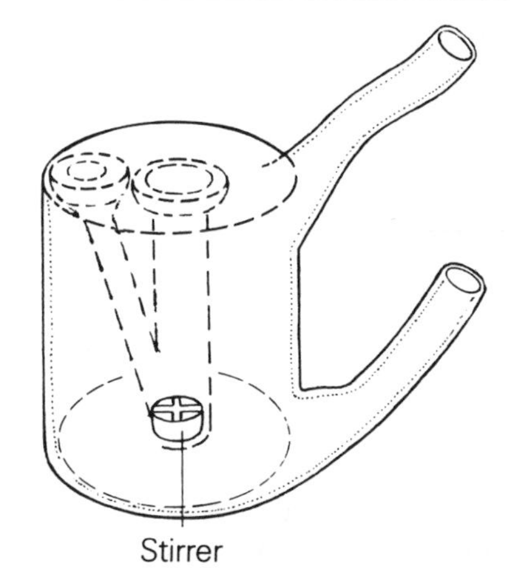

Fig. 18.2. A glass reaction vessel for measurement of proton flux using a pH electrode.

by promoting the useless escape of protons from the thylakoid interior, with consequent partial or complete loss of ability to sustain a transmembrane pH gradient and thus drive photophosphorylation. The accelerating effect on electron transport of a fully uncoupling concentration of, for example, NH_4Cl provides a reference rate against which the uncoupling activity of an unknown may be judged.

In principle, the effect of uncouplers on H^+ flux can also be observed directly, by measuring the extent and kinetics of pH changes in a thylakoid suspension that is subjected to cycles of illumination and darkness. However, these light-induced pH changes are difficult to detect in the suspending medium, owing to its large volume relative to that of the thylakoids. Laboratory pH meters do not provide sufficient amplification and offset to present such small pH responses on a chart recorder. This problem can be overcome using the amplifier circuit shown in Fig. 18.1, which is cheap to construct.

A stirred, thermostatically controlled reaction vessel that can be made from glass is shown in Fig. 18.2. It is designed to accommodate a magnetic follower (stirrer), a thin combination pH electrode and about 2 ml of sample. The narrow side arm is used for making additions and cleaning the chamber, without removing the electrode. A suction line connected to a length of thin polyethylene tubing facilitates cleaning.

This equipment is useful not only in monitoring transmembrane proton fluxes, but also in measuring the rate of photophosphorylation and electron transport. Its virtues are cheapness and availability. If appropriate laboratory facilities are available, phosphorylation can be measured under a wider range of conditions using the radioisotope ^{32}P [1]. The reduction of terminal electron acceptors is also conveniently studied using a spectrophotometer equipped with a suitable light source and complementary blocking filters.

18.2.1 Measurement of proton flux

Intact chloroplasts are freshly shocked in the reaction vessel: add 0.9 ml 10 mM $MgCl_2$ in 1 mM HEPES buffer (pH 6.5), and while slowly stirring, add 0.1 ml chloroplasts (containing about 0.1 mg chlorophyll). After 2 min, add 1.0 ml 0.66 M sorbitol and adjust the pH to 6.5 ± 0.1 with 10 mN NaOH or HCl. An artificial electron acceptor should be provided, such as 50 μM pyocyanine or 20 μM phenazine methosulphate. Illuminate with strong light from a slide projector, focused by means of a water-filled round-bottom flask (250 ml) and filtered through orange or red Perspex (white light will introduce an artefact). Observe the effect of uncouplers and unknowns on the extent of the response and its rate of relaxation in the dark (expressed as a half-time). Compare the effects of addition (i) before illumination, (ii) in the illuminated steady state, and (iii) immediately after turning off the light. Calibrate the proton uptake by adding small volumes of standard NaOH to match the light-induced alkalization within ±20%. Express the extent as microequivalents H^+ taken up per micromole chlorophyll (note that 1 μmol chlorophyll is equivalent to about 0.9 mg.)

An uncoupler will speed the collapse of the response, whether added in dark or light and lower the extent of the pH shift in the light.

18.2.2 Measurement of photophosphorylation

Proceed as above, but after osmotic rupture in unbuffered 10 mM $MgCl_2$, add 1 ml 0.66 M sorbitol, 20 mM KCl, 1 mM Tris, 2 mM Na phosphate (pH 8.2); then add 10 μl 0.1 M ADP (pH 6.7–7.0). After adjusting the pH to 8.2 with 50 mM NaOH, measure the linear rate of alkalisation in the light. Calibrate by back-titration with standard HCl and calculate the ATP formed (μmol ATP mg Chl^{-1} h^{-1}).

Uncouplers which might be tested in the above systems are final concentrations of 2 mM NH_4Cl or 1 μM carbonyl cyanide *m*-chlorophenylhydrazone (CCCP).

Inhibition of phosphorylation, though most commonly due to uncoupling, arises in rare instances from interference with the operation of coupling factor. Substances having this activity are called energy transfer inhibitors; they inhibit coupled phosphorylation, electron flow and the ATPase activity of CF_1. Quercetin (40 μM) is an interesting example of this class [2]. In alkaline suspensions, it may increase the extent of the light-induced proton gradient, by blocking the normal pathway for its utilization.

18.2.3 Measurement of electron transport

The reduction of electron acceptors which support O_2 evolution, such as ferricyanide or $NADP^+$, leads to acidification of the suspending medium:

$$NADP^+ + H_2O \rightarrow NADPH + H^+ + \tfrac{1}{2}O_2$$

In the above case, one H^+ is released for every two electrons transported. If ferricyanide is reduced, the stoichiometry is given by $H^+:e^- = 1.0$, as described here:

$$2Fe(CN)_6^{3-} + H_2O \rightarrow 2Fe(CN)_6^{4-} + 2H^+ + \tfrac{1}{2}O_2$$

Figure 18.3 shows how the transmembrane proton uptake and this *scalar* proton release can be kinetically distinguished and their ratio obtained [3]. The resulting value (2.0), for protons taken up per electron transported via two photosystems, is of fundamental importance in our understanding of the coupling mechanism and the efficiency of photosynthesis.

18.3 PARTIAL ELECTRON TRANSPORT REACTIONS ASSAYED WITH THE OXYGEN ELECTRODE AND A CONVENTIONAL RECORDING SPECTROPHOTOMETER

The reactions described below are illustrated in Fig. 18.4. Intact chloroplasts are freshly shocked in the electrode vessel by dilution in 50 mM Tricine–KOH, 50 mM KCl, 5 mM $MgCl_2$ (pH 7.6) to a final chlorophyll concentration of 20–50 μg ml^{-1}. Additions are given below. Other media may be substituted provided that Mn is not included. Electron transport reactions catalysed by methyl viologen may be of indeterminate stoichiometry; consult Allen and Hall [4] on this complex topic. See Chapter 10 for details of the oxygen electrode.

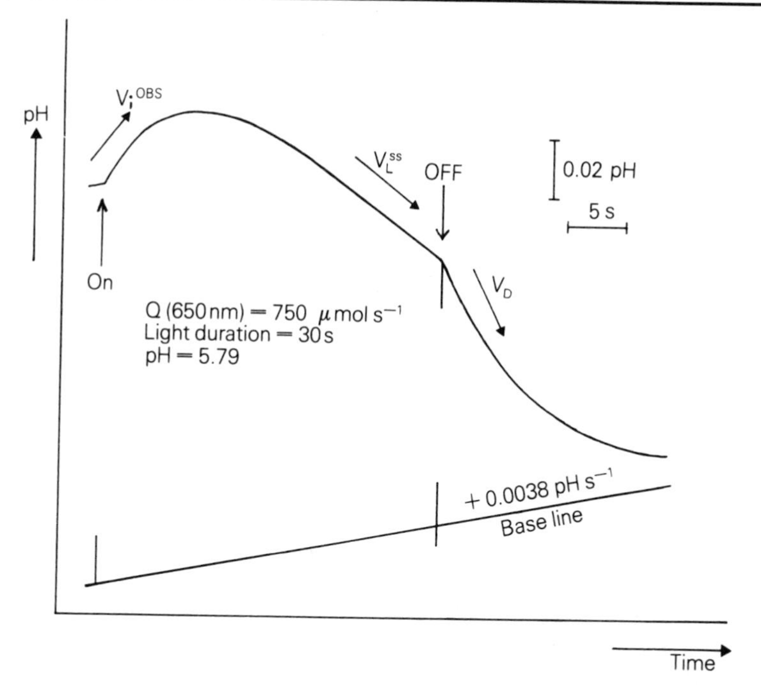

Fig. 18.3. Recorder trace obtained by M. Schwartz [3] showing light-induced pH changes in a suspension of spinach thylakoids. After turning on the light, the observed inward proton flux (V_i^{OBS}) gradually saturates within 20 s; this phase is suppressed or absent in the presence of an uncoupler. The steady decrease in pH that follows (V_L^{SS}) results from the reduction of ferricyanide and is a direct measurement of its rate (see text). The coupling efficiency (H^+/e^-) is given by V_D/V_L^{SS}, where V_D is the rate immediately after turning off the light. In this example, $H^+/e^- = 2.0$.

Figure 18.4 shows diagrammatically the following partial reactions:

1. Water to methyl viologen:
Activity assayed: whole chain electron transport excluding ferredoxin and FNR. The reaction medium also contains 50 μM methyl viologen (or flavin mononucleotide), 5 mM NH_4Cl and 2 mM sodium azide. The end product is H_2O_2; the stoichiometry is four electrons transported per O_2 consumed.

2. Dichlorophenolindophenol (DCPIP) to methyl viologen:
Activity assayed: PS I, including plastocyanin. The reaction medium also contains 50 μM methyl viologen, 5 mM NH_4Cl, 2 mM sodium ascorbate, 2 mM sodium azide, 50 μM DCPIP and 5 μM DCMU. One electron is transported per O_2 consumed.

3. Water to ***p****-phenylenediamine:*
Activity assayed: PS II, including the DCMU-sensitive site. Additions to the reaction medium are 5 mM NH_4Cl, 4 mM potassium ferricyanide and 1 mM p-phenylenediamine. Four electrons are transferred per O_2 evolved.

4. Water to silicomolybdate:
Activity assayed: PS II, excluding DCMU-sensitive site. The Tricine in the stock reaction medium should be replaced with 50 mM HEPES–KOH, pH 7.0; also added are 0.5 mM potassium ferricyanide, 0.1 mM silicomolybdic acid (*Pfaltz and Bauer, 375*

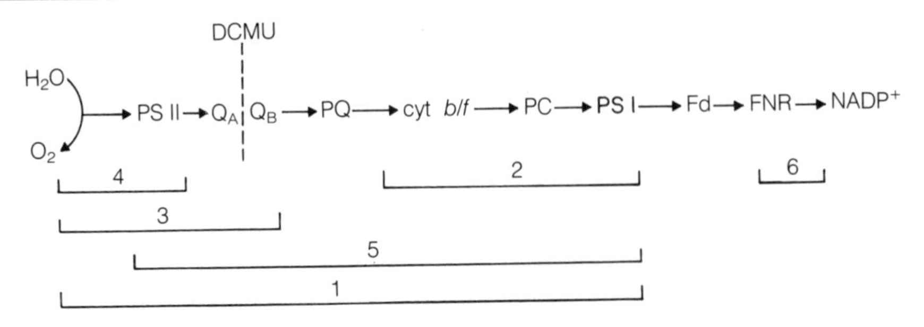

Fig. 18.4. Partial electron transport reactions described in the text. Q_A, Q_B, primary electron acceptors; PQ, plastoquinone; PC, plastocyanin; Fd, ferredoxin.

Fairfield Ave., Stamford, CT 06902, USA) and 5 µM DCMU. Four electrons are transferred per O_2 evolved.

5. *Diphenylcarbazide (DPC) to methyl viologen:*
Activity assayed: PS I and II, excluding water-splitting complex. The normal pH 7.6 reaction medium is used, supplemented with 5 mM NH_4Cl, 0.5 mM DPC, 2 mM sodium azide and 50 µM methyl viologen. DPC is prepared as a 0.1 M stock solution in dimethylsulphoxide. Electron flow from water splitting is inhibited by incubation of the chloroplasts for 2 minutes at 50°C. One electron is transported per O_2 consumed.

6. *Assay for FNR using a recording spectrophotometer:*
Activity assayed: FNR diaphorase, independent of ferredoxin. The reaction buffer contains 50 mM Tris, 100 µM potassium ferricyanide, adjusted to pH 9.0 with NaOH; 2 ml are loaded into a spectrophotometer cuvette followed by 50 µl of sample (equivalent to approx. 50 µg chlorophyll). The wavelength is set at 420 nm. A baseline is registered, then the reaction started by addition of 20 µl 0.1 M NADPH (dissolved in 0.1 M Tricine, pH 8.0). Scaling down these proportions to conserve NADPH is possible by use of narrow cuvettes. The extinction coefficient (*E*) of ferricyanide is 1.0 $(mM.cm)^{-1}$. The pH used in this assay gives high rates that are not influenced by binding of FNR to the thylakoid membrane.

18.4 ANALYSIS AND QUANTIFICATION OF THYLAKOID PIGMENT COMPLEXES

The absorption spectrum of the thylakoid is dominated, in the visible region, by the combined contributions of the light-harvesting components: chlorophylls, carotenoids and other accessory pigments. This spectrum gives only a crude representation of the component pigments owing, for the most part, to distortion caused by scattering of the measuring beam, as well as the considerable overlap between absorption bands. Furthermore, the chromophores are complexed with specific polypeptides that determine their function and location in the membrane. Chlorophyll *a*, for example, may be found in over a dozen different protein complexes in a higher plant thylakoid. Analysis of such complexes may involve (i) determination of the macrocomplex (photosystem) in which the pigment–protein functions, (ii) isolation of the pigmented complex with minimal loss of associated pigments, and (iii) analysis of the bound pigments. Non-ionic detergents have proved most effective in resolving the photosystems and their component complexes, while minimizing pigment release [5–7]. One- or two-dimensional electrophoresis of detergent extracts is used to discriminate complexes on the basis of molecular mass and charge.

Analyses of free pigments are preferably done on appropriate extracts (see Section 16.7) prepared with due regard to protection from light and oxidative destruction. Accessory pigments (e.g. carotenoids, xanthophylls, phycobilins) are best resolved and quantified by high pressure liquid chromatography [8–10]. Column, paper or thin layer chromatography are used for qualitative surveys and with care can be made quantitative [10–12].

18.4.1 Electron transport components

The concentration of reaction centres and electron carriers is commonly about 0.1–0.4% of the total chlorophyll in higher plants. Plastoquinone is unusual in being the most abundant carrier in the electron transport chain (6 mol/P700 [13]) and not attached to a protein. Along with phylloquinone, it can be assayed by direct extraction [14,15] though with both quinones, an inactive pool is present. The active plastoquinone pool can be estimated indirectly from dye reduction [13] or fluorescence (Chapter 12).

A sensitive spectrophotometer, preferably capable of operating in dual wavelength mode, is required for quantitative assay of most other electron carrying components, along with an appropriate protocol for selectively oxidizing and reducing the component of interest. One popular instrument is the *SLM DW* series UV-VIS spectrophotometer (*SLM Instruments Inc., 810 W. Anthony Drive, Urbana, IL 61801, USA*) which can operate either in split beam or dual wavelength mode and has adapters for actinic illumination and for low temperature spectroscopy.

Only in a few instances is the kinetic turnover of a membrane component slow enough to be monitored in a conventional spectrophotometer. Special purpose instruments with flash excitation are widely used in kinetic studies, but the commercial development of a portable kinetic spectrophotometer for field use is still awaited. Kramer and Crofts [16] have described a portable double-flash kinetic spectrophotometer with adequate sensitivity to measure the redox reactions of cytochromes *b* and *f* in intact leaves. This instrument has been successfully used in a field study of the thioredoxin-modulated activation of coupling factor (ATP synthase) by the proton-motive force [17]. A commercially available version of this instrument is now much needed for both field and laboratory research.

A modified version of the PAM fluorimeter (*H. Walz, Effeltrich, Germany*), developed specifically for monitoring P700, is now available. P700 is the specialized chlorophyll *a* molecule forming the PS I reaction centre that bleaches upon oxidation. The corresponding decrease in absorbance at 700 nm, or accompanying increase above 800 nm, offer a convenient measure of reaction centre (P700) concentration. P700 content is a valuable benchmark in quantification of major and minor thylakoid components. It is preferred to total chlorophyll as a reference point since this latter includes accessory pigment whose abundance can vary widely with growth conditions, genetic background, developmental stage, etc. The stoichiometry of P700 to PS II reaction centres usually lies within the range 0.5–1.0 [18]. Three methods of P700 determination are in general use: chemical or photochemical oxidation with detection at 700 nm, and photochemical oxidation with spectrophotometric detection at 800–840 nm. The former approaches are well documented [19] and apply to isolated thylakoids or subfractions thereof. In contrast, the ED 800 T probe of the PAM fluorometer detects absorbance or reflectance change at 830 nm, and can monitor P700 turnover in leaves [20].

PS II reaction centres (P680) can be quantified in thylakoids or intact leaves by measuring the flash-induced oxygen yield [21]. Growth in high light tends to increase the P680 content relative to P700, though the PS II antenna size is correspondingly smaller.

18.5 POLYACRYLAMIDE GEL ELECTROPHORESIS (PAGE)

PAGE has become a routine procedure for protein analysis, and a detailed account of PAGE methodology for chloroplast proteins is available [22]. A principal virtue of this method is that protein components are directly visualized. Vertical slab gels, in which the polyacrylamide is supported as a sandwich between glass plates, are the most popular form. The standard size gel is approx. 18 × 16 cm and 0.7–3.0 mm in thickness; up to 15 samples can be loaded on this size of gel. Important variations in physical design include (i) extra-length (height) gels, used to increase resolution; (ii) extra-width gels for comparison between large numbers of samples (e.g. fractions from chromatographic elution of a column) or for isolation of individual polypeptides, in which case the sample is applied as a continuous band across the width of the gel; (iii) 'mini-gels', approx. 8 × 8 cm, useful when detailed resolution is less important than a speedy analysis, e.g. when verifying the effectiveness of an ongoing protein purification procedure.

Electrophoresis on polyacrylamide gels can be performed under conditions which maintain the native form of soluble proteins; enzymes will retain some, if not all activity, and pigment–protein complexes will retain colour (chlorophyll–protein complexes appear as distinct greenish bands, for instance). Relative migration of proteins on non-denaturing gels is a function of molecular mass and charge.

Most usually, the sample is denatured with the detergent sodium dodecylsulphate (SDS), added before and during electrophoresis. SDS binds to protein and its strong negative charge overcomes interactions between component polypeptide chains, causing them to unfold. The sample may also be pretreated with mercaptoethanol to rupture covalent — S — S —) bonds within and between polypeptides, and may be heated to encourage unfolding. The unfolded, negatively-charged chains migrate in an electric field at a rate which decreases as a logarithmic function of their molecular weight. After staining with a dye, an SDS gel reveals polypeptide components in the original sample as a series of bands that are in order of molecular mass and have stain intensities indicative of the relative concentration of individual components.

Because many enzymes or pigment-proteins are oligomeric (made up of more than one polypeptide chain), the pattern of polypeptide bands on denaturing and non-denaturing gels is usually very different. These approaches yield complementary information, so it is advisable to analyse an unknown under both denaturing and non-denaturing conditions, especially if pigments are present.

Results from PAGE are usually presented or published directly by photographing the gels. In consequence, the placement of samples on the gel lanes requires forethought, so that appropriate experimental controls and molecular weight standards adjoin the samples of interest. Sharp, straight bands are also desirable, but are not always obtained. Bands will be indistinct if too much protein is loaded in the sample well and will curve upwards at the edges if the gel overheats through electrophoresis at excessive currents. A power supply that delivers constant current or power to the gel is advantageous, as are apparatus designs that circulate water over the glass plates holding the gel slab. Non-denaturing gels should be electrophoresed in a cold room to minimize heat-denaturation. Denaturing gels can also be run in the cold if lithium dodecylsulphate replaces SDS, which precipitates at low temperatures. Sharper resolution of polypeptides smaller than 25,000 daltons (25 kDa), and more even spacing of bands, may be achieved using gels in which the polyacrylamide forms a concentration gradient from high (e.g. 18%) at the bottom to low (e.g. 10%) at the top of the gel.

18.5.1 Preparation of solutions

Rubber gloves should be worn while preparing and handling solutions. Ensure that ventilation of the work area is adequate. Acrylamide monomer is very toxic. Replace 10% mercaptoethanol with 0.1 M dithiothreitol, if available, to minimize risk of allergic sensitization to thiols.

Acrylamide/Bis stock:
Dissolve 300 g acrylamide completely in less than 1.0 litre of water, add 4.0 g Bis (N,N′-methylenebisacrylamide), adjust to 1.0 litre, decolorize with 0.5 g activated charcoal, filter, store in refrigerator.

3.0 M Tris–Cl, pH 8.8:
Dissolve 36.3 g Tris in about 60 ml water, adjust to pH 8.8 with HCl, make up to 100 ml.

0.5 M Tris–Cl, pH 6.7:
Dissolve 3 g Tris in about 30 ml of water, adjust to pH 6.7 with HCl, make up to 50 ml.

Other solutions needed:
10% w/v SDS; 0.2 M Na_2EDTA; 10% w/v ammonium persulphate (freshly made); 50% w/v glycerol.

Resolving gel, sufficient for three mini-gels:

acrylamide/Bis	15.0 ml*
3 M Tris-Cl, pH 8.8	3.8 ml
10% SDS	0.3 ml
50% glycerol	6.0 ml
0.2 M EDTA	0.3 ml
water	4.3 ml

Evacuate and agitate in small Buchner flask, to remove air. Immediately before use, gently mix in:

TEMED**	30 µl
10% ammonium persulphate	0.3 ml

*This gives a 15% acrylamide gel. Adjust acrylamide/Bis and water proportions to give other percentage gels.
** TEMED (N,N,N′,N′-tetramethylethylenediamine) is an accelerator; slower polymerization occurs in its absence.

Stacking gel:

acrylamide/Bis	3.0 ml
0.5 M Tris-Cl, pH 6.7	4.5 ml
10% SDS	0.2 ml
0.2 M EDTA	0.1 ml
water	10.0 ml

Immediately before use add:

TEMED	20 µl
10% ammonium persulphate	0.2 ml

Running buffer, 10 × concentrated stock:

Tris	30 g
glycine	144 g
10% SDS	100 ml
0.2 M EDTA	100 ml

Make up to 1.0 litre with water. The pH should be 8.6.

Destain solution:
Methanol : glacial acetic acid : water (40 : 7 : 53 by volume).

Stain solution:
Dissolve 0.25% w/v Coomassie Brilliant Blue R250 in methanol : glacial acetic acid : water (50 : 7 : 43 by volume), then filter.

18.5.2 Procedure for making slabs

Assemble the glass sandwich according to the manufacturer's instructions. If these are unavailable, take a notched glass plate and border the straight sides with three plastic spacer strips that have been coated on both sides with a continuous bead of petroleum jelly. Complete the sandwich with an unnotched plate and hold together with clamps or binder clips spaced along the three unnotched edges. Hold upright, fill to the notch with water and examine for leaks (particularly at the bottom corners). Decant the water and dry inside the sandwich with filter paper strips. Measure the distance between notch and bottom spacer; using a marker, draw a horizontal line on the notched plate, about one-eighth this distance from the notch. Fill the sandwich to this level with *resolving gel*. Carefully cover the surface with a water layer to give an even, horizontal boundary. After polymerization (10 minutes) decant the water. Rinse the surface of the resolving gel with a few millilitres of *stacking gel* solution, decant, then fill almost to the notch with stacking gel. Insert the comb (lightly coated with petroleum jelly) until the teeth are 1–2 mm short of being fully submerged in gel, taking care not to trap air bubbles below the teeth. After the gel has set (10 minutes), it may be used directly or stored for less than 3 days in a sealed plastic bag in a refrigerator.

18.5.3 Preparation of samples

Sample preparation affects the results markedly. For denaturing gels, protein may be precipitated with 6–10% w/v cold trichloroacetic acid. After centrifugation, the precipitated protein is washed in 1:1 ethanol: ether and dissolved in approx. 15 µl 0.1 M Na_2CO_3, 10% v/v mercaptoethanol using agitation or sonication if needed; then 15 µl 25% w/v glycerol, 4% w/v SDS, 0.04% w/v Bromophenol Blue is added. Better results may be obtained if acid precipitation is omitted. The effect of heating the samples for 1 minute at 85°C should be investigated, as resolution may be enhanced or diminished.

Extra caution is needed in sample preparation for non-denaturing electrophoresis since acid precipitation cannot be used. Divalent cations and high salt concentrations (especially K^+) interfere and must be removed, e.g. by dialysis. Chlorophyll–protein complexes will retain pigment if first extracted from the membrane using a non-ionic detergent [23] and electrophoresed at <0.05% SDS, or if extracted and electrophoresed in the cold using lithium dodecylsulphate [24].

Each lane (or 'well') on the gel will be loaded with 10–100 µl of fluid consisting of a control, a molecular weight marker or a solubilized sample containing 10–50 µg of protein. The lower values are used for 'mini-gels', or when only a few polypeptides are expected to be present. It follows that some method is needed for determining approximate protein content (Section 18.7.3); in the case of thylakoids, this can be estimated assuming 1.0 µg of chlorophyll represents 5 µg protein. It is advisable to run at least two different loadings of unknown samples, to resolve detail in both the major and minor constituents. Outside lanes should be reserved for blanks (e.g. solubilization buffer) and molecular weight markers. Take careful notes of the lane assignments and details of sample preparation.

18.5.4 Loading the samples

Remove the comb carefully and rinse the sample wells with *running buffer* (diluted from 10 × stock). Remove the bottom clamps and the spacer from inside the sandwich. Apply petroleum jelly in a bead around the sides and bottom of the notch or use a length of thin silicone tubing to form a gasket, and clamp the sandwich to the electrophoresis apparatus with the notch facing the upper reservoir. Fill the upper reservoir with running buffer until the sample wells formed by the stacking gel teeth are at least 1.0 cm deep. Test that buffer does not leak past the seal, then fill the bottom reservoir with running buffer also. Dislodge any bubbles that are trapped under the sandwich.

Glycerol in the sample buffer provides a dense solution that can be layered under the running buffer. The same total volume of sample should be applied to each well, a microsyringe being convenient for this purpose. After loading the wells, connect the power supply (positive to the bottom reservoir, negative to the upper reservoir terminal). Switch on, and adjust the current to 15 mA for a mini-gel or 30 mA for the regular size. Observe the progress of the tracking dye, Bromophenol Blue, and turn off the power when this has almost reached the bottom of the gel. Disconnect the electrodes.

18.5.5 Processing the gel slab

Remove the sandwich from the apparatus. Insert a straight edge between the glass plates at the bottom and gently prise the plates apart. Rinse the gel with water and transfer to a plastic box. Wear gloves when doing this. Cover the slab with *stain solution* and gently agitate for about 3 hours. Pour off the stain (this can be filtered and re-used), rinse the slab and box in water, then cover the slab with enough *destain solution* to half-fill the box. Agitate as before for 1 hour, decant and discard the solution and agitate with destain solution for a further 2 hours. Decant and rinse with destain solution. Mount on a gel-drying apparatus, with the slab up, cover with cellulose acetate film and apply vacuum.

If a microwave oven is available, processing time and stain concentration can be reduced. Cover the gel in stain solution diluted 20-fold with destain solution. Microwave (700 W) for 3 minutes then agitate gently for 30 minutes at room temperature. Repeat this entire step using fresh, diluted stain solution. Finally repeat this step a third time but omit the stain from the destain solution. The gel will store indefinitely in destain solution.

Staining with silver offers the highest sensitivity of detection, but many published procedures give disappointing results. The novel method of Blum *et al.* [25] consistently gives good band intensity against a clear background. The original paper should be consulted as it contains a wealth of detail. The high sensitivity of silver-based methods (1 ng of protein or less) requires scrupulous cleanliness and wearing of sterile gloves throughout.

18.5.6 Troubleshooting

1. The standards run well, but other samples show a diffuse smear from the well downwards.
   (a) The samples were not fully stabilized before loading: try increasing the SDS:protein ratio and/or microcentrifuging particulate material out of the solution.
   (b) Too much salt in the sample: K^+ salts are worst because dodecyl sulphate can precipitate; replace with Na^+. If salt cannot be lowered to <0.1 M before loading the gel, start running the gel at about 25% of the normal current for 1 h, then increase to the normal operating current. Salts retard the movement of the marker dye, which is thus a good indicator of this problem.
2. Some or all of the bands are not sharp.
   (a) Proteins have hydrolysed before or during sample preparation; perhaps the samples were frozen and thawed too often (as is common with standards) or were allowed to age (this occurs even on ice). Try adding 1 mM phenylmethylsulphonyl fluoride (PMSF), diluted from a 40 mM stock solution in isopropanol, to the buffers used in sample preparation.
   (b) The sample was overheated during solubilization. In extreme cases, coagulated protein will not enter the gel and the symptoms are as for (1) above.
3. The bands curve upwards at the edges. 'Smiling' results from overheating the gel during electrophoresis: reduce the current (by decreasing the applied voltage).
4. One or more bands appear in every lane, irrespective of the sample loaded.
   (a) Only applies to bands in the range 50–65 kDa, usually on silver-stained gels.) This could be an artefact from DTT or mercaptoethanol in the solubilization buffer: try omitting thiol or removing

the artefactual bands by the procedure of Beis and Lazou [26].
   (b) Gross contamination of the equipment: be sure to thoroughly clean the syringes or pipettes used for sample loading.
   (c) Bacterial/fungal growth in reagents used to prepare the sample.
5. Opaque white areas appear on the gel, or white powder encrusts the upper reservoir.
   This usually indicates that SDS has come out of solution: check concentrations and temperature. PAGE cannot be run in the cold without special precautions: replace SDS with lithium dodecylsulphate [23] or reduce the concentration of SDS in the upper reservoir to 0.04%.
6. Marker dye is moving slowly or not at all. **Caution: this could accompany an electrical shock hazard; turn off power and disconnect before touching the apparatus.**
   (a) Check for removal of the bottom spacer and trapped air bubbles under the sandwich.
   (b) Check that the positive terminal of the power supply output is connected to the bottom reservoir of the apparatus, and the negative to the top reservoir.
   (c) Ensure that enough buffer is in the top reservoir to contact the sample wells (this should have been done before loading the samples).
   (d) No voltage: some power supplies indicate a set voltage which is not effective until a timer is activated. Consult the instruction manual.

18.6 PROTECTION OF CELL PROTEINS FROM PHENOLS AND PHENOL OXIDASE ACTIVITY

The homogenization of plant tissues is a first step in the isolation of individual protein components for biochemical and functional analysis, and in the study of membrane reactions in chloroplasts, mitochondria, etc. However, the contents of cell compartments such as the vacuole may include substances that denature proteins directly, or may promote undesirable enzyme reactions that complicate protein purification or electrophoresis (Section 18.5). A common problem arises from the presence of *o*-diphenols. Nearly all cells contain some form of phenol oxidase (monophenol oxidase, polyphenol oxidase, tyrosinase, etc.), a class of copper-containing enzymes that can oxidize a diphenol to an *o*-quinone [27–30]. This may react in turn with -SH or -NH_2 groups of proteins, or may polymerize to yield heterogenous dark-coloured compounds (as is evident when fruits, tubers, mushrooms, etc. are cut or bruised). The consequences are:

1. Inactivation of an enzyme;
2. Precipitation of protein, i.e. making soluble enzymes appear particulate;
3. Changing physical characteristics such as redox potential, UV absorption spectrum.

To reduce such effects to a minimum during any homogenization of plant tissues, or to help preserve plant products:

1. Keep preparations cold (0–4°C);
2. Remove O_2 (i.e. under N_2);
3. Add inhibitors (copper chelators, e.g. diethyl dithiocarbamic acid);
4. Add reducing compounds, e.g. ascorbate;
5. Add thiols or mercapto compounds, e.g. thioglycollate, mercaptoethanol, dithiothreitol;
6. Add an excess of another protein, e.g. bovine serum albumen (BSA);
7. Add polymers which complex phenols, e.g. polyvinylpyrrolidone (PVP), polyethylene glycol, polyamid, nylon 66.

The level of phenol oxidase in a plant extract, and the efficiency with which it is combatted using these compounds can be followed using the oxygen electrode [27].

18.6.1 Demonstration of phenol oxidase activity and design of protective media

During grinding of most plant tissue, both phenols and the enzyme phenol oxidase are liberated, forming toxic quinones and resulting in the inactivation of the protein that may be required for experimental purposes. The basic reaction is as follows:

σ-diphenol —Phenol oxidase ($\frac{1}{2}O_2$ → H_2O)→ Quinone

Since this reaction results in the uptake of oxygen it can easily be monitored using an oxygen electrode (Chapter 10).

Materials:
Oxygen electrode, pestle and mortar or blender, 10 mM catechol, chlorogenic acid or DL-DOPA (a concentrated instant coffee/water extract can be used as a crude substitute), DIECA (diethyldithiocarbamic acid), thiols, mercaptans, polymers, made up in 10 × concentration series from 1.0 μM to 10 mM.

Method:
Ensure the oxygen electrode is calibrated and operational (Chapter 10). Homogenise mature, deveined leaves of local plants such as maize, bean, sunflower, cassava, castor bean, etc.; say, 20 g leaves in 150 ml of water; filter homogenate through eight layers of muslin and keep on ice. Measure the rate and extent of oxygen uptake using, initially, 1.0 ml homogenate plus 1.0 ml water, and adjusting as appropriate within a constant final volume of 2 ml (smaller quantities of homogenate may be required for more active extracts). When endogenous phenols have all been oxidized, add microlitre quantities of a substrate such as catechol, chlorogenic acid, DOPA, coffee or tea extract. (N.B. coffee and tea extracts make interesting substrates, but they should not be used for routine assays!) Vary the amount of homogenate to evaluate the endogenous phenol concentration. Check the effect of inhibitors, reducing agents and thiol compounds on the reaction. A control of boiled homogenate should be run, with subsequent additions of fresh homogenate. Protective reagents such as PVP (polyvinyl pyrrolidene) and BSA (bovine serum albumen) should be tested in the homogenisation medium.

Results:
Plot observed rates as a function of concentration of enzyme in homogenate added inhibitor (express as nmol O_2 uptake per minute per gram [protein, chlorophyll, fresh or dry weight]). Also calculate amount of endogenous phenol.

Discussion:
The oxidation of catechol or DOPA in the presence of phenol oxidase can also be monitored in a recording spectrophotometer at 410–460 nm. Furthermore, phenol oxidases may survive non-denaturing SDS-PAGE (Section 18.5), so catechol can be used as an activity stain on the gel (resulting in brown bands). For maximum sensitivity, immerse the gel in 12.5 mM DOPA dissolved in 25 mM Tris-Cl^- at pH 7.5. Following colour development (30 minutes, 25°C) the gel may be stained with Coomassie Blue but *not* with silver. Thus phenol oxidase activity can be assayed by three independent procedures: oxygen electrode, spectrophotometer and activity gel staining.

Such experiments can suggest conditions for minimizing phenol oxidase activity and designing appropriate protective media for tissue preservation or homogenization.

18.7 DETERMINATION OF PROTEIN CONTENT

18.7.1 Lowry method

This is the most widely used method for protein determination. The Folin–Ciocalteau reagent reacts with certain parts of the protein molecule, producing a dark blue colour. Reagents are prepared as follows:

A: 2% Na_2CO_3 (anhydrous) in 0.1 M NaOH.
B: 0.5% $CuSO_4$.
C: 1.0% $NaKC_4H_6O_6 \cdot 4H_2O$ (sodium potassium tartrate).
D: 48 ml of Reagent A + 1 ml Reagent B + 1 ml Reagent C, mixed immediately before use.
E: Folin–Ciocalteau reagent, diluted with water (usually 1:1) to give a solution 1.0 N in acid.

Mix 0.5 ml protein suspension (containing up to 500 µg protein) with 5.0 ml Reagent D. Allow to stand for 15 minutes at room temperature. Add 0.5 ml of Reagent E and mix at once. Stand for a further 30 minutes at room temperature, and measure the absorbance at 700 or 750 nm. It is necessary to prepare a calibration graph using known amounts of BSA, since the colour produced is not directly proportional to the amount of protein present.

18.7.2 Kjeldahl method

This is a measurement of the protein nitrogen content. A sample containing approximately 1.0 mg nitrogen (about 6 mg protein) is digested with 2 ml concentrated sulphuric acid in a long-necked flask. To this is added 1.0 g catalyst mixture (prepared by grinding to a powder in a mortar 80 g K_2SO_4, 20 g $CuSO_4 \cdot 5H_2O$ and 0.34 g sodium selenate). After gentle refluxing for about 8 h, during which time the protein nitrogen is quantitatively converted into ammonium sulphate, the flask is allowed to cool and the contents carefully diluted with 25 ml of distilled water.

The solution is transferred quantitatively to a distillation apparatus with an outlet tube immersed in a flask containing 10 ml 2% boric acid (A.R.) solution. 10 ml 30% NaOH is added to the digestion chamber, and steam passed through the mixture. Ammonia liberated from the ammonium sulphate under the alkaline conditions distils over into the boric acid. About 20 ml distillate is collected, and the ammonia titrated with 0.01 M HCl. A solution of 0.2% methyl red and 0.1% methylene blue is used as an indicator; this shows a colour change from purple to grey when all the alkali has been titrated. An ammonium sulphate control is usually run: 1.0 ml 0.01 M HCl corresponds to 0.14 mg nitrogen.

This method measures total nitrogen, whether in protein or otherwise. For this reason, contaminating nitrogenous substances such as ammonium sulphate must be absent from the sample. Most proteins have a nitrogen content of about 16%; therefore to convert nitrogen content into weight of protein, multiply by a factor of 6.25.

18.7.3 Bradford method

Coomassie Brilliant Blue G (100 mg) is dissolved in 50 ml 95% ethanol. This solution is mixed with 100 ml 85% (w/v) phosphoric acid, and diluted with water to a final volume of 1 litre. The resulting suspension is filtered to remove undissolved material (use Whatman No. 1 paper), and stored in a darkened bottle [31].

To prepare a standard curve with BSA stock solution (1 mg ml^{-1}) follow the method:

1. Take aliquots of the stock solution to give the following amounts of protein: 0, 1, 2, 5, 10, 20, 25, 50 µg BSA. Make up to 100 µl.
2. Add 1 ml of reagent to the sample and shake (with a vortex mixer).
3. Leave the solution in the dark for 5 minutes.
4. Read the absorbance at 595 nm (use a 1 ml cuvette)

Note: Some detergents (Triton X-100, sodium dodecyl sulphate) may interfere with the assay, so always include a reagent blank in the protein determinations. When testing an unknown sample, add an appropriate amount (1–100 µl) to give an absorbance reading in the linear part of the standard curve.

REFERENCES

1. Avron, M. (1960) Photophosphorylation by Swiss Chard chloroplasts. *Biochim. Biophys. Acta* **40**, 257–272.
2. Shoshan, V., Y. Shahak and N. Shavit (1980) Quercetin interaction with the chloroplast ATPase complex. *Biochim. Biophys. Acta* **591**, 421–433.
3. Schwartz, M. (1968) Light induced proton gradient links electron transport and photophosphorylation. *Nature* **219**, 915–919.
4. Allen, J.F. and D.O. Hall (1974) The relationship of oxygen uptake to electron transport in Photosystem 1 of isolated chloroplasts: the role of superoxide and ascorbate. *Biochem. Biophys. Research Communications* **58**, 579–585.
5. Green, B.R. (1988) The chlorophyll-protein complexes of higher plant photosynthetic membranes. *Photosynth. Research* **15**, 3–32.
6. Chitnis, P.R. and J.P. Thornber (1988) The major light-harvesting complex of photosystem II: aspects of its molecular and cell biology. *Photosynth. Research* **16**, 41–63.
7. Allen, K.D. and L.A. Staehelin (1991) Resolution of 16–20 chlorophyll protein complexes using a low ionic strength native green gel system. *Analyt. Biochem.* **194**, 214–222.
8. Young, A., P. Barry and G. Britton (1989) The occurrence of β-carotene-5,6–epoxide in the photosynthetic apparatus of higher plants. *Zeitschrift Naturforschung* **44c**, 959–965.
9. Braumann, T. and L.H. Grimme (1981) Reversed phase high-performance liquid chromatography of chlorophylls and carotenoids. *Biochim. Biophys. Acta* **637**, 8–17.
10. Demmig-Adams, B. (1990) Carotenoids and photoprotection in plants: a role for the xanthophyll zeaxanthin. *Biochim. Biophys. Acta* **1020**, 1–24.
11. Bryant, D.A. and G. Cohen-Bazire (1981) Effects of chromatic illumination on cyanobacterial phycobilisomes. *European J. Biochem.* **119**, 415–424.
12. Jeffrey, S.W. (1968) Quantitative thin-layer chromatography of chlorophylls and carotenoids from marine algae. *Biochim. Biophys. Acta* **162**, 271–285.
13. Graan, T. and D.R. Ort (1984) Quantitation of the rapid electron donors to P700, the functional plastoquinone pool, and the ratio of the photosystems in spinach chloroplasts. *J. Biol. Chem.* **259**, 14003–14010.
14. Schoeder, H.-U. and W. Lockau (1986) Phylloquinone copurifies with the large subunit of photosystem I. *FEBS Letters* **199**, 23–27.
15. Okayama, S. (1976) Redox potential of plastoquinone A in spinach chloroplasts. *Biochim. Biophys. Acta* **440**, 331–336.
16. Kramer, D.M. and A.R. Crofts (1990) Demonstration of a highly-sensitive portable double-flash kinetic spectrophotometer for measurement of electron transfer reactions in intact plants. *Photosynth. Research* **23**, 231–240.
17. Kramer, D. M., R.R. Wise, J.R. Frederick, D.M. Alm, J.D. Hesketh, D.R. Ort and A.R. Crofts (1990) Regulation of coupling factor in field-grown sunflower: a redox model relating coupling factor activity to the activities of other thioredoxin-dependent chloroplast enzymes. *Photosynth. Research* **26**, 213–222.
18. Mauzerall, D. and N.L. Greenbaum (1989) The absolute size of a photosynthetic unit. *Biochim. Biophys. Acta* **974**, 119–140.
19. Marsho, T.V. and B. Kok (1980) P700 detection. *Methods Enzymol.* **69**, 280–289.
20. Schreiber, U., C. Klughammer and C. Neubauer (1988) Measuring P700 absorbance changes around 830 nm with a new type of pulse modulation system. *Zeitschrift Naturforschung* **43c**, 686–698.
21. Chow, W.S., A.B. Hope and J.M. Anderson (1989) Oxygen per flash from leaf disks quantifies Photosystem II. *Biochim. Biophys. Acta* **973**, 105–108.
22. Piccioni, R., G. Bellemare and N-H. Chua (1982) Methods of polyacrylamide gel electrophoresis in the analysis and preparation of plant polypeptides. In: *Methods in Chloroplast Molecular Biology* (M. Edelman, R.B. Hallick and N-H Chua, eds.) Elsevier Biomedical Press, Amsterdam. p. 985.
23. Camm, E.L. and B.R. Green (1980) Fractionation of thylakoid membranes with the nonionic detergent octyl-β-D-glucopyranoside. *Plant Physiol.* **66**, 428.
24. Delepelaire, P. and N-H. Chua (1979) Lithium dodecyl sulphate/polyacrylamide gel electrophoresis of thylakoid membranes at 4° C: characterizations of two additional chlorophyll a-protein complexes. *Proc. Natl. Acad. Sci. USA* **76**, 111.
25. Blum, H., H. Beier and H.J. Gross (1987) Improved silver staining of plant proteins, RNA and DNA in polyacrylamide gels. *Electrophoresis* **8**, 93–99.
26. Beis, A. and A. Lazou (1990) Removal of artifactual bands associated with the presence

of 2–mercaptoethanol in two-dimensional polyacrylamide gel electrophoresis. *Analyt. Biochem.* **190**, 57–59.

27. Baldry, C.W., C. Bucke and J. Coombs (1970) Phenols, phenol oxidase and photosynthetic activity in chloroplasts isolated from sugar cane and spinach. *Planta* **94**, 107–123; 124–129.
28. Golbeck, J.H. and K.V. Cammarata (1981) Spinach thylakoid polyphenol oxidase: isolation, activation and properties of the native chloroplast enzyme. *Plant Physiol.* **67**, 977–984.
29. Meyer, H-U. and B. Biehl (1982) Relation between photosynthetic and phenolase activities in spinach chloroplasts. *Phytochemistry* **21**, 9–12.
30. Yu, H-F., S.M. Newman, N.T. Emnetta, M.I. Fisher and J.C. Steffens (1991) Cloning of tomato polyphenol oxidase cDNAs and enzyme conservation betweeen polyphenol oxidases and tyrosinases. *Plant Physiol.* **96** (Suppl.), 5.
31. Bradford, M.M. (1976) Rapid and sensitive method for quantitation of microgram quantities of protein utilising principle of protein-dye binding. *Analyt. Biochem.* **72**, 248–254.

19

Nitrogen fixation and nitrate reduction

P. Lindblad and M.G. Guerrero

19.1 INTRODUCTION

Nitrogen (N) is one of the most abundant elements in plants, and constitutes 0.5–5% of the plant dry weight (8–10% in the case of microalgae). The atmosphere contains almost 80% nitrogen in the form of dinitrogen gas (N_2). Only certain prokaryotic organisms are able to utilise this N_2 by a process called biological nitrogen fixation – amounting to about 10^8 tonnes of N_2-N per year. In the case of legumes and some other higher plants, nitrogen-fixing bacteria are contained in nodules in the roots. A large number of other symbiotic associations also exist. The N_2 is fixed into ammonia which can be assimilated and metabolised (Fig. 19.1; Chapter 20). Biological nitrogen fixation was first reported in 1888, and since then this process has attracted a lot of interest in both its basic and applied aspects. The first part of this chapter will introduce the process of biological nitrogen fixation, the requirements, and its basic physiology and biochemistry.

For most plants in their natural environment, nitrate is the primary source of nitrogen. The assimilation of nitrate by higher plants and algae is a fundamental photosynthetic process in which a highly oxidised form of inorganic nitrogen is reduced to ammonium, which becomes in turn combined with carbon compounds to form the various nitrogenous components of the cell (Chapter 20). The process is known as assimilatory nitrate reduction to differentiate it from nitrate reduction of the respiratory type carried out by various bacteria which, under microaerophilic or anaerobic conditions, use nitrate as an electron acceptor in place of molecular oxygen. It has been estimated that the plant kingdom assimilates about 10^{10} tonnes of nitrate-N per year.

Photosynthesis and Production in a Changing Environment: a field and laboratory manual
Edited by D.O. Hall, J.M.O. Scurlock, H.R. Bolhàr-Nordenkampf, R.C. Leegood and S.P. Long
Published in 1993 by Chapman & Hall, London.
ISBN 0 412 42900 4 (HB) and 0 412 42910 1 (PB).

19.2 NITROGEN FIXATION

Biological nitrogen fixation is catalysed by the enzyme complex nitrogenase, and can be summarized by the following equation:

$$N_2 + 8H^+ + 8e^- + 16MgATP \xrightarrow[\text{nitrogenase}]{} 2NH_3 + H_2 + 16MgADP + 16P_i$$

As seen above, the fundamental requirements for active biological nitrogen fixation are:

1. the enzyme complex nitrogenase;
2. a powerful reductant; and
3. ATP.

The following criteria also need to be fulfilled for active nitrogenase activity:

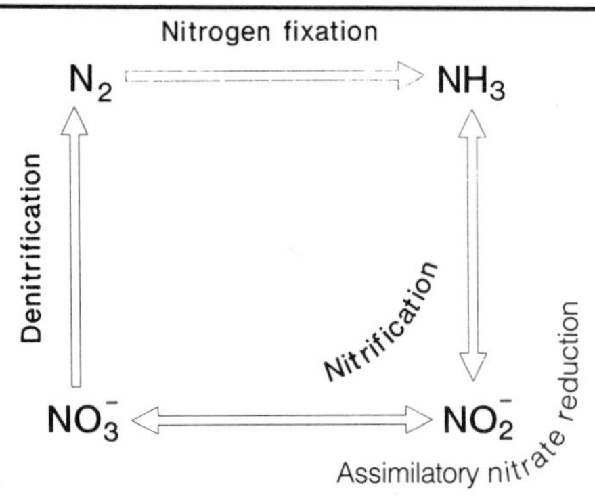

Fig. 19.1. A schematic representation of the biological cycle of inorganic nitrogen.

4. low or near-zero oxygen level and
5. an efficient removal/assimilation of the fixed nitrogen (ammonia) into amino compounds.

19.2.1 Nitrogenase

The nitrogenase complex consists of two enzymes: a reductase (dinitrogenase reductase, Fe-protein, Component II), which provides electrons with high reducing power, and a nitrogenase (dinitrogenase, MoFe-protein, Component I) which uses these electrons to reduce N_2 to ammonia. Both enzymes are oxygen sensitive, and different organisms have developed different strategies to lower the oxygen tension around the active nitrogenase enzyme. The genes for nitrogen fixation are called the *nif*-genes, and besides the three structural genes, at least 14 more *nif*-genes have been described.

(i) Dinitrogenase reductase

The dinitrogen reductase, also called the Fe-protein or Component II, is the smaller of the two enzymes. The enzyme consists of two identical sub-units (α_2) with molecular weights in the range of 30–36 kilodaltons (kDa) and has one 4Fe-4S iron-sulphur cluster per sub-unit. The structural gene is called *nif*H.

(ii) Dinitrogenase

The larger dinitrogenase, also called the MoFe-protein or Component I, has two different sub-units, each being a dimer of identical sub-units ($\alpha_2\beta_2$). The metal content can be summarized as $Mo_2Fe_{30-52}S_{approx.28}$. The molecular weights of the different sub-units vary around 50–60 kDa. The structural genes are called *nif*K and *nif*D.

(iii) Alternative nitrogenase

At least 2 alternative nitrogenases have been described recently. The alternative nitrogenases have very much in common with the 'normal' or conventional nitrogenase although recent studies have shown that they are coded for by separate structural genes. In one of the alternative nitrogenases the Mo in the dinitrogenase enzyme is replaced by V, whereas a second one does not have any Mo or V.

19.2.2 Hydrogenase

Nitrogen-fixing organisms may contain at least three enzymes involved in dihydrogen (H_2) metabolism: (i) Nitrogenase evolves H_2 concomitantly with the fixation of N_2 to NH_3 at a rate of at least 25% of the electron-flow through nitrogenase (Section 19.2.1); (ii) a membrane-bound uptake hydrogenase, oxidising the H_2 produced by the nitrogenase; and (iii) a soluble reversible hydrogenase, the physiological function of which has been suggested to evolve H_2.

The presence of an active uptake hydrogenase, catalysing the net reaction $H_2 \rightarrow 2H^+ + 2e^-$, is believed to improve the efficiency of the nitrogen fixation process in several ways. The electrons are 'passed down' an electron transport pathway to oxygen, and ATP is regained by oxidative phosphorylation. The efficient removal of H_2 from the vicinity of the nitrogenase enzyme complex protects the enzyme from hydrogen inhibition, as well as lowering the oxygen level.

Table 19.1. Organisms capable of biological nitrogen fixation.

(1) Free-living nitrogen-fixing bacteria.
- (a) Archaebacteria (e.g. *Methanosarcina* and *Methanococcus*).
- (b) Eubacteria (e.g. *Clostridium, Desulphovibrio, Desulfotomaculum, Klebsiella, Azotobacter, Azospirillum, Rhodospirillum, Rhodopseudomonas, Chromatium* and *Chlorobium*).
- (c) Cyanobacteria (e.g. *Gleothece, Synechococcus, Oscillatoria, Trichodesmium, Anabaena, Nostoc*).

(2) Nitrogen-fixing bacteria in symbiotic association.
- (a) *Rhizobium* symbioses.
  - (i) Symbioses with legumes (e.g. *Phaseolus, Trifolium, Medicago, Vigna, Glycine*).
  - (ii) Symbioses with non-legumes (*Parasponia*).
- (b) *Frankia* symbioses (e.g. *Alnus, Casuarina, Dryas, Myrica*).
- (c) Cyanobacterial symbioses.
  - (i) Symbioses with fungi (lichens, e.g. *Collema, Peltigera* and *Lobaria*).
  - (ii) Symbioses with bryophytes (e.g. *Blasia, Cavicularia* and *Anthoceros*).
  - (iii) Symbioses with pteridophytes (*Azolla spp.*).
  - (iv) Symbioses with gymnosperms (cycads, approx. 150 species).
  - (v) Symbioses with angiosperms (*Gunnera*, approx. 50 species).

19.3 ORGANISMS CAPABLE OF BIOLOGICAL NITROGEN FIXATION

The ability to carry out nitrogen fixation is restricted to prokaryotic organisms, which can be either free-living or in symbiosis with another organism (Table 19.1).

19.3.1 Free-living nitrogen-fixing bacteria

(i) Archaebacteria

All archaebacteria are strict aerobes, and are generally divided into three main groups; the methanogens, halophiles, and themo-acidophiles. A few species within the methanogens, *Methanosarcina* and *Methanococcus*, have recently been shown to be able to fix nitrogen.

(ii) Eubacteria

Nitrogen fixation is widespread among different eubacteria, ranging from the strict anaerobic Gram-positive *Clostridium*, the sulphur-reducing *Desulphovibrio* and *Desulfotomaculum*, the facultative anaerobic *Klebsiella* (including the common *K. pneumoniae* which is used to study the physiology, biochemistry and molecular biology of nitrogen fixation), to the strict aerobic *Azotobacter* and *Azospirillum*.

Among the photosynthetic bacteria, nitrogen fixation occurs in the non-sulphur genera *Rhodospirillum* and *Rhodopseudomonas*, and is found, though rarely, in both the purple sulphur bacteria (*Chromatium*) and the green sulphur bacteria (*Chlorobium*).

(iii) Cyanobacteria

Cyanobacteria (blue-green algae) are Gram-negative prokaryotic photosynthetic organisms. A large number of cyanobacterial genera, ranging from unicellular (e.g. *Gleothece, Synechococus*) to filamentous, both without (e.g. *Oscillatoria, Trichodesmium*) and with heterocysts (e.g. *Anabaena, Nostoc*), are capable of fixing atmospheric N_2. Cyanobacteria may produce vegetative cells, heterocysts and akinetes (spores). Their primary functions are photosynthesis, nitrogen fixation and survival under harsh environments, respectively. The ammonia produced by nitrogen fixation is assimilated by the enzymes glutamine synthetase and glutamate synthase (GOGAT) (Chapter 20). In heterocystous cyanobacteria, the glutamine is transported from the heterocysts to the neighbouring vegetative cells, where it is converted to glutamate by GOGAT. Nitrogen-fixing cyanobacteria are, in contrast to the majority of other nitrogen-fixing organisms, capable of photoautotropic growth. As in eukaryotic organisms Rubisco is the primary CO_2-fixing enzyme. It is undesirable that nitrogen fixation (which is O_2 sensitive) and photosynthesis (which evolves oxygen) occur together, but different cyanobacteria have evolved different strategies to separate these two processes, in time, in space or metabolically.

19.3.2 Nitrogen-fixing bacteria in symbiotic association

Prokaryotic nitrogen-fixing bacteria have established symbioses with a high number of higher plant families and genera. The economically most important symbioses are those between nitrogen-fixing *Rhizobium* and legumes in both tropical regions (e.g. soybeans, ground nuts, chick-peas) and temperate regions (e.g. beans, peas, clover, lucerne, lupin). The actinomycete *Frankia* and filamentous cyanobacteria also are found in symbioses with a number of eukaryotic organisms.

(a) Rhizobium *symbioses*

(i) Symbioses with legumes:
The legumes comprise three major subgroups; Papilionaceae, Mimosaceae and Caesalpinaceae. Within the larger group, the Papilionaceae, as much as 80–90% of the species form nodules containing nitrogen-fixing rhizobia. However, in both Mimosaceae and Caesalpinaceae approximately one quarter of the species contain these nitrogen-fixing structures. With the exception of the tropical legume *Sesbania*, which has nodules on the stem, the nodules are formed on the roots. The rhizobia, Gram-negative motile rods, are classified into two major groups; the fast-growing *Rhizobium*, mainly establishing symbioses with temperate legumes (e.g. beans (*Phaseolus spp.*), clovers (*Trifolium spp.*), lucerne (*Medicago sativa*) and peas (*Pisum sativum*)), and the slow-growing *Bradyrhizobium* which mainly are found in symbioses with tropical legumes (e.g. acacia, cowpea (*Vigna*), lupins, peanut, and soya-beans (*Glycine*)). Both the leguminous plants, and the species of *Rhizobium* and *Bradyrhizobium* show a certain degree of specificity, although not very strict because the groups do overlap in their infection and the establishment of functional symbiosis. Generally, the tropical/sub-tropical legumes, establishing symbiosis with slow-growing *Bradyrhizobium*, develop determinate nodules from which the fixed nitrogen is exported as ureides (allantoin and allantoic acid). Temperate legumes, establishing symbioses with the fast-growing *Rhizobium*, form indeterminate nodules, and the fixed nitrogen is exported in the form of glutamine and asparagine. The above distinction clearly divides the tropical/sub-tropical and the temperate legumes into two different groups. However, there are exceptions and some genera can be infected by both *Rhizobium* and *Bradyrhizobium* strains. In mature infected regions of the nodule, almost all the host cells contain large numbers of intracellular symbionts enclosed within membrane-bound vesicles. The symbionts are usually termed bacteroids. Leghaemoglobins, comprising up to 20% of the soluble protein content (soybean), are invariably present in effective nodules, confined to the bacteroid-containing cells, and making the nodules appear pink. Ineffective nodules are white. The main function of leghaemoglobin, analogous to the role of haemoglobin in mammalian blood, is to bind and carry oxygen.

(ii) Symbiosis with non-legumes:
The *Parasponia* (Ulmaceae) root nodules are the only example of symbiosis between a

non-legume and rhizobia. Structurally and physiologically this symbiosis shares features both with legume and actinorhizal nodules.

(b) Frankia *symbioses:*

The actinomycete *Frankia* can establish symbioses with a broad range of plants from 8 families and 25 genera (e.g. *Alnus, Casuarina, Dryas, Myrica*). Collectively these plants are called actinorhizal plants. The rates of nitrogen fixation in actinorhizal symbiosis are comparable to those in legumes. The actinorhizal plants develop nodules which contain nitrogen-fixing *Frankia*. The extracellular symbioses are formed through a root hair infection and by cell to cell growth of the *Frankia* hyphae. Vesicles, the site for nitrogen fixation, are differentiated from swellings of some hyphal tips, and are localized in the periphery in an infected host cell. However, in *Allocasuarina* and *Casuarina*, where *Frankia* does not form vesicles, nitrogenase is believed to be localized in the hyphae.

(c) Cyanobacterial symbioses:

Filamentous heterocystous cyanobacteria develop symbioses with all major plant groups. However, only a few examples exist within each group. The enzyme nitrogenase is localized exclusively in the heterocysts.

(i) Symbioses with fungi:
Out of about 13,500 known lichen species, approximately 10% contain extracellular nitrogen-fixing cyanobacteria. The lichens containing cyanobacteria can either be of the bipartite (fungus + cyanobacteria) or tripartite type (fungus + green algae + cyanobacteria).

(ii) Symbioses with bryophytes:
Cyanobacteria develop endophytic symbioses with a few genera of both liverworts (e.g. *Blasia, Cavicularia*) and hornworts (e.g. *Anthoceros*). Moreover, epiphytic cyanobacteria have been described in a number of mosses and liverworts. The intercellular symbiotic *Nostoc* filaments within the liverwort/hornwort thallus are localised in distinct colonies or units.

(iii) Symbioses with pteridophytes:
Azolla is a small water-fern with seven species, all of which contain nitrogen-fixing filamentous cyanobacteria within the dorsal leaf cavities. *Azolla* is used extensively in rice plantations in Asia as a biological nitrogen fixer, where under optimal conditions it can accumulate as much as 10 kg nitrogen per hectare per day.

(iv) Symbioses with gymnosperms:
The cycad – *Nostoc* symbiosis is the only example of a symbiosis between a gymnosperm and cyanobacteria, in fact the only symbiosis between a gymnosperm and a nitrogen-fixing organism. All cycads examined contain intercellularly located filamentous, heterocystous cyanobacteria within cycad coralloid roots. The heterotrophic cyanobacteria fix nitrogen in darkness, and the fixed-N is then transferred to the cycad. Further translocation within the cycad is in the form of either glutamine or a combination or glutamine and citrulline.

(v) Symbioses with angiosperms:
The angiosperm genus *Gunnera*, with approximately 50 described species, contain intracellular nitrogen-fixing *Nostoc* filaments within specialized glands along the stems. It is the only known example of symbioses between angiosperms and cyanobacteria.

19.4 METHODS TO MEASURE NITROGEN FIXATION

Dinitrogen (N_2) is a very inert molecule. Nitrogenase is capable of reducing many types of complexes, including N_2 which has a dissociation energy of 945 kJ mol^{-1}. Among other complexes that are reduced by nitrogenase are nitrous oxide, azide, cyanide, cyclopropene and acetylene. Nitrogen fixation can be measured using (i) total nitrogen content, (ii) isotopes or (iii) by assaying the reduction of acetylene to ethylene. The most commonly used isotope is the naturally occurring and stable ^{15}N. A definitive test for nitrogen fixation is the incorporation of ^{15}N from $^{15}N_2$ in the gas phase. The radioactive isotope ^{13}N has been used in short-duration studies. However, its very short half-life and the need for extensive precautions has limited the use of this isotope.

19.4.1 Acetylene reduction activity

When the nitrogenase enzyme complex converting N_2 and H^+ into ammonia and H_2 (Section 19.2) is exposed to 10% acetylene, the enzyme will reduce acetylene into ethylene at a rate which is proportional to the total activity of the nitrogenase enzyme complex. This ability is used in the acetylene reduction technique to measure total nitrogenase activity. The method is based on quantification of the ethylene produced using a gas chromatograph:

$$N_2 \xrightarrow[\text{nitrogenase}]{} \text{ammonia}$$

$$C_2H_2 \xrightarrow[\text{nitrogenase}]{} C_2H_4$$

The following schedule is used for measuring nitrogenase activity using the acetylene reduction technique:

1. Harvest the biological material (e.g. whole plant, root system, legume nodules, cyanobacteria, etc.).
2. Distribute the biological material into incubation vials or gas-proof plastic bags. Place a damp piece of filter-paper in the incubation vial in order to retain moisture and prevent desiccation.
3. Seal the incubation vials/plastic bags with a tight gas-proof serum stopper made of good rubber (this is important, otherwise leaks of C_2H_4 and/or H_2 may cause problems).
4. Withdraw 10% of the air.
5. Add C_2H_2 (same volume as above). In the field C_2H_2 may be generated from calcium carbide (CaC_2) and water (Fig. 19.2):

   $$CaC_2 + 2H_2O \rightarrow C_2H_2 + Ca(OH)_2$$

   (N.B. Calcium carbide is highly reactive and should be handled with care.)
6. Incubate the biological material together with the acetylene.
7. Sample for C_2H_4 by withdrawing, for example, 3 × 1 ml gas samples from each incubation at several time intervals after the start of the incubation.
8. Run C_2H_4 standards, for example by injecting 5 × 1 ml 100 ppm C_2H_4 into a gas chromatograph. (Other dilutions are possible using pure (100%) C_2H_4.)
9. Analyze the gas samples for C_2H_4 content.
10. Determine the amount of biological material (e.g. fresh weight/dry weight/chlorophyll).
11. Calculate the C_2H_4 standards. Assuming that C_2H_4 is an ideal gas, the amount of gas (in moles) = pV/RT, where:
    p = Atmospheric pressure (kPa);
    V = Volume (dm^3);
    R = Universal gas constant (0.08205);
    T = Room temperature (K).

    therefore 1 ml 100% C_2H_4 = X mol C_2H_4 and 1 ml 100 ppm C_2H_4 = X × 10^{-6} mol C_2H_4

    Convert 'standard readings' to mol C_2H_4:

    1 unit (from the gas chromatograph) = Y mol C_2H_4
12. Calculate the amount of C_2H_4 in the 1 ml gas samples.
13. Multiply the above figure (12) by the total gas volume in the incubation vessel/plastic bag.
14. Express the results as mol C_2H_4 per unit amount of biological material per unit time.

If possible, it is useful to estimate hydrogen (H_2) evolution under the same conditions as used for acetylene reduction in order to estimate the efficiency of the biological material used. The relative efficiency (RE) of a

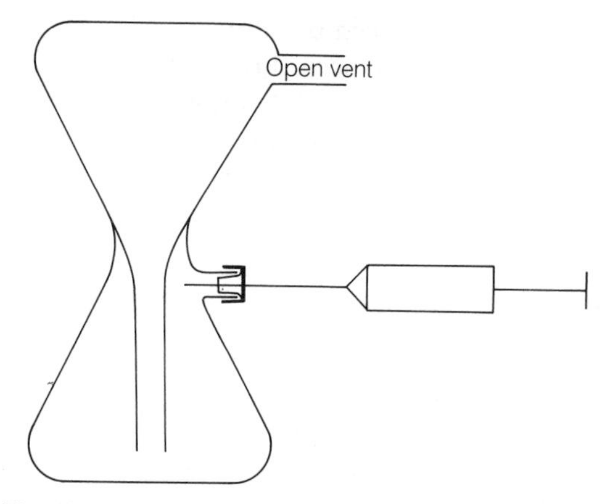

Fig. 19.2. A simple apparatus for the generation of acetylene using CaC_2 and water. Fill the lower reservoir completely with distilled water. Drop in two CaC_2 pellets, and wait a few seconds before adding a serum cap (to allow the water to be purged of air). Withdraw a portion of the acetylene using a syringe via the serum cap. Clean the vessel with HCl after use to remove $Ca(OH)_2$.

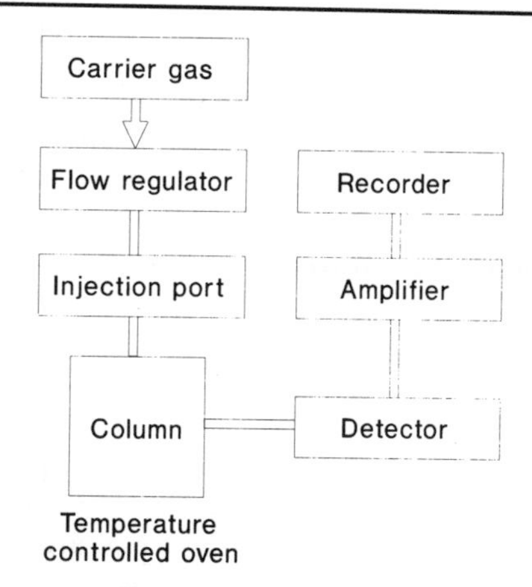

Fig. 19.3. Schematic diagram of a gas chromatograph.

given biological nitrogen-fixing system can be expressed as:

$$RE = 1 - \frac{H_2 \text{ evolved}}{C_2H_2 \text{ reduced}}$$

19.5 GAS CHROMATOGRAPHY

A schematic drawing of a gas chromatograph is shown in Fig. 19.3. Carrier gas, e.g. from a cylinder, is passed through a flow regulator and an injection port before it picks up the sample for analysis. Injection should be carried out as quickly as possible. The carrier gas and the sample then pass through a column in a thermostatic oven where the individual components of the sample are separated. The columns, column packings and detectors used vary widely depending on the nature of the material being analysed. The components are detected, for example with a flame ionisation detector or a semiconductor gas sensor, and the signal amplified before it is recorded on a meter and/or a printer. A single substance will give rise to a single peak.

The volume of carrier gas passed between the point of injection of the sample and the peak maximum, measured at the average pressure and temperature of the column, is termed the true retention volume of the substance (V_r). V_r is related to the retention time (t_r), usually used to characterize a given substance, by the equation $t_r = V_r/f_v$, where v = volumetric gas flow rate. The retention time of solvent, air or other component which moves with the carrier gas equals the volume of gas in the column (V_g) and may be called the dead volume of the column. The net retention volum V'_r is the true retention volume less the dead volume:

$$V'_r = V_r - V_g$$

Since the retention volume is measured at the temperature and pressure of the laboratory (or in the field), the true retention volume has to be obtained from the experimental retention volume V_{exp} by the formula:

$$V_r = V_{exp}\,(Tp_r)/(T_rp),$$

where T and p are column temperature and pressure, and T_r and p_r are laboratory/field temperature and pressure, respectively. This volume is then compared with that obtained using a standard sample.

The area of the peak depends on the amount of the substance present, the detector efficiency, and the degree of amplification used. If the latter factors are held constant, the recorded peak area is a direct measure of the amount of substance present in the sample. Methods of determining area vary, from measurement by cutting out and weighing the recorder paper to electronic integration by a microprocessor.

Packing material:

Discussion is restricted here to the *Porapak* column packing materials, which are porous polymer beads (*Dow Chemical Company, Texas, USA*). Eight different types are available, identified in order of increasing polarity as types P, Ps, Q, etc., to T. They are relatively robust, but should not be used at temperatures above 250°C. A typical separation of gases using a small portable gas chromatograph packed with *Porapak* T is shown in Fig. 19.4.

19.6 NITRATE REDUCTION

Assimilatory nitrate reduction takes place in higher plants and algae, as well as in yeasts, fungi and many bacteria. The 8-electron reduction of nitrate to ammonium occurs in two steps:

$$\overset{(+5)}{NO_3^-} \xrightarrow[\text{nitrate reductase}]{+2e} \overset{(+3)}{NO_2^-} \xrightarrow[\text{nitrite reductase}]{+6e} \overset{(-3)}{NH_4^+}$$

This involves the sequential participation of two metalloproteins: nitrate reductase and nitrite reductase. The physiological source of electrons are reduced pyridine nucleotides or reduced ferredoxin according to the source or the type of enzyme. ATP is not required for nitrate or nitrite reduction. Both reactions take place with a large decrease in free energy [1].

19.6.1 Enzymology

In eukaryotes [2,3], nitrate reductase is a homodimeric enzyme complex (molecular weight about 200 kDa) containing flavin (FAD), heme (cytochrome b_{557}) and a molybdenum-pterin cofactor as redox-active prosthetic groups. Electrons provided by NAD(P)H (from photosynthesis or carbohydrate oxidation) are transferred to nitrate through the electron transport chain of the enzyme:

$$H^+ + NAD(P)H \rightarrow NAD(P)^+ \quad \longrightarrow [FAD \rightarrow \text{cyt } b \rightarrow Mo] \longrightarrow \quad NO_3^- \rightarrow NO_2^- + H_2O$$

In addition to catalysing the reduction of nitrate by reduced pyridine nucleotides, the NAD(P)H: nitrate reductase exhibits a variety of partial activities that involve only part of the overall electron-transport capacity of the enzyme. Diaphorase activities result in the reduction by NAD(P)H of various 1- and 2-electron acceptors (cytochrome *c*, ferricyanide and other oxidants). Terminal nitrate reductase activities consist of the NAD(P)H-independent reduction of nitrate by reduced flavin nucleotides or viologens:

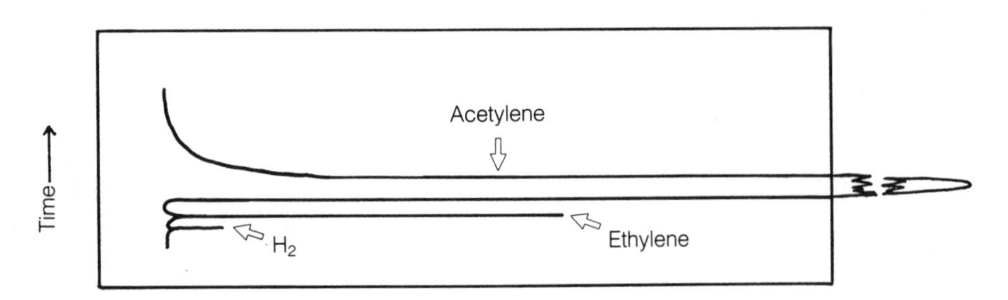

Fig. 19.4. Typical separation of gases using a small portable gas chromatograph equipped with a smoke detector (*Figaro TGS 812*). The column (300 mm × 4 mm) was packed with *Poropak T* (80–100 mesh), using air as the carrier gas and a column/detector temperature of about 60°C.

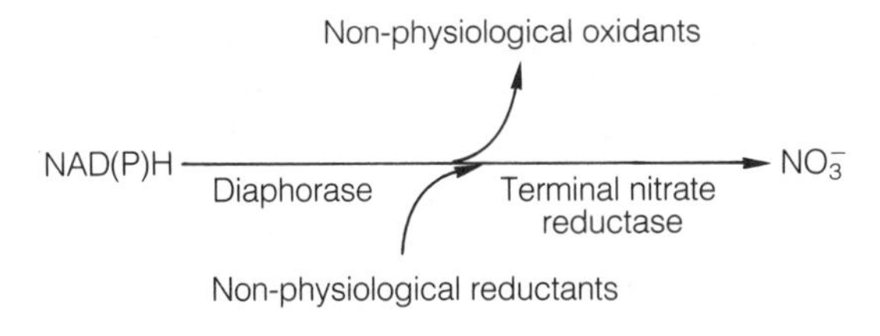

The intracellular localization of nitrate reductase in eukaryotic algae and photosynthetic tissues of higher plants has not yet been firmly established. Although most evidence points to a cytosolic location, an association of nitrate reductase with the chloroplast has been found by some workers. In prokaryotic photosynthesizing organisms, such as blue-green algae (cyanobacteria) [4], nitrate reductase is a smaller protein (MW 75–85 kDa) containing molybdenum and two binuclear iron-sulphur centres, but no flavin or heme. The enzyme is tightly bound to thylakoids and uses photosynthetically reduced ferredoxin, but not NAD(P)H, as the electron donor.

A variety of environmental and nutritional factors influence the amount of assimilatory nitrate reductase. The enzyme from different organisms, either prokaryotes or eukaryotes, is turned over rapidly and is present in high amounts when cells are fed with nitrate, but is repressed if the cells or plants are grown on media containing ammonium ions. Significant positive correlations have been found between nitrate reductase and the nitrogen status of some higher plant systems. This enzyme is considered to be an important limiting factor for plant growth and development. Growth, yield or protein content are sometimes correlated with the nitrate reductase activity level of seeds or leaves.

The reduction of nitrite to ammonium in photosynthesising eukaryotic and prokaryotic cells is catalysed by ferredoxin: nitrite reductase [3,4], which accepts electrons from photosynthetically reduced ferredoxin but not from reduced pyridine nucleotides.

The enzyme protein (MW about 63 kDa) contains a tetranuclear iron-sulphur centre and siroheme. The electron-transport chain from reduced ferredoxin to nitrite may be depicted as follows:

$$8H^+ + 6Fd_{red} \rightarrow 6Fd_{ox} \quad \rightarrow [(Fe_4 - S_4^*) \rightarrow \text{Sirohaem}] \rightarrow \quad NO_2^- \rightarrow NH_4^+ + 2H_2O$$

Nitrite is directly reduced to ammonium, without liberation of free intermediates.

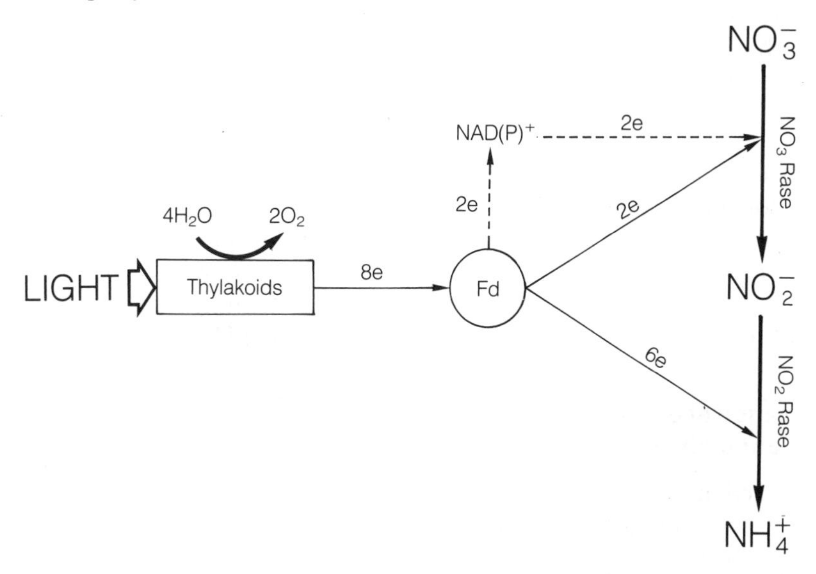

Fig. 19.5. The reduction of nitrate to ammonium by photosynthetically generated reductant. Fd, ferredoxin.

Photosynthetic nitrite reductase is located within chloroplasts or thylakoids; a ferredoxin-dependent nitrite reductase is also present in root tissue.

19.6.2 Photosynthetic nature of nitrate reduction

Assimilatory nitrate reduction in green cells represents a genuine photosynthetic process [1,5], which consumes photogenerated reductant and is coupled to oxygen evolution (Fig. 19.5). The photoreduction of nitrate to ammonia with water as the electron donor can in fact be considered as one of the simplest examples of photosynthesis. In cyanobacteria, with the two enzymes being ferredoxin-dependent, nitrate reduction is intimately linked to photosynthesis, and its coupling with the photolysis of water has been conclusively demonstrated [4]. In eukaryotic green algae and in leaves of higher plants, nitrite reductase also makes direct use of photogenerated reductant (reduced ferredoxin), whereas NAD(P)H nitrate reductase accepts electrons indirectly from ferredoxin, through pyridine nucleotide [1,5]. The assimilation of nitrate in actively photosynthesising cells consumes a significant fraction of the photosynthetically generated assimilatory power. In microalgae and young leaves, with a C/N ratio below 10, the proportion of electrons used in nitrate assimilation might represent as much as one-fourth of that required for the fixation of CO_2.

In leaf tissue in darkness, and in non-green tissues, such as roots, the reducing power required for nitrate and nitrite reduction is provided by carbohydrate oxidation.

19.6.3 Determination of enzymatic activities (nitrate reductase and nitrite reductase)

Nitrate reductase activity is usually estimated by measuring the amount of nitrite produced from nitrate, and nitrite reductase activity by the disappearance of nitrite from the reaction mixture.

The method used for measuring nitrite [6] is based on the formation of a diazonium salt during the reaction of nitrite with sulphanilamide in acid medium, and the further establishment of a coloured complex as the diazonium salt reacts with N-(1-naphtyl) ethylenediamine (NNEDA). The resulting complex is pink in colour and has an absorption maximum at 540 nm.

(a) Nitrite estimation:
Add 1 ml 1% (w/v) sulphanilamide in 2.4 N HCl to 3 ml of the sample containing nitrite (10–100 nmol). Add 1 ml 0.02% (w/v) N-(1-naphtyl)ethylenediamine dichloride to each tube. Mix (Vortex mixer) and measure absorbance at 540 nm (A_{540}) after 10 minutes.

$$E^{1\,\text{mM}}_{1\,\text{cm}} \text{ nitrite complex (540 nm)} = 55$$

where E is the extinction at the given wavelength.

(b) Nitrate reductase activity determination:

(i) In vivo assay of nitrate reductase in plant leaf tissue [7]:
Nitrate is reduced to nitrite with endogenously generated reductant. Isopropanol is used to increase the permeability of the tissue to nitrate and nitrite. The reaction is carried out in darkness to limit reduction of nitrite to ammonia. Anaerobic conditions prevent competition with oxygen for endogenously generated reduced pyridine nucleotide.

Procedure:
Cut the leaves into pieces (squares or discs of about 5 mm^2). Add 0.2 g to 5 ml of incubation mixture containing: 0.1 M potassium phosphate buffer, pH 7.7, 0.1 M KNO_3, and 1% (v/v) isopropanol. Use suba-seal stoppers. Vacuum infiltrate for 2–3 minutes and then flush with argon (or N_2) for about 5 minutes. Incubate in the dark at 30°C. Remove aliquots (0.2–1 ml) with a hypodermic syringe for nitrite determination at zero time and after 15–30 minutes.

This method is also sometimes used for estimating the nitrite content of plant tissues, following the same procedure but omitting nitrate from the reaction mixture. Allow the reaction to proceed until reaching a constant concentration of nitrite (1–2 hours).

(ii) In situ assay of nitrate reductase activity in algal cells:

Green algae:
For *Chlorella* use the freezing–thawing procedure [8,9] as follows:

(1) Collect cells by centrifugation (500 *g* for 2 minutes).
(2) Wash the cells with 0.067 M phosphate buffer, pH 6.8.
(3) Resuspend the cells in 0.1 M Tris–HCl buffer, pH 8.5, to reach a concentration of about 2 mg dry weight cells per ml buffer.
(4) Freeze overnight (about 16 hours at −20°C).
(5) Thaw at about 3°C, centrifuge and resuspend in a total volume of 1.5 ml of the reaction mixture which contains:

Tris–HCl buffer, pH 8.5	100 μmol
KNO_3	10 μmol
Malate	75 μmol
NADH	0.03 mg

Note: NADH is not required when the species of *Chlorella* is *C. fusca*.

(6) Incubate for 10 minutes at 30°C.
(7) Add 0.25 ml 25% (w/v) $ZnSO_4$ and 0.25 ml 1 M NaOH. Let the mixture stand for 5 minutes at 0°C, and centrifuge.
(8) Use an aliquot (1 ml) of the supernatant for nitrite estimation as indicated above.

For assays with cells of other strains of green algae (*Ankistrodemus braunii* and *Chlamydomonas reinhardii*) see [8].

Blue-green algae (Cyanobacteria): [10,11]

(1) Add 20 μl toluene to 1 ml of cell suspension (either in buffer or in culture medium) containing about 10 μl packed cell volume cells (70–80 μg chlorophyll).
(2) Shake vigorously (Vortex mixer) for 90 s.
(3) Add a 0.1 ml aliquot of the suspension to 0.8 ml of standard assay mixture for ferredoxin-nitrate reductase activity determination (dithionite-reduced methyl viologen (MV) as reductant) [8].

0.5 M $NaHCO_3/Na_2CO_3$ buffer, pH 10.5	6.25 ml
KNO_3	63.1 mg
methyl viologen (MV)	32.2 mg
water to a total volume of	25 ml

(4) Mix gently and add 0.1 ml of freshly prepared dithionite solution containing 20 mg $Na_2S_2O_4$ in 0.3 M $NaHCO_3$.
(5) Mix gently and incubate for 5 minutes at 30°C.
(6) Stop the reaction by shaking in Vortex mixer until blue colour (reduced MV) disappears.
(7) Add 0.1 ml 25% $ZnSO_4$ and 0.1 ml 1 M NaOH.
(8) Centrifuge (in a bench centrifuge) and determine nitrite as described above using an aliquot (0.6 ml) of the supernatant.

A modification of this assay involves the inclusion of 75 μg ml^{-1} mixed alkytrimethylammonium bromide (MTA) in the reaction mixture [11], thus avoiding the pretreatment of the cells with toluene.

(iii) In vitro assays of nitrate reductase activity:
Nitrate reductase from microalgae and plant tissues is usually very unstable [3]. Precautions should be taken during breakage of cells and extraction of the enzyme. The presence of EDTA and dithioerythritol in the extraction buffer is recommended, as well as FAD (about 20 μM) in the case of NAD(P)H-dependent enzymes.

Ferredoxin-nitrate reductase: from blue-green algae, with reduced MV as reductant. Use the same reaction mixture and assay conditions as described above for the *in situ* assay (omit toluene treatment or MTA addition).

NAD(P)H-nitrate reductase: for the enzyme from green algae and higher plants [12], with NADH (or NADPH) as reductant.

(1) Mix

0.5 M Tris–HCl buffer, pH 7.5	0.2 ml

0.2 M KNO_3	0.1 ml
3 mM NADH (or NADPH)	0.1 ml
enzyme preparation + water	0.3 ml

(2) Start the reaction by addition of the enzyme preparation.
(3) Incubate at 30°C for 5–15 minutes.
(4) Stop the reaction by addition of 0.1 ml 0.2 mM phenazine methosulphate to oxidise the remaining NAD(P)H. After 5 minutes add 0.1 ml 25% $ZnSO_4$ and 0.1 ml 1 M NaOH. Centrifuge and remove 0.65 ml of the supernatant for nitrite estimation.

In many cases the presence of FAD in the reaction mixture (20 µM final concentration) is required for achieving maximal activity.

The reaction can also be followed spectrophotometrically by recording the oxidation of NAD(P)H at 340 nm. Use 3 ml cuvettes containing 0.6 ml of the Tris buffer + 0.1 ml 0.2 M KNO_3 + 0.3 ml 1 mM NADH (or NADPH) + 1.7 ml water. Start the reaction by adding 0.3 ml of enzyme preparation. Follow the decrease in absorbance at 340 nm. Run a blank without KNO_3 to correct for any nitrate-independent oxidation of NAD(P)H.

$$E_{1\,cm}^{1\,mM}\,\mathrm{NAD(P)H}\ (340\,\mathrm{nm}) = 6.2$$

(c) Nitrite reductase activity determination:
For the enzyme from algae and higher plants [13], with dithionite-reduced MV as reductant.

Measure disappearance of nitrite using dithionite-reduced MV as electron donor.

Reaction mixture:

0.5 M Tris–HCl buffer, pH 7.5	6.25 ml
$NaNO_2$	4.32 mg
Methyl viologen (MV)	6.01 mg
Water	up to 25 ml

Procedure:
Add 0.3 ml of the enzyme preparation to 1.5 ml of the reaction mixture. Run a blank without enzyme. Start the reaction by adding 0.2 ml of a recently prepared dithionite solution (25 mg ml^{-1} $Na_2S_2O_4$; 25 mg ml^{-1} $NaHCO_3$). Incubate for 15 minutes at 30°C. Stop the reaction by vigorously shaking (Vortex mixer) until blue colour disappears. Use a 20 µl aliquot for nitrite determination. Estimate the amount of nitrite which has disappeared using as a reference the assay without enzyme.

This can also be used as an *in situ* assay with intact microalgae by including an adequate amount of mixed MTA in the reaction mixture. In the case of blue-green algae, the optimum amount of the detergent is 750 µg per ml reaction mixture to which an amount of cells containing about 25 µg chlorophyll is added [11].

For assays with the natural reductant (reduced ferredoxin), the assay is performed as described above, but with 0.5 mM ferredoxin instead of MV [3].

19.6.4 Other analytical procedures useful for the study of nitrate assimilation

Sometimes it is useful to follow the disappearance of nitrate or the appearance of ammonium in a cell suspension or in a sample of soil, or even to determine the concentration of these compounds in plant tissues or algal cells. Chemical methods are described for the determination of nitrate and ammonium. Ammonium can also be determined enzymatically with glutamate dehydrogenase, and nitrate with nitrate reductase. In both cases the oxidation of NAD(P)H, which is dependent upon the presence of the respective substrates, can be followed spectrophotometrically by measuring the decrease in absorbance at 340 nm. A method for the enzymatic determination of ammonium is also described. Methods for the determination of small amounts of nitrate and ammonium (in the picomole range) by high performance liquid chromatography are also available [14,15].

(a) Determination of nitrate:
This method (after Cawse, [16]) is based on

measurement of the ultraviolet (210 nm) absorption due to nitrate, after the removal of interference due to other ions (mainly nitrite) by treatment of the sample with perchloric and sulphamic (amidosulphuric) acids.

Add 0.1 ml of 10% (w/v) sulphamic acid to 1.5 ml of the sample containing nitrate (10–200 nmol). Shake (Vortex mixer), let stand for 2 minutes, shake again and add 0.4 ml 20% (v/v) perchloric acid. Shake again, and measure the absorbance at 210 nm using silica cuvettes:

$$E_{1\,cm}^{1\,mM} \text{ nitrate (210 nm)} = 7.4$$

(b) Chemical determination of ammonium and dissolved ammonia:
This method (after Solorzano, [17]) is based on the formation of indophenol after the reaction, at high pH, of ammonia, phenol and hypochlorite. Possible interference due to calcium or magnesium are eliminated by treatment with citrate. Under the final assay conditions, indophenol exhibits a blue colour with an absorbance maximum at 640 nm.

Solutions:

(I) Phenol–alcohol solution: 0.1% (w/v) phenol in 95% (v/v) ethanol.
(II) Nitroprusside solutions: 0.5% (w/v) sodium nitroprusside (store in amber bottle for not longer than 1 month).
(III) Alkaline solution: 20% (w/v) trisodium citrate in 1% (w/v) NaOH solution.
(IV) Oxidising solution: 16 ml of solution (III) + 0.4 ml of commercial hypochlorite solution (1.5 N at least) + 3.6 ml of water. This oxidising solution should be prepared just before using it (use it within the same day).

Procedure:
To 1.6 ml of the sample containing ammonia (up to 80 nmol) add 0.2 ml of solution I, then 0.2 ml of solution II, and finally 0.5 ml of solution IV, mixing thoroughly after each addition. The colour is allowed to develop for 60–90 minutes at room temperature in the dark. Finally, absorbance at 640 nm is measured.

$$E_{1\,cm}^{1\,mM} \text{ indophenol-nitroprusside (640 nm)} = 18.5$$

Precautions:
Newly distilled water must be used, and the glassware must be cleaned by washing with warm diluted HCl and rinsing thoroughly with distilled water.

In either nitrate or ammonium determinations, blanks without the corresponding N-compound should be run in parallel. Run also standards containing about 50 nmol of the corresponding N-compound.

It may be necessary to increase the amount of hypochlorite in solution IV if this reagent is more than a few months old.

(c) Enzymatic determination of ammonium and dissolved ammonia:
With glutamate dehydrogenase, by estimating the oxidation of reduced pyridine nucleotide [18].

Solutions:

(A) 0.15 M triethanolamine buffer, pH 8.6.
(B) 0.3 M α-ketoglutarate (2–oxoglutarate).
(C) 30 mM ADP.
(D) 2.4 mM NADPH.

Mix 1.45 ml of solution A, 0.1 ml of solution B, 0.1 ml of solution C and 0.2 ml of solution D. Add up to 1.1 ml of the sample containing ammonium (up to 300 nmol) and make up to 2.95 ml with water. Measure absorbance at 340 nm (this should be about 1.0). Add 50 μl of an ammonium-free preparation of glutamate dehydrogenase containing about 400 units ml^{-1}. Wait 10 minutes. Measure A_{340} again. Repeat the measurement every 5 minutes till the value of absorbance does not change more than 0.01–0.02.

A blank without ammonium should be run in parallel, and also a standard containing 100 nmol ammonium.

$$E_{1\,cm}^{1\,mM} \text{ NADPH (340 nm)} = 6.2$$

Therefore:

$$\frac{(A_{340} \text{ sample} - A_{340} \text{ blank}) \times 3}{6.2}$$

= μmol ammonium in the assay

REFERENCES (NITRATE REDUCTION)

1. Guerrero, M.G., J.M. Vega and M. Losada (1981) The Assimilatory nitrate-reducing system and its regulation. *Ann. Rev. Plant Physiol.* **32**, 169–202.
2. Solomonson, L.P. and Barber, M.J. (1990) Assimilatory nitrate reductase: functional properties and regulation. *Ann. Rev. Plant Physiol. Plant Mol. Biol.* **41**, 225–253.
3. Wray, J.L. and Fido, J. (1990) Nitrate reductase and nitrite reductase. In: *Methods in Plant Biochemistry*, Vol. 3, Enzymes of Primary Metabolism (P.J. Lea, ed.) Academic Press, London. pp. 241–256.
4. Guerrero, M.G. and Lara, C. (1987) Assimilation of inorganic nitrogen. In: *The Cyanobacteria* (P. Fay and C. Van Baalen, eds.) Elsevier, Amsterdam. pp. 163–186.
5. Lara, C. and Guerrero, M.G. (1989) The photosynthetic assimilation of nitrate and its interactions with CO_2 fixation. In: *Techniques and New Developments in Photosynthesis Research* (J. Barber and R. Malkin, eds.). Plenum, London.
6. Snell, F.D. and C.T. Snell (1949) *Colorimetric Methods of Analysis* Van Nostrand, New York. p. 804.
7. Scott, D.B. and C.A. Neyra (1979) Glutamine synthetase and nitrate assimilation in sorghum (*Sorghum vulgare*) leaves. *Canadian J. Bot.* **57**, 754–758.
8. Hipkin, C.R. and P.J. Syrett (1977) Nitrate reduction by whole cells of *Ankistrodesmus braunii* and *Chlamydomonas reinhardii*. *New Phytol.* **79**, 639–648.
9. Syrett, P.J. and E.M. Thomas (1973) The assay of nitrate reductase in whole cells of *Chlorella*: strain differences and the effect of cell walls. *New Phytol.* **72**, 1307–1310.
10. Herrero, A., E. Flores and M.G. Guerrero (1981) Regulation of nitrate reductase level in the cyanobacteria *Anacystis nidulans*, *Anabaena sp.* strain 7119, and *Nostoc sp.* strain 6719. *J. Bacteriol.* **145**, 175–180.
11. Herrero, A., and M.G. Guerrero (1986) Regulation of nitrite reductase in the cyanobacterium *Anacystis nidulans*. *J. Gen. Microbiol.* **132**, 2463–2468.
12. Hageman, R.H. and A.J. Reed (1980) Nitrate reductase from higher plants. *Methods Enzymol.* **89**, 270–280.
13. Vega, J.M., J. Cárdenas and M. Losada (1980) Ferredoxin-nitrite reductase. *Methods Enzymol.* **69**, 255–270.
14. Thayer, J.R. and R.C. Huffaker (1980) Determination of nitrate and nitrite by high-pressure liquid chromatography: comparison with other methods for nitrate determination. *Analyt. Biochem.* **102**, 110–119.
15. Corbin, J.L. (1984) Liquid chromatographic-fluorescence determination of ammonia from nitrogenase reaction: a 2 minute assay. *Applied Environmental Microbiology* **47**, 1027–1030.
16. Cawse, P.A. (1967) The determination of nitrate in soil solutions by ultra-violet spectroscopy. *Analyst* **92**, 311–315.
17. Solorzano, L. (1969) Determination of ammonium in natural waters by the phenol-hypochlorite methods. *Limnology and Oceanography* **14**, 799–801
18. Bergmeyer, H.U. (1974) *Methoden der Enzymatischen Analyse*, 2nd edn. Verlag Chemie, Weinheim.

FURTHER READING (NITROGEN FIXATION)

Burris, R.H. (1991) Nitrogenases. *J. Biol. Chem.* **266**, 9339–9442.

Dixon, R.O.D. and C.T. Wheeler (1986) *Nitrogen Fixation in Plants*. Blackie, London.

Greshoff, P.M. (ed.) (1990) *Molecular Biology of Symbiotic Nitrogen Fixation*. CRC Press Inc., Boca Raton, Florida, USA.

Holfeld, H.S., C.S. Mallard and T.A. LaRue (1979) A portable gas chromatograph. *Plant and Soil* **52**, 595–598.

Packer, L. and A.N. Glazer (eds.) (1988) *Methods in Enzymology*. Vol. 167, Cyanobacteria. Academic Press, New York.

Peoples, M.B., A.W. Faizah, B. Rertasem and D.F. Herridge (eds.) (1989) *Methods for Evaluation of Nitrogen Fixation by Nodulated Legumes in the Field*. ACIAR Monograph No. 11, Canberra, Australia.

Postgate, J.R. (1987) *Nitrogen Fixation*. 2nd edn. Edward Arnold, London.

Rai, A.N. (ed.) (1990) *Handbook of Symbiotic Cyanobacteria*. CRC Press Inc., Boca Raton, Florida, USA.

Sprent, J.I. and P. Sprent (1990) *Nitrogen-fixing Organisms; Pure and Applied Aspects*. Chapman and Hall, London.

Stacey, G., R.H. Burris and H.J. Evans (eds.) (1992) *Biological Nitrogen Fixation*. Chapman and Hall, London.

20

Ammonia assimilation, photorespiration and amino acid biosynthesis

P.J. Lea and R.D. Blackwell

20.1 AMMONIA ASSIMILATION

20.1.1 Introduction

In the preceding chapter, the mechanism by which nitrogen gas and nitrate can be converted to ammonia have been described. There is now good evidence that ammonia is also released within the plant by a number of reactions [1].

1. Conversion of glycine to serine in the photorespiration process.
2. Metabolism of nitrogen transport compounds, e.g. asparagine, arginine and ureides.
3. Metabolism of proteins and subsequent deamination of amino acids via glutamate dehydrogenase, e.g. in seed germination and senescence.
4. Synthesis of lignins, etc., via the enzyme phenylalanine ammonia lyase.

The rate of ammonia production by the four reactions described above is over 10 times the rate of primary nitrogen uptake. On a quantitative basis, reaction 1 (which will be described in Section 20.3) is by far the most important in C_3 plants. Ammonia is toxic to all living organisms but plants have a limited capacity to store ammonia in the acidic vacuole. However, in order to ensure that there is an adequate supply of amino acids for protein synthesis and to prevent any build up of the potentially toxic metabolite, it is essential that all ammonia is assimilated into an organic form as rapidly and efficiently as possible.

The reaction is catalysed by the enzyme glutamine synthetase (GS), which has a very high affinity for ammonia (K_m = 5 μM). In the leaf the majority of the GS is present in the chloroplast, but a second form, under separate gene control, is also present in the cytoplasm [1]. The ammonia is initially assimilated into the amide position of glutamine, but this needs to be transferred to the 2-amino position if it is to become a major constituent of protein amino acids. The enzyme glutamate synthase (sometimes written as GOGAT; glutamine, 2-oxoglutarate amido transferase) carries out the transfer of the amide nitrogen to 2-oxoglutarate yielding glutamate. The enzyme requires reducing power which is either supplied as reduced ferredoxin or NADH. Thus a twostep re-

Photosynthesis and Production in a Changing Environment: a field and laboratory manual
Edited by D.O. Hall, J.M.O. Scurlock, H.R. Bolhàr-Nordenkampf, R.C. Leegood and S.P. Long
Published in 1993 by Chapman & Hall, London.
ISBN 0 412 42900 4 (HB) and 0 412 42910 1 (PB).

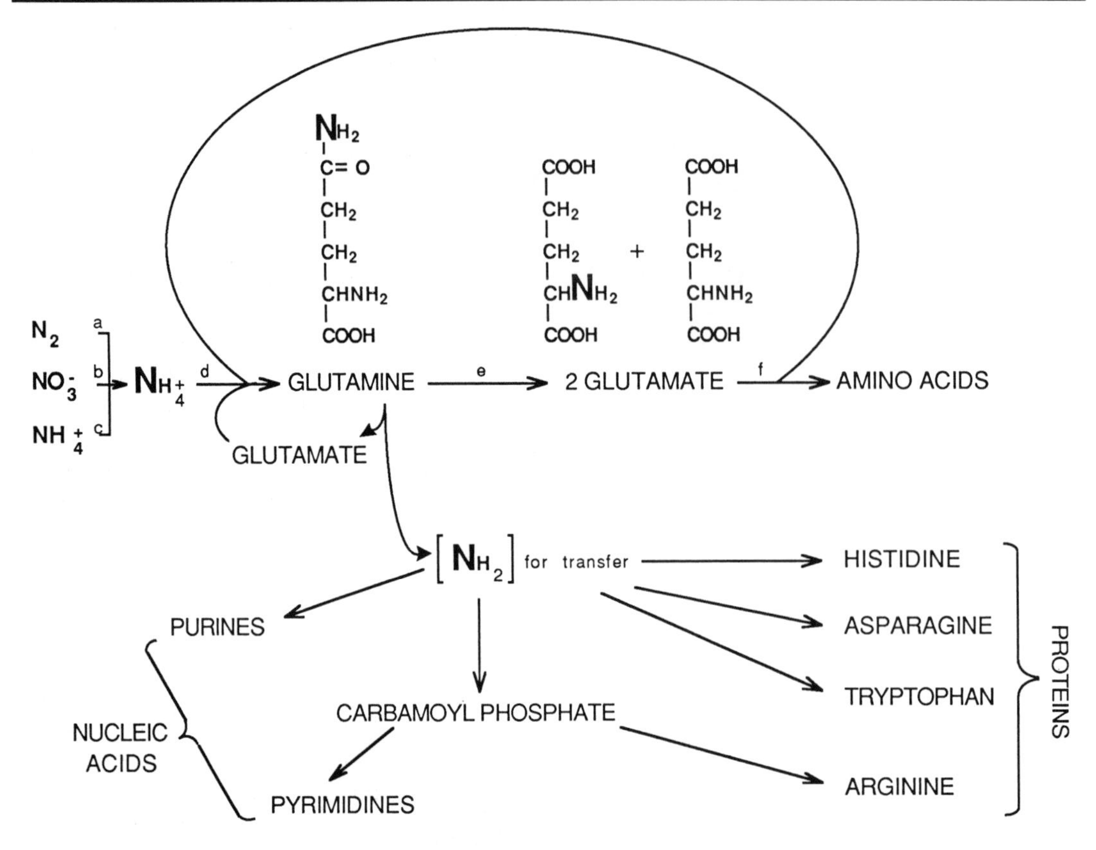

Fig. 20.1. A summary of the flow of nitrogen into proteins and nucleic acids via the glutamate synthase cycle. Enzymes: (a) nitrogenase; (b) nitrate reductase; (c) internal generation of ammonia as described in Section 20.1.1; (d) glutamine synthetase; (e) glutamate synthase; (f) aminotransferases.

action carries out the process of ammonia assimilation in higher plants, the reactions are jointly known as the glutamate synthase cycle (Fig. 20.1). In the leaf the ammonia is assimilated in the chloroplast, where the ATP and reduced ferredoxin are generated directly from light energy. This ammonia assimilation can be said to be a true **photosynthetic** process. Glutamine and glutamate act as a source of nitrogen for all the protein amino acids, as well as nucleic acids amines and ureides (Fig. 20.1).

Most textbooks, and even some research papers, suggest that the enzyme glutamate dehydrogenase also plays an important role in ammonia assimilation in plants. The enzyme which catalyses the direct amination of 2-oxoglutarate to yield glutamate is found in all plant tissues and is localised predominantly in the mitochondria. There is now a wide body of evidence based on: (1) enzyme kinetics, (2) ^{15}N and ^{13}N labelling data, (3) inhibitor studies and (4) mutants lacking GS and glutamate synthase that **the enzyme does not play a significant role in ammonia assimilation**. Recent studies now suggest that the enzyme may operate in the direction of glutamate breakdown in times of shortage of 2-oxo acids required for the Krebs cycle [1].

20.1.2 Enzyme isolation

Unless absolutely essential it is not recommended that any assays are carried out on totally crude extracts. Plant extracts contain a large number of inhibitors (in particular oxidised phenols) which will result in an underestimate (or zero determination) of the enzyme activity being measured. It must be remembered that just because an enzyme cannot be detected in an extract under one set of conditions, it does not mean that the enzyme is not present. It should also be noted that a set of conditions that is optimum for one enzyme is unlikely to be correct for another. Thus, it is not always possible to assay a range of enzymes using the same isolation procedure or assay medium. It is very important for each new enzyme and plant material that optimum conditions of the extraction and assay are determined on each occasion. When crude extracts have been prepared the simplest form of purification is to desalt them using Sephadex G-25 columns.

20.1.3 Glutamine synthetase (E.C. 6.3.1.2)

A reliable method for the isolation of GS is to grind 200 mg of liquid nitrogen frozen material in 2 ml of buffer containing 50 mM Tris–HCl (pH 7.8), 1 mM EDTA, 1 mM dithiothreitol, 10 mM $MgSO_4$, 5 mM sodium glutamate (it is **essential not** to use the ammonium salt in experiments on GS) and 10% ethanediol.

Glutamine synthetase is frequently assayed in the presence of glutamine, hydroxylamine, ADP, arsenate and Mn^{2+}. This so-called 'transferase' assay is **not recommended** as the physiological relevance to GS is not known. The assay is, however, widely used mainly because very high activities may be determined.

A more physiological, but by no means perfect, assay is the 'synthetase' assay which employs hydroxylamine in place of ammonia:

$$\text{glutamate} + \text{hydroxylamine} + \text{ATP} \rightarrow \gamma\text{-glutamyl-hydroxamate} + \text{ADP} + P_i.$$

The γ-glutamyl-hydroxamate may be rapidly quantified by a simple reaction with acidified ferric chloride to yield a brown-coloured product that may be determined spectrophotometrically at 540 nm. For the determination of GS activity the 500 µl assay mixture contains: 50 mM sodium glutamate (pH 7.0), 5 mM hydroxylamine hydrochloride (pH 7.0), 50 mM $MgSO_4$, and 20 mM ATP in 100 mM Tris–HCl (pH 7.8). The ATP solution must be prepared immediately before use and the pH adjusted carefully to just above 7.0. The reaction is started by the addition of 200 µl of enzyme extract and incubated at 30°C for 30 minutes. The reaction is terminated by the addition of 700 µl ferric chloride reagent (0.67 M ferric chloride, 0.37 M HCl and 20% (w/v) trichloroacetic acid, TCA). Control tubes should be incubated comprising all of the components except ATP. After termination of the assay the reaction tubes are centrifuged at 10,000 *g* for 5 minutes to remove the precipitated protein and the absorbance of the supernatant read on a spectrophotometer at 540 nm. A standard curve of γ-glutamyl-hydroxamate may be prepared with up to 3 µmol per 1 ml of assay mixture. The rate of enzyme activity may then be calculated. For very active enzyme preparations it may be necessary to dilute the enzyme or use very short assay times as the production of hydroxamate is not linear at very high rates. It is not known in crude extracts how close an agreement there is between the true rate of glutamine synthesis and the production of the hydroxamate complex. Two other assays may be used in purer extracts: (1) the determination of the phosphate liberated from ATP, and (2) the determination of ADP based on a spectrophotometric assay involving pyruvate kinase and lactate dehydrogenase [2].

It is well established that two isoenzymic forms of GS occur in leaves, one present in

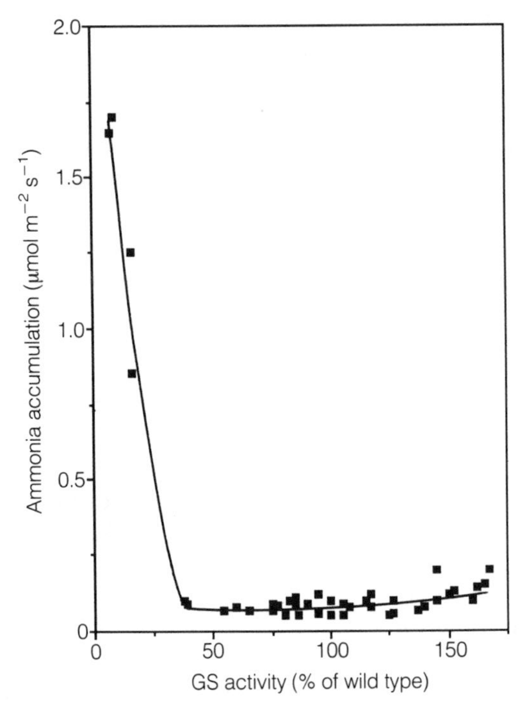

Fig. 20.2. Correlation between glutamine synthetase (GS) activity and ammonia content in a range of mutant barley lines, following exposure to air.

the chloroplast and one in the cytosol. These forms can be readily separated by ion-exchange chromatography. However, the distribution of the two forms varies considerably between different plants. A detailed survey has been able to identify four groups of plants: (1) parasitic plants containing only the cytosolic enzyme (e.g. *Orobanche* sp.); (2) plants containing 50% each of the cytoplasm and chloroplastic enzyme (e.g. *Zea, Pennisetum*); (3) plants containing a small proportion of the cytosolic enzyme (e.g. wheat, barley, pea); (4) plants containing solely the chloroplast enzyme (e.g. spinach, tobacco, lupin) [1]. The reason for the differences between the plant species is not clear. However, it is known that the two isoenzymes, although catalysing the same reaction, may have slight differences in their kinetic constants, molecular weights and stability, and are under separate gene control [1].

As part of a study identifying mutants of barley lacking enzymes of the photorespiratory carbon and nitrogen cycle (Section 20.3), Blackwell *et al.* [3] related the total GS activity to the ability of each plant to assimilate ammonia (Fig. 20.2). It was only when the leaves contained less than 40% of wild-type levels of GS activity, that ammonia accumulated following exposure to air. This demonstrates that in barley there is at least a 2-fold excess of GS activity *in vivo* in air, to allow for any sudden increase in the rate of photorespiration. Mutant plants containing 10% of the normal GS activity were shown to be completely devoid of the chloroplastic isoenzyme. Such plants could grow normally at elevated levels of CO_2 but showed severe symptoms of stress following exposure to air.

Inhibition of glutamine synthetase:
A frequently used piece of evidence for the operation of the glutamate synthetase cycle is that ammonia assimilation in higher plants and algae is blocked by L-methionine-S-sulphoximine (MSO):

MSO	PPT
CH_3	CH_3
|	|
O=S=NH	O=P—OH
|	|
CH_2	CH_2
|	|
CH_2	CH_2
|	|
$CHNH_2$	$CHNH_2$
|	|
COOH	COOH

MSO **PPT**

The compound acts as an inhibitor by binding to the active site of glutamine synthetase in place of the activated γ-glutamyl phosphate. A more potent inhibitor is phosphophinothricin (PPT) which is marketed as a non-selective herbicide by Hoechst as the ammonium salt of gluphosinate under the trade name 'Basta' [4].

The effect of MSO on plant metabolism can be determined relatively simply. Leaves of a C_3 cereal plant (e.g. wheat, barley or rice) should be cut under water into 10 cm lengths. The cut ends are placed into 2 ml of MSO solutions of concentrations varying from 0 to 1.0 mM in a 5 ml test-tube. The leaves and tubes are then placed in (1) the dark, (2) low light, (3) high light, (4) high light plus elevated CO_2. A high level of CO_2 can readily be generated in an enclosed glass or plastic chamber by the addition of a beaker of an acidified sodium carbonate solution. Sufficient leaves should be available to allow a time course for the different light and MSO treatments to be carried out over a period of 0–6 h.

The ammonia content of the leaves can then be readily determined as follows:

Grind the weighed leaf in a mortar with 1 ml of 0.1 M HCl into a uniform paste and centrifuge the extract for 5 minutes. Place 50 µl of each extract into a 1.5 ml plastic microfuge tube (any small glass or plastic tube will do as long as it is **clean**), and add 25 µl of 1 N H_2SO_4 and 25 µl of 10% (w/v) sodium tungstate. Mix and centrifuge for a further 5 minutes to remove the precipitated protein. Using a pasteur pipette transfer the whole of the supernatant to another clean microfuge tube and add 0.5 ml of phenol: nitroprusside reagent (1.0 g phenol, 5 mg sodium nitro-prusside per 100 ml of deionized water) and 0.5 ml of alkaline hypochlorite reagent (0.5 g NaOH, 2 g anhydrous Na_2HPO_4 and 1 ml of commercial sodium hypochlorite (bleach) per 100 ml of deionized water). Mix the solutions well and incubate in a water bath at 30°C for 30 minutes. An intense blue colour denotes the presence of ammonia, which can be determined quantitatively by reading the absorbance at 640 nm. A standard curve utilizing 0, 10, 15, 20, 25 and 30 µl of a 0.5 mM solution of ammonium sulphate should be constructed.

Typical results obtained with barley are shown in Table 20.1. These data clearly show that ammonia is generated in a light-dependent reaction that is prevented by the presence of the elevated levels of CO_2. The liberation of ammonia is not linear with time as MSO has been shown to inhibit the rate of photosynthetic CO_2 fixation in air, but not at elevated levels of CO_2.

Table 20.1. Ammonia content of barley leaves following the feeding of 1 mM MSO in the xylem stream for 1 h

Treatment	Ammonia content ($\mu mol\ g^{-1}$ fresh weight)
Light, 800 $\mu mol\ m^{-2}\ s^{-1}$	0.8
Dark + 1 mM MSO	1.4
Light, 200 $\mu mol\ m^{-2}\ s^{-1}$ + 1 mM MSO	3.2
Light, 800 $\mu mol\ m^{-2}\ s^{-1}$ + 1 mM MSO	10.8
Light, 800 $\mu mol\ m^{-2}\ s^{-1}$ + 1 mM MSO in 0.7% CO_2	1.9

The dramatic effect of MSO and PPT on plant metabolism can be explained by the following:

1. MSO inhibits GS activity and blocks all ammonia assimilation, from whatever source.
2. In the light with leaves in the air, the major source of ammonia is the conversion of glycine to serine in the photorespiratory nitrogen cycle (Section 20.3). This reaction does not occur in the dark or at levels of CO_2 sufficient to saturate Rubisco.
3. If ammonia is not reassimilated following the conversion of glycine to serine, the pool of amino acids becomes depleted and there are insufficient amino donors to convert more glycine to serine.
4. The level of glyoxylate increases, this is a metabolite that has been shown to prevent the activation of Rubisco (at micromolar concentrations). The amount of glycerate

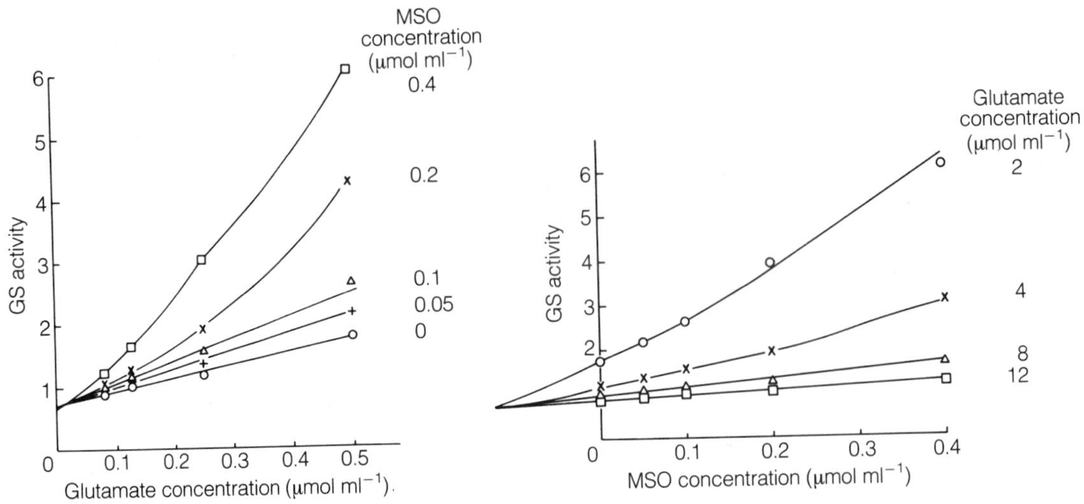

Fig. 20.3. Inhibition of pea leaf glutamine synthetase (GS) by methionine sulphoximine (MSO): (a) Lineweaver–Burke plot; (b) Dixon plot.

returning to the chloroplast is reduced. Either response will cause a lowering in the level of Calvin cycle intermediates and a subsequent inhibition of the rate of photosynthetic CO_2 assimilation.

The interruption of the photosynthetic and photorespiratory pathways by inhibitors of GS has clearly identified the close relationship between carbon and nitrogen metabolism. The ability of MSO to inhibit GS activity can be used to demonstrate simple enzyme inhibitor kinetics. A Lineweaver–Burke plot of GS activity against glutamate concentration will yield a K_m value of 5–10 mM depending on the plant material. MSO concentrations of 0 to 0.4 mM can be tested at different glutamate concentrations varying from 2 to 12 mM. The resulting data can be plotted out using the standard Lineweaver–Burke or Dixon method (Fig. 20.3). At low concentrations of MSO the plots are linear, but at higher levels an upward curve can be detected. There is now sufficient data available to show that for MSO and PPT there is an initial reversible binding to the active site which is competitive with glutamate. There then follows a phosphorylation of the inhibitor with ATP that causes irreversible inhibition. K_i values for MSO with GS isolated from higher plants range between 100 and 160 μM, whereas the equivalent K_i values for PPT are 6–8 μM [4].

There has recently been considerable interest in the production of plants resistant to the herbicide 'Basta' using genetic engineering. An enzyme has been isolated from *Streptomyces* that can acetylate the amino group of PPT and render it inactive. The gene coding for the enzyme has now been transferred to a number of experimental plants.

20.1.4 Glutamate synthase (Ferredoxin-, EC 1.4.7.1; NADH-, EC 1.4.1.13)

For the extraction of both forms of the enzyme for *in vitro* activity measurements of crude extracts, a phosphate buffer (pH 7.5) is normally used [2]. The components of the extraction buffer vary according to the source of the enzyme (i.e. leaves, cotyledons or root nodules). A suitable extraction buffer for leaves, however, comprises 50 mM KH_2PO_4-KOH (pH 7.5), 100 mM KCl, 5 mM EDTA,

12.5 mM mercaptoethanol, 1 mM PMSF, 2 mM 2-oxoglutarate, 20% (v/v) ethanediol and 0.05% (v/v) Triton X-100.

Ferredoxin-glutamate synthase:
The electron donor ferredoxin (Fd) can be replaced in the *in vitro* assay by reduced methyl viologen. Provided the methyl viologen is present in saturating amounts, no significant difference in activity with ferredoxin or methyl viologen has been found for the barley leaf enzyme [2]. For the determination of Fd-glutamate synthase activity the reaction mixture normally consists of 100 µl 100 mM glutamine in 500 mM KH_2PO_4-KOH (pH 7.5), 100 µl 100 mM 2-oxoglutarate in 500 mM KH_2PO_4-KOH (pH 7.5), 100 µl 150 mM methyl viologen and 400 µl of crude enzyme preparation. After pre-incubation at 30°C for 30 minutes the reaction can be started by the addition of 100 µl of reductant (47 mg Na_2S_2O4 and 50 mg $NaHCO_3$ dissolved per 1 ml of distilled water). After 20 minutes incubation at 30°C the reaction is terminated by the addition of 1 ml ethanol, with vigorous shaking to oxidize any remaining methyl viologen and dithionite. A blank without dithionite is run for each reaction. The reaction mixture for Fd-glutamate synthase determination in other plant tissues differs only slightly [2]. The transaminase inhibitor amino-oxyacetate should be added to the reaction mixture if there is any evidence of reductant-independent glutamate synthesis.

Fd-glutamate synthase activity is determined by the quantitative measurement of glutamate, the product of the reaction between glutamine and 2-oxoglutarate. Glutamate, therefore, has to be separated from the mixture after termination of the assay. Several different procedures have been applied [2], the simplest being paper chromatography. After centrifugation, 200 µl of the reaction mixture is spotted onto Whatman No. 4 chromatography paper. The separation of glutamate may be carried out by chromatography in 25% (w/v) phenol in the presence of a small beaker of concentrated ammonia. The papers should be dried thoroughly after use and sprayed with a freshly prepared specific ninhydrin reagent [5]. The reagent comprises 0.05 g cadmium acetate, 1.0 ml acetic acid and 5.0 ml H_2O in 50 ml acetone containing 0.5 g ninhydrin. The papers should be left in the dark in an enclosed container for a period of 12–18 h in the presence of a small beaker of concentrated sulphuric acid. The glutamate spots may then be cut out and the dark-red colour eluted with 8 ml of a reagent containing 600 ml ethyl acetate, 600 ml water, 600 ml methanol, 18 ml glacial acetic acid and 18 g cadmium acetate. The absorbance of the coloured solution may be determined at 500 nm. A standard curve of glutamate concentrations should be made, and an internal standard of 20 µl of a 1 mg ml^{-1} solution should always be run on each chromatogram, to allow for variations in relative humidity and temperature during development. There is a linear relationship between absorbance at 500 nm and the glutamate applied over a range of 0–50 µg.

Glutamate may also be separated from glutamine using Dowex-1 columns. The Dowex-1–chloride form may be converted to the acetate form by washing thoroughly with 1 M NaOH, 1 M acetic acid followed by excess distilled H_2O.

The assay mixture is completely transferred to the Dowex-1-acetate column (1 × 8, 200–400 mesh size, 12 mm diameter × 35 mm length). Glutamine is eluted from the column with 15 ml of distilled water. Residual water is removed by centrifuging the column for 2 minutes at 1500 rpm. Glutamate is eluted from the column with 5 ml of 3 M acetic acid and the centrifugation step is repeated. Quantification of glutamate is best carried out by a ninhydrin assay similar to that used for paper chromatography. An aliquot of 500 µl of the Dowex-1-acetate eluate is added to 1 ml ninhydrin reagent (0.4 g ninhydrin, 80 ml 95% ethanol, 1 g $CdCl_2$, 10 ml acetic acid, 20 ml H_2O). After incubation at 80°C for precisely 10 minutes the samples are cooled and the absorbance measured at 506 nm.

NADH-glutamate synthase:
The enzyme in roots, root nodules and developing fruits may be assayed by measuring the decrease in absorbance of NADH at 340 nm in a recording spectrophotometer. For measurement of NADH oxidation the assay mixture contains 0.1 mM NADH, 10 mM glutamine, 10 mM 2-oxoglutarate, 100 mM KH_2PO_4-KOH (pH 7.5) and 200 µl of crude extract. The rate of reaction for this enzyme is relatively low and care should be taken to avoid artefacts due to the presence of contaminating NADH oxidases and GDH. This is achieved by running separate controls with glutamine and 2-oxoglutarate each being omitted from the reaction medium. The reaction is followed at 340 nm for about 10 minutes at 30°C. Activity is determined by measuring the difference between the rate of NADH oxidation in the presence and absence of the substrates. The actual rate of the enzyme reaction can be calculated, as the extinction coefficient (*E*) of 1 mM NADH is 6.2 at 340 nm.

The use of inhibitors (e.g. methionine sulphoximine (MSO) for GS, and azaserine for glutamate synthase) in the study of nitrogen metabolism is well established [4]. However, it is the use of plants that lack GS and/or glutamate synthase [3] that have demonstrated the very close interdependence of CO_2 fixation with nitrogen metabolism. Although CO_2 fixation rates of mutants lacking GS, glutamate synthase and both enzymes were indistinguishable from wild type under non-photorespiratory conditions, a dramatic decrease in this rate could be seen for the mutant plants on transfer to air (Fig. 20.4). A similar situation has been found in plants treated with MSO, where there is a fall in the rate of photosynthetic CO_2 assimilation and an accumulation of ammonia. Close inspection of the change in the rate of CO_2 fixation with time exhibited by the mutants on transfer to air revealed that the rate detected in plants lacking GS fell more slowly than those lacking glutamate synthase. Simultaneous measurements of the ammonia content indicated that plants deficient in GS accumulated much more of this potentially toxic compound than similarly treated plants lacking glutamate synthase. Further investigation indicated that in air the fixation rate was progressively inhibited by increasing the light intensity above 400 $\mu mol\ m^{-2}\ s^{-1}$, while the content of ammonia remained virtually unchanged. It may be concluded that the ammonia may accumulate in a compartment, perhaps the vacuole, away from the chloroplast and has no direct effect on the photosynthetic machinery.

20.1.5 Glutamate dehydrogenase

As mentioned previously the enzyme can be readily detected in all higher plant tissues [1]. A standard extraction buffer is 50 mM Tris-acetate (pH 8.2), 10% (v/v) glycerol, 0.5 mM EDTA and 5 mM mercaptoethanol. Activity can be determined by measuring the rate of NADH oxidation at 340 nm in a spectrophotometer. The reaction mixture contains in a final volume of 1.0 ml: 150 mM ammonium acetate, 80 mM NADH, 12.5 mM 2-oxoglutarate, 0.8 mM calcium chloride and 50 mM Tris–acetate buffer (pH 8.2) plus enzyme extract. All solutions are adjusted to pH 8.2 with Tris. The absolute activity is determined by subtracting blank values of NADH oxidation run in the absence of 2-oxoglutarate and ammonia. Two important points to note are:

1. the concentration of the ammonium ion must be high to ensure that the enzyme is fully saturated;
2. there must be excess of a divalent cation (normally calcium) to activate the enzyme completely. It is not possible to make an accurate determination of glutamate dehydrogenase activity in the presence of EDTA alone, as the majority of the enzyme would be inhibited.

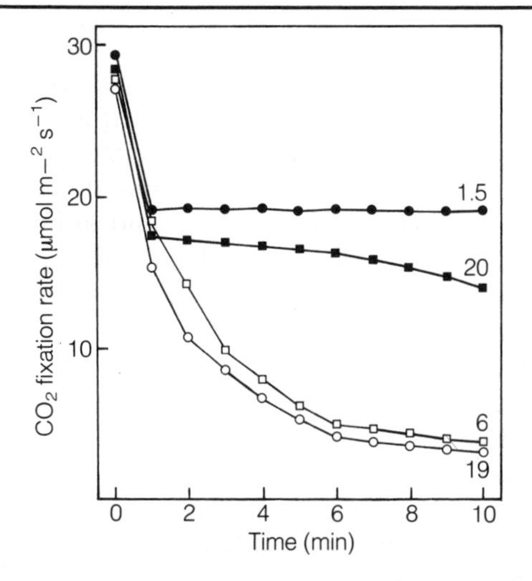

Fig. 20.4. Rate of CO_2 fixation of detached leaves of wild-type barley (●), mutants deficient in glutamine synthetase (■), glutamate synthase (□) and both activities (○), after transfer to air from 1% O_2. Figures after 10 minutes denote the ammonia content of the leaf in μmol per g fresh weight.

Glutamate dehydrogenase may also be assayed in the reverse direction by a similar procedure. The reaction mixture should contain: 15 mM glutamate, 2 mM NAD, 0.8 mM calcium chloride and 100 mM Tris–acetate buffer (pH 9.0). A buffer suitable for use at pH 10.0 may be employed if required. This assay mixture may be utilised to visualise glutamate dehydrogenase activity on polyacrylamide gels, following electrophoresis, in the presence of 0.3 mM nitroblue tetrazolium and 0.081 mM phenazine methosulphate. Frequently, seven bands of glutamate dehydrogenase activity are detected using this procedure.

20.1.6 Aminotransferases

Aminotransferases catalyse the transfer of the amino group of an amino acid to a 2-oxo acid yielding a different amino acid and 2-oxo acid. Assay methods may involve the measurement of the amino acid product by high performance liquid chromatography (HPLC), thin layer chromatography (TLC) or paper chromatography the keto acid product. Alternatively, may be determined by coupling to another reaction, usually one that involves NADH or NADPH, which permits assay by monitoring absorbance change at 340 nm. The 2-oxo product can also be determined by radioactive means or by derivatisation with dinitrophenyl-hydrazine or other reagents [6]. Probably the most extensively studied of the plant aminotransferases is aspartate aminotransferase (glutamate : oxaloacetate aminotransferase or aspartate: 2-oxo-glutarate aminotransferase, E.C. 2.6.1.1).

$$\text{Aspartate} + \text{2-oxoglutarate} \rightarrow \rightarrow \text{oxaloacetate} + \text{glutamate}$$

Pyridoxal phosphate (PLP) is a co-enzyme in aminotransferase reactions but in plants appears to be very tightly bound to the enzyme. Therefore, the addition of PLP to the assay mixture is usually necessary only for highly purified extracts or unusual cases, although PLP requirement should be checked for each enzyme under investigation.

Extraction is usually made into a buffer comprising 50 mM Tris-HCl (pH 7.5), 2 mM EDTA and 20% (v/v) glycerol. Probably the simplest method of testing for aminotransferase is to incubate the enzyme with 5 mM 2-oxo acid and 5 mM amino acid in 50 mM Tris–HCl (pH 7.5) buffer for varying times and stopping the reaction with an equal volume of ethanol. After centrifuging the protein, the extract may be chromatographed on paper or TLC plates in a solvent that gives a good separation of the initial and end-product amino acids (e.g. *n*-butanol : acetone : diethylamine : water (70 : 70 : 14 : 35 by volume) gives a good separation of the amino acids aspartate and alanine). The rate of synthesis of the product amino acid may be determined by the method of Atfield and Morris [5] described in the previous section.

A second method is to incubate the amino acid with a very small amount of the 2-oxo acid and determine the product 2-oxo acid by formation of a dinitrophenylhydrazone. Alanine aminotransferase may be assayed by incubating the enzyme with 100 mM alanine and 2 mM 2-oxoglutarate in 100 mM Tris-HCl (pH 7.4) buffer. The pyruvate formed may be determined by reaction with 2,4-dinitrophenylhydrazine and measuring the colour at 546 nm. A standard curve of varying pyruvate concentrations must be constructed.

A third more refined but expensive method is to couple the 2-oxo acid produced to NADH oxidation by an added enzyme. A standard reaction mixture may be set up containing 2.5 mM 2-oxoglutarate, 2 mM EDTA, 1 mM NADH and 40 µl enzyme extract. If aspartate aminotransferase is to be measured the product is oxaloacetate which is rapidly converted to malate by malic dehydrogenase (MDH) with the subsequent oxidation of NADH, which may be measured on a spectrophotometer at 340 nm. The final reaction medium in the spectrophotometer cell should also include 2.5 mM aspartate, 0.5 ml 100 mM HEPES (pH 8.0) buffer, 5 units MDH and 280 µl H_2O. In crude extracts of plants there is usually sufficient MDH already present to drive the reaction without further addition of the enzyme. If alanine aminotransferase is to be measured the product is pyruvate, which may be converted to lactate by lactate dehydrogenase. In this case the final reaction medium should also include 10 mM alanine, 0.5 ml 100 mM HEPES (pH 7.5) buffer, 100 µl 100-fold diluted commercial lactate dehydrogenase and 200 µl H_2O. The activity of the enzyme may be calculated in both cases using a blank reaction cuvette containing no 2-oxoglutarate.

20.2 TRANSPORT OF NITROGENOUS COMPOUNDS

20.2.1 Compound utilised

If ammonia is formed in the roots it has to be transported to the leaves and developing fruits in a non-toxic form. Nitrate on the other hand may be transported directly to the leaves in the xylem as the unmetabolised ion. The most common forms of transport compound are the amides glutamine and asparagine [1]. Arginine is frequently found in tree species and citrulline, and although originally isolated from the fruit of the water-melon (*Citrullus*), is utilised for nitrogen transport in alder (*Alnus*) species. Certain tropical legumes (e.g. *Glycine, Phaseolus* and *Vigna*) have the capacity to synthesise large amounts of the ureides, allantoin and allantoic acids in nitrogen-fixing nodules. There is now good evidence to show that ureides are synthesised by the oxidative catabolism of purines [7].

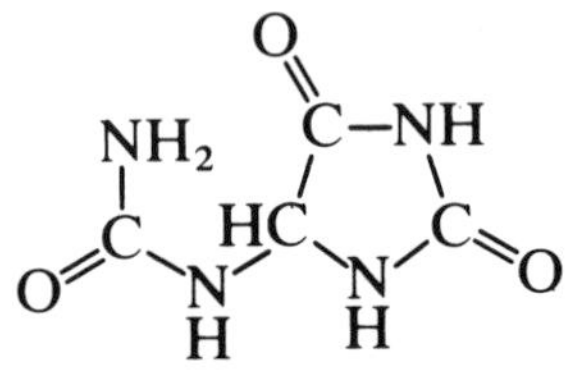

Allantoin

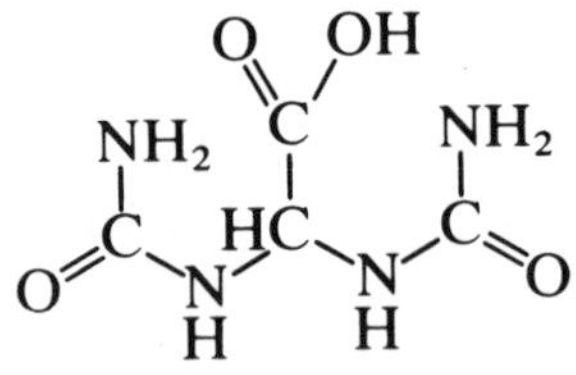

Allantoic acid

20.2.2 Collecting samples for analysis

The simplest method is to cut a plant close to ground level with a sharp razor blade and immediately place over the stump a silicon rubber (or other suitable plastic) tube of the same diameter. The root bleeding sap (RBS) exuding under root pressure into each piece of tubing may be sampled after 15–20 minutes using a Pasteur pipette or syringe. The method works well with glasshouse-grown plants, but sampling may be difficult in the field, particularly in dry areas. Watering the plants in advance of sampling will stimulate root bleeding, but this may cause unreliable results through its effects on the soil environment.

Vacuum-extracted xylem sap (VES) may be obtained from freshly detached stems by applying a mild vacuum (60–70 kPa) to the stem base and progressively cutting off small (3–5 cm) sections from the top of the shoot. Exudate may be collected in a 5 ml 'vacutainer' connected in series to the shoot and pump. If samples are to be transported long distances, they should be packed in ice to inhibit bacterial growth. If a vacuum apparatus is not available, the shoot sections may be collected and oven-dried at 80°C. The samples are then ground to a fine powder and extracted in boiling water. Such a procedure is rather harsh and may cause the breakdown of glutamine.

20.2.3 Assay of transport compounds

(a) Nitrate:

The nitrate content of the RBS or VES may be analysed by methods described in the previous chapter or by the salicylic acid technique [8].

(b) Amino Acids:

A qualitative determination of the amino acid content may be obtained by spotting 10–20 μl of the sap onto a cellulose TLC plate. The chromatogram is developed in n-butanol:acetone:diethylamine:water (70:70:14:35 by volume) and the plate sprayed with ninhydrin. The procedure gives a very good separation of aspartate, glutamate, glutamine, asparagine, glycine, alanine and serine. The individual amino acids may be quantified by the method of Atfield and Morris [5] as described previously.

More frequently it is only necessary to establish the total α-amino nitrogen content of the sap. Herridge [9] has described a procedure which involves the addition of 0.5 ml of citrate buffer (citric acid 16.8%, sodium hydroxide 6.4% in water) and 1.2 ml of ninhydrin reagent (ninhydrin 0.96% w/v and ascorbic acid 0.033% w/v in 2-methoxyethanol) to 0.05 ml of the sap in a 2.5 cm diameter glass test-tube. The tubes are placed in a boiling water bath for 25 min, allowed to cool and diluted with 3 ml of aqueous (60%) ethanol. The absorbance at 570 nm is determined in a spectrophotometer. In order to allow for the amino acid content of the sap, solutions of a 1:1 mixture of glutamine and asparagine are used as standard.

(c) Ureides:

The ureide content of the extracts may be qualitatively determined by spraying TLC plates with Ehrlich's reagent (1% solution of 4-dimethylamino benzaldehyde in 96% ethanol). Allantoin and allantoic acid give brown colours when the plates are exposed to HCl vapour.

The best quantitative method of ureide analysis involves the measurement of glyoxylate after hydrolysis. The allantoin in the sap is converted to allantoic acid by alkaline (pH 12) hydrolysis at 100°C for 5 minutes. The sap is then heated in 0.05 M HCl for 2 minutes at 100°C and immediately cooled on ice. The hydrolysed samples are then incubated with phenylhydrazone hydrochloride at a concentration of 0.66 mg ml^{-1} at 30°C for 15 minutes. The mixture should be cooled in a salt-ice bath and the phenylhydrazone oxidised to form the diphenylformazan by reaction with 3 M HCl and 3.3 mg ml^{-1} potassium ferricyanide. This forms a coloured 1,5-diphenylformazan, which absorbs strongly between 520 and 540 nm with a molar extinction coefficient of 51,000. Standard curves of allantoic acid must be prepared at least in duplicate before any determination can be made.

$$\text{Allantoin} \xrightarrow{\text{base } 100^\circ\text{C}} \text{Allantoic acid} \xrightarrow{\text{acid } 100^\circ\text{C}}$$

$$\text{Glyoxylate} \xrightarrow{\text{phenylhydrazine}} \text{Phenylhydrazone}$$

$$\xrightarrow[\text{strong acid}]{\text{ferricyanide}} \text{Diphenylformazane}$$

A second method involves the separation of the ureides from amino acids on a strongly acidic cationic exchange resin. The ureides are then oxidised to ammonia, which may be determined by the phenol hypochlorite assay described in Section 20.1.3.

20.2.4 Determination of Nitrogen Fixation

Allantoin and allantoic acid are synthesised in the nodule and are the major products of recently fixed nitrogen in tropical legumes. Within 30 minutes of exposure of nodulated roots to $^{15}N_2$, labelled ureides were detected in the xylem sap of cowpea. After 2 h, ureides contained over 90% of the ^{15}N in the xylem exudate.

Herridge [9] has established that there is a very strong correlation between the relative abundance of ureide nitrogen in the RBS or VES and the proportion of plant nitrogen derived from nitrogen fixation. This correlation was not affected by the plant genotype used or the strain of rhizobia. It is relatively simple using the techniques described in Section 20.2.3 to calculate the relative abundance of ureide-N:

$$\text{Relative ureide-N}(\%) = \frac{400a}{4a + b + c}$$

where a, b and c are, respectively, the molar concentrations of ureides, nitrate and α-amino acids. The equation takes into account the fact that ureides contain four atoms of nitrogen per molecule.

It is therefore possible, by collecting samples of RBS or VES in the field or in the glasshouse, to establish the contribution made by nitrogen fixation to the growth of a plant. This procedure does not require the use of an expensive gas-liquid chromatograph for acetylene reduction assays (described in Chapter 19) or a mass spectrometer for the determination of ^{15}N content.

20.3 PHOTORESPIRATORY CARBON METABOLISM

20.3.1 Introduction

Photorespiration is the light-dependent evolution of CO_2, a process that is carried out in the leaves of C_3 plants at the same time as photosynthetic CO_2 assimilation. The process is inhibited by elevated levels of CO_2 but stimulated by high light and O_2 concentrations. Photorespiration can be traced back to the dual substrate specificity of Rubisco (Chapter 16). When CO_2 acts as the substrate, phosphoglycerate is the product, which is then metabolised via the Calvin cycle. If oxygen is the substrate then only one molecule of phosphoglycerate is formed, plus the two carbon molecule, phosphoglycolate. The photorespiratory carbon and nitrogen cycle shown in Fig. 20.5 is simply a mechanism for regenerating phosphoglycerate from phosphoglycolate. This pathway is complex and requires the interaction of three types of organelle: chloroplasts, peroxisomes and mitochondria. Within the leaf these bodies are in close proximity, allowing rapid diffusion between the organelles. The key reaction in the cycle is the conversion of glycine to serine (reactions 6 and 7), which results in the liberation of equal amounts of ammonia and CO_2. The ammonia is rapidly reassimilated via the glutamate synthase cycle (Section 20.1), but the CO_2 is lost to the atmosphere.

It has been calculated that for every oxygenation of RuBP in air, there are 2.5 carboxylation reactions [10]. Methods for determining the rate of photorespiration are complex and subject to error due to the continuous reassimilation of CO_2 in the light. There is little doubt that in C_3 plants there is a major loss of carbon through the photorespiratory cycle, with estimates of the loss of

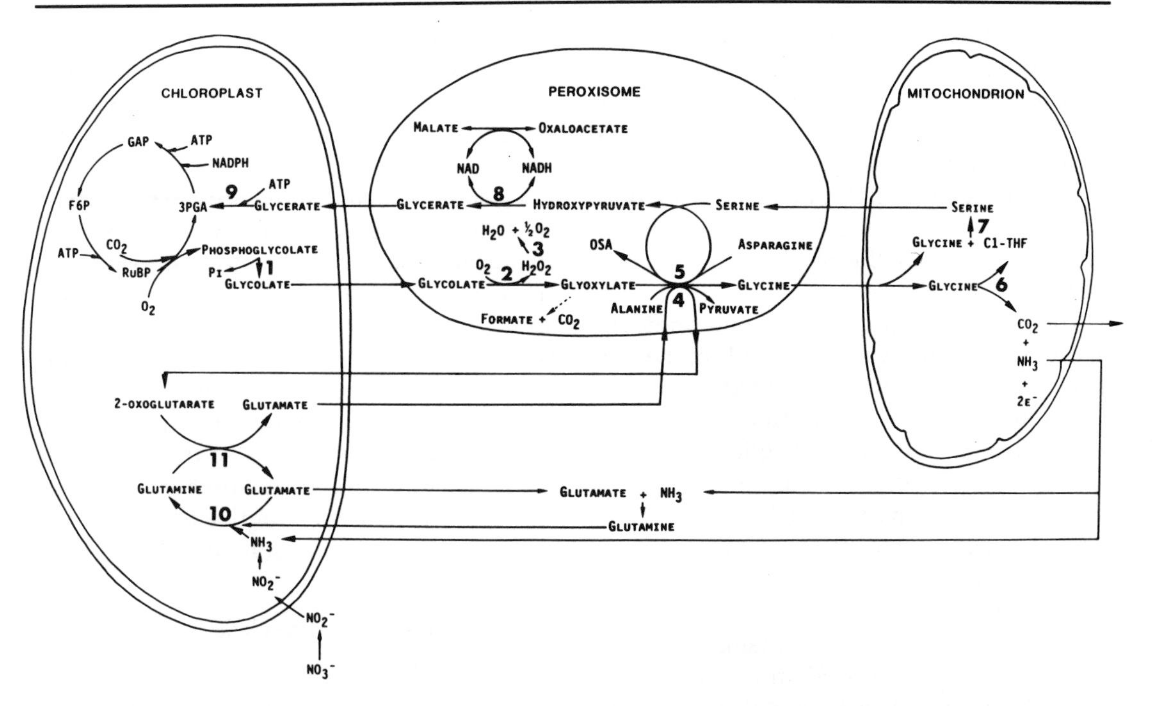

Fig. 20.5. The photorespiratory carbon and nitrogen cycle. Enzymes: (1) phosphoglycolate phosphatase; (2) glycolate oxidase; (3) catalase; (4) glutamate:glyoxylate aminotransferase; (5) serine:glyoxylate aminotransferase; (6) glycine decarboxylase; (7) serine transhydroxymethylase; (8) hydroxypyruvate reductase, (9) glycerate kinase; (10) glutamine synthetase; (11) glutamate synthase.

productivity ranging from 25% to 40% of the net photosynthetic rate [10].

Due to the potential increases in yield that could be obtained by the prevention of photorespiration, there is now a considerable amount of interest in the genetic manipulation of the enzyme Rubisco to prevent the oxygenase reaction. So far this has not proved possible.

A number of mutant lines of different plants have been isolated that are lacking one of the 11 enzymes of the photorespiratory carbon and nitrogen cycle (Fig. 20.5). These plants have the capacity to grow normally at elevated levels of CO_2, when the photorespiratory pathway does not operate. However when the mutant plants are transferred to air, there is a rapid fall in the rate of photosynthetic CO_2 assimilation due to a lesion in the photorespiratory cycle. Such mutant lines have been of considerable value in increasing our understanding of the photorespiratory process, and have highlighted the important role played by ammonia and amino acids.

All enzymes of the photorespiratory carbon pathway can be assayed following extraction in 40 mM Tricine (pH 7.8), 2 mM $MgCl_2$, 1 mM EDTA, 2% (w/v) insoluble polyvinylpyrollidone, 1 mM PMSF, 1 mM aminocaproic acid and 14 mM DTT [11].

20.3.2 Phosphoglycolate phosphatase

The simplest method to detect phosphoglycolate phosphatase (enzyme 1) activity is a two-step method which produces glycolate and inorganic phosphate (P_i) from phosphoglycolate, and then measures the P_i produced.

The 360 µl reaction mixture to produce the

glycolate and P_i contains 10 mM $MgSO_4$, 40 mM cacodylic acid (pH 6.3) and 10 mM phosphoglycolate. The reaction is initiated by the addition of 20 µl of crude extract and terminated after 5 minutes incubation at 30°C by the addition of 120 µl of 10% (w/v) TCA. The mixture is centrifuged to remove precipitated protein and the supernatant used to determine P_i.

The reagents required for P_i determination are 0.5% (w/v) TCA and acid molybdate reagent. The ammonium molybdate is made up as follows; prepare 100 ml of 4% (w/v) ammonium molybdate in distilled water and separately 100 ml of 2.5 M sulphuric acid. Mix these two solutions and make up to 250 ml with distilled water. Before use, $FeSO_4$ is added to this solution to make a final concentration of 5% (w/v).

The reaction mixture contains 0.5 ml of 0.5% TCA, 0.5 ml of acid molybdate/$FeSO_4$ reagent and 0.5 ml of supernatant from the phosphoglycolate phosphatase assay. This mixture is incubated for 15 minutes at 20°C and the absorbance of the blue colour is measured at 740 nm. Phosphate concentration and hence enzyme activity is then calculated by reference to a calibration graph prepared using phosphate standards made up in 2.5% (w/v) TCA.

20.3.3 Glycolate oxidase and catalase

Glycolate oxidase (enzyme 2) activity is most simply measured by monitoring the rates of oxygen uptake using an oxygen electrode with the following assay conditions.

The reaction mixture of 800 µl is made up with 50 mM Tris/HCl (pH 8.3), containing 1 mM sodium azide and 0.1 mM flavine mononucleotide with crude leaf extract and the reaction initiated by adding 100 µl of 10 mM glycolate in 50 mM Tris–HCl (pH 8.3). Sodium azide is essential to inhibit catalase activity. Methods of assaying the enzyme have also been developed based on the use of glycolate-dependent production of H_2O_2 by oxidation of dyes. Glyoxylate production can also be quantified by monitoring the phenylhydrazone derivative. However, this reaction is not totally specific for glyoxylate, since acids such as L-lactic acid may also react.

Catalase (enzyme 3) activity can be assayed spectrophotometrically by monitoring the change in absorbance at 240 nm. Into a quartz cell place 3.0 ml of 10 mM Tris-HCl (pH 8.5) and 0.1 ml of 0.88% H_2O_2 in 100 mM Tris–HCl (pH 8.5). Start the reaction by adding 0.2 ml crude extract and monitor the change in A_{240} against a blank of Tris-HCl and extract. It is important to use fresh reagents for this assay and to keep them at room temperature. Inhibition of this activity at 4°C (or kept on ice) has been reported.

20.3.4 Glycine decarboxylase

Glycine is decarboxylated in the mitochondrial matrix by the glycine cleavage system (E.C. 2.1.2.10; enzyme 6) to yield 5,10-methylene tetrahydrofolate, CO_2 and NH_3 (Fig. 20.6). The former compound reacts with a second molecule of glycine to form serine in a reaction catalysed by serine hydroxymethyltransferase. It is extremely difficult to measure the conversion of glycine to serine, except in isolated intact mitochondria. However it is possible to measure the enzyme *in vitro* following the exchange between the carboxyl carbon of glycine and ^{14}C-bicarbonate.

The reaction mixture, made up in 20 mM MOPS–KOH (pH 7.0) containing 0.1 mM pyridoxal phosphate, 20 mM glycine, 2 mM DTT and crude enzyme extract, is initiated by the addition of 4 µmol $NaH^{14}CO_3$ (55 mCi mol^{-1}). This is incubated for 30 minutes at 30°C and terminated with 25 µl of glacial acetic acid.

After termination, dissolved $^{14}CO_2$ is driven off by drying the samples under an infra-red lamp and acid-stable ^{14}C-glycine estimated by scintillation counting.

Glycine decarboxylation *per se* can be measured by trapping $^{14}CO_2$ liberated in an

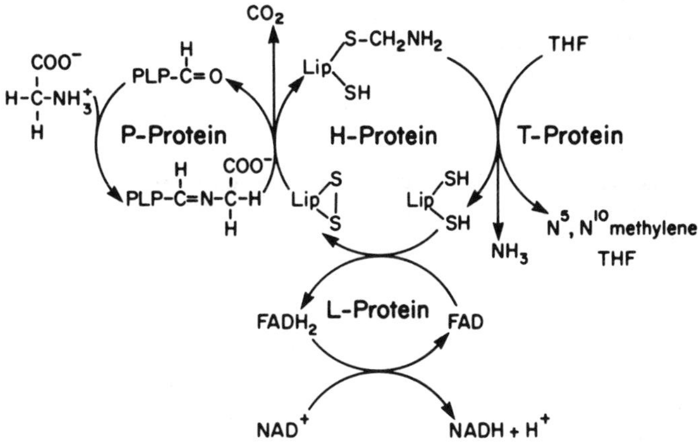

Fig. 20.6. The mitochondrial glycine decarboxylase complex showing the role played by the four individual proteins (kindly supplied by D.J. Oliver).

assay system supplied with ^{14}C-glycine. Crude enzyme extract is incubated with 8 mM ^{14}C-glycine (3.5 μCi $μmol^{-1}$) in a suitable assay buffer (20 mM MOPS–KOH, pH 7.0, 30 mM pyridoxal phosphate, 0.5 mM tetrahydrofolate, 1 mM NAD and 2 mM DTT). The reaction mixture is placed in small cups suspended over 10% (v/v) triethanolamine in 20 ml scintillation vials, and the reaction initiated by the addition of glycine. The reaction is terminated after 30 minutes by the addition of 100 μl of 6 M acetic acid and allowed to stand for 16 h to permit trapping of all $^{14}CO_2$. The amount of trapped ^{14}C radioactivity can then be determined by scintillation counting.

20.3.5 Serine transhydroxymethylase

The simplest estimation of serine transhydroxymethylase (enzyme 7) activity is based on the production of a radioactive C-1 unit in 5,10-methylene tetrahydrofolate from [3-^{14}C]-serine using the reverse reaction of serine transhydroxymethylase.

The complete assay system made up in 75 mM potassium phosphate buffer (pH 7.4) comprises 0.25 mM [3-^{14}C]-serine (6.56 μCi $μmol^{-1}$), 0.25 mM pyridoxal phosphate, 2 mM DL-tetrahydrofolate, 10 mM mercaptoethanol and enzyme extract.

The reaction mixture (except serine) is incubated for 5 minutes at 37°C and the reaction initiated by the addition of [3-^{14}C]-serine. After 15 minutes incubation at 37°C the reaction is terminated by the addition of 0.3 ml of 1 M sodium acetate (pH 4.5). This is followed by the addition of 0.2 ml of 0.1 M formaldehyde and 0.3 ml of 0.4 M dimedon (in 50% ethanol) and boiled for 5 minutes to accelerate formation of the formaldehyde/dimedon derivative. After cooling, the dimedon compound is extracted by shaking with 5 ml toluene at 20°C and the phases separated by centrifugation. The upper phase is removed for quantification of radioactivity in 5,10-methylene tetrahydrofolate by scintillation counting.

20.3.6 Hydroxypyruvate reductase

The assay for NADH or NADPH-dependent hydroxypyruvate reductase (enzyme 8) activity follows the oxidation of either NADH or NADPH spectrophotometrically at 340 nm.

The assay mixture (1 ml) contains 0.5 mM hydroxypyruvate, 25 mM phosphate buffer (pH 6.3), 0.2 mM NADH or NADPH and varying amounts of enzyme extract. Reactions are initiated by the addition of hydroxypyruvate and control assays contain the components except hydroxypyruvate to correct for non-specific oxidation of NADH or NADPH and the change in A_{340} is recorded.

20.3.7 Glycerate kinase

Glycerate kinase (enzyme 9) activity can be measured spectrophotometrically by linking 3-PGA and/or ADP formation to NADH oxidation in a coupled assay.

The most suitable set of coupling enzymes use a reaction mixture (volume 1.1 ml) placed in a cuvette containing 0.2 mM NADH (pH 7.0), 5 mM ATP (pH 7.0), 10 mM $MgCl_2$ and 100 mM Tris–HCl (pH 7.8), 5 units of phosphoglycerate kinase and 5 units of glyceraldehyde 3-phosphate dehydrogenase and varying amounts of crude enzyme extract. The reaction is initiated by adding 0.1 ml of 100 mM glycerate and the change in A_{340} measured.

20.3.8 Mutagenesis

A considerable amount of early biochemistry involved the use of mutant micro-organisms that lacked key enzymes of metabolic pathways. In this chapter we have shown how mutants of barley lacking GS and glutamate synthase have been used to study photorespiration: in a later section mutants with alterations in the feedback sensitivity of aspartate kinase will be described. There is little doubt that mutant selection in conjunction with genetic engineering will play an important role in improving the productivity of plants in the future.

The treatment described here for sodium azide can be used for most dry seeds. Other mutagens such as nitrosomethylurea (NMU) and ethyl methane sulphonate (EMS) have been used just as successfully with barley, *Arabidopsis*, pea and tomato. The method of mutagenesis for these chemicals is the same as for azide with alterations in concentration and incubation time. A simple test as to the effectiveness of a mutagen is the appearance of a high proportion of albino plants (5–10%) when the seeds are germinated [12].

Seeds should be soaked overnight in water at 4°C and then aerated vigorously for 8 h at 20°C. Incubation with sodium azide (1 mM) is carried out in 0.1 M phosphate buffer (pH 3.0) for 2 h. The seed is then washed in 2 volumes of distilled water and running tap water in the cold for 30 minutes. The seed should be dried and then planted out in the field or in pots. The seed must be grown through a complete generation to produce the M_2 seed that is used for the selection of mutants.

Safety Note:

Azide is a volatile mutagen and respiratory poison, coupled with which it may form an explosive mixture if poured down the sink, due to a reaction with heavy metals, e.g. lead and mercury. All handling of azide must therefore be carried out in a fume cupboard, and the original solution and first two washes must be treated with 15% (w/v) ammonium ceric nitrate before disposal, after which treatment azide is decomposed. This method is completely safe if reasonable precautions are taken.

20.4 THE BIOSYNTHESIS OF AMINO ACIDS

20.4.1 Introduction

It is still generally accepted that only 20 amino acids are incorporated into protein. However, there may be considerable post-translational modification (e.g. the formation of N-methylysine, 4-hydroxyproline and 4-carboxyglutamate). Other amino acids (e.g. homoserine, ornithine and diaminopimelic acid) are formed during metabolic interconversions. However, at least 200 other 'non-protein' amino acids occur in plants, some of which have a restricted taxonomic distribution.

The protein amino acids can be divided into separate 'families' depending on the precursor compound common to their synthesis [13].

*Aspartate (asparagine, **lysine, isoleucine, methionine** and **threonine**):*
Asparagine is synthesised as a nitrogen transport compound and at times of nitrogen excess by the glutamine-dependent amination of aspartate. The enzymology of this reaction has remained particularly elusive in higher plants, but recently it has proved possible to isolate the gene coding for the enzyme [1]. The synthesis of the other amino acids will be discussed in Section 20.4.2.

*Glutamate (glutamine, arginine and **proline**):*
The synthesis of glutamine in the primary assimilation of ammonia has been discussed in Section 20.1. Arginine is synthesised via ornithine, citrulline and arginosuccinate in a similar pathway to the urea cycle. Proline is synthesised via the cyclisation of glutamate semialdehyde to yield pyrroline-5-carboxylate which is then reduced. Only the last enzyme pyrroline-5-carboxylate reductase has been characterised in higher plants. There is evidence that arginine can be catabolised via ornithine to yield glutamate [13].

*Phosphoenol pyruvate (the aromatic amino acids **phenylalanine, tyrosine** and **tryptophan**):*
Phosphoenol pyruvate and erythrose-4-phosphate are converted by a seven-step series of reactions to yield chorismate. At this stage, chorismate serves as the branch point for the synthesis of tryptophan or phenylalanine and tyrosine. The sixth enzyme in the pathway, enol-pyruvylshikimate phosphate (EPSP) synthase has been the focus of considerable interest due to its involvement in the action of the herbicide glyphosate. Inhibition of EPSP synthase results in the accumulation of shikimate phosphate or shikimate a process that can be reversed by the addition of the aromatic amino acids. Chorismate may either be converted to anthranillate which is a precursor of tryptophan or to prephenate. Two alternate pathways are available for the conversion of prephenate to phenylalanine and tyrosine. The precise route taken in individual plants has not yet been clarified.

*Pyruvate (alanine, **leucine** and **valine**):*
Pyruvate may be transaminated directly to yield alanine: glutamate is the normal amino donor. The synthesis of leucine and valine requires two molecules of pyruvate, whilst isoleucine is derived from one molecule of pyruvate and 2-oxobutyrate, formed by the deamination of threonine. The pathways of isoleucine, leucine and valine synthesised are considered to be biochemically parallel, with the four reactions being catalysed by enzymes possessing dual substrate specificity. Acetolactate synthase the first common enzyme in the pathway is sensitive to inhibition by two classes of herbicides, the sulphonylureas and the imidazilinones. The herbicides act by starving the plants of the three branched chain amino acids, isoleucine, leucine and valine.

*3-Phosphoglycerate (serine, glycine and **cysteine**):*
The majority of textbooks state that phosphoglycerate can be metabolised to serine via hydroxypyruvate. Serine may then act as a precursor of cysteine following acetylation and reaction with sulphide. The conversion of serine to glycine by serine hydroxymethyltransferase is also necessary for the synthesis of one-carbon moieties which are utilised in a variety of biosynthetic reactions.

However, as can be seen in Section 20.3, glycine and serine are major intermediates in the photorespiratory carbon and nitrogen cycle. It is likely that in C_3 plants, and to a lesser extent, in C_4 plants, the majority of glycine and serine is synthesised in the leaves in the light and transported to other organs of the plant.

Ribose-5-phosphate:
Ribose-5-phosphate is the precursor of *histidine* in micro-organisms in a 10-step pathway. Little work has been carried out on histidine synthesis in higher plants. Plants

and bacteria are able to synthesise all 20 protein amino acids. Animals, however, are unable to synthesise the eleven amino acids marked above in bold italics. These are normally termed 'essential' amino acids. Animals do have the capacity to convert phenylalanine to tyrosine and methionine to cysteine. Thus cysteine and tyrosine are not considered essential provided there is an adequate supply of the other amino acids. There is now a wealth of evidence to suggest that those amino acids that animals require in their diet, are synthesised solely inside the chloroplasts in higher plants.

20.4.2 The aspartate family

A pathway that has been extensively studied is the synthesis of lysine, threonine, methionine and isoleucine from aspartate (Fig. 20.7). These amino acids frequently limit the quality of plant foodstuffs; in particular, cereal seeds are deficient in lysine and legume seeds deficient in methionine. The ability of chloroplasts to synthesise these amino acids can be readily demonstrated using intact chloroplasts prepared by the method described in Chapter 17.

Chloroplasts containing 100 µg chlorophyll should be resuspended in 0.5 ml buffer containing sorbitol (300 mM), EPPS (pH 8.3) and 30 mM KCl, with 2 µCi of ^{14}C-aspartate for 20 minutes in a bright light (800 µmol $m^{-2} s^{-1}$). Controls should be run in complete darkness. The reaction is stopped by the addition of an equal volume of ice-cold 10% TCA. The reaction products can be separated by two-dimensional chromatography on paper or TLC plates.

Amino acids can be identified by co-chromatography with authentic standards. Radioactivity in each amino acid is determined by scraping each spot into a scintillation vial. A full description of this technique is given in Mills *et al.* [14]. Pea leaf chloroplasts readily convert aspartate to lysine, threonine and homoserine.

Initial experiments suggested that chloroplasts could also synthesise methionine, but later works showed that homocysteine was the chloroplast product and the last step of the methionine synthesis took place in the cytoplasm. It is interesting that the requirement for methionine in animals can be met by homocysteine, and that animals can also carry out the final methylation step.

The complex pathway to lysine, threonine, methionine and further on to isoleucine requires a considerable amount of energy input in the form of NADPH and ATP. If these reactions are carried out in the chloroplast the energy is derived directly from light. The light dependence of the reactions can be demonstrated by keeping the reaction tubes in the dark, when little conversion of aspartate to other amino acids takes place. As these reactions utilize NADPH and ATP directly from the electron transport chain of the light process, they are true photosynthesis in the same manner as CO_2 fixation. If amino acid synthesis is carried out in the root or maturing seed then the energy required for the synthesis must be derived from previously fixed carbon via oxidation in the mitochondria. It is possible to calculate the amount of glucose required to synthesise 1.0 g of methionine in the root from nitrate and sulphate as 2.13 g. If the reaction had gone on in the chloroplast using light energy then the cost would only be 0.51 g. Thus there is a considerable saving in energy if plants carry out their synthetic reactions in the chloroplast, and there could be a considerable increase in total yield if plants were selected which carried out more of their metabolism in the leaf rather than in the root or seed.

The metabolic interconversions in the different pathways are subject to strict feedback regulatory control by the end-product amino acids. Lysine almost totally inhibits its own synthesis at levels above 1.0 mM, and also that of homoserine and threonine. Threonine, however, only inhibits its own synthesis and that of homoserine. The feedback inhibition can be readily explained by the known properties of the enzymes involved in the path-

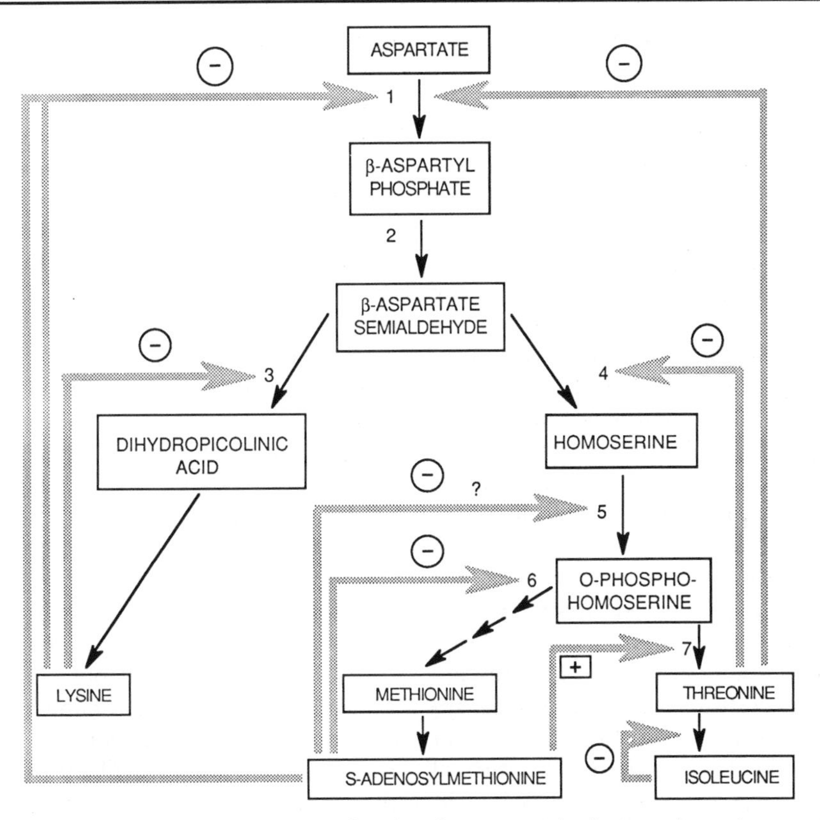

Fig. 20.7. The biosynthesis of the aspartate family of amino acids (lysine, threonine, methionine and isoleucine) showing the key regulatory steps. Enzymes: (1) aspartate kinase; (2) aspartate semialdehyde dehydrogenase; (3) dihydrodipicolinate synthase; (4) homoserine dehydrogenase; (5) homoserine kinase; (6) β-cystathionine synthase; (7) threonine synthase. (−) inhibition of enzyme activity apart from on (6) which is probably through repression. (+) stimulation of enzyme activity. (kindly supplied by R. Azevedo).

way. Aspartate kinase which catalyses the first step, the formation of aspartyl phosphate, is inhibited strongly by lysine and to a lesser extent by threonine. The inhibition by lysine and threonine is additive suggesting the presence of at least two separate enzymes. Lysine is also able to inhibit the first enzyme unique to its own synthesis, dihydrodipicolinate synthase, and in a similar manner threonine inhibits homoserine dehydrogenase (Fig. 20.7). The plant is thus able to carefully regulate the flow of metabolites along the complex pathways of amino acid synthesis. High levels of amino acids do not build up, and the rate of synthesis is then determined by the rate of incorporation into protein. A similar system operates in bacteria, where the regulation of metabolite flow is even more important. Bacteria are also able to regulate amino acid synthesis by 'repressing' the amount of a particular enzyme synthesised within the cell. In most cases such a system does not operate in higher plants. However, recent evidence now suggests that the first enzyme unique to methionine synthesis (which is not feedback regulated), cystathionine-γ-synthase, is repressed by methionine [13].

20.4.3 Selection of amino acid metabolism mutants

In Fig. 20.7 it can be seen that lysine and threonine are able to inhibit the passage of carbon from aspartate to homoserine by inhibiting both aspartate kinase and homoserine dehydrogenase.

If they are added together to a growing plant they are able to kill the plant by preventing the formation of methionine due to the prevention of homoserine synthesis. Under normal conditions it is not possible to add lysine and threonine to growing plants: (1) due to the large cost of the amino acids required to feed plants growing in a pot or in the field; (2) due to the stimulation of bacterial and fungal growth which will remove the amino acids and interfere with the natural growth of the plants. This problem was initially overcome by growing plants in tissue culture, but it was not always possible to regenerate healthy plants after the tests had been carried out. Studies at Rothamsted over the last few years by Dr. S.W.J. Bright have developed a technique for growing young barley plants in sterile culture, in such a way that the action of various compounds can be tested, and viable plants can be obtained at the end of the experiment [15].

Method:
Barley seeds should be dehusked in 50% (v/v) H_2SO_4 for 3 hours, washed in tap water and three changes of distilled water before being soaked overnight at 5°C. A wash in saturated $CaCO_3$ solution for 20 minutes after the acid treatment is also recommended. The embryos may then be dissected by hand from the seeds and allowed to dry on filter paper at 25°C for at least 24 hours. Petri dishes should be prepared containing agar ($6\,g\,l^{-1}$), Murashige and Skoog's medium without hormones or casein hydrolysate and sucrose ($30\,g\,l^{-1}$) which has been autoclaved previously. If amino acids are to be tested they should be sterile filtered and added to the mixture whilst it is still liquid; 20 ml of medium may be added to a standard 9 cm plastic petri dish.

Dry embryos should be sterilised by shaking for 15 minutes at 25°C in the supernatant of $70\,g\,l^{-1}$ calcium hypochlorite with a small amount of detergent added as a wetting agent. The embryos are then rinsed in sterile distilled water, washed for 5 minutes in sterile 0.01 M HCl and finally rinsed with 400 ml sterile distilled water. Embryos should be placed on the agar medium under sterile conditions so that the scutellum faces downwards. The plates should be incubated for 7 days at 25°C under white light (300 $\mu mol\,m^{-2}\,s^{-1}$) with a 16-hour day. The embryos develop roots and leaves, and examples of the plants at different stages of growth can be seen in Fig. 20.8. After the leaves reach 6 cm long the plants may be successfully transplanted to a small pot of compost and provided they are maintained in a humid atmosphere for a further period until the roots establish, they will grow into healthy mature plants.

If the plants are grown in the presence of 2 mM lysine and 2 mM threonine then there is an 80% reduction in growth. The inhibition in growth can be prevented by the addition of 0.5 mM methionine to the medium, suggesting that the action of lysine + threonine is to prevent methionine synthesis (Fig. 20.7). If normal embryos are grown in the presence of lysine and threonine then none of them will show any significant leaf development. However, if the seeds are first treated with a mutagen then there is a large increase in the probability of obtaining a mutant plant that is in some way resistant to the toxic action of lysine + threonine.

The majority of mutants examined have alteration in the feedback regulation of the enzyme aspartate kinase (E.C. 2.7.24). The enzyme which converts aspartate into aspartyl phosphate can be separated into three isoenzymic forms in barley, one sensitive to inhibition by threonine and two sensitive to lysine [14]. Mutations were found

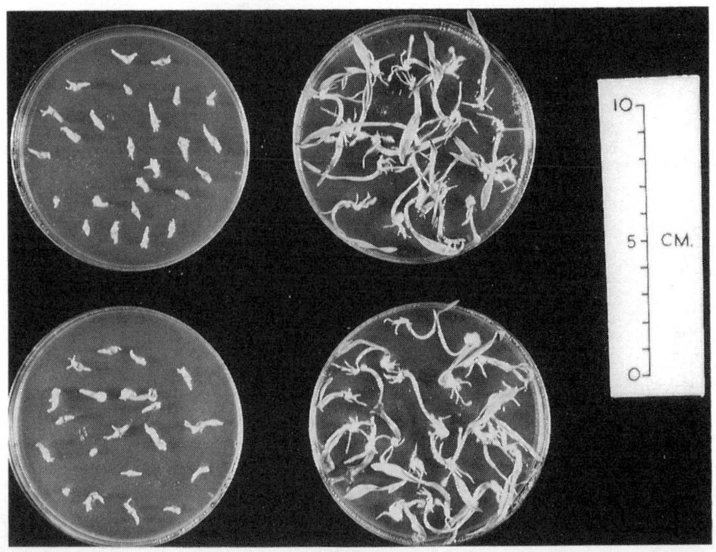

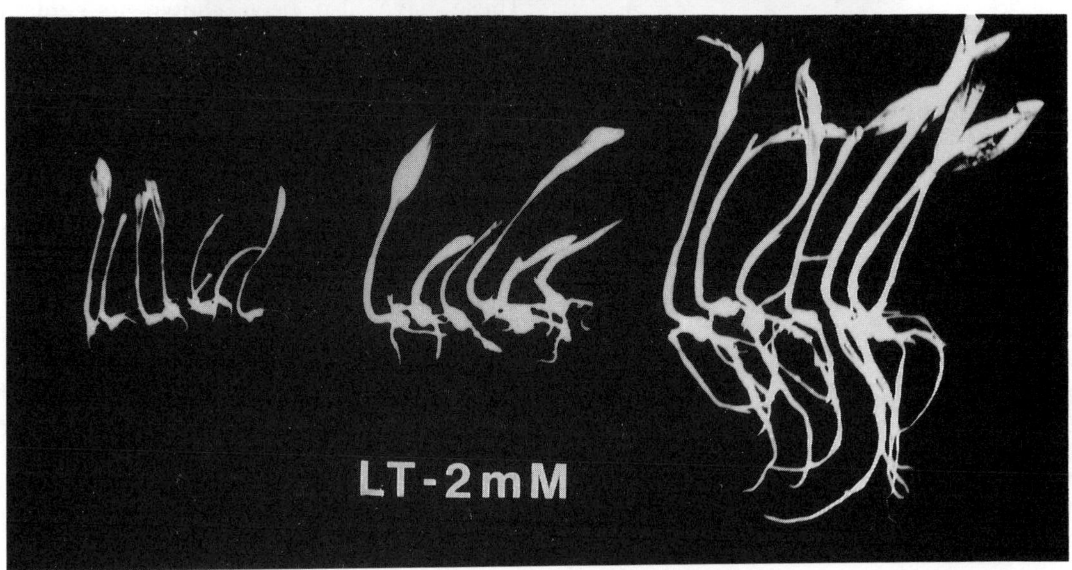

Fig. 20.8. (a) Embryos of barley after 2 days (left) and 7 days (right) growth on Murashige and Skoog's medium. (b) Embryos of maize after 7 days growth on 2 mM lysine plus 2 mM threonine medium showing from left to right sensitive (lt/lt), heterozygous (Lt/lt) and resistant (Lt/Lt) genotypes [16].

in one or other of the lysine-inhibited isoenzymes such that the enzyme activity had a greatly reduced sensitivity to lysine. In Fig. 20.9 the sensitivity of aspartate kinase II (cv. Bomi) is shown; the enzyme is about 80% inhibited by 1 mM lysine. However the enzyme from the homozygous mutant line is totally insensitive to lysine, whilst the heterozygous line shows intermediate sensitivity.

The soluble levels of lysine, threonine and methionine in the mutants analysed so far at Rothamsted are shown in Table 20.2. It is

Table 20.2. Threonine, lysine and methionine content of the free amino acid pools of mature grains from four barley mutants resistant to lysine plus threonine.

Plant	Amino acid content (nmol per g dry weight)			
	Threonine	Lysine	Methionine	Total
Bomi	118	84	24	10 400
R2501	6129	1266	149	28 200
R2506	9032	1620	72	20 400
Bomi	127	79	19	6 900
R3004	2376	107	23	12 400
R3202	125	109	19	11 700
Double mutant R3004 × R3202	689	129	21	10 100

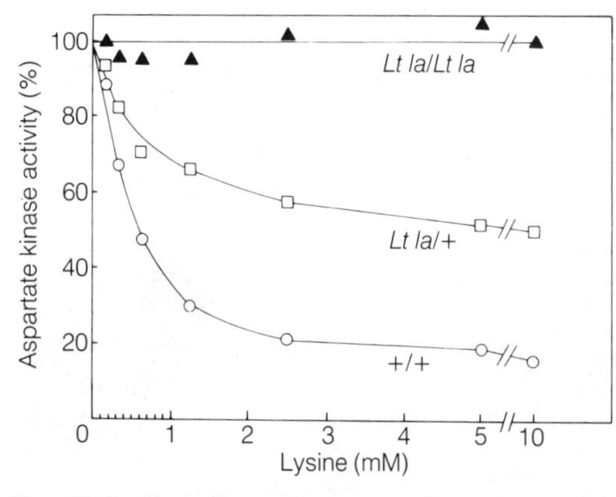

Fig. 20.9. The effect of lysine on the activity of the barley isoenzyme of aspartate kinase AKII. (▲) homozygous mutant plant resistant to lysine plus threonine, (□) heterozygous mutant plant, (○) wild type plant, sensitive to the inhibitory action of lysine plus threonine [15].

clear that an alteration to the feedback sensitivity of aspartate kinase can cause a build up in the seed of the levels of threonine and to a lesser extent lysine. The quantities are sufficient to eliminate the second limiting amino acid status of the threonine in the barley grain and to make a significant contribution to reducing the requirement for lysine supplementation. Similar attempts have been made to select mutants of maize that are resistant to lysine and threonine. These have also been shown to contain aspartate kinase with altered feedback inhibition and elevated levels of threonine [16–19].

20.4.4 Isolation and characterisation of aspartate kinase

A suitable extraction buffer is 25 mM potassium phosphate (pH 7.5), 2 mM $MgCl_2$, 2 mM EDTA, 15% (v/v) glycerol and 0.2% (v/v) 2-mercaptoethanol. Prior to assay the extract should first be precipitated with 65% saturation ammonium sulphate and passed through Sephadex G-25. The assay is similar to that used for glutamine synthetase. The mixture should contain 75 mM L-aspartate, 15 mM $MgCl_2$, 30 mM ATP and 225 mM hydroxylamine hydrochloride at pH 7.4; zero time and minus-aspartate blanks should be used. The formation of aspartyl-hydroxamate should then be assayed using the ferric chloride reagent described in Section 20.1.3. A more refined assay using ^{14}C-aspartate is also available [20].

In order to separate the three isoenzymes of aspartate kinase in barley, the enzyme should be subjected to DEAE–Sephacel chromatography. The extract should be loaded in the presence of 50 mM KCl on to a 1 × 11 cm column and washed with 30 ml of extraction buffer plus 75 mM KCl; this will elute the threonine-sensitive isoenzyme AKI. The first lysine-sensitive isoenzyme, AKII, can then be eluted with 15 ml of extraction buffer plus 120 mM KC1. The second lysine-sensitive isoenzyme, AKIII, requires 20 ml of extraction buffer plus 220 mM KCl for complete elution. The feedback sensitivities of the different fractions can then be tested by carrying out the aspartate kinase assays in increasing concentrations of lysine as shown in Fig. 20.9.

20.4.5 Proline and drought stress

When subject to a reduction in the water potential of tissues or cells, the majority of plants accumulate the cyclic imino acid proline [13,21]. Examples of estimates of the lower limit to the threshold water potential for the response are barley (−0.7 MPa), cotton (−1.2 MPa) and sorghum (−2.4 MPa). The concentration of proline may rise to 10% of the total leaf dry weight although most values are in the region of 20–30 mg g^{-1} dry weight. Most studies have examined the proline concentration in shoot or leaves but the amino acid will also accumulate in the root and other organs. Proline accumulation may also be stimulated by high or low temperatures, salinity and abscisic acid.

A number of workers have attempted to establish a correlation between the ability of a plant to accumulate proline and drought resistance. The subject has caused some controversy and it is difficult to assess the importance of proline accumulation in the complexity of factors which determine crop resistance to water deficit in the field.

Assay for proline:

Approximately 200 mg of tissue should be extracted successively with 0.5 ml 3% (w/v) 5-sulphosalicylic acid four times in a glass pestle and mortar. The pooled homogenate is centrifuged at 500 g for 10 minutes and reacted with 2 ml glacial acetic acid and 2 ml of acid ninhydrin (1.25 g ninhydrin warmed in 30 ml glacial acetic acid plus 20 ml 6 M phosphoric acid; this mixture is stable at 4°C for 24 hours) for 1 hour at 100°C in a boiling tube. The mixture is then cooled in an ice bath and 4 ml of toluene added. After vigorous mixing the two layers separate out and the pink-red colour at the top may be removed with a Pasteur pipette. The top colour-containing layer should be allowed to warm up to room temperature and the absorbance read at 520 nm using toluene as a blank in a spectrophotometer. A standard curve should always be obtained at the same time. The absorbance is linear up to 40 μg of proline per 2 ml sample.

Note:

It is important for this assay that either fresh tissue or that which has been rapidly frozen (e.g. in liquid nitrogen) is used. If tissue is allowed to dry after it has been removed from the plant (even for a very short period), proline will accumulate and give spurious results.

REFERENCES

1. Lea, P.J., S.A. Robinson and G.R. Stewart (1990) The enzymology and metabolism of glutamine and asparagine. In: *The Biochemistry of Plants*, Vol. 16 (B.J. Miflin and P.J. Lea, eds.) pp. 121–159. Academic Press, San Diego.
2. Lea, P.J., R.D. Blackwell, F-L. Chen and U. Hecht (1990) Enzymes of ammonia assimilation. In: *Methods in Plant Biochemistry*, Vol. 3, (P.J. Lea, ed.) pp. 257–276. Academic Press, London.
3. Blackwell, R.D., A.J.S. Murray, P.J. Lea and K.W. Joy (1988) Photorespiratory amino donors, sucrose synthesis and N metabolism in a photorespiratory mutant barley deficient in chloroplastic glutamine synthetase and/or ferredoxin dependent glutamate synthase. *J. Exp. Bot.* **39**, 845–858.
4. Lea, P.J. and S.M. Ridley (1989) Glutamine synthetase and its inhibition. In: *Herbicides and Plant Metabolism* (A.D. Dodge, ed.) pp. 137–167. Cambridge University Press, London.
5. Atfield, G.N. and C.J.O.R. Morris (1961) Analytical separation by high voltage electrophoresis of amino acids in protein hydrosylates. *Biochemical J.* **81**, 606–613.
6. Ireland, R.J. and K.W. Joy (1990) Aminotransferases. In: *Methods in Plant Biochemistry*,

Vol. 3, (P.J. Lea, ed.) pp. 277–286. Academic Press, London.

7. Schubert, K.R. and M.J. Boland (1990) The Ureides. In: *The Biochemistry of Plants*, Vol. 16, (B.J. Miflin and P.J. Lea, eds.) pp. 197–282. Academic Press, San Diego.
8. Cataldo, D.A., M. Haroon, L.E. Schrader, V.L. Youngs (1975) Rapid colorimetric determination of nitrate in plant tissues by nitration of salicylic acid. *Communications in Soil Science and Plant Analysis* **6**, 71–80.
9. Herridge, D.F. and M.B. Peoples (1990) Ureide assay for measuring nitrogen fixation by nodulated soybean calibrated by ^{15}N methods. *Plant Physiol.* **93**, 495–503.
10. Keys, A.J. (1986) Rubisco, its role in photorespiration. *Phil. Trans. Roy. Soc. Lond. B.* **313**, 325–336.
11. Blackwell, R.D., A.J.S. Murray and P.J. Lea (1990) Enzymes of the photorespiratory carbon pathway. In: *Methods in Plant Biochemistry*, Vol. 3, (P.J. Lea, ed.) pp. 129–144. Academic Press, London.
12. Kleinhofs, A., R,L. Warner, F.J. Muehlbauer and R.A. Nilan (1978) Induction and selection of specific gene mutations in *Hordeum* and *Pisum*. *Mutation Research* **51**, 29–35.
13. Bryan, J.K. (1990) Advances in the biochemistry of amino acid biosynthesis. In: *The Biochemistry of Plants*, Vol.16 (B.J. Miflin and P.J. Lea, eds.) pp. 161–195. Academic Press, San Diego.
14. Mills, W.R., P.J. Lea and B.J. Miflin (1980) Photosynthetic formation of the aspartate family of amino acids in isolated chloroplasts. *Plant Physiol.* **65**, 1166–1171.
15. Arruda, P., S.W.J. Bright, J.S.H. Kueh, P.J. Lea and S.E. Rognes (1984) Regulation of aspartate kinase isoenzymes in barley mutants resistant to lysine plus threonine. *Plant Physiol.* **76**, 442–446.
16. Azevedo, R.A., J.L. Arana and P. Arruda (1990) Biochemical genetics of the interaction of the lysine plus threonine resistant mutant *Ltr l* with *opaque-2* maize mutant. *Plant Science* **70**, 81–90.
17. Dotson, S.B., D.A. Frisch, D.A. Somers and B.G. Gengenbach (1990) Lysine-insensitive aspartate kinase in two threonine-overproducing mutants of maize. *Planta* **182**, 546–552.
18. Shaul, O. and G. Galili (1992) Increased lysine synthesis in tobacco plants that express high levels of bacterial dihydrodipicolinate synthase in their chloroplasts. *Plant J.* **2**, 203–210.
19. Frankard, V., M. Ghislain and M. Jacobs (1992) Two feedback-insensitive enzymes of the aspartate pathway in *Nicotiana sylvestris*. *Plant Physiol.* **99**, 1285–1293.
20. Bonner, P.L.R. and P.J. Lea (1990) Enzymes of lysine synthesis. In: *Methods in Plant Biochemistry*, Vol. 3. (P.J. Lea, ed.) pp. 297–313. Academic Press, London.
21. Hanson, A.D. and W.D. Hitz (1982) Metabolic responses of mesophytes to plant water deficits. *Ann. Rev. Plant Physiol.* **33**, 163–203.

21

Microalgae: laboratory growth techniques and the biotechnology of biomass production

A. Vonshak

21.1 INTRODUCTION

Microalgae are photosynthetic micro-organisms which contain at least one type of chlorophyll – chlorophyll *a*. They are considered to be one of the most versatile groups of organisms in terms of their size, form and ecological function. Microalgae may vary in shape from small single cells of less than 1 μm in diameter to branched filamentous multicellular types. They are found in fresh-water habitats, marine environments and salt marshes, in hot springs and under ice.

Microalgae, including the prokaryotic types known as cyanobacteria (blue-green algae), play an important part in both aquatic and soil environments. For example, the nitrogen-fixing capability of cyanobacteria plays a crucial role in providing nitrogen fertiliser in rice fields and a number of other soils, and many green algae help to maintain soil structure. Furthermore, the role of algae in both water purification and pollution is increasingly being recognised: there are opportunities to improve their participation in water treatment and in abating the problems of toxins produced in polluted water. In the aquatic food chain, primary production by microalgae leads to the growth and production of many fish and other seafood products. The importance of and demand for microalgae have therefore expanded with the growing industry of aquaculture.

The importance of algal culture in physiological and biochemical research, and the fact that algae are among the most efficient converters of solar energy to useful form, have led to increased interest in techniques of algal culture. Algal-culturing techniques can be divided into two categories: the first applies to laboratory conditions and a controlled environment, and the second to outdoor conditions for the large-scale production of biomass.

This chapter aims to provide basic techniques and concepts of algal culture and to present a state of the art report on algal biomass production, its problems and achievements. It is not the intention to provide a complete comprehensive manual of the techniques of algal cultivation, and the reader is encouraged to make use of the

Photosynthesis and Production in a Changing Environment: a field and laboratory manual
Edited by D.O. Hall, J.M.O. Scurlock, H.R. Bolhàr-Nordenkampf, R.C. Leegood and S.P. Long
Published in 1993 by Chapman & Hall, London.
ISBN 0 412 42900 4 (HB) and 0 412 42910 1 (PB).

references at the end of the chapter. However, I will try to present an overall approach to algal growth techniques and to point out problems which the scientist should be aware of when planning his or her own growth system and choosing the parameters to be measured.

In this chapter, I discuss only aquatic microalgae which grow photoautotrophically (i.e. requiring only light, CO_2 and inorganic nutrients) and reproduce solely by cell division.

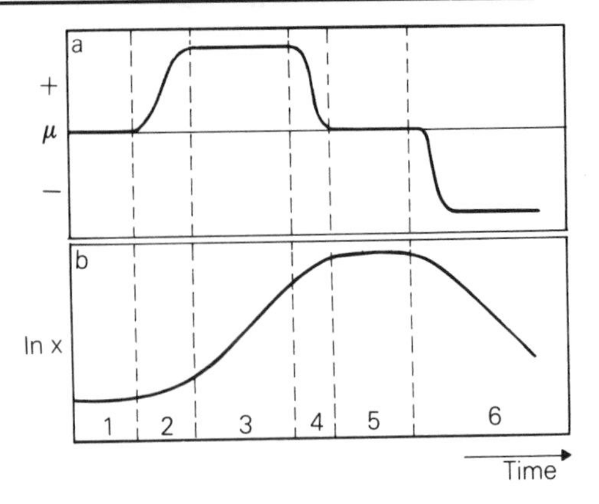

Fig. 21.1. Schematic representation of changes in (a) growth rate (μ) and (b) biomass (x) as a function of time in batch culture. Numbers refer to the various growth phases described in the text.

21.2 GROWTH OF MICROALGAE: TECHNIQUES AND KINETICS

The algae dealt with here behave as simple microorganisms, undergoing a non-sexual life cycle and multiplying only by cell division. Hence common growth techniques and mathematical analyses used in bacteriological studies may also be applied to these algal cultures. Algae may be grown in batch cultures or in continuous culture.

21.2.1 Batch culture

A batch culture is initiated by the transfer of a small portion of a culture into a new culture medium, resulting in growth and an increase in biomass. Biomass concentration can be measured in many ways: as cell number, dry weight, packed cell volume, or in terms of any convenient biochemical component or parameter. The rate of increase in biomass concentration is generally expressed by the specific growth rate (μ), which is calculated according to the following formula:

$$\mu = \frac{1}{x} \cdot \frac{dx}{dt}$$

where x is the biomass concentration. Thus μ represents the increase in biomass per unit time per average biomass. Units are reciprocal time, i.e. s^{-1} or d^{-1}.

The changes in specific growth rate during the development of a culture are shown in Fig. 21.1a. Fig. 21.1b shows the increase of biomass concentration (x) with time (t) in such a batch culture. These figures indicate that the following phases can be distinguished in the growth of a batch culture: (1) lag phase, (2) accelerating phase, (3) logarithmic phase (balanced growth), (4) decelerating phase, (5) stationary phase, and (6) death phase. Each growth phase is a reflection of a particular metabolic state of the cell population at any given time. These phases of growth are considered in more detail below.

Lag phase:
A newly transferred culture may have a lag phase for several reasons.

(a) The population transferred may have been in a metabolically 'bad' ('shifted down') state. This case occurs when the inoculum is taken from the stationary or death phase of the parent culture.
(b) The freshly inoculated batch culture has first to become conditioned to the culture medium (e.g. through the chelation of metals by excretion products).
(c) The measured biomass parameter does not take the non-viable portion of the population into account, and therefore, the biomass production of the small but vigorously growing portion of viable cells is masked by the non-viable cells

representing the major part of the population.

Acceleration phase:
In this phase different biomass parameters increase sequentially: the first component to increase is RNA, followed by protein and then dry weight. Cell number is usually the last parameter to show an increase. This phase may also be referred to as 'shift up'.

Logarithmic phase:
During this period the growth rate remains constant and the biomass concentration changes according to the formula:

$$x_2 = x_1 \exp[\mu(t_2 - t_1)]$$

Thus:

$$\mu = \frac{\ln x_2 - \ln x_1}{t_2 - t_1}.$$

The mean doubling time (t_d) or generation time (g) can thus be determined as follows:

$$g = t_d = \frac{\ln 2}{\mu} = \frac{0.693}{\mu}$$

During this phase of growth the concentration ratio of the different biochemical components stays constant as shown in Fig. 21.2. This pattern of growth is termed 'balanced growth'. It may also be referred to as exponential growth.

Deceleration phase:
During this phase the biochemical composition changes in a sequence opposite to that in the acceleration phase. This may also be termed 'shift down'.

Stationary phase:
During this phase biomass remains constant. Other parameters may increase or decrease. The final concentration of biomass reached at the stationary phase is usually taken either as a function of the depletion of some essential nutrient in the medium which becomes limiting, or due to limitations resulting from gas exchange. Other factors that might determine the biomass concentration at the stationary

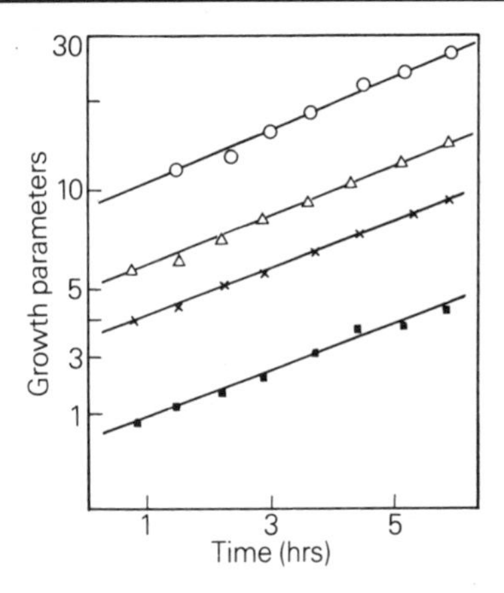

Fig. 21.2. Increase in several growth pararmeters with time in a culture of *Anacystis nidulans* during exponential or balanced growth: (O-O) chlorophyll ($\mu g\ ml^{-1} \times 10$); (△-△) dry weight ($\mu g\ ml^{-1} \times 10^{-1}$); (■-■) cell number ($ml^{-1} \times 10^{-7}$); (x-x) absorbance ($A_{560} \times 10$).

phase include excretion products that inhibit further growth, possibly due to changes in pH of the medium, or light limitation due to self-shading in dense algal cultures.

Death phase:
A decline in biomass concentration occurs as a result of an increase in the ratio of respiration to photosynthesis greater than unity, or as a result of cell death and lysis.

21.2.2 Continuous culture

Semicontinuous cultures:
Semicontinuous culture is a type of batch culture which is diluted at frequent intervals (Fig. 21.3a). The biomass concentration in such cultures has to be monitored so that the frequency of dilution and the dilution ratio (volume of culture:volume of new medium) can be estimated.

Turbidostat culture:
This type of culture (Fig. 21.4) is similar in nature to a semicontinuous culture, except that the monitoring of the biomass is not performed manually but by an optical device that controls the dilution rate, maintaining the culture at a preset optical density (Fig.

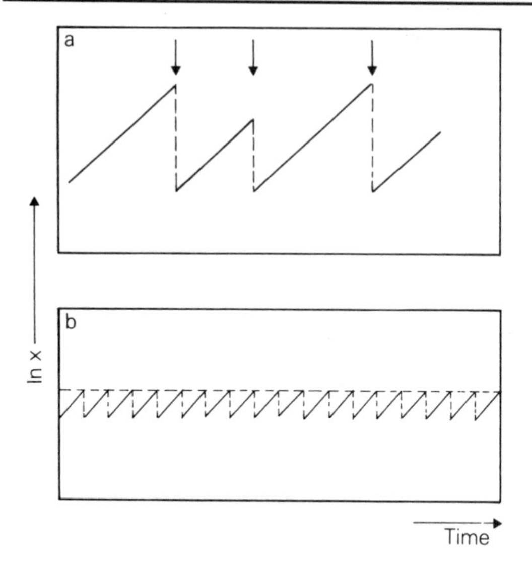

Fig. 21.3. Growth patterns in (a) semicontinuous and (b) turbidostat cultures. The arrows in (a) represents the points at which the culture was diluted. The dotted line in (b) represents the preset turbidity value. x = biomass concentration measured as absorbance.

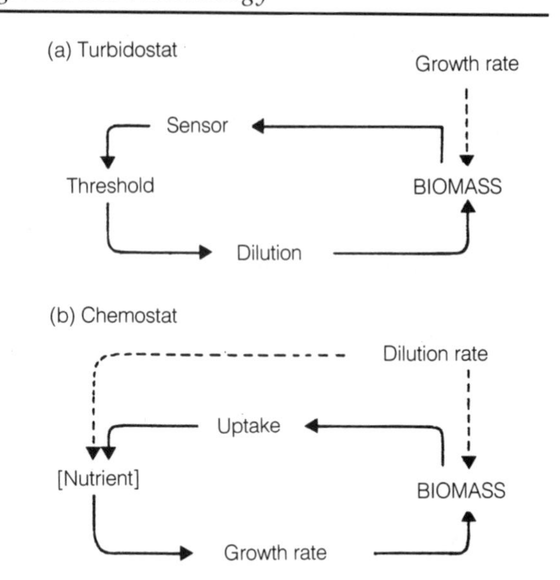

Fig. 21.5. The control cycles in (a) a turbidostat, and (b) a chemostat.

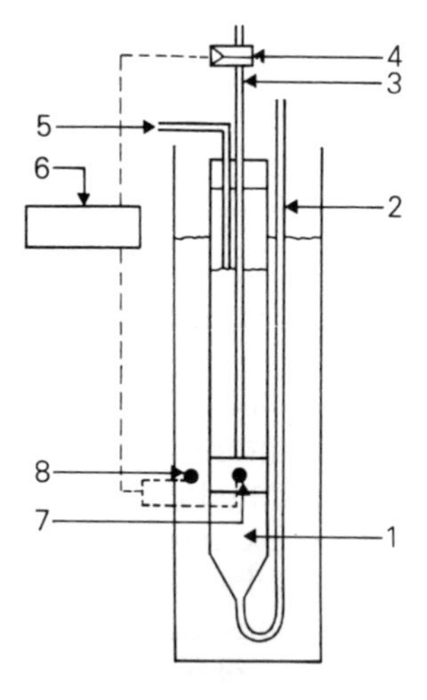

Fig. 21.4. Schematic diagram of a turbidostat: (1) culture vessel; (2) air + CO_2 inlet; (3) medium inlet; (4) magnetic valve; (5) air and medium outlet; (6) electronic control device; (7,8) measuring and reference photocells.

21.3b). In this case turbidity is used as the measure of biomass concentration. In such turbidostat cultures growth is such that nutrients are always non-limiting. Light, on the other hand, may be a limiting factor. A schematic representation of the control cycle for a turbidostat culture is shown in Fig. 21.5a.

21.3 CHEMOSTAT CULTURES

The principle of chemostat culture is based on the relationship between the specific growth rate and a limiting nutrient concentration that regulates the growth rate in such a way that it matches a preset constant dilution rate. The chemostat represents the most simple of the continuous culture devices, because the regulatory mechanism is dependent on the metabolism of the organism. The chemostat is the only culture method from which data can be obtained on the relationship between growth rate and nutrient limitations of the organisms. The control cycle for a chemostat is shown in Fig. 21.5b.

The chemostat system (Fig. 21.6) consists of a reservoir containing the culture medium which has a constant concentration of the limiting nutrient, a constant-flow device (pump, capillary with constant hydrostatic pressure) and a culture vessel with constant volume (maintained by an overflow). Culture fluid leaves the vessel at the same rate as the new medium is fed into the vessel by the flow device. The culture has to be homogeneously mixed in the culture vessel. This mixing may be achieved by the aeration stream, or by mechanical means (using an impeller or, with a smaller culture vessel, a magnetic stirrer).

In both chemostats and turbidostats the rate of production of cells through growth is equal to the rate of loss of cells through the overflow. Hence, when the flow rate (pump rate) f = dv/dt (where v is the culture volume) and D, the dilution rate, = f/v, then at a steady state:

$$\mu x = \frac{dv}{dt} \cdot \frac{x}{v} = Dx$$

hence

$$\mu = D$$

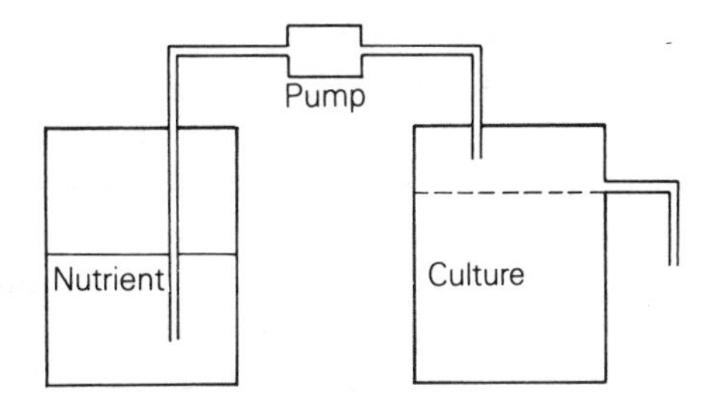

Fig. 21.6. Schematic diagram of a chemostat.

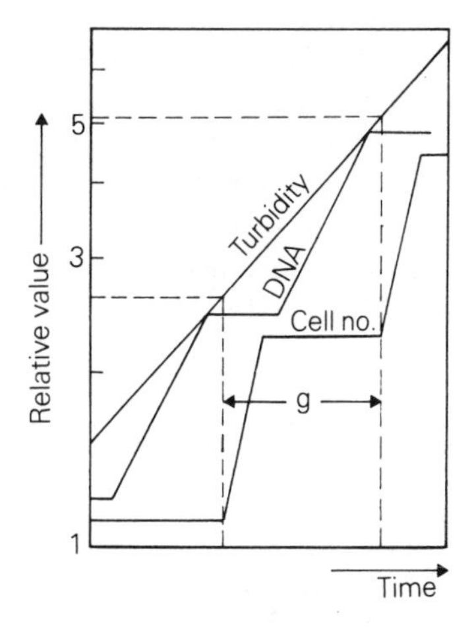

Fig. 21.7. The changes in some growth parameters in a synchronised culture; g, generation time.

21.4 SYNCHRONISED CULTURES

The three techniques described above are used for studying the growth pattern of populations of microorganisms which contain cells at all stages of the life cycle. In contrast, synchronised cultures are used for studies of individual growth patterns of cells in culture, since in this type of culture the cells are all at the same stage of the cell cycle at any given time. The pattern of growth and the changes in some of the growth parameters are shown in Fig. 21.7. Synchronised cultures can be obtained using either of the following techniques: (i) induction of synchrony by repeated shifts in the environmental or nutritional conditions; (ii) physical separation of cells from a random population and subsequent reculturing of those that are at the same stage of the cell cycle.

21.5 NUTRIENT-LIMITED GROWTH

Fig. 21.8 illustrates some hypothetical results from a chemostat culture showing the relationships between biomass concentration (as determined by measuring the amount of limiting nutrient taken up), the limiting nutrient concentration and the production of biomass as a function of the dilution rate (observed growth rate/maximum growth rate). The production of biomass (which equals flow rate times biomass concentration – the units being weight of biomass produced per time interval) exhibits a clear maximum at a growth rate close to the maximal growth rate. This picture neglects the dependence of the relative yield (biomass production per unit of nutrient used) on growth rate. The ratio of biomass parameters (other than those associated with the limiting nutrient) to the limiting nutrient incorporated into the organism changes in characteristic ways at different growth rates. For substrates providing energy (organic nutrients such as sugar, etc.) the amount of biomass produced per unit of nutrient increases with increasing growth rate (Fig. 21.9). This is a result of a change in the ratio of anabolism to catabolism. If the limiting nutrient is an inorganic salt or an essential vitamin the amount of biomass produced per unit of limiting nutrient decreases with increasing growth rate (Fig. 21.10).

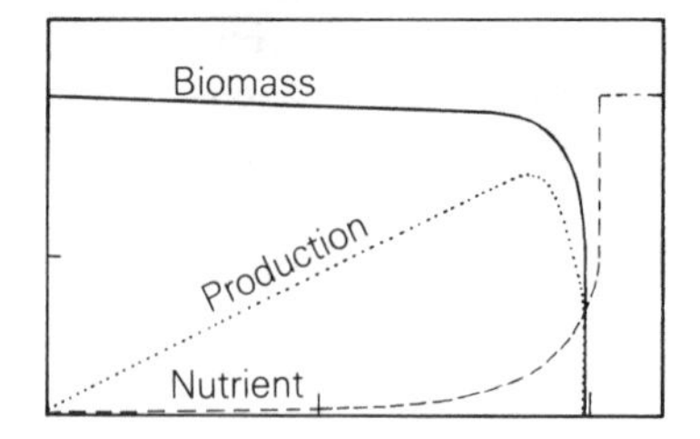

Fig. 21.8. Schematic representation of changes in biomass concentration, nutrient concentration and production as a function of dilution rate.

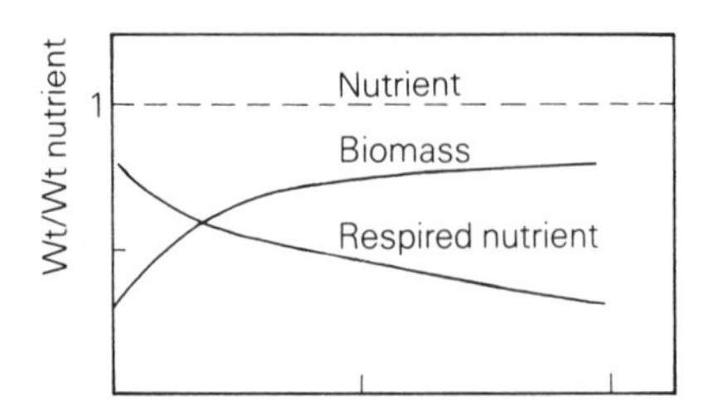

Fig. 21.9. The relationship between biomass production or loss of energy substrate through respiration as a function of the fractional growth rate.

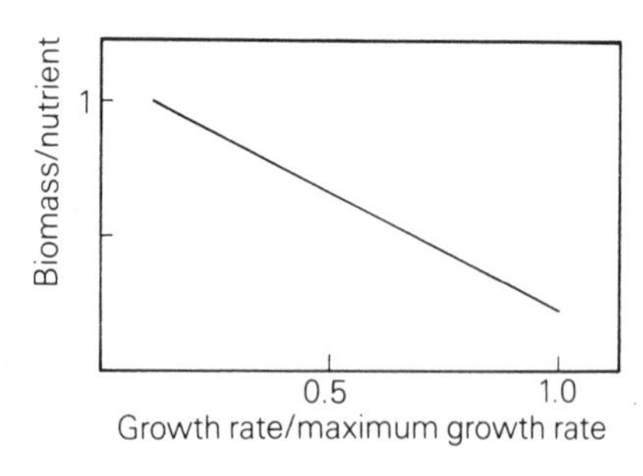

Fig. 21.10. The production of biomass on a nutrient basis (yield) for a 'non-energy' nutrient as a function of the fractional growth rate.

21.6 ANALYTICAL TECHNIQUES

21.6.1 Cell counting

Direct Microscopic Counting:
This procedure is commonly used to determine algal growth. For counting natural dilute populations, an inverted microscope and settling chambers are recommended. The major difficulty in microscopic enumeration is reproducibility; thus, adequate attention should be given to sampling, diluting, and filling the chamber, as well as choosing the right counting chamber, microscope magnification, and range of right counting chamber, microscope magnification, and range of cell concentration. Recommended counting chambers are listed in Table 21.1 and depend on the cell size and concentration. In general, counting chambers need to be cleaned and dried before use. The algal cells should settle for 4–6 minutes before counting, especially if small cells are counted in a deep chamber. The use of two counting chambers is recommended, so that the cells can settle in one while the other is being counted. After the chamber is filled, it should be examined at low magnification to ensure satisfactory distribution and concentration of cells.

Electronic Counting:
The Coulter counter and similar instruments are very useful for counting non-filamentous round algal cells, but very difficult to apply

Table 21.1. Cell-counting devices

Commercial name	Chamber volume (ml)	Depth (mm)	Objectives for magnification	Cell size (μm)	Cell concentration counted easily
Redgwick Rafter	1.0	1.0	2.5–10	50–100	$30–10^4$
Palmer Malony	0.1	0.4	10–45	5–150	$10^2–10^5$
Speirs-Levy hemacytometer	4×10^{-3}	0.2	10–20	5–75	$10^4–10^7$
Improved Neaubouer	2×10^{-4}	0.1	20–40 (phase)	2–30	$10^5–10^7$
Petroff Houser	2×10^{-5}	0.02	40–100 (phase)	0.5–5	$10^5–10^8$

to rod-shaped or other irregular cells. The principle of electronic counting is as follows.

A diluted algal suspension is pumped through a very small orifice between two fluid-filled compartments. Electrodes in each measure the electrical resistance of the system. The orifice is small enough to ensure that its electrical resistance is high. When a particle passes through the orifice, the resistance increases further since the conductivity of the algal cells is less than that of the medium. The change in resistance is converted into an electrical pulse which is counted. Two main problems exist in the outline use of this device: (a) failure of daughter cells to separate after cell division, and (b) clogging of the orifice. Thus highly pure reagents and careful cleaning procedures are recommended. Useful information and detailed operating procedures are usually provided with the instruction manuals of the instruments.

21.6.2 Light scattering (turbidity)

Light-scattering methods are the most general techniques used to follow the growth of pure cultures. They can be very powerful but can lead to erroneous results. The major advantages are that they can be performed quickly and non-destructively. They mainly give information about increase in biomass (dry weight) and not directly about the number of cells. The common practice is to use any available colorimeter or spectrophotometer to measure turbidity. At low turbidities there is a simple geometric relationship between the number of cells in the light path and the measured light intensity, since the intensity of the unscattered light decreases exponentially as the number of cells increases. Of the many instruments available, the *Klett-Summersen* colourimeter is the most popular, probably as a result of its stability and ease of use.

21.6.3 Dry weight measurement

Measuring the increase in dry weight is one of the most direct ways to estimate biomass production and involves the following steps.

Sampling. Taking representative aliquots from the algal culture is one of the most crucial steps. Special attention should be paid to achieving (a) a well-stirred culture, (b) rapid pipetting of samples to avoid settling, and (c) a large enough sample.

Separation. After sampling, cells are separated from the medium by membrane filtration or by centrifugation. The cells have to be washed to remove salts and other contaminants, usually with diluted medium or buffer. Marine algae must not be washed with distilled water, to avoid plasmolysis and bursting of the cells.

Drying. A wide range of temperatures (70–110°C) is mentioned in the literature. It is best to optimise the drying temperature to the particular organism according to some basic rules: (a) avoidance of excessive heat, (b) good reproducibility, and (c) identical weights obtained for the same sample taken at intervals of 1 hour.

Expression of results. The results of dry weight determination are expressed per unit of volume or in outdoor ponds per illuminated area of algal culture.

While the growth rate of higher plants is linear under constant conditions, in microalgae growth takes place logarithmically and thus resembles growth of microorganisms rather than higher plants. Therefore, in order to demonstrate the uninterrupted growth process in microalgae, one should present the growth curve on a semi-logarithmic scale.

21.7 MEASUREMENT OF PHOTOSYNTHESIS AND PHOTOINHIBITION IN ALGAL CULTURES USING AN OXYGEN ELECTRODE

A significant part of the studies related to photosynthesis and photoinhibition have been conducted using microalgae as model organisms. It is now common practice to use the oxygen electrode to measure changes in O_2 evolution rate as an indication of changes in the photosynthetic activity of microalgae, in response to modifications in nutritional or environmental conditions. The basic set-up of equipment and experimental conditions for using this methodology is described below.

21.7.1 Type of electrode and measuring cell

The basic concept of O_2 measurement by an electrode is described in Chapter 10. There are two kinds of Clark-type electrode available. One is the disk-form type and is usually used as the bottom part of the measuring cell. The whole system is available from a number of manufacturers, one of the most common being the *Hansatech* type (Hansatech Ltd., King's Lynn, UK). This system has several advantages, of which the most important is the ability to measure O_2 evolution rates in small volumes (<1 ml). The system is highly sensitive and reliable. The electrode is somewhat delicate and needs careful handling and maintenance. Although detailed instructions on use and maintenance are provided, some practice is required before good performance of the electrode is achieved. Another technical requirement is a good fit between the disk and the bottom of the cell in order to avoid leakage. The main problem with this kind of system is the price (about US$ 700); unless full use is to be made of its specific advantages, a much simpler device can be constructed.

The alternative is a stick-shaped electrode that can be immersed in a cell of any size, shape or volume. The most well-known manufacturer of such electrodes is *Yellow Springs* (Yellow Springs Instrument Co. Inc., Yellow Springs, Ohio, USA), although other makes are available in the price range US$ 100–150. This manufacturer supplies details for constructing a simple electronic amplifier to (a) provide the electrode with the polarization current required, and (b) adjust the output signal so that the changes in oxygen concentration can be recorded on a chart recorder with a sensitivity of 1–5 mV full scale.

The problem with this kind of set-up is that the minimal working volume is about 3–5 ml, depending on the cell's shape. A recommended cell is illustrated in Fig. 18.2. Since constant mixing is important in this system, careful attention should be paid to the mixing device and the shape of magnetic bar used.

For specific applications, a micro-electrode with a small diameter enabling the use of smaller volumes is also available on the market. Details of manufacturers and their full addresses may be found in Appendix A.

The following are some basic features that must be taken into consideration when constructing the system:

Recorder. A good chart recorder with a sensitivity range of 1–5 mV and a chart speed in the range of 0.5–2.0 cm min^{-1}.

Temperature. The electrical signal, as well as O_2 solubility in water, is highly dependent on temperature. Thus it is essential to have a good temperature control system, usually achieved by a water bath with a good circulation pump. Reaction cells must have double walls to allow cooling by circulation of water. When high light intensity is required, a water filter of 0.01% $CuSO_4$ is used to prevent infrared radiation from direct heating of the measuring cell.

Light source. A variety of light sources are available on the market, unless specific application of light is needed. In most cases a simple slide projector with a concentration lens will provide enough light for carrying out most kinds of measurements. Modification of light intensity for obtaining a light response curve should not be carried out by modification of voltage, since in most cases this results in a significant change in the light spectrum. When modification of light intensity is required the use of neutral density filters is highly recommended; tissue paper or fine glass fibre sheet is an acceptable substitute.

Preparation of samples. In order to obtain reliable and reproducible rates, the following must be considered when preparing algal samples for the oxygen electrode measurement:

(a) Constant concentration of chlorophyll should be used. Although in most cases results are expressed on the basis of chlorophyll, big differences in chlorophyll concentration between the samples will cause a difference in light availability to each cell (self-shading phenomenon).
(b) Whenever possible, it is recommended to harvest the algal cells and resuspend them in fresh growth medium to avoid nutritional limitation.
(c) Ensure sufficient supply of carbon. Some of the algae, mainly the marine strains, are grown in medium containing very little, if any, inorganic carbon in the form of carbonate. Supplying carbon to support growth is achieved by continuous bubbling of a CO_2/air mixture. It is recommended to have $NaHCO_3$ at a minimal concentration of 20 mM so as to assure sufficient carbon supply.

21.7.2 Performing the measurement

Calibration:
In order to be able to convert the recorder traces to rates of O_2 evolution, two reference points must be determined. One is the 100% point for air saturation obtained by bubbling air through fresh growth medium. The second point is usually the zero point obtained by bubbling N_2 through the medium or by using sodium dithionite. When dithionite is used, extensive rinsing of the electrode is essential before adding the algal sample.

Cell Concentration:
In order to measure the maximal potential of the photosynthetic activity, it is important to perform the measurements under non-limiting conditions. As already mentioned, self-shading may be a problem with algae. Preliminary measurements of O_2 evolution rates should be performed with several samples varying in their chlorophyll concentration; the recommended range is 0.5–5 μg ml Chl^{-1}. Plotting the rate of O_2 evolution as a function of chlorophyll concentration will give a good estimate of the chlorophyll concentration that should be used for unlimited (light-saturated) rate measurement.

Taking the measurement:
Before starting the measurement, samples should be allowed to adjust to the temperature of the cell. Depending on previous growth conditions and the species of algae used, an induction period of up to 1–2 minutes may be observed before a linear

rate of O_2 evolution is observed. In order to obtain a reliable measurement, it is recommended to measure the rates over a (linear) period of 1–2 minutes. Results are usually expressed as $\mu mol\ O_2\ h^{-1}\ mg\ Chl^{-1}$.

21.7.3 Measurement of photoinhibition

Photoinhibition is a very common phenomenon studied in algal systems, so this experiment is useful for both research and teaching. A common way to assay photoinhibitory damage is by measuring the decrease in the O_2 evolution rate during exposure to strong light (high photon flux density). A few experimental points have to be taken into consideration when performing this kind of experiment.

(a) High light is usually associated with production of heat. Water filters containing $CuSO_4$ are very efficient in preventing over-heating of samples by removal of infrared radiation.
(b) Neutral density filters (or even layers of white tissue paper) should be used to derive response curves of photosynthesis versus light intensity.
(c) Good stirring of the illuminated sample is important in order to avoid build-up of high O_2 concentrations in the illuminated cultures.
(d) Carbon supply: It is recommended to add some bicarbonate to the cultures before the photoinhibitory treatment. When using concentrated cultures for a long treatment, cultures may be deprived of carbon and the reduction in photosynthetic activity may reflect a carbon limitation rather than true photoinhibitory damage.

21.8 GROWTH, MAINTENANCE AND PRESERVATION OF ALGAL CULTURES

Once algal cultures have been obtained by direct isolation from nature or from a culture collection, it is recommended that a locally maintained culture is established.

Sterilisation:
This is mandatory even for algae that are not bacteria-free, so as to avoid contamination from other sources.

Illumination:
After a small amount of algal cells is transferred to fresh medium, it is placed under cool-white fluorescent tubes (2150–4300 lux). In liquid cultures for rapid multiplication 5400 lux are used. After good growth is obtained (1–2 weeks) the cultures are transferred to lower illumination (540–1100 lux) for slower growth and storage.

Algae stocks on agar media are illuminated after transfer with an intensity of 2700 lux for 6 or 7 days until good growth has been obtained. Such stocks are then moved to areas with an illumination level of 540–800 lux.

Temperature:
For preservation, most algae do well at room temperature (15–20°C). Only a few may prefer higher temperatures for survival.

Frequency of transfer:
Cultures are transferred at different intervals depending on the maintenance conditions and species. Unicellular and filamentous non-motile species may be transferred once every 6–12 months. Flagellated species may require more frequent transfers. With some species long preservation under liquid nitrogen has been successful.

21.9 CULTURE MEDIA

There are many recipes for media for the cultivation of microalgae under laboratory conditions. Most of them are modifications of previously published formulae and some are derived from analysis of water in the native habitat and ecological considerations. The main considerations in developing a nutrient recipe for algal cultivation are as follows:

(a) The total salt concentration, mostly dependent on the ecological origin of the organism.
(b) The composition and concentration of major ionic components such as potassium, magnesium, sodium, calcium, sulphate and phosphate.
(c) Nitrogen sources: nitrate, ammonia and urea are widely used as the nitrogen sources, mainly dependent on the species performance and the pH optimum. Growth is highly dependent on nitrogen availability. Most microalgae contain 7–9% nitrogen per dry weight. Thus for the production of 1.0 g cells in 1.0 litre of culture, a minimum of 500–600 mg l^{-1} KNO_3 will be required.
(d) Carbon source: inorganic carbon is usually supplied as CO_2 gas in a 1–5% mixture with air. Another means of supplying carbon is as bicarbonate. The preference is highly dependent on the pH optimum for growth.
(e) pH: usually, acidic pH values are used to prevent precipitation of calcium, magnesium and some of the trace elements.
(f) Trace elements: these are usually supplied in a mixture at concentrations previously found to be effective (of the order of micrograms per litre). Nevertheless, the necessity of such components for growth has not always been demonstrated. For stability of the mixture of trace elements, chelating agents such as citrate and EDTA are used.
(g) Vitamins: many algae require vitamins such as thiamine and cobalamin for growth.

The following recipes for culture media may be useful.

Allen's medium:

For each 1000 ml of medium required, add the following to 966 ml of glass-distilled water:

	To add	Stock solutions (g per 200 ml H_2O)
$NaNO_3$	1.5 g	–
K_2HPO_4	5 ml stock	1.5
$MgSO_4.7H_2O$	5 ml stock	1.5
Na_2CO_3	5 ml stock	0.8
$CaCl_2$	10 ml stock	0.5
$Na_2SiO_3.H_2O$	10 ml stock	1.16
Citric acid	1 ml stock	1.2
Trace elements (detailed below)	1 ml	–

Adjust pH to 7.8. Solidify, if desired, with agar (10 g l^{-1}).

Medium for *Anacystis nidulans*:

	Stock (g l^{-1})	Add to medium (ml l^{-1})
$MgSO_4.7H_2O$	246.5	0.5
$CaCl_2.2H_2O$	147.0	0.1
NaCl	233.7	0.5
$Na_2MoO_4.2H_2O$	10.0	0.1
KNO_3	202.2	10.0
K_2HPO_4 (see note below)	174.2	12.0

plus 1.0 ml per litre of Solution A_5 and Fe-EDTA solution as detailed below.

Medium for *Chlamydomonas reinhardtii*:

	Minimal medium	High salt minimal medium
NH_4Cl	0.05 g	0.50 g
$MgSO_4.7H_2O$	0.02 g	0.02 g
$CaCl_2.2H_2O$	0.01 g	0.01 g
K_2HPO_4 (see note below)	0.72 g	1.44 g
KH_2PO_4	0.36 g	0.72 g
Hutner's trace elements soln. (detailed below)	1 ml	1 ml
Distilled water	1 litre	1 litre

Medium for *Chlorella*:

	$(g\ l^{-1})$
KNO_3	1.25
KH_2PO_4	1.25
$MgSO_4.7H_2O$	1.00
$CaCl_2$	0.0835
H_3BO_3	0.1142
$FeSO_4.7H_2O$	0.0498
$ZnSO_4.7H_2O$	0.0882
$MnCl_2.4H_2O$	0.0144
MoO_3	0.0071
$CuSO_4.5H_2O$	0.0157
$Co(NO_3)_2.6H_2O$	0.0049
EDTA	0.50

The pH of the medium is 6.8.

Medium for *Euglena*:

To 1000 ml of Pyrex-distilled water add:

Sodium acetate	1.0 g
Beef extract	1.0 g
Tryptone	2.0 g
Yeast extract	2.0 g
$CaCl_2$	0.01 g

If desired, this may be solidified by adding 15 g agar.

Medium for *Spirulina*:

	$(g\ l^{-1})$
NaCl	1.0
$MgSO_4.7H_2O$	0.2
$CaCl_2$	0.04
$FeSO_4.7H_2O$	0.01
EDTA	0.08
K_2HPO_4 (see note below)	0.5
$NaNO_3$	2.5
K_2SO_4	1.0
$NaHCO_3$	16.8

plus 1 ml per litre of solutions A_5 and B_6 as below:

Artificial sea-water medium for *Porphyridium*:

	Add to 1 litre H_2O
NaCl	27.0 g
$MgSO_4.7H_2O$	6.6 g
$MgCl_2.6H_2O$	5.6 g
$CaCl_2.2H_2O$	1.5 g
KNO_3	1.0 g
KH_2PO_4	0.07 g
$NaHCO_3$	0.04 g
Tris/HCl (1.0 M, pH 7.6)	20 ml
Microelements solution (detailed below)	1 ml

Enriched sea-water medium:

	Stock	Add to 1 litre filtered and sterilised sea-water
KNO_3	$10.0\ g\ l^{-1}$	2 ml
KH_2PO_4	$13.61\ g\ l^{-1}$	2 ml
$NaHCO_3$	$84.0\ g\ l^{-1}$	5 ml
Na_2SiO_3	$30.0\ g\ l^{-1}$	5 ml
$Na_2EDTA.2H_2O$ + $FeCl_3.6H_2O$	30 mM 25 mM	1.5 ml
Tris–HCl	1.0 M (pH 7.6)	20 ml
Trace elements II (detailed below)		10 ml

Trace elements solution II:

	$(g\ l^{-1})$
$CaSO_4.5H_2O$	0.98
$ZnSO_4.7H_2O$	2.2
$CaCl_2.6H_2O$	1.0
$MnCl_2.4H_2O$	18.0
$Na_2MoO_4.2H_2O$	0.63

Hutner's trace elements solution:

EDTA	50.0 g
$ZnSO_4.7H_2O$	22.0 g
H_3BO_3	11.4 g
$MnCl_2.4H_2O$	5.1 g
$FeSO_4.7H_2O$	5.0 g
$COCl_2.6H_2O$	1.6 g
$CuSO_4.5H_2O$	1.6 g
$(NH_4)_6Mo_7O_{24}.4H_2O$	1.1 g
Distilled water	750 ml

Boil, cool slightly, and bring to pH 6.5–6.8 with KOH (do not use NaOH). The clear solution is diluted to 1000 ml with distilled water and should have a green colour which changes to purple on standing. It is stable for at least one year.

For heterotrophic acetate mutants the media may be supplemented with sodium acetate at a concentration in the medium of 0.20%.

Microelements solution:

	Add to 1 litre H_2O
$ZnCl_2$	4.0 mg
H_3BO_3	60.0 mg
$CaCl_2.2H_2O$	4.0 mg
$MnCl_2.4H_2O$	40.0 mg
$(NH_4)_2SO_4.4H_2O$	37.0 mg
Fe-EDTA solution (100 ml 0.05 M Na_2EDTA, pH 7.6, + 240 mg $FeCl_3.4H_2O$)	1.0 ml

Solution A_5:

	$(g\,l^{-1})$
H_3BO_3	2.86
$MnCl_2.4H_2O$	1.81
$Na_2MoO_4.2H_2O$	0.252 (or 0.015 g MoO_3)
$ZnSO_4.7H_2O$	0.222
$CuSo_4.5H_2O$	0.079

Solution B_6:

	$(mg\,l^{-1})$
NH_4NO_3	22.96
$K_2Cr_2(SO_4)_4.24H_2O$	96.0
$NiSO_4.7H_2O$	47.85
$Na_2SO_4.2H_2O$	17.94
$Ti(SO_4)_3$	40.0
$Co(NO_3)_2.6H_2O$	43.98

Fe-EDTA solution:

Dissolve 16 g EDTA (free acid) and 10.4 g KOH in 186 ml of distilled water. Mix this solution with another containing 13.7 g $FeSO_4.7H_2O$ (with low Mn content) in 364 ml of distilled water. Bubble air through the mixture to oxidise Fe^{2+} to Fe^{3+} (3–4 h). Final pH should be about 3.

Note on K_2HPO_4 solution:

Place 12 ml 1.0 M K_2HPO_4 (174.2 g l^{-1}), to which 0.84 g $NaHCO_3$ has been recently added, in a separate tube. Sterilise (20 min, 1 atm.) and leave to cool. Mix both solutions under aseptic conditions.

21.10 DEFINITION OF COMMON UNITS AND TERMS

Dilution rate (D) = flow rate (f) divided by culture volume (V). Critical dilution rate (D_c) occurs when specific growth rate is maximal and any increase in flow rate will result in wash-out of the culture.

Mean generation time (g) in a continuous culture is equal to retention time (t_d). In a homogeneously mixed culture, on average:

$$t_d = \frac{\ln 2}{\mu} = \frac{0.693}{\mu}$$

$$\text{Specific growth rate } (\mu) = \frac{1}{x} \cdot \frac{dx}{dt}$$

$$\text{Growth yield (Y)} = \frac{dx}{ds}$$

= increase in biomass resulting from consumption of a given amount of substrate.

Productivity = rate of production of biomass (dry weight per unit time per unit volume).

21.11 TECHNICAL PROBLEMS

The user of chemostat or turbidostat cultures should be aware of the following technical problems which may arise.

Chemostats:

(a) Non-exponential wash-out can occur due to poor mixing or growth of cells at high density on solid supports within the vessel, e.g. walls, stirrers.
(b) Variation in the flow rate due to perishing of the tubing in peristaltic pumps, formation of deposits in cylinder pumps, or fluctuation in temperature and hydrostatic pressure in capillary tubing.
(c) Heterogeneous biomass due to differentiation into sexual or resting stages.

Turbidostats:

(a) Wall growth as mentioned above, or agglomeration of cells in the light path.
(b) Fluctuation in the performance of flow devices.

In general:

The air supply may become contaminated, in particular by oil drops where a pressurised air supply is used. This problem can be overcome by using suitable filters.

21.12 WHERE TO OBTAIN ALGAL CULTURES

The following are addresses of recommended culture collections:

1. Culture Center of Algae and Protozoa (CCAP)
   Freshwater Biology Association
   The Ferry House
   Ambleside
   Cumbria LA22 0LP
   UK
2. University of Texas Culture Collection (UTEX)
   Department of Botany
   University of Texas
   Austin, TX 78712
   USA
3. American Type Culture Collection (ATCC)
   12301 Parklawn Drive
   Rockville, MD 20852
   USA
4. Sammlung von Algenkulturen
   Pflanzenphysiologysches Institute
   Universität Gottingen
   18 Nikolausbergerweg
   Gottingen D-3400
   Germany
5. Microbial Culture Collection (NIES Collection)
   National Institute for Environmental Studies
   Onogawa 16-2 Tsukuba
   Ibaraki 305
   Japan
6. SERI Microalgal Culture Collection
   Solar Energy Research Institute FTLB
   1617 Cole Boulevard
   Golden, CO 80401
   USA

21.13 ALGAL BIOMASS PRODUCTION

During the past 20 years or so, much progress has been made in developing the biotechnology of algal mass culture. The potential uses of algal biomass for the benefit of mankind have been extensively reviewed [1].

Microalgae are aquatic microorganisms with a photosynthetic process similar to that of higher plants. The main advantages of culturing microalgae as a source of biomass are as follows:

(a) Algae are considered to be the most efficient biological system for harvesting solar energy and for the production of organic compounds via the photosynthetic process.
(b) The entire biomass is available for harvest and use, as most algae are non-vascular plants and lack complex reproductive organs.
(c) Many species of algae can be induced to produce particularly high concentrations of specific compounds (proteins, carbohydrates, lipids, and pigments) which are of commercial value.
(d) Genetic selection and strain screening are relatively easy and quick since algae are microorganisms without a sexual stage, undergoing simple cell division and completing their life cycle within a few hours. This also allows much faster development and demonstration of viable production compared with conventional agricultural processes.
(e) Microalgae can be grown using marginal water (brackish or sea water). This is of particular interest in increasing productivity and securing a basic protein supply in regions of low productivity with shortages of fresh water or poor soils.
(f) Algal biomass production systems can be easily adapted to various levels of operational skills and investment, from simple, labour-intensive production units to fully automated, capital-intensive operations.

The major inputs and potential uses of algal biomass are presented in Fig. 21.11.

The biotechnology of mass cultivation of algae, like other branches of biotechnology, is a combination of the biological know-how required for the optimization of algal growth

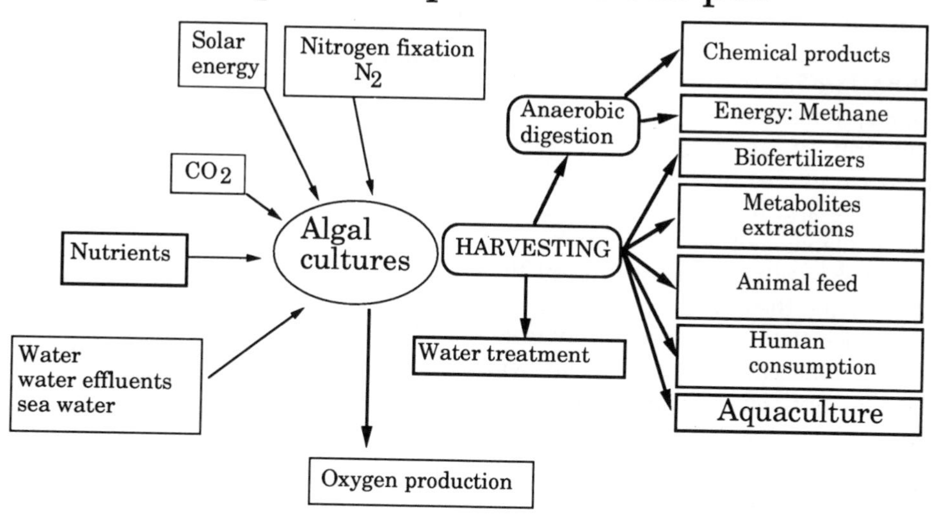

Fig. 21.11. Algal biomass production: inputs and potential outputs (Courtesy of Dr. S. Boussiba).

and the technology required for the design of large-scale ponds and the subsequent harvesting and processing of the resultant algae biomass. Both aspects are briefly described below.

21.13.1 Biological know-how: optimization of algal growth

The first step in growing algae combines the biological know-how of growing photo-autotrophic microorganisms with the special requirements for designing an appropriate reactor for the process.

Biological know-how involves understanding the interactions of environmental factors (such as light and temperature, as well as salinity, photoinhibition and dark respiration) on algal growth and productivity. These parameters have to be considered in developing an operational protocol for pond management which also includes nutrient levels and pH, in order to establish a continuous culture for sustained production and to avoid development of predators and contamination by other algae. The parameters have been reviewed in detail for *Spirulina* [2] and for *Dunaliella* [3].

21.13.2 Biological problems

Water source:
Algae growing in brackish or saline water in ponds are often faced with the problem of the increasing salinity caused by evaporation. Thus, when the water source is brackish or saline the organism of choice will be one that is capable of growing successfully in a wide range of saline conditions and thus of adapting rapidly to changes in salinity.

Optimisation of growth:
It is quite clear that the economics of algal biomass production under outdoor conditions can be improved by increasing the amount of biomass obtained per unit area.

This aim can be achieved by optimising growth conditions. While the nutritional requirements of the algae culture can easily be met by an excess supply of nutrients, environmental limitations, such as light and temperature, are more difficult to control in outdoor systems. The productivity of the system, in terms of dry weight produced per unit time per unit area, is a function of both the growth rate and the algal concentration, while the availability of solar radiation to a single cell in the culture is a function of the total radiation and the cell concentration. In a culture in which there are no other environmental limitations, the maximal yield will depend on the optimal cell concentration, so regulation of cell concentration is thus of the utmost importance. The optimal cell concentration is a function of the algal species, the total radiation, the degree of induced turbulence and other environmental limitations, and is in particular influenced by the depth of the algal culture.

Contamination:
In open systems, contamination by other algal species may take place and, therefore, the product may not be uniform. This problem can sometimes be solved by choosing growth conditions that give a selective advantage to the required algal species.

Choice of algal species:
Although there are about 20 000 species of algae, only a few of them have been studied under outdoor conditions. The possibility of cultivating other algal species out of doors and their potential use for the production of biomass should be investigated. The organism of choice should exhibit the following characteristics:

(i) fast growth rate;
(ii) wide range of tolerance to temperature and radiation extremes;
(iii) high content of proteins, lipids or carbohydrates, or selective accumulation of a specific metabolite (e.g. glycerol);
(iv) ease of harvesting.

Other requirements may be added to this list depending on the specific approach and requirements of the system.

21.13.3 Uses of algal biomass

Today it appears that the use of algal biomass as a protein source for general human consumption is not practicable. However, it is thought that biomass production systems should be developed for such diverse purposes as: waste-water treatment; production of fish food; production of commercial chemicals (glycerol, mannitol, lipids); production of biochemicals and pigments such as natural dyes; and energy production by biomass fermentation.

The following microalgae are currently used or appear potentially useful:

Chlorella	Health food (claimed to contain a special growth factor)
Spirulina	Vitamin-rich health food Source of γ-linoleic acid Source of blue pigment
Dunaliella	Source of β-carotene Source of glycerol
Haematococcus	Source of Astaxantin, a red pigment used in aquaculture
Porphyridium	Source of polysaccharides Source of arachidonic acid
Chlamydomonas	Source of polysaccharides Soil conditioner
Various cyanobacteria (blue-green algae)	Nitrogen fixation Production of ammonia Soil conditioner

Reviewing the past 30 years of research and the current state of the algal biomass

Fig. 21.12. *Spirulina* cultivation ponds (each 5000 m^2) at Earthrise Farms, Inc., California.

production industry, it seems that a regional approach, i.e. the exploitation of local resources to solve local problems such as waste-water treatment or the production of high-value products (pigments, food supplements), is a more logical strategy than trying to enter global markets, where requirements vary considerably and may change rapidly depending on consumer preferences.

21.13.4 Technology: reactor design and harvesting processes

Most commercial reactors used in large-scale production of algal biomass are based on shallow raceways in which algal cultures are mixed in a turbulent flow sustained by a paddle wheel. The size of commercial ponds varies according to the width and length of the raceways, usually covering 1000–5000 m^2 (Fig. 21.12).

Algal cultures grown photoautotrophically outdoors are very dilute, and for maximal productivity the concentration of biomass is usually kept in the range 0.5–0.7 $g\,l^{-1}$. The harvesting process requires an intensive step of water removal; thus using an efficient harvesting device with low energy input is highly desirable.
The various techniques used in algal harvesting are as follows.

(a) Filtration. This is the preferred method whenever filamentous algae (such as *Spirulina*) are being harvested, and may be designed with either vibrating or static nets. As long as no problems in clogging of the filters are encountered, this is a relatively inexpensive and efficient harvesting method.
(b) Centrifugation. This technique has been widely used in the harvesting of

microalgae, mainly unicellular, such as *Chlorella*. It is considered to be a very efficient but relatively expensive process due to the high investment cost and high energy inputs.

(c) Flocculation and Sedimentation. These techniques are used mainly in the removal of algal biomass in waste-water treatment, and are usually induced by the addition of chemicals such as aluminium sulphate, ferric sulphate or lime, or by the modification of growth conditions such as pH. These procedures are relatively inexpensive, but the main problem is that the flocculated product cannot be directly used as food since complete removal of the flocculant chemical is required.

The final processing step is highly dependent on the type of product and its marketing. Whenever the final product is aimed at the health food market, the harvested biomass is dried using a spray drier. If specific chemicals or pigments are being extracted, the biomass is usually used in a wet form for the extraction process, as in the cases of *Dunaliella* and its valuable extract, β-carotene.

REFERENCES

1. Richmond, A. (ed.) (1986) *Handbook of Microalgal Mass Culture*. CRC Press, Boca Raton, Florida.
2. Vonshak, A. and Richmond, A. (1988) Mass production of *Spirulina* – an overview. *Biomass* **15**, 233–248.
3. Ben-Amotz, A. and Avron, M. (1989) The biotechnology of mass culturing of *Dunaliella* for products of commercial interest. In: *Algal and Cyanobacterial Technology* (Cresswell, R.D., Rees, T.A.V. and Shah, N., eds.). Longman Scientific and Technical Press, London. pp. 90–114.

FURTHER READING

Abeliovich, A. (1986) Algae in wastewater oxidation ponds. In: *Handbook of Microalgal Mass Culture* (A. Richmond, ed.). CRC Press, Boca Raton, Florida.

Carr, N.G. and Whitton, B.A. (1982) *The Biology of Cyanobacteria*. Blackwell Scientific Publishers, Oxford.

Coleman, A.W., Coff, L.J. and Stein, J.R. (eds.) (1989) *Algae as Experimental Systems*. Alan and Liss, Inc., New York.

Falkowski, P.G., ed. (1980) *Primary Productivity in the Sea*. Brookhaven Symposia in Biology, No. 31. Plenum Press, New York and London.

Fogg, G.E. (1975) *Algal Cultures and Phytoplankton Ecology*. University of Wisconsin Press, Madison.

Gerhardt, P., ed. (1981) *Manual of Methods for General Bacteriology*. American Society for Microbiology.

Lobban, C.S., Chapman, D.J. and Kremer, B.P., eds. (1981) *Experimental Phycology: A Laboratory Manual*. Cambridge University Press.

Megnell, G.G. and Megnell, J.E. (1970) *Experimental Bacteriology*. Cambridge University Press.

Pirt, S.J. (1975) *Principles of Microbe and Cell Cultivation*. Blackwell, Oxford.

Raymont, J.E.G. (1980) *Plankton and Productivity in the Ocean*. Vol. 1, Phytoplankton, 2nd edn. Pergamon Press, Oxford.

Richmond, A. and A. Vonshak, eds. (1991) Special Issue on Algal Biotechnology. *Bioresource Technology* **38**, 83–252.

Shelef, G., Soeder, C.J. and Balaban, M, eds. (1980) *Algal Biomass*. Elsevier, North Holland.

Stanier, R.Y., J.L. Ingraham, M.L. Wheelis and P.R. Painter (1987) *General Microbiology*, 5th edn. Prentice-Hall Inc., New Jersey, USA, and Macmillan Education, London.

Stewart, W.D.P. (1974) *Algal Physiology and Biochemistry*. University of California Press, Berkeley.

UNESCO (1973) *A Guide to the Measurement of Marine Primary Productivity under some Special Conditions*. Monographs on oceanographic methodology, No. 3. UNESCO, Paris.

Vollenweider, R.A., ed. (1974) *A Manual for Measuring Primary Production in Aquatic Environments*. IBP Handbook No. 12, 2nd edn. Blackwell Scientific Publications, Oxford.

Handbook of Phycological Methods. Cambridge University Press.
Vol. 1. *Culture Methods and Growth Measurements* (1973) ed. J.R. Stein

Vol. 2. *Physiological and Biochemical Methods* (1978) eds. J.A. Hellebust and J.S. Craigie
Vol. 3. *Cytological and Developmental Methods* (1979) ed. E. Gantt
Vol. 4. *Ecological Field Methods: Macroalgae* (1985) eds. M.M. Littler and D.S. Littler

Appendix A
Equipment for plant physiology research in a changing environment

M.J. Bingham and S.P. Long

Introduction

The following tables are extracted from Bingham, M.J. and S.P. Long (1992) Equipment for Crop and Environmental Physiology: specifications, sources and costs, 4th edn., University of Essex, Colchester, UK.

When establishing or extending a research facility, much time may be consumed in identifying suitable equipment and its commercial availability. The research worker often has to rely upon personal contacts, the chance discovery of advertisements, or incomplete descriptions of equipment in research publications. This can produce a narrow and out-dated view of what is available and usually prohibits an objective assessment of the 'best buy'. The following tables compare the specifications of different manufacturers' equipment for field and laboratory research on the physiology of plants in crop and natural environments. Further tables include equipment necessary for describing the environment in which plants grow, with respect to microclimate, atmospheric composition and soil conditions.

In choosing an item of equipment for a research task, you should first draw up a specification. What precision will you require? There is little point in purchasing an instrument with a precision of $\pm 1\,\mu g\,cm^{-3}$ when $\pm 1\,mg\,cm^{-3}$ is all that is necessary. This will almost certainly increase the size and the price of the instrument. At the same time you must ensure adequate precision.

Similarly, consider the space available for the instrument, and whether or not it will be used in the field. If the latter, you should ask whether it needs to be weatherproof, whether mains power will be conveniently available, whether it will be used in very hot or cold environments, whether it will be too heavy to be carried around all day. Once you have decided upon a specification, then the tables provided here should enable you to identify the commercially available items which meet or approach your requirements. However, the tables do not provide complete information on each item listed. When you have reached a short-list, you should contact the manufacturer for full and current information before making a purchase.

In compiling these tables, we have attempted to demonstrate the range of instruments availables, and to include a number of manufacturers from different countries. It should be emphasised that the lists are not fully inclusive and do not feature the only instruments or manufacturers of such equipment. We would welcome information from or about other manufacturers so that we may

Photosynthesis and Production in a Changing Environment: a field and laboratory manual
Edited by D.O. Hall, J.M.O. Scurlock, H.R. Bolhàr-Nordenkampf, R.C. Leegood and S.P. Long
Published in 1993 by Chapman & Hall, London.
ISBN 0 412 42900 4 (HB) and 0 412 42910 1 (PB).

consider inclusion of their equipment in future versions of these lists.

N.B. Inclusion in these tables does not constitute a recommendation or guarantee, in any respect, of any piece of equipment. There is no significance in the numerical order in which the instruments appear.

The tables are compilations of specifications as quoted by the manufacturers (*Note: they are not independent tests*) and all tables have been returned to the manufacturers or their agents for correction. Whilst data were to the best of our knowledge correct at the time of manuscript submission, equipment is frequently improved and lines updated. We advise readers to contact the manufacturer for the latest specifications before making any purchase.

Accuracy

Most items of equipment listed in these tables are said to have a certain 'accuracy'. We have conformed to the manufacturers' usage of the word in constructing the tables. However, it should be stressed that except where an item is a standard, this should be regarded as the precision or repeatability, since where the instrument has to be calibrated (e.g. infra-red gas analysers) the accuracy depends upon the exactness of the calibration procedure as well as the repeatability. These figures should therefore be regarded as the best accuracy that could be obtained, given perfect calibration.

Abbreviations

Each group of equipment has tended to use specialist and sometimes personalised units of measurement; we have endeavoured to convert these to S.I. units for consistency of comparison, but inevitably this has meant some degree of approximation where exact conversions are not possible. Codes used within a table are explained in the footnotes to each table. Other abbreviations have been used throughout, *viz.* DC – direct current; FSD – full scale deflection (of a meter); n.a. – not applicable; RH – relative humidity; VAC – volts, alternating current.

Prices

With continually fluctuating currency exchange rates, it would be pointless to quote an exact price for a piece of equipment, so we have used coded price ranges as a guide to the comparative prices of different items of equipment. The range given is for a 'typical system', since there are often many options which alter the total price. Price ranges are based on the export price in US$ for December 1991. Note that prices may vary considerably depending on agency overhead charges (which may often double the manufacturer's price) and local taxes. The letter codes used in the tables represent the following price ranges:

A = US$0–100
B = 100–500
C = 500–1000
D = 1 000–2000
E = 2 000–3000
F = 3 000–5000
G = 5 000–10 000
H = 10 000–20 000
I = 20 000–50 000
J = 50 000–10 0000
K = US$100 000+

The full names and addresses of manufacturers (and where available, telephone, fax and telex numbers) can be found at the end of the tables. For large companies, the address of the corporate headquarters or regional headquarters in the UK has been given; this should be contacted for details of branches, agents or distributors in other countries.

We are grateful to the companies listed here for their cooperation and, in many cases, for their encouragement and helpful suggestions which will allow us to improve future editions of this equipment review.

Contents

The tables are grouped as follows:

	FLUORIMETERS			
INSTRUMENT	1	2	3	4
Measurements:	Fo;Fm;Fv; t/2;qp*;qnp*	Fo;Fm:Fv;Fv/Fm; Area over curve	Fo;Fm;Fv;qp; qn;Fv/Fm	Fo;Fm;Fv;qp; qn;Fv/Fm
Light source {1}	Y	R	R	R
Wavelength (nm)	583	650	655	655
Modulation frequency (kHz){2}	4.8	N	1.6/100	0.6/20
Actinic beam (μmol/m²/s):				
Accessory LS2 light source	0-1500	0-3500	2000	0-400(0-3500Halogen lamp)
Saturating beam (μmol/m²/s):				
Accessory PLS1 light source	5000	n.a.	7500	12000 (internal lamp)
Photodetector {3}	Ph	Ph	Ph	Ph
Filter {4}	V	P	V	B
Time resolution (mS)	1	0.01	0.1	0.1
P700+ measurement accessory kit {5	A/B	n.a.	E	
Light source {1}	I	n.a.	I	
Wavelength (nm)	820	n.a.	810-840	
LCD display	+	+	+	+
Output : V	+	+	+	+
Computer interface {6}	B	RS	B	B
Manufacturers software available	+	included	+	included
Dimensions [main unit] (cm) :				
Length	26	24	28	
Width	24	12	53	
Depth	10	10	15	
Weight (kg) *Inc. integral battery	5	1.5*	6	

Battery life (h)	30	24		
Operating environment :				
Maximum temp. (°C)	50	50	40	
Minimum temp. (°C)	0	0	0	
Humidity (%RH)	80	80	80	
Price range (U.S.$)§ :				
Single channel only	F	G		
Twin channel only	G			
Single channel + LS2/PLS1 light so	G			
Twin channel + LS2/PLS1 light sou	H			
P700 + accessory kit	D			
INSTRUMENT	1	2	3	4

INSTRUMENTS :

1.Modulated Chlorophyll Fluorimeter/Absorbsiometer (single & twin versions) :Hansatech Instruments Ltd.

2.Plant Efficiency Analyser :Hansatech Instruments Ltd.

3.PAM-101Chlorophyll Fluorescence Measuring System :Walz Mess- und Regeltechnik

4.Portable Fluorimeter :Walz Mess- und Regeltechnik

*Requires PLS1 light source

{1}Light source :Y=Yellow LED ;R=Red LED ;I=Infra-red LED

{2}Modulation frequency :N=Non-modulated

{3}Photodetector :Ph=Photodiode

{4}Filter :V=Various ;B=Balzers DT Cyan ;P=IR photographic

{5}P700+ measurement :A=Twin channel version ;B=IBM compatible board ;E=ED800T emitter-detector unit

{6}Computer interface :B=IBM compatible board ;RS=RS232

{7}Power supply :1=110/120VAC 50-60Hz ;220/240VAC 50-60Hz ;D=DC ;R=Rechargeable batteries

§ See Introduction

PORTABLE PHOTOSYNTHESIS SYSTEMS

INSTRUMENT	1	2	3
MEASUREMENTS :			
CO2 Concentration :			
Sensor {1}	N	N	N
Range (ppm)	0-1500	0-1600	Differential +/-50 Absolute 0-1000
Absolute measurement	+	+	+
Differential measurement	+	+	+
Accuracy (%FSD)	0.2	0.2	<1
Sample cell dimensions (cm):			
Diameter	1.3	0.4	1.8
Length	15.2	6	20
Leaf Temperature :			
Sensor {2}	Te	Ti	Te
Range (°C)	0 - 50	0 - 50	0 - 50
Accuracy (°C)	1	1	0.2
Chamber & Air Temperature :			
Sensor {2}	Ti	Ti	Pt-100
Range (°C)	0 - 50	0 - 50	0 - 50
Accuracy (°C)	0.5	0.3	0.1
Relative Humidity :			
Sensor {3}	T	T	T
Stomatal Conductance/Resistance :	+	+	+
Apparent Photosynthesis :	+	+	+
PAR :			
Sensor {4}	Si	Se	Si
Range (μmol/s/m^2)	0 - 2500	0 - 3000	0 - 4000
Accuracy (%)	5	10	5
OUTPUT :			
V		+	+
Digital display	+	+	+
RS232C	+	+	+
Printer	+		+
DIMENSIONS (cm) :			
Analyser :			
Width	39	28	10.5
Depth	15	18.5	30.6
Height	10.5	21	12.9
Weight (kg)	5	6	5

Leaf Chamber * :	A	B	C	D	E	F	G	V	Cl
Width/diameter	14.2	20.7	11.8	12	7	16	13.5	6	10
Depth	17	14.5	7.3	9	9	9	9		2.5
Height	18.5	4.5	4.5	37	41	37	43	9	

Width	23.4		42
Depth	15		26
Height	14.5		21
Weight (kg)	2.7		11
Sensor Housing :			
Width/diameter	11.1		3.3
Depth	4.3		26
Height	5.3		
Weight (kg)	0.7		1.1
Power supply {5}	1;2;D;R	1;2;D;R	1;2;D;R
Battery life (h)	2 - 4	5	4 - 8
Operating environment :			
Maximum temp. (°C)	50	45	45
Minimum temp. (°C)	0	0	5
Humidity (%RH)	80	80	80
Price range (U.S.$)§	H	H	
INSTRUMENT	1	2	3

INSTRUMENTS :

1.LI-6200 Portable Photosynthesis System :LI-COR,inc.
2.LCA-3 Portable Leaf Chamber Analyser & Parkinson Leaf Chamber :Analytical Development Co. Ltd.
3.Compact CO2/H2O Porometer :Walz Mess- und Regeltechnik

{1}Sensor :N=Non-dispersive infra red gas analyser
{2}Sensor :Te=Thermocouple ;Ti=Thermistor ;Pt-100=Platinum resistance thermometer
{3}Sensor :T=Thin film capacitance
{4}Sensor :Si=Silicon ;Se=Selenium
*Leaf chamber :A=4l chamber
B=1l chamber
C=0.25l chamber
D=Cereal / Grass leaf model
E=Broad leaf model
F=Conifer chamber
G=P.O.D. chamber
V=Various chambers available on request
Cl=Climate unit :eliminates glasshouse effect caused by sun radiation
{5}Power supply :1=110/120VAC 50-60Hz
2=220/240VAC 50-60Hz
D=DC
R=Rechargeable batteries
§ See Introduction

	GAS EXCHANGE SYSTEMS WITH TEMPERATURE CONTROLLED CHAMBER
INSTRUMENT	I
Configuration :	
Open system	Based on real differential measurement
Closed system	
CONTROLS :	
Chamber temperature :	
Range (°C)	-5.....+45
Attainable temp. difference (K)	+/- 15 (at 10-30°C ambient temp.)
Accuracy (°C)	0.15
Input humidity :	
Range (°C dew point)	-5.....+45
Attainable temp. (°C dew point)	30 (below ambient temp.)
Accuracy (°C dew point)	0.15
Illumination :	
Range (μmol/s/m²)	0.....1500
Accuracy (%)	5
CO2 Concentration :	
Range (ppm)	250-2500 at 4l/min flow rate
Accuracy (%)	1
MEASUREMENTS :	
CO2 concentration :	
Sensor {1}	N
Range (ppm)	+/-50 differential ;0-1000 absolute
Accuracy (%FSD)	<1
Sample cell dimensions (cm) :Diameter x Length	1.8 x 20
H2O Concentration :	
Sensor {1}	N
Range (ppm)	0-5000 differential ;0-3% absolute
Accuracy (%FSD)	<1
Sample cell dimensions (cm) :Diameter x Length	1.8 x 20
Leaf Temperature :	
Sensor {2}	Te
Range (°C)	-5....+45
Accuracy (°C)	+/-0.2
Chamber Temperature :	
Sensor {2}	Ti
Range (°C)	-5....+45
Accuracy (°C)	+/-0.15
External Temperature :	
Sensor {2}	Ti
Range (°C)	-5....+45
Accuracy (°C)	+/-0.15
Relative Humidity in Chamber :	
Sensor {3}	T
Range (%RH)	0-100
Accuracy (%RH)	+/-1
External PAR :	
Sensor {4}	Si
Range (μmol/s/m²)	[illegible]

Internal Radiation :			
Sensor {4}	Si		
Range (relative)	0-4000		
Accuracy (%)	5		
Flow Rates :			
Sensor {5}	M		
Range (cm^3/min)	250-2250		
Accuracy (%)	2		
OUTPUT :			
mV/V	+		
Digital display	+		
RS232C	+		
Printer	+		
DIMENSIONS (cm) :			
Temperature Controlled Chamber* :	A	B	C
Width	24	10	15.5
Depth	37	40	22.5
Height	21	14	18
Weight (kg)	5.3	3.5	3.8
System Console :			
Width	52		
Depth	60		
Height	30		
Weight (kg)	40		
POWER SUPPLY {6}	1;2;D		
OPERATING ENVIRONMENT :			
Maximum temp. (°C)	45		
Minimum temp. (°C)	0		
Humidity (%RH)	80		
PRICE RANGE (U.S.$)§			
INSTRUMENT	1		

INSTRUMENTS :

1.Compact Minicuvette System :Walz Mess- und Regeltechnik

{1}Sensor :N=Non-dispersive infrared gas analyser
{2}Sensor :Te=Thermocouple ;Ti=Thermistor
{3}Sensor :T=Thin film-capacitance
{4}Sensor :Si=Silicon
{5}Sensor :M=Mass flow meter
{6}Power supply :1=110/120VAC 50-60Hz
2=220/240VAC 50-60Hz
D=DC

*Leaf chambers :A=GK-022 Climate unit with flanged cuvette (conifer model)
B=GK-0235 Climate unit with flanged cuvette (grass model)
C=KK-02 Measuring chamber (broad leaf model)

§ See Introduction

INSTRUMENT	1
Sensor:	
Type{1}	C
Disc diameter (cm)	4
Disc height (cm)	1.3
Temperature control{2}	T
Control box:	
Output (mV)	0-1999
Analogue	+
Precision (digit)	1
Dimensions (cm):	
Width	20
Depth	17
Height	7.8
Weight (kg)	2
Casing{3}	M
Accessories available{4}	Cu,Ch,L,Co,Fl
Power supply{5}	D
Battery life (h)	200
Operating temperatures (°C):	
Upper limit	50
Lower limit	-5
Price range (U.S.$)§:	
Disc only	B
Control box	C

INSTRUMENT	1

INSTRUMENTS:

1.S1 Disc Electrode & CB1-D Control Box :Hansatech Instruments Ltd.
This electrode is designed for incorporation into both temperature controlled solution vessels and leaf disc chambers for measurement of photosynthesis and respiration.

{1}Type : C=Clark
{2}Temperature control : T=disc is operated in temperature controlled cuvettes.
{3}Casing : M=metal
{4}Accessories available : Cu=Temperature controlled stirred cuvette for chloroplast,algal and cell suspensions.
Ch=Temperature controlled chamber for leaf disc oxygen exchange measurement.
L=Light source for Ch/Cu.
Co=Computer controlled system for automatic photosynthetic light response curve measurement
Fl=Chlorophyll fluorimeter attachment for Cu and Ch.
{5}Power supply : 1=110/120VAC 50-60Hz
2=220/240VAC 50-60Hz
D=DC ; R=Rechargeable batteries
§See Introduction

POROMETERS		
INSTRUMENT	1	2
Type{1}	S	Y
Diffusion resistance range(s/cm)	0.5-50	0-30
Resolution(s/cm)	0.05-0.1	0.05
Relative humidity sensor{2}	V	V
Leaf temperature sensor{3}	T	Th
Cuvette temperature sensor{3}	Th	Th
Light measurement{4}	Q	Q
Other displays available{5}	C,N	C
Data output:		
Cassette	+	
RS232C	+	(+)
Calibration{6}	A	A/P
Sensor head dimensions(cm):		
Width/Diameter	6.3	3
Depth	11.5	11
Height		3
Weight(kg)	1	0.1
Meter dimensions(cm):		
Width	25.4	30
Depth	15	14
Height	14.5	22
Weight(kg)	4.6	3
Power supply{7}	1/2/R	1/2/R
Battery life(h)	6 - 8	16
Operating environment:		
Maximum temp.(°C)	50	50
Minimum temp:(°C)	0	0
Humidity(%RH)	95	
Price range (U.S.$)§		F
INSTRUMENT	1	2

INSTRUMENTS :

1.LI-1600 Steady State Porometer :LI-COR,inc.
2.AP4 Porometer :Delta-T Devices Ltd.

{1}Type : Y=Dynamic ; S=Steady state
{2}R.H. sensor : V=Vaisala thin film capacitance type
{3}Leaf/Cuvette sensor : Th=Thermistor ; T=Thermocouple
{4}Light measurement : Q=Quantum sensor
{5}Other displays : C=Stomatal conductance ; N=Transpiration
{6}Calibration : A=Automatic ; P=Plate
{7}Power supply : 1=110/120 VAC 50-60Hz
2=220/240 VAC 50-60Hz
R=Rechargeable batteries
§ See Introduction

PRESSURE CHAMBERS			
INSTRUMENT	1	2	3
Chamber material {1}	A	S	B
Chamber dimensions (cm) :			
Height	12.7*	12.7*	2-15*
Diameter	6.3*	6.3*	5-10*
Cover type {2}	L	L	Sc
Maximum range (MPa)	4	7	4.1
Precision (%FSD)	0.25	0.25	0.2
Weight (kg)	6.1	11.5	
Case dimensions (cm) :			
Width	33	33	
Depth	25	25	
Height	20	20	
Casing {3}	A/F	A/F	
Price Range (U.S.$)§	D	E	D
INSTRUMENT	1	2	3

INSTRUMENTS :

1.Model 600 :P.M.S. Instrument Co.
2.Model 1000 :P.M.S. Instrument Co.
3.Model SKPM 1400 Plant Moisture Vessel :Skye Instruments Ltd.

{1}Chamber material : A=Aluminium ; S=Stainless steel ; B=Brass
*Volume reducer to save gas
{2}Cover type : L=Lug type ; Sc=Screw type
{3}Casing : A=Aluminium ; F=Fibreglass
§See Introduction

LEAF AREA & ROOT LENGTH METERS

NSTRUMENT	1	2	3	4	5	6	7	8
'ype {1}	V	L	V		L	V	V	V
:esolution (mm² or mm)	0.01	1	0.1/1	1	0.15-0.5	0.01-1	0.15	0.01
.ccuracy (%)	1	2		1		1 - 4	10 - 20	1 - 3
/lax. sample dimensions (cm):								
Width		12.7	25.4	12.7	11.8	30	10	40/25
Length (U=unlimited)	U	100	U	U	100	U	10	37
Thickness		0.8	2.5	0.75	0.8			
'ortable		+			+			
Jse with microscope	+					(+)	(+)	(+)
:S232 interface	+	+			+	+		+
)ther applications {2}	T,F,I,S					T,F	T	T,F,I,S
:onveyer belt speed (cm/s)			6.7-8	5.4-6.3				6 - 19
.ight source {3}	Fl	Fl	Fl			Fl	Fl	Fl
'ower supply {4}	1;2	1;2;R	1;2	1;2	R	1;2	1;2	1;2
Battery life (h)		15			30			
)perating environment :								
Maximum temp. (°C)		55	55	55				
Minimum temp. (°C)		0	15	0				
Humidity (%RH)								
'rice range (U.S.$)§	F					F/G	G#	E* F/G#
NSTRUMENT	1	2	3	4	5	6	7	8

NSTRUMENTS:

1.Computer-based Analysis System :Skye Instruments Ltd.
2.LI-3000A Portable Area Meter :LI-COR,inc.
3.LI-3100 Area Meter :LI-COR,inc.
4.LI-3050A Transparent Belt Conveyor :LI-COR,inc.
5.LAM-100 Leaf Area Meter :C.I.D.inc.(from ADC Ltd.)
6.Area Meter :Delta-T Devices Ltd.
7.MK2 Area & Root Lenght Measurement System :Delta-T Devices Ltd.
8.DIAS Digital Image Analysis System :Delta-T Devices Ltd.

1}Type :V=Video camera scan ;L=Light emitting diode
2}Other applications :T=Root lenght ;F=Fish size ;I=Image analysis ;S=Seed counting
3}Ligth source :Fl=Fluorescent tube
4}Power supply :1=110/120VAC 50-60Hz ;2=220/240VAC 50-60Hz ;D=DC ;R=Rechargeable batteries

Card and software only
Complete basic system inc.computer software

See Introduction

PLANT CANOPY STRUCTURE ANALYSERS

INSTRUMENT	1	2	3
Measurement {1}	G	G,P	G
Derived parameter {2}	L,M,A	L,A	L
Sensor type {3}	Si	Ph	
Probe dimensions (cm):			
Length	61	80/40	
Diameter	2.9	1.5	
Memory capacity (data sets)		1000	670
RS232	+	+	+
Processing software available	+		+
Unattended measurement		+	
LCD display	+	+	+
Waterproof		+	
Control unit dimensions (cm):			
Length	21	16	
Width	11.4	8	
Depth	6.9	5.5	
Weight (kg)	1.8	1	
Power supply {4}	D	D	R
Battery life (h)	260	8500	
Operating environment :			
Maximum temp. (°C)	50		
Minimum temp. (°C)	-15		
Humidity (%RH)	100		
Price range (U.S.$)§		E	
INSTRUMENT	1	2	3

INSTRUMENTS :

1.LAI-2000 Plant Canopy Analyser :LI-COR,inc.
2.Sunfleck Ceptometer :Decagon (from Delta-T Devices Ltd.)
3.C.S.I.R.O. Demon :Assembled Electrics

{1}Measurement : G=Gap fraction ;P=PAR (Photosynthetically active radiation)
{2}Derived parameter :L=LAI (Leaf area index);M=Mean tip angle ;
A=Leaf angle distribution
{3}Sensor type : Ph=Photodiode ;Si=Silicon "fish-eye" optical sensor
{4}Power supply :D=DC ;R=Rechargeable batteries
§ See Introduction

INFRA-RED GAS ANALYSERS

INSTRUMENT	1	2	3	4	5	6	7	8
Differential capability	+	+	+			+		
Measurement range(0- ppm)	3000	3000	1500	2000 - 50000	2000 - 50000	1000		
No. of ranges	1	1	1	2	2	1	2	4
Precision (%FSD)	0.2	0.2	0.2	0.4	0.4	0.05	3	1
Detector :								
Type{1}	S	S	S	S	S	S	S	M
Configuration{2}	P	P	P			(P)		
Sample tube :								
Volume (cm³)	11.9	11.9	11.9	1.6	1.6	15		
Length (cm)	15.2	15.2	15.2	8	8	15	0.1-20	0.015-20
Output :mA		+	+				+	+
mV/V	+	+	+	+	+	(+)	+	+
RS232		+	+	+	+			
Digital display	+	+	+		+	+		+
Dimensions (cm) :								
Width	13	13	13	13	15.8	30	23	50
Depth	24	24	24	18	21	20	13	29
Height	33.5	33.5	33.5	4.5	6.2	20	35	27
Weight (kg)	3	3.5	3.5	0.4	1.6	6.5	2.5	18
Casing {3}		M	M		P	M	M	M
Weatherproof					(+)			
Power supply {4}	1;2;D;R	1;2;D;R	1;2;D;R	D	R;(2)	D	2	2
Battery life (h)	2 - 4	2 - 4			6			
Operating environment :								
Maximum temp.(°C)	50	50	50	50	50	45	45	45
Minimum temp.(°C)	0	0	0	0	0	0	10	0
Humidity (%RH)	90	90	90	90	90	90		
Price range (U.S.$)§	G	G	G	C	E	F		
INSTRUMENT	1	2	3	4	5	6	7	8

INSTRUMENTS :

1.LI-6251 CO2 Analyser :LI-COR,inc.
2.LI-6252 CO2 Analyser :LI-COR,inc.
3.LI-6262 CO2/H2O Analyser :LI-COR,inc.#
4.SBA-1 OEM CO2 Analyser :P.P.Systems
5.EGM-1 Environmental Gas Analyser :P.P.Systems
6.CI-301 Portable IRGA :C.I.D. Inc.
7.Model QGD-07 Portable IRGA :Beijing Analytical Instrument Factory
8.Model QGS-08 IRGA :Beijing Analytical Instrument Factory

{1}Detector type: M=Microphone; S=Solid state
{2}Configuration: P=Parallel
{3}Casing :M=Metal; P=Plastic
{4}Power supply: 1=110/120VAC 50-60Hz ;2=220/240VAC 50-60Hz
D=DC ;R=Rechargeable batteries
# Includes simultaneous infra-red H2O vapour measurement with a range of 0-75mB and accuracy of 1% FSD.
§See Introduction

NITROGEN OXIDES ANALYSERS		
INSTRUMENT	1	2
Detection method {1}	C	C
Gases measured	NO2;NO;NOx	NO2;NO;NOx
Maximum range (ppm)	0 - 1	0 - 100
Minimum range (ppm)	0 - 0.1	0 - 0.5
Lowest detectable limit (ppm)	0.002	0.002
Precision (%)		1
Output :		
mV/V	+	+
mA/A	+	+
Digital display	+	+
RS232C		(+)
Dimensions (cm) :		
Width	43	43.2
Depth	55	50.8
Height	22.1	17.8
Weight (kg)	30	22
Power supply {2}	1;2	1;2
Unattended operation	+	+
Operating environment :		
Maximum temp. (°C)	40	40
Minimum temp. (°C)	0	5
Humidity (%RH)		95
Price range (U.S.$)§		
INSTRUMENT	1	2

INSTRUMENTS :

1.APNA-350E Ambient NOx Monitor :Horiba Ltd.

2.Model 2108 Chemiluminescent Nitrogen Oxides Analyser :Dasibi Environmental Corp.

{1}Detection method :C=Chemiluminescence

{2}Power supply :1=110/120VAC 50-60Hz
2=220/240VAC 50-60Hz

§ See Introduction

OXYGEN ELECTRODE PROBES & METERS							
INSTRUMENT	1	2	3	4	5	6	7
SENSOR :							
Type {1}		E	G/C	C	C	M	C
Dimensions (cm):							
Diameter	1.5		1.2	1.5		1.3	2.5
Length	19		10.4	14		16.8	15
Temperature compensated	+	+	+	+	+	+	+
Range (0- ppm or mg/l)	20	20	15	50	19.9	19.99	19.9
Precision (%FSD)	1	1	1	1	2	1	0.3
Steam sterilisable			+				
Output :							
mV	+	+	+	+	+	+	+
Analogue		+	+				(+)
Digital	+			+	+	+	(+)
BCD							(+)
METER-Dimensions (cm):							
Width	5.5	24.5	22	8.7	6	23	28
Depth	4	23	23	5.5	2.6	22	10
Height	13.5	12	7	17.1	15	9	22
Weight (kg)	0.25	1.5	1.5	0.3	0.14	2.3	1.4
Casing {2}	P		M	P	P		
Power supply {3}	V	V;(2)	1;2	D	D	1;2;D	1;2;D
Battery life (h)				200		100	100
Operating temperatures (°C):							
Upper limit	40		50	50		50	50
Lower limit	0		5	-5	0	-5	-10
Price range (U.S.$)§	B		C	B			C
INSTRUMENT	1	2	3	4	5	6	7

INSTRUMENTS :

1.Checkmate D.O. Set :Ciba Corning Analytical
2.Model PE-130 Dissolved Oxygen Analyser :Elico Private Ltd.
3.E5 Series Oxygen Electrode & K500 Series Meter*:Uniprobe Instruments Ltd.
4.Oxi 191 Battery Oxygen Meter :Wissenschaftlien-Technische Werkstätten (W.T.W.)
5.9070 Dissolved Oxygen Meter :Jenway Ltd.
6.Model 7135 Dissolved Oxygen Meter :Kent Industrial Measurements
7.YSI 54ARC Dissolved Oxygen Sensor & Model 5739 Dissolved Oxygen Indicator :Yellow Springs Instrument Co.

{1}Sensor type :C=Clark type (polargraphic) ;M=Mackereth type ;G=Galvanitic ;E=Electrolytic
{2}Casing :M=Metal ;P=Plastic
{3}Power supply :1=110/120VAC 50-60Hz
220/240VAC 50-60Hz
D=DC
R=Rechargeable batteries
V=9V PP3
*E500 Series are multifunctional monitoring & control instruments with RS232 interface.
Each will accept 1/2 transducer inputs in any combination of :temperature,dissolved oxygen, gaseous oxygen,pH,redox,conductivity & ion selective.
§See Introduction

OZONE MONITORS/ANALYSERS							
INSTRUMENT	1	2	3	4	5	6	7
Detection method {1}	U	U	U	U	U	U	U
Measurement range (ppm O3):							
Maximum	1	2	50	60000	50	1	1
Minimum	0	0	0	0	0	0	0
Precision (ppm)	0.001	0.002	0.001	1%	0.001	0.001	0.002
Measurement interval (s)	12or24		10	12or24	10	12or24	
Flow rate (l/m)	1.5 - 3	1 - 2	2	1 - 3	1 - 2	1.5 - 3	
Built-in ozone generator					+	+	
Output :							
mV/V	+	+	+	+	+	+	+
mA	(+)	(+)	(+)	(+)	(+)	(+)	+
Digital display	+	+	+	+	+	+	+
BCD	(+)		(+)	(+)	(+)	(+)	
RS232C		(+)	(+)		(+)		
Dimensions (cm) :							
Width	43.7	43.2	43.7	43.7	43.7	43.7	43
Depth	57.2	50.8	57.2	57.2	57.2	57.2	55
Height	13.3	17.8	13.3	13.3	13.3	13.3	22.1
Weight (kg)	13.6	16	12.7	13.6	13.6	16	20
Power supply {2}	1;2	1;2	1;2	1;(2)	1;2	1;2	1;2
Unattended use	+	+	+	+	+	+	
Portable	+	+	+	+	+	+	
Operating environment :							
Maximum temp.(°C)	45	45	45	50	45	50	40
Minimum temp.(°C)	0	5	0	0	0	0	0
Humidity (%RH)							
Price range (U.S.$)§							
INSTRUMENT	1	2	3	4	5	6	7

INSTRUMENTS :

1. Model 1003-AH Ambient Air Quality Ozone Monitor :Dasibi Environmental Corp.
2. Model 1108 U.V. Photometric Ozone Analyser :Dasibi Environmental Corp.
3. Model 1008-AH U.V. Photometric Ozone Analyser :Dasibi Environmental Corp.
4. Model 1003-HC High Concentration Ozone Analyser :Dasibi Environmental Corp.
5. Model 1008-PC U.V. Photometric Ozone Analyser with Built-in Ozone Generator :Dasibi Environmental Corp.
6. Model 1003-PC Ozone Analyser with Built-in Ozone Generator :Dasibi Environmental Corp.
7. APOA-350E Ambient Ozone Monitor :Horiba Ltd.

{1}Detection method :U=U.V. absorption

{2}Power supply :1=110/120VAC 50-60Hz ;2=220/240VAC 50-60Hz

§See Introduction

MASS FLOW CONTROLLERS/METERS							
INSTRUMENT	1	2	3	4	5	6	7
Maximum range (std.l/min)*	50	30	500	1500	10	15	50
Minimum range (std.cm^3/min)*	5	10	50		50	10	0
Control valve type {1}	S	S	S	S	PM	S	
Accuracy (%FSD)	1	1	1	1	0.8	1	1.2
Output :							
mA	+					(+)	+
V	+	+	+	+	+	+	+
Dimensions (cm):							
Lenght	7.7	7.6	13.6	35.6	17.4	13	3.1
Width	2.5	3.2	5	6.5	3.8	2.5	1.3
Height	11	13.4	15.2	20.3	14	11.8	5.1
Minimum differential pressure for full performance (kPa)	5	70	70	70	10#		
Readout meter/power supply{2}	C	S	S	S	S/C	C	S
Power supply {3}	1;2;D	1;2;D	1;2;D	1;2;D			1;2
Power requirement :							
V	15	15	15	15	15		
mA	300	50-200	50-200	50-200	400		
Operating temperatures (°C) :							
Maximum	70	50	50	50	40	65	70
Minimum	-10	0	0	0	25	0	0
Price range (U.S.$)§	C	C	D	E	D		
INSTRUMENT	1	2	3	4	5	6	7

INSTRUMENTS :

1.Hi-Tec MFC 201 Mass Flow Controller :Bronkhorst High-Tech B.V.
2.Hastings HFC-202 Series Mass Flow Controllers :Teledyne Hastings-Raydist
3.Hastings HFC-203 Series Mass Flow Controllers :Teledyne Hastings-Raydist
4.Hastings HFC-204 Series Mass Flow Controllers :Teledyne Hastings-Raydist
5.Type 1259A Mass Flow Controller :MKS Instruments
6.Model 5850 TR Mass Flow Controller :Brooks Instruments
7.Accu-mass Series 730 Mass Flowmeter :Sierra Instruments

*Overall values for a range of instruments.
{1}Control valve type :S=Solenoid ; PM=Proportioning magnet
#Value for flowmeter only.Conductance is ruled by that of a 50std.l/min orifice at 100mbar ΔP.
{2}Readout meter/Power supply : S=Separate ; C=Combined separate unit
{3}Power supply :1=110/120 VAC 50-60Hz
2=220/240 VAC 50-60Hz
D=DC
§ See Introduction

VARIABLE AREA FLOWMETERS

INSTRUMENT	1	2	3	4	5	6	7	8
Maximum flowrate (l/min)*	50	67	100	60.7	35	54	40	71
Minimum flowrate (cm^3/min)*	4	4.8	5	46	10	50	2	0.5
Accuracy (%FSD)	2 - 10	1 - 2	1.25	3(1)	10	2	1.6	
Float type {1}	C/S	S	C	S	S	S	C	S
Needle valve	(+)	+		+	(+)		+	+
Overall dimensions (cm) :								
Width/diameter	4.3		3.2	3.2	2.5	10.2		2.8
Depth	3.8	5	3.3	4.9	2.5	2.9	5	4.2
Height	36.2	26.3	18.4	24.9	9.5	27.9	35	15
Scale length (cm)	24	10 - 13	10	25	4	15.2	17	10
Operational limits :								
Maximum temp. (°C)	60	120	100	120	80	121	130	100
Maximum pressure (kPa)	600	1800	350	1400	700	1700	2000	1700
Price range (U.S.$)§		A	A		A	B		A
INSTRUMENT	1	2	3	4	5	6	7	8

INSTRUMENTS :

1.Series 1100 Variable Area Glass Tube Flowmeters :KDG Flowmeters
2.Series 10A320 Purgemeter :Fischer & Porter
3.Small Gapmeter Type GS :Platon Flowbits
4.Sho-Rate Purgemeter Model 1357 :Brooks Instruments
5.Poddymeters (short scale) :Comoy Components
6.Low Flow Meters :Wallace & Tiernan
7.Rotameters (with standard metering tubes) :Rota
8.DK800 Miniature Flowmeter :Krohne

*Values given refer to the top and bottom of the range available and not individual instruments.
{1}Float type :C=Conical ;S=Spherical
§ See Introduction

PLANT GROWTH CABINETS			
INSTRUMENT	1	2	3
Temperature range (°C) :			
Upper limit	45	45	45
Lower limit *	10	5	5
Normal range of variation	0.5	0.75	1
Humidity range (%RH) :			
Upper limit	85-95	90	90
Lower limit		55	40
Illumination :			
Max. no. fluorescent lights	16	12	14
Max. no. incandescent lights	12	4	3
Max. no. metalhalide lights	8		
Photon flux density at plant level (lux)	80000	25000	45000
Control system {1}	M	C	M
Programmable features {2}	R;E;B;A;S;F P;L;H		A;E
Alarm system	+	+	+
Interior exposed surface {3}	E		E
External dimensions (cm) :			
Width	242		236
Depth	90		89.5
Height	194		179
Total weight (kg)	694		510
Growth area (m^2)	1.4	0.72	1.2
Power supply {4}	1;2	2	3
Made to buyers specifications	+		(+)
Operating environment :			
Maximum temp. (°C)	Average:		
Minimum temp. (°C)	21		
Humidity (%RH)	50		
Price range (U.S.$)§			
INSTRUMENT	1	2	3

INSTRUMENTS :

1.E15 Plant Growth Chamber :Controlled Environments Ltd.

2.Fi-totron 600H Growth Cabinet :Galenkamp Industrial

3.Heraphyt HPS 1500 :Heraeus-Vötsch GmbH

*With all lights on

{1}Control system : M=Microprocessor CMP3244 ;C=Clock

{2}Programmable features :R=Real time programming ;E=Editing ;B=Back-up ;
A=Automatic ramping ;S=Security code ;F=Defrost ;
P=Barometric pressure ;L=4 level lighting ;H=Service history

{3}Interior exposed surface material :St=Stainless steel ;E=Enamel

{4}Power supply :1=110/120VAC 50-60Hz
2=220/240VAC 50-60Hz
3=380V,3/N,50Hz
D=DC

§See Introduction

AUTOMATIC WEATHER STATIONS										
SYSTEM	1	2	3	4	5	6	7	8	9	10
No. inputs	10	4	17 - 62	14	20	9	12 - 26	16	20	20
Sensors available :										
Wind speed	+		+	+	+	+	+	+	+	+
Wind direction	+		+	+	+	+	+	+	+	+
Air temperature	+	+	+	+	+	+	+	+	+	+
Relative humidity	+		+	+	+	+	+	+	+	+
Atmospheric pressure	+		+	+	+	(+)	+	+	+	+
Precipitation		+		+	+	+	+	+	+	+
Rain amount	+		+	+	+	+	+	+	+	+
Evaporation				+	+		+	+		+
Net/solar radiation	+	+	+	+	+	+	+	+	+	+
Sun detector							+			
Soil temperature	+	+	+	+	+	+	+	+	+	+
Soil moisture				+	+		+		+	+
Heat flow	+			+	+					
Dew point	+			(+)	(+)	+			+	+
Water temperature	+		+	+	+			+	+	+
Others {1}			L	L;G	L;G	L		L;G		
Output :										
Analogue	+			(+)	+	+	+		+	
Digital		+		+	+		+	+	+	+
RS232C	+	+	+	+	+	+	+	+	+	+
BCD				+						
C-type cassette				+	+				+	+
Solid state data recorder	+			+	+	+	+	+	+	+
Chart	+									
Data transmission options :										
Current loops				+	+		+		+	
Modem			+	+	+	+	+	+	+	+
UHF/VHF radio				+	+		+	+	+	+
Satellite				+	+			+	+	
Meteor scatter				+	+			+	+	
Power supply options :										
Mains	+	(+)	+	+	+		+	+	+	+
24/28 VAC power unit				(+)	(+)				+	
Solar generator			+	+	+		+	+	+	+
Wind generator				+	+			+	+	

Housing dimensions (cm) :										
Width	49	19	28	30	30	22	38.1	27	36	51
Depth	37	19	22	17	17	12	20.9	18	14	28
Height	15	56	14	20	20	29	38.1	19	48	61
Weight (kg)	8	6	2.7	9.4	11.3	3.2	13.5	4.5		10
Housing material {2}	M/Pl	W	Pl/M	M/F	M/F	Pl	M	F	M	M
Mast type {3}	Pr/T		P	P/T/Pr	P/T/Pr	P	T	P/Pr	P/S	S/T
Operating environment :				*	*					
Maximum temp. (°C)	65	55	60	50	50	50	50	55	50	55
Minimum temp. (°C)	-30	-25	-20	-25	-25	-25	-30	-25	-30	-25
Price range (U.S.$)§	F	D/E	G	F	F	F	F	E/F	G	
SYSTEM	1	2	3	4	5	6	7	8	9	10

SYSTEM :

1.R500/SQW Anemograph/Logger :Vector Instruments
2.LI-1200 Minimum Data Set Recording System :LI-COR,inc.
3.Delta-T Weather Station :Delta-T Devices Ltd.
4.Automatic Weather Station (with CR10 Logger) :Campbell Scientific Ltd.
5.Automatic Weather Station (with 21X Logger) :Campbell Scientific Ltd.
6.EMS-10 :P.P.Systems
7.Easy Logger Recording System :Omnidata International Inc.
8.Environdata Automatic Weather Station :Environdata
9.AWS 1 Automatic Weather Station :Didcot Instrument Co.Ltd.
10.Automatic Weather Station :Qualimetrics

{1}Other sensors :L=Leaf wetness ;G=Grass temperature
{2}Housing material :M=Metal ;Pl=Plastic ;W=Wood ;F=Fibreglass
{3}Mast type :Pr=Portable ;T=Tripod ;P=Pole ;S=Stacked
* Extended temperatures to special order
§See Introduction

INSTRUMENT	ANEMOMETERS										
	1	2	3	4	5	6	7	8	9	10	11
Type{1}	C	P	T	C	H	C	C	V	P	P	P
Maximum speed (m/s)	75	100	60 (80)	75	30	45	75	17	20	60	
Threshold (m/s)	0.15	1	1	0.2	3	0.22	0.9	0.8	0.3	0.15	0.8
Accuracy (%)	1	0.5	2	1		1	2	2	2.3		1
Output :mA											
mV	+			+	+	+					+
pulse	+	+				+	+		+	+	+
Remote	+	+	+	+		+	+			+	+
Handheld					+			+	+		
Direction indicator	+	+	+							+	+
Sensor dimensions (cm):											
Rotor diameter	15.2	18		16.4		19.4	8.6	7	2	140	39.5
Body diameter	5.5		17.8	4.5		7	6.5		1.9	58	
Height	19.5	37	107	29		30.5	30		30	9	75.9
Weight (kg)	0.35	1	15	0.6		1.1	0.87				5.7
Readout meter * :											
Analogue display	+		(+)	+	+			+	+		
Digital display	+	+	+	+							
Counter											
Chart				+							+
Dimensions (cm) :											
Width/diameter	13	22		9.6	6.2			10	9.2		48.2
Depth	25	17		0.7	12.3			10	3.3		24.8
Height	16	7		4.4	19.3			10	15.9		17.7
Weight (kg)	2.7			1.5	1				0.45		4.1
Power supply{2}	1;2;D;R	1(2)	1;2	D	R				D	2	1
Operating environment :											
Maximum temp.(°C)	70	40	40	60	50	60	55	95	50	50	65
Minimum temp.(°C)	-50	-40	-40	-30	0	-40	-40		0	-35	
Humidity (%RH)	100		100	100			93		70	100	
Price range (U.S.$)§	B			A	B	B					F

1.A100 Porton Anemometer & D600 Series Indicator :Vector Instruments
2.Model 05103 Wind Monitor & 04503 Wind Speed/Direction Indicator :R.M.Young Company
3.PTA-1 Pitot Tube Anemometer/Wind Direction Sensor (&CR10 Measurement & Control Module) :Campbell Scientific Ltd.
4.DWR204 & 204G Anemometer :Didcot Instrument Co.
5.AVM521 Air Velocity Meter :Prosser Scientific Instruments
6.Model 2030 Micro Response Anemometer :Qualimetrics
7.WAA 12 Opto Electronic Anemometer :Vaisala
8.Portable Air Meter :Casella
9.Electronic Thermo-Anemometer :Haenni
10.NEZ 11 3-Dimensional Wind Sensor :Alcyon Data
11.Type LD Digital Wind Set :Belfort Instrument Co.

{1}Type :C=Cup ;P=Propeller ;V=Vane ;H=Hot wire ;T=Pitot tube
*Where available together
{2}Power supply :1=110/120VAC 50-60Hz ;2=220/240VAC 50-60Hz
D=DC ;R=Rechargeable batteries
§See Introduction

ACCESSORIES FOR METEOROLOGICAL MEASUREMENTS	
INSTRUMENT	1
Type of accessory {1}	R
Sensor mounting for :	
Temperature	+
Dewpoint	+
Relative humidity	+
Variable sensor mounting	+
Material {2}	P
Accuracy :	
Radiation error *(°C)	
@ wind speed of 3m/s	0.4
@ wind speed of 1m/s	1.5
Mounting {3}	U
Overall dimensions (cm) :	
Diameter	12
Height	27
Plate dimensions (cm) :	
Thickness	0.2
Spacing	1.1
Weight (kg)	0.7
Price range (U.S.$)§	B
INSTRUMENT	1

INSTRUMENT :
1.Gill Multi-Plate Radiation Shield :R.M.Young Co.

{1}Type of accessory :R=Radiation sheild
{2}Material :P=White thermoplastic (UV stabilised)
*Radiation error is dependant on wind speed (ventilation rate) and is measured under radiation intensity of 1080W/m².
{3}Mounting :U=U-bolt to fit vertical pipe 25-50mm diameter
§ See Introduction

WIND DIRECTION INDICATORS

INSTRUMENT	1	2	3	4	5	6
Measurement range (°)	360	360	360	360	360	360
Threshold windspeed (m/s)	0.6	1	0.9	0.22	0.6	0.5
Maximum windspeed (m/s)	75	100	45	45	75	75
Accuracy (°)	2	3	3	2	1	1
Output :						
Digital						
Analogue	+	+		+	+	+
Overall dimensions (cm):						
Height	23.0	37.0	23.0	30.5	39.0	29.0
Fin radius	15.0	38.0	27.0	45.0	11.4	9.7
Body diameter	5.5	5.0	5.7	7.0	3.9	3.4
Weight (kg)	0.3	1.0	0.9	1.1	1.0	0.6
Power supply {1}	D	D	1 ; 2		D	D
Operating environment :						
Maximum temp. (°C)	70	50	50	60	60	60
Minimum temp. (°C)	*-50	-50	-10	-40	-30	-30
Price range (U.S.$)§	B	C	B	B	C	B
INSTRUMENT	1	2	3	4	5	6

INSTRUMENTS :

1.W200P Potentiometer Windvane :Vector Instruments
2.Model 05103 Wind Monitor :R.M.Young Co.
3.Wind Minder Vane :Qualimetrics
4.Micro Response Vane :Qualimetrics
5.DWD/103 Potentiometer Wind Direction Sensor :Didcot Instrument Co.
6.DWD/104 Budget Wind Direction Sensor :Didcot Instrument Co.

{1}Power supply: 1=110/120VAC 50-60Hz
2=220/240VAC 50-60Hz
D=DC

* with de-icing heater
§ See Introduction

METERS FOR RADIATION & LIGHT SENSORS

INSTRUMENT	1	2	3	4	5	6	7	8	9
Type of instrument {1}	Q,R,P,I	I	Q,R,P	Q,R	Q,R,P	I	Q,R,P	Q,R,P,I	Q
Maximum range (μmol/s/m²)	20000		19999	3000	20000	40000	30000	20000	2000
Maximum range (W/m²)	2000	2000	1999	1500	2000	40000	3000	2000	
No. of ranges	8	1	3	1	3	3	11	7	3
Precision (%FSD)	1	1	0.3	0.01	0.2	0.2	1.5	1	0.01
Output : mV/V	+				+	+	+	+	+
mA	+							+	
Digital display	+	+	+	+	+	+	(+)	+	+
Integrator	+	+				+		+	
Other features {2}	(T,Pr)	(T)				Pr/C		(Pr)	
Dimensions (cm):									
Width	22	8	7.7	8	8	15	11.5	10	7
Depth	24	5.8	3.8	13	3.5	24	6	4.3	14
Height	9	14	14	15	14.5	11	19	18.5	3
Weight (kg)	2.3	0.4	0.26	0.3	0.28		1	0.6	0.1
Casing {3}	P	P		P	P	P	P	P	
Weatherproof		+				+			
Power supply {4}	1/2/D/R	D	D	R	D/R	D/R	R	D	
Battery life (h)	8 - 10	4320	200	100	250	1200	50	57	
Operating environment :									
Maximum temp. (°C)	40	60	55	50	60	60	50	40	
Minimum temp. (°C)	0	-20	0	-5	-10	-15	0	0	
Humidity (%RH)	75	100	95	90	90	100	80	75	
Price range (U.S.$)§	D	B		B	B	C	C	C	B

1.IL1700 Radiometer :International Light,inc.
2.MV2 Microvolt Integrator :Delta-T Devices Ltd.
3.LI-189 Quantum Radiometer Photometer :LI-COR,inc.
4.ESR-1 Sensor Readout Unit :P.P.Systems
5.Radiation Sensor Measuring Unit :Skye Instruments Ltd.
6.Integrating Printing Logger :Skye Instruments Ltd.
7.Q101 Quantum Radiometer Photometer :Macam Photometrics
8.IL1400A Radiometer :International Light,inc.
9.Control Unit :Hansatech Instruments Ltd.

{1}Type of instrument :Q=Quantum meter ;R=Radiometer ;P=Photometer ;I=Integrator
{2}Other features :T=Temperature integrator board ;Pr=Printer ;C=Multichannel
{3}Casing :P=Plastic
{4}Power supply :1=110/120VAC 50-60Hz
220/240VAC 50-60Hz
D=DC
R=Rechargeable batteries
§ See Introduction

RADIATION SENSORS

INSTRUMENT	1	2	3	4	5	6	7	8	9	10	11
Measurement{1}	R,Q	R	R	R	R	Q	Q	Q	R	Q	Q
Sensor surface{2}	P	L	L	L	P	P	L	P	P	P	P
Underwater use	(+)										
Detector{3}	Si	Th	Th	Th	Si	Si	Si	Si	Th	Si	Si
Wavelength range (nm):											
Maximum	700	2500	2500	10000	1000	700	700	700	2700	700	750
Minimum	400	350	350	300	400	400	400	400	300	400	550
Cosine correction	+	+	+	+	+	+		+	+	+	+
Sensitivity *	1A/2000B	15C	15C	15C	10.8C	10D	D	333D	666C	3A	3A/10D
Output :mV/μV		+	+	+	+	+		+	+	+	+
mA/μA	+										
Response time (μs or s)	60μs	40	5	9	10μs	10μs	0.1	10μs	30	10μs	10μs
Sensor dimensions(cm):											
Diameter/width	4.2	2.6	1.1	2.9	1.5	1.5	1.5	4	5.5	4.5	4.5
Height/length	3.5	97	38	99	3.8	3.8	80/40	3.3	3	5.8	5.8
Weight (kg)	1.25	0.4	0.2	0.8	0.02	0.02	1	0.33	0.33	0.16	0.16
Housing{4}	Al	P	P	P	M	M	P	P	Al		
Weatherproof	+	+	+	+	+	+	+	+	+	+	+
Suitable meter #	1	2	2	2	2	2		4	4	9	9
Price range (U.S.$)§	B	B	B	B	B	B	E	B	B	B	B

1.SED(L)033 Quantum Detector :International Light,inc.
2.Large Tube Solarimeter :Delta-T Devices Ltd.
3.Miniature Tube Solarimeter :Delta-T Devices Ltd.
4.Large Tube Net Radiometer :Delta-t Devices Ltd.
5.ES Silicon Cell Energy Sensor :Delta-T Devices Ltd.
6.Type QS Silicon Cell Quantum Sensor for PAR :Delta-TDevices Ltd.
7.CEP Sunfleck Ceptometer :Decagon(from Delta-T Devices Ltd.)
8.PAR-1 PAR Sensor :P.P.Systems
9.SRM-1 Solarimeter :P.P.Systems
10.QSPAR Large Area PAR Quantum Sensor :Hansatech Instruments Ltd.
11.QSRED Large Area RED Quantum Sensor :Hansatech Instruments Ltd.

{1}Measurement :Q=Photon flux ;R=Radiation flux
{2}Type of sensor :P=Point ;L=Line
{3}Detector :Si=Silicon photocell (blue enhanced) ;Th=Thermopile
{4}Housing :M=Metal ;P=Plastic ;Al=Aluminium
*Figures given have various units :

A=μA/1000μmol/m^2/s
B=μA/1000W/m^2
C=mV/1000W/m^2
D=mV/1000μmol/m^2/s
E=1mV/100W/m^2

#See"Meters for Radiation & Light Sensors" table
§See Introduction

RADIATION SENSORS cont'd.

INSTRUMENT	·12	13	14	15	16	17	18	19	20	21	22
Measurement{1}	Q	Q	Q	Q	R	R	R	Q	Q	Q	Ruv
Sensor surface{2}	P	L	P	S	P	P	P	P	P	P	P
Underwater use			+	+						(+)	+
Detector{3}	Si	Si	Si	Si	Si	Si	Si	Si	Si	Si	G
Wavelength range (nm):											
Maximum	700	700	700	700	1100	700	1100	700	700	700	378
Minimum	400	400	400	400	400	400	300	400	400	400	282
Cosine correction	+	+	+	+	+	+	+	+	+	+	+
Sensitivity *	8A	3A	3A	3A	80B	55B/10C	56B/10C	3A	7A	100A/10D	1E
Output :mV/μV	+	+	+		+	+	+	+	+	+	+
mA/μA						+	+	+	+	+	+
Response time (μs)	10	10	10	10	10	10	10	10	10	10	10
Sensor dimensions(cm):											
Diameter/width	2.4	2.5	3.2	6.1	2.4	3.3	3.3	3.3	3.3	1.5	5.2
Height/length	2.5	116	46.2	10.7	2.6	4	4	4	4	3.6	5.2
Weight (kg)	0.03	1.8	0.23	0.14	0.28	0.13	0.13	0.13	0.13	0.02	0.25
Housing{4}	M	M	M	M	M	P	P	P	P	M	M
Weatherproof	+	+	+	+	+	+	+	+	+	+	+
Suitable meter #	3	3	3	3	3	5 / 6	5 / 6	5 / 6	5 / 6	1	
Price range (U.S.$)§	B	D	B	B	B	B	B	B	B	B	C

INSTRUMENTS:

12.LI-190SA Quantum Sensor :LI-COR,inc.
13.LI-191SA Line Quantum Sensor :LI-COR,inc.
14.LI-192SA Underwater Sensor :LI-COR,inc.
15.LI-193SA Spherical Sensor :LI-COR,inc.
16.LI-200SA Pyranometer Sensor :LI-COR,inc.
17.SKE510 Energy Sensor :Skye Instruments Ltd.
18.SKS1110 Pyranometer Sensor :Skye Instruments Ltd.
19.SKP215 Quantum Sensor :Skye Instruments Ltd.
20.SKP210 PAR "Special" Sensor :Skye Instruments Ltd.
21.SD101Q Quantum Sensor :Macam Photometrics Ltd.
22.PD104AB Ultra-Violet Sensor :Macam Photometrics Ltd.

{1}Measurement :Q=Photon flux ;R=Radiation flux ;Ruv=Ultra-violet radiation
{2}Type of sensor :P=Point ;L=Line ;S=Spherical
{3}Detector :Si=Silicon photocell (blue enhanced) ;G=GaAs Photodiode
{4}Housing :M=Metal ;P=Plastic ;Al=Aluminium
*Figures given have various units :

A=μA/1000μmol/m^2/s
B=μA/1000W/m^2
C=mV/1000W/m^2
D=mV/1000μmol/m^2/s
E=1mV/100W/m^2

#See"Meters for Radiation & Light Sensors" table
§See Introduction

SPECTRORADIOMETERS & OTHER SPECIALIST RADIATION MEASURING

INSTRUMENT	1	2	3	4	5	6	7
Detector {1}	S	S	P	Pt	Pt	S/L	S/L
Maximum range (nm)	1100	1100	1100	800	800	3200	3200
Minimum range (nm)	350	300	360	240	200	280	200
Accuracy (nm)	2	2		1.5	0.2	0.5	0.5
Scan rate (nm/s) :							
Maximum	2	40		20	70	70	70
Minimum		20		4	1	1	1
Maximum irradiance level (W/cm²/nm)	4	$5x10^{-3}$		2x10-5	$1x10^{-3}$	$1x10^{-3}$	$1x10^{-3}$
Optical input {2}	C(F,I)	C,F,I,T	C,T	C,F,(I)	I,(F)	I,(F)	I,(F)
Output :							
mV/V	+	+		+	+		
RS232C	+	+	+		+		
BCD				(+)		+	+
X-Y recorder	+	+	+	+		+	+
Digital display	+	+	+	+	+	+	+
Dimensions (cm) :	#						
Width	15.8	36	30	48	25	33	53
Depth	12.1	20.1	7	36	26	28	46
Height	6.4	16.3	22	20	17	35	25
Weight (kg)	7	6.4	6	10	9	15	27
Weatherproof		+			(+)		
Power supply {3}	1;2;D;R	1;2;D;R	R	1;2;D	R	1;2	1;2
Battery life (h)	5 - 10	4 - 8	10 - 12		4 - 8		
Operating environment :							
Maximum temp. (°C)	40	45	50	50	40	40	40
Minimum temp. (°C)	0	0	0	0	5	5	5
Humidity (%RH)	75	100	90	80	90	90	90
Price range (U.S.$)§	G	H	F	G	I	I	I
INSTRUMENT	1	2	3	4	5	6	7

1.IL 1700/586D Spectroradiometer System :International Light,Inc.

2.LI-1800 Portable Spectroradiometer :LI-COR,inc.

3.Cropscan Multispectral Radiometer & Data Aquisition System :Skye Instruments Ltd.

4.SR 3010 Digital Spectroradiometer :Macam Photometrics

5.OL 752 High Accuracy UV-Vis-NIR Spectroradiometer :Optronic Laboratories,Inc.

6.OL 746 Infrared Spectroradiometer :Optronic Laboratories,Inc.

7.OL 746D Infrared Spectroradiometer Double Monochromator :Optronic Laboratories,Inc.

{1}Detector : S=Silicon ;P=Photodiodes ;Pt=Photomultiplier tube ;L=Lead sulphide

{2}Optical input : C=Cosine collector ;F=Fibre optic probe ;I=Integrating sphere ;T=Telescope

{3}Power supply : 1=110/120 VAC 50-60Hz

2=220/240 VAC 50-60Hz

D=DC

R=Rechargeable batteries

Sensor head only

§ See Introduction

SPECTRAL COMPARATORS for 660/730nm		
INSTRUMENT	1	2
Sensor:		
Sensor type {1}	Si	Si
Cosine corrected	+	+
Dimensions (cm):		
Diameter/width	3.3	7.5
Depth		
Height	5.5	14.2
Temperature range (°C):		
Maximum	75	40
Minimum	-25	0
Readout unit :		
Max. range (μmol/m²/s)	200	(+)
Max. range (W/m²)		+
No. ranges	3	2
V/mV output	+	+
Digital display	+	+
Other features{2}		L
Dimensions (cm):		
Width	8	16.5
Depth	3.5	19
Height	14.5	6.8
Weight (kg)	0.2	0.67
Power supply{3}	D/R	R
Battery life (h)	250	
Price range (U.S.$)§	C	E/F
INSTRUMENT	1	2

INSTRUMENT:

1.SKR110 660/730 Radiation Detector & Measuring Unit :Skye Instruments Ltd.
2.SRM4 Red/Far Red 660/730nm Spectral Ratio Meter :Macam Photometrics*

{1}Sensor type :Si=Silicon photocell
{2}Other features :L=Datalogger
{3}Power supply :1=110/120VAC 50-60Hz
2=220/240VAC 50-60Hz
D=DC
R=Rechargeable batteries
§See Introduction
*The 4-channel SRM4 measures incident radiation (cosine corrected irradiance) and reflected radiation (radiance with a 10°-30° field of view) in either red/far red spectral bands or 660/730nm narrow spectral bands.

WET & DRY BULB PSYCHROMETERS			
INSTRUMENT	1	2	3
SENSOR :			
Type {1}	Pt		Th
Range :			
Max.humidity (%RH)	100	100	99
Min.humidity (%RH)	0	10	1
Max.temperture (°C)	65	150	99.9
Min.temperature (°C)	-20	-30	-30
Accuracy (°C)	0.2	0.5-2	0.1
99% Response time (s)	50		
Dimensions (cm):			
Diameter	16.5		5
Length	25.5		32
Airflow (m/s)	4		2.5
Water capacity (cm^3)	500		8
Readout meter dimensions (cm):			
Width	9.6	6	12.5
Depth	16.7	2.6	5.5
Height	9.6	15	17
Weight (kg)	1.4	0.16	3.5
Overall dimensions (cm):			
Width	16.5		
Depth	18.3		
Height	25.5		
Casing {2}	M/P	P	
Output : mV	+	+	
Digital display	+	+	+
Power supply {3}	1/2/3/R	D	R
Operating environment :			
Maximum temp. (°C)	65		60
Minimum temp. (°C)	-20		-10
Price range (U.S.$)§	C		
INSTRUMENT	1	2	3

INSTRUMENTS :

1.H301 Psychrometer /DUH203 Display :Vector Instruments
2.5100 Digital Psychrometer :Jenway Ltd.
3.Type 461 Digital Psychrometer :Adolf Thies

{1}Type : T=Thermocouple ; Th=Thermistor ; Pt=Platinum resistance
{2}Casing : M=Metal ; P=Plastic
{3}Power supply : 1=110/120VAC 50-60Hz
2=220/240VAC 50-60Hz
3=12VAC 50-60Hz
D=DC
R=Rechargeable batteries
§See Introduction

DEW POINT GENERATORS

INSTRUMENT	1	2	3	4	5	6
Generator type {1}	B	B	B	B	B	B
Gas conduits :	2*	2*	2*	1	1	1**
Maximum gas flow (l/min)	1.5	1.5	3	10	25	1.7
Range (°C dew point) :	-5...+45	-5...+45	0...50	0...50	0...50	0...50
Accuracy (°C dew point)	0.2	0.2	0.2	0.2	0.2	0.2
Attainable temp. (°C dew point)	30 below ambient	30 below ambient	30 below ambient	30 below ambient	30 below ambient	35 below ambient
Generator dimensions (cm) :						
Width	7	7	12.5	12.5	21.5	21
Depth	17.5	17.5	23.5	23.5	46	28.5
Height	19	19	25.5	25.5	25	23.5
Weight (kg)	1.2	1.2	3.2	3.2	10.5	8
Set points :	any,by user	any,by user	any,by user	any,by user	any,by user	any,by user or mV source
Output :						
mV	+	+	+	+	+	+
mA	+	+	+	+	+	
Instrument dimensions (cm) :						
Width	26	42	36	36	48	21
Depth	26.5	26	26.5	26.5	27	28.5
Height	15	21	15	15	13.5	23.5
Weight (kg)	6	9	7	7	12.5	8
Casing {2}	M	M	M	M	M	M
Rack mounting	+		+	+	+	
Power supply {3}	1;2	1;2;D;R	1;2	1;2	1;2	1;2;D;R
Battery life (h)		8				
Operating environment :						
Maximum temp. (°C)	45	45	50	50	50	50
Minimum temp. (°C)	-5	-5	0	0	0	0
Humidity (%RH)	80	80	80	80	80	100
Price range (U.S.$)§						F

1.Measuring Gas Cooler MGK-1 with Temperature Controller KR-K-9/2 :Walz Mess- und Regeltechnik
2.Measuring Gas Cooler Unit MGE-130 :Walz Mess- und Regeltechnik
3.Measuring Gas Cooler Unit MGK-4 with Temperature Controller KR-KW-12/2 :Walz Mess- und Regeltechnik
4.Cold Trap KF-18/2 with Temperature Controller KR-KW-12/2 :Walz Mess- und Regeltechnik
5.Cold Trap KF-24/6BM with Temperature Controller KR-KW-42/6T :Walz Mess- und Regeltechnik
6.LI-610 Portable Dew Point Generator :LI-COR,inc.

{1}Generator type :B=Peltier cooled block
*Pneumatically separated
**One inlet,2 outlets
{2}Casing :M=Metal
{3}Power supply :1=110/120VAC 50-60Hz ;2=220/240VAC 50-60Hz ;
D=DC ;R=Rechargeable batteries
§ See Introduction

DEW POINT HYGROMETERS								
INSTRUMENT	1	2	3	4	5	6	7	8
Sensor type{1}	M	M	C	M	L	M	C	M
Measurement range (°C)*:								
Upper limit	50	60	50	80	42	100	115	100
Lower limit	-40	-20	-60	-40	-12	-40	-80	-60
Accuracy (°C)	0.25	0.1	0.5	0.3	0.9	0.3	0.5	0.4
Response time (°C/s)				2	0.02	2		2
Probe dimensions (cm):								
Diameter	3.8	5	1.9		0.94	73	1.9	3.9
Length	11	7	20		9.4	6.7	20	15.5
Output :								
mV/V	+	+		+	+	+	+	+
mA		+		+		(+)		+
Digital display	+	+	+	+		+	+	+
BCD			+			+		
RS232C	(+)		+					
Multichannel			+					
Temperature readout	+	+	+	+	(+)	+	+	+
Instrument dimensions (cm):								
Width	11	10.5	22	38	25	22	19	19
Depth	27	18	23	25	19	26	26	16
Height	8	13	7	18	13	16	16	10
Weight (kg)	1.5	1.6	1.8		3	6.8	1.2	4
Casing {2}	M	M	P		P		P	M
Rack mounting		+		+		(+)	(+)	
Power supply {3}	2;(1;D)	1;2	D;R	1;2	1;2	1;2	1;2;D;R	1;2

Probe-Max. temp.(°C)	60	60	55	80		70	115	110
Min. temp.(°C)	-20	-20	-10	-40		-50	-40	-40
Inst.-Max. temp.(°C)	50	60	50	50		50	55	50
Min. temp.(°C)	0	-20	-5	0		4	5	0
Price-range (U.S.$)§	F		D		C			
INSTRUMENT	1	2	3	4	5	6	7	8

INSTRUMENTS :

1.DP 989M Dewpoint Meter :Protimeter plc.
2.MTS Dew-Point Mirror Measuring System :Walz Mess- und Regeltechnik
3.HTL-1 Humidity Temperature Probe & Logger :P.P.Systems
4.System 1100DP Dew Point Hygrometer :General Eastern
5.YSI Model 91 :Yellow Springs Instrument Co.
6.Model 911 Dew-All Digital Humidity Analyser :E.G.&G. Environmental Equipment
7.HMI 33 Humidity,Temperature & Dewpoint Indicator :Vaisala
8.DP14 Dewpoint Instrument :MBW Elektronic

{1}Sensor type :M=Chilled mirror ;C=Capacitor ;L=Lithium chloride ;S=Silicon chip

*Measurement range :Where a range of sensors are available the figures given are for the total range and not necessarily an individual sensor.

{2}Casing :M=Metal ;P=Plastic

{3}Power supply :1=110/120VAC 50-60Hz
2=220/240VAC 50-60Hz
D=DC ; R=Rechargeable batteries

§See Introduction

ELECTRICAL CONDUCTIVITY HUMIDITY SENSORS & METERS

INSTRUMENT	1	2	3	4	5	6	7	8	9
SENSOR:									
Type{1}	C	C		C	C	E	C	W	C
Probe dimensions(cm):*									
Diameter/width	8	1.3	10	1.9	8.5	B	V	V	2.3
Length	14.5	9	20	20	3				18
Weight (kg)	0.3			0.15					0.16
Maximum measurement(%RH)	100	98	95	90	90	95	100	100	97
Minimum measurement(%RH)	0	20	5	0	15	10	0	10	0
No. of ranges		1		1		1	1	1	1
Accuracy (%FSD)	2	3	4	2	3	2	2	2 - 3	2
Hysteresis (%RH)	<0.5	<1		1		n.a.	1		1
METER:									
Output : mV/V	+			(+)				+	+
mA	(+)						+		
Digital display		+	+	+	+		(+)	+	+
Temperature readout	+	+	+	+		+	+	+	+
Dimensions (cm) :									
Width		8	7.5	8	3.5	7	24	7	9.2
Depth		2.5	25	14.4	2.2	4	23	3.6	4
Height		16		3.5	20	26.5	10	13	25.2
Weight (kg)		0.23	1	0.3	0.17	0.25	3	0.24	0.59
Casing {2}	P	P		P	P	P	P	P	P
Mounting {3}				C		C	B/C	C	B/C
Power supply {4}		D	D	D;R	D	D	1;2;D	2;D	2;D

INSTRUMENT	1	2	3	4	5	6	7	8	9
Probe-Max.temp.(°C)	70	50	100	55	60	50	70	70	
Min. temp.(°C)	-30	-10	0	-10	0	-10	-10	0	
Meter-Max.temp.(°C)		40	-30	50	60	50		40	50
Min.temp.(°C)		-10	0	-5	0	-10		0	0
Price range (U.S.$)§		B		B				C	B

INSTRUMENTS :

1.2000 Series Humidity & Temperature Probe :Skye Instruments Ltd.
2.Digital Hygrometer :Protimeter plc.
3.Infrared AG Multimeter :Everest Interscience Inc.
4.HTP-2 Probe & ESR-1 Readout :P.P.Systems
5.5060 Stick Humidity Meter :Jenway Ltd.
6.MIK 2000 :Novasina
7.Humilab 24 Portable Hygrometer :Lee Integer
8.Series 5500 Digital Humidity & Temperature Sensor :Channel Electronics
9.KM8004 Digital Humidity Meter :Kane-May

{1}Sensor type :C=Thin film capacitance; E=Electrolytic resistor; W=Wet & dry bulb
*Probe dimensions :B=Built-in ;V=Various
{2}Casing :P=Plastic
{3}Mounting :B=Bench mounted ;C=Portable
{4}Power supply :1=110/120VAC 50-60Hz ;2=220/240VAC 50-60Hz ;D=DC ;R=Rechargeable batteries
§ See Introduction

EVAPORATION MEASURING EQUIPMENT

INSTRUMENT	1	2	3	4	5	6
Sensing method{1}	W	L	F	F	H	L/W
Evaporation surface{2}	E	D	E	P	E	E
Range (mm)	20	110	160	10	100	290
Accuracy (mm)			0.4	0.3	0.02	0.2
Area (cm^2)	250	200	11433	8.81		169.7
Recording method{3}	C	C	C	C(B)		E/B
Unattended period (days)	1	1 ; 7	1 ; 7	1 ; 7	1	14
Overall dimensions (cm):						
Width/diameter	28	32	47	32	122	27
Depth	14	21		14	122	32
Height	25	87	61	20	35.4	55
Weight (kg)	3.2	17	29	4	58	20
Power requirement{4}	n.a.	D	n.a.	(D)	n.a.	R
Official specification{5}			+			
Price range (U.S.$)§			D	C		
INSTRUMENT	1	2	3	4	5	6

INSTRUMENT :

1. Evaporation Recorder :Adolf Thies
2. Evaporation Recorder (acc. to Czeratzki) :Adolf Thies
3. Evaporation Recorder & Pan :Qualimetrics
4. Evaporograph :Qualimetrics
5. Evaporation Pan :Eijkelkamp
6. Evaporation Pan :Didcot Inst. Co.

{1}Sensing method :W=Weight ;L=Water level ;F=Float ;H=Hook gauge evaporimeter
{2}Evaporation surface :E=Evaporation pan ;D=Ceramic disc ;P=Filter paper
{3}Recording method :C=Spring wound chart ;B=Battery run chart ;E=Electrical pulse(data logger)
{4}Power requirement :D=DC ;R=Rechargeable batteries ;n.a.=not applicable
{5}Built to national weather bureau specifications of the country of origin.
§ See Introduction

SURFACE EVAPORATION MEASUREMENT

INSTRUMENT	1
Measurement principle :	Bowen ratio
Parameters measured :	Air temperatures Dew points Soil heat flux Average soil temperature Net radiation
Recording/storage method :	21X Micrologger
Mounting system :	Steel tripod & mast
Power supply :	DC ;Rechargeable batteries
Price range (U.S.$)§	
INSTRUMENT	1

INSTRUMENT :

1.Bowen Ratio System :Campbell Scientific Ltd.

RAINFALL GAUGES

INSTRUMENT	1	2	3	4	5	6	7	8
Sensor type {1}	Ca	T	T	T	W	T	M	F
Collection area (cm²)	100	507	200	342	323	203.2	200	130
Capacity (cm³) *	U	U	4500	U	9600	U	1200	U
Resolution (mm)	1	0.2	0.1;(0.25)	0.1	0.5	0.2;0.5	0.05	
Measurement/recording method{2}	D	D	S	D;S;N;C	C	D	Ms	C
Recording period (days)*	U	U	31	1;7;31	1;2;4;7;36	n.a	1	1;7;8
Self-levelling mechanism	n.a	+						
Overall dimensions (cm) :								
Diameter	14	25.4	37	21	36	22.5	14	30
Height	65	35	1200	44.5	90	40	45	64
Weight (kg)	2.5	2.1	21	3.6	11.4	4.9	2.5	8.2
Recorder drive :	n.a	n.a				n.a		+
Clockwork				+	+			
Battery			+	+	+			
Heating system	+		+	+		+		
Power supply {3}	D	n.a	2	1;2	n.a	3;D	n.a	
Price range (U.S.$)§	C	B		C	D	B		B

INSTRUMENT	1	2	3	4	5	6	7	8

1.Model 50202 Precipitation Gauge :R.M.Young Company
2.Type RG1 Tipping Bucket Raingauge :Delta-T Devices Ltd.
3.Precipitation Recorder :Adolf Thies
4.Model 6021A Electrically Heated Rain & Snow Gauge :Qualimetrics
5.Model 6032 Weighing/Recording Rain & Snow Gauge :Qualimetrics
6.Type DRRG/3 Heated Tipping Bucket Raingauge :Didcot Instrument Co.
7.No.1500 Rain Gauge,Hellmann type :Wilh. Lambrecht
8.Natural Siphon Rainfall Recorder (small size) :Casella

{1}Sensor type :T=Tipping bucket ;M=Measuring can ;Ca=Capacitative sensor
W=Weighing bucket ;F=Float
*U=Unlimited
{2}Measurement/recording method :D=Data logger ;S=CMOS semi-conductor memory
C=Chart ;N=Counter ;Ms=Measuring cylinder
{3}Power supply :1=110/120VAC 50-60Hz
2=220/240VAC 50-60Hz
3=24VAC 50-60Hz
D=DC
n.a=not applicable
§ See Introduction

TEMPERATURE SENSORS								
INSTRUMENT	1	2	3	4	5	6	7	8
Application{1}	A	A	A	A/G	A	S/A	V	Re
Type of sensor{2}	Pl	Pl(Si,Th)	Pl/Th	I	Th	Th	Si	Py
Upper temperature limit(°C)	65	70	70	100	50	80	150	500
Lower temperature limit(°C)	-20	-50	-30	-30	0	-20	-30	-50
Accuracy @ 20°C (°C)	0.1	0.5-2	0.2	0.2-0.5	0.2	0.2	0.5	
Resolution @ 20°C (°C)	0.05	0.1		0.1	0.1		0.1	0.1
Other features{3}	C	C	C	Df;H;V S;K;W	Pa;H	C		C
Output :mA		+	(+)					
mV					+			
Digital display				+			+	+
RS232C				+				
Dimensions (cm):								
Width/diameter	27.5	7.5	8	7.5	3	0.48	3.5	8.2
Depth			8.2	25	25	12.5	2.2	12.6
Height	23.5	12	14.5				12	9
Weight (kg)	1	0.25	0.3	1	0.25		0.13	0.72
Casing{4}	P/Gl/M	P/M	P		Al	M	P	
Weatherproof	+	+	+					
Power supply{5}	3/D	D		R	D		D	(1)/D
Battery life (h)								
Operating environment:								
Maximum temp.(°C)	65	70		50	50	80	45	50
Minimum temp.(°C)	-20	-50		0	0	-20	-5	0
Humidity(%RH)		100		95	95	100		
Price range (U.S.$)§	C	A			C	A		C

INSTRUMENT 1 2 3 4 5 6 7 8

INSTRUMENTS:

1.T302 Precision Air Temperature Sensor :Vector Instruments
2.T351 Outside Air Temperature Sensor :Vector Instruments
3.2000 Series Temperature Sensor :Skye Instruments Ltd.
4.Infrared AG Multimeter Plant Stress Monitor :Everest Interscience Inc.
5.HTR-1 RH/T/PAR Measuring Probe :P.P.Systems
6.Type ST 1 Soil Temperature Probe :Delta-T Devices Ltd.
7.Model 2060 Temperature Meter :Jenway Ltd.
8.Cyclops Compac 3 :Land Infrared Ltd.

{1}Application :A=Air temperature ;S=Soil temperature ;G=General purpose ;Re=Remote sensing ; V=Various probes available including :rapid response surface probe,hypodermic probe & plumb line probe for water depth profile.
{2}Type of sensor :Pl=Platinum resistance ;Si=Silicon semiconductor ;Th=Thermistor ;I=Infra red ; Py=Pyroelectric cell
{3}Other features :C=Compatible data processor available; Df=Differential measurement;Pa=PAR/Lux measurement;H=Humidity V=Vapour pressure deficit;S=Solar radiation;K=Canopy temp.;W=Crop water stress index
{4}Casing :M=Metal; P=Plastic; Gl=Glass-fibre
{5}Power supply :1=110/120VAC 50-60Hz ;2=220/240VAC 50-60Hz ;3=12VAC 50-60Hz ; D=DC ;R=Rechargeable batteries
§ See Introduction

	ATOMIC ABSORPTION SPECTROMETERS		
INSTRUMENT	1	2	3
Optics :			
Light source {1}	H	H	H
Single beam	+		
Double beam	(+)	+	+
Wavelength range (nm)	180 - 860	180 - 860	180 - 860
Detector {2}	Ph	Ph	Ph
Monochromator :			
Type {3}	E	E	E
Focal length (mm)	174	174	174
Grating (lines/mm)	1200	1200	1200
Spectral bandwidth (nm)	0.2;0.5;1.0	0.2;0.5;1.0	0.2;0.5;1.0
Wavelength accuracy (nm)	1	0.3	0.3
Furnace :			
No. elements analysed		16	16
Typical sensitivity eg.Pb			
Hg			
Na			
K			
Ca			
Measurement time (s)		0.1 - 200	0.1 - 200
Minimum sample size (μl)		200	200
Output :mV/V	+	+	+
Digital display	+		
RS232C		+	+
VDU		+	+
Data station	(+)	+	+
Options available with data station {4}	(S);(G);(A);(C)	A;C;G;(E)	A;(C)
Power supply {5}	1;2	1;2	1;2
Dimensions (cm) :			
Width	80.5	117	117
Depth	49.5	64	64
Height	47.2	61.5	61.5
Weight (kg)	50	120	120
Operating environment :			
Maximum temp.(°C)			
Minimum temp. (°C)			
Humidity (%RH)			
Price range (U.S.$)§			
INSTRUMENT	1	2	3

INSTRUMENTS :

1.PU9100X Atomic Absorption Spectrometer :Unicam Ltd.
2.PU9200X Atomic Absorption Spectrometer :Unicam Ltd.
3.PU9400X Atomic Absorption Spectrometer :Unicam Ltd.

{1}Light source :H=Hollow cathode lamp
{2}Detector :Ph=Side window photomultiplier
{3}Monochromator type :E=Temperature compensated Ebert diffraction grating
{4}Options with data station :S=Sample changer ;G=Graphite furnace ;A=furnace autosampler ; C=Continuous flow vapour system ;E=Electrothermal atomiser
{5}Power supply :1=110/120VAC 50-60Hz ;2=220/240VAC 50-60Hz
§ See Introduction

FLAME PHOTOMETERS

INSTRUMENT	1	2	3
Detector {1}		P	P
Sensitivity (ppm) :			
Sodium	3	2	1
Potassium	3	5	2
Lithium	5	2	
Calcium	50	20	15
Barium	200		
Sample size (cm³)		3	
Fuel supply {2}	Pr;B;B/Pr	Cg;Bu	Cg;Bu
Other features {3}	I;(S)		
Output :			
mV/V	+		
Digital display	+		+
Analogue display		+	
Dimensions (cm) :			
Width	39		40
Depth	34.5		21
Height	51		21
Weight (kg)	9.5		18
Power supply {4}	1;2	2	2
Operating environment :			
Maximum temp. (°C)			
Minimum temp. (°C)			
Humidity (%RH)			
Price range (U.S.$)§			
INSTRUMENT	1	2	3

INSTRUMENTS :

1.410 Flame Photometer :Ciba Corning Analytical

2.Model CL-22A Flame Photometer :Elico Pvt. Ltd.

3.Model CL-22D Digital Flame Photometer :Elico Pvt.Ltd.

{1}Detector :P=Photosensitive device

{2}Fuel supply :Pr=Propane ;B=Butane ;Cg=Calgas ;Bu=Burshane

{3}Other features :I=Auto-ignition ;S=Auto-sample dilution

{4}Power supply :1=110/120VAC 50-60Hz

2=220/240VAC 50-60Hz

§See Introduction

GAS CHROMATOGRAPHS					
INSTRUMENT	1	2	3	4	5
Capacity (no. complete chromatography channels)					2
Detectors available {1} :	F;E;T;N;P	Ph;(F;T;U; E;C;Np)	Ph	Ph;F;T;U; E;C;Np	F;A;E;Ph; C
Injectors available :					
Split	+			+	+
Splitless	+			+	+
PTV	+				
Capillary-on-column	+	+	+	+	+
Packed-on-column	+	+		+	
Gas sampling		+		+	
Pyrolysis					
Automatic injection	(+)			+	
Injector oven :					
Capacity (no. injectors)					
Temperature range (°C)					
Accuracy (%)					
Column oven :					
Capacity (no. columns)					
Volume (cm³)		4916		18000	
Temperature range (°C)		Ambient to 300	200	35 to 400	Ambient+5 to 350
Accuracy		0.1°C			1.5°C
Detector oven :					
Capacity (no. detectors)				3	
Temperature range (°C)					
Output : mV/V	+				
Digital display	+	+	+		+
RS232C	+		+		+
Data processing available {2}	D	C	C		I;Pr
Dimensions (cm) :					
Width	58.2				66
Depth					45
Height					30

Weight (kg)					
Power supply {3}	2;(1)				
Portable		+	+		
Field use			+		
Operating environment :					
Maximum temp.(°C)					
Minimum temp.(°C)					
Humidity (%RH)					
Price range (U.S.$)§					
INSTRUMENT	1	2	3	4	5

INSTRUMENTS :

1.PU4600 Series Gas Chromatograph :Unicam Ltd.
2.300 Series Compact Gas Chromatograph :HNU Systems
3.311 Portable Gas Chromatograph :HNU Systems
4.421 Research Gas Chromatograph :HNU Systems
5.Nordion Micromat HRGC 412 Analysis System :HNU Systems

{1}Detectors available :F=Flame ionisation ;E=Electron capture ;T=Thermal conductivity ;N=Nitrogen selective ; P=Flame photometric ;U=Far UV absorbance ;Ph=Photoionization ;A=Alkali thermoionization C=Electrolytic conductivity ;Np=Nitrogen phosphorus

{2}Data processing available :D=Data system ;I=Integrator ;C=Chart recorder ;Pr=Printer/plotter

{3}Power supply :1=110/120VAC 50-60Hz ;2=220/240VAC 50-60Hz ;D=DC ;R=Rechargeable batteries

§ See Introduction

FTIR SPECTROMETERS

INSTRUMENT	1	2
Detector {1}	D	D;M;P;Pb;In;S;G;IGa
Spectral range (cm-1) :		
Maximum	6000	15800;(6000);(450)
Minimum	400	4000;(400);(10)
Resolution (cm-1)	2;4;8;16	
Fastest scan time (min)		
%T Accuracy		
Output :		
mV/V	+	+
Digital display	+	
Liquid crystal spectral display	+	
Built-in recorder	+	
Data station	(+)	(+)
Dimensions (cm) :		
Width	77.4	61
Depth	61.9	62.3
Height	21	30
Weight (kg)	35	
Power supply {2}	1;2	1;2
Operating environment :		
Maximum temp.(°C)		
Minimum temp.(°C)		
Humidity (%RH)		
Price range (U.S.$)§		
INSTRUMENT	1	2

INSTRUMENTS :

1.Mattson 1000 Series FT-IR Spectrometers :Unicam Ltd.

2.Galaxy FT-IR Spectrometers :Unicam Ltd.

{1}Detector :D=DTGS (L-Alanine doped deuterated tri-glycine sulphate detector) ;M=MCT P=PbSe ;Pb=PbS ;In=InSb ;S=Si ;G=Ge ;IGa=InGaAs

{2}Power supply :1=110/120VAC 50-60Hz ;2=220/240VAC 50-60Hz

§ See Introduction

HIGH PRESSURE LIQUID CHROMATOGRAPHS

INSTRUMENT	1	2
Detectors available {1}	U;F;R;E;M;D	U;F;R;E;C;M;P
Column oven :		
Capacity - No.of columns		4
External diameter of columns(cm)		1.27
Length of columns (cm)		30 or 50
Maximum temperature (°C)		100
Minimum temperature (°C)		Room temp. + 10°C
Autosampler :		
Variable volume		+
Fixed loop injection	+	+
Refrigerated		(+)
Pump :		
Flowrate - Preparative (ml/min)	0.1 - 25	0.1 -10;(20);(50)
-Analytical (μl/min)	10 - 5000	
Accuracy		
Data capture & storage options :		
Data station	+	
Integrator	+	+
Analytical recorder	+	
Overall dimensions (cm) :		
Width		
Depth		
Height		
Weight (kg)		
Power supply {2}	1 ; 2	
Operating environment :		
Maximum temp. (°C)	40	
Minimum temp. (°C)	5	
Humidity (%RH)	20 -80	
Price range (U.S.$)§		
INSTRUMENT	1	2

INSTRUMENTS :

1.4100M HPLC System :Unicam Ltd.

2.LC-800 Series High Performance Liquid Chromatograph :Jasco Inc.

{1}Detectors available :U=UV/VIS ;F=Fluorescence ;R=Refractometer ;E=Electrochemical ;
C=Chemiluminescence ;M=Multi-channel UV ;D=Diode array ;P=Polarimeter

{2}Power supply :1=110/120VAC 50-60Hz ;2=220/240VAC 50-60Hz

§ See Introduction

MERCURY DETECTION	
INSTRUMENT	1
Detection method {1}	F
Range (ppb) :	
Maximum	0 - 500
Minimum	0 - 0.5
Number of ranges	4
Detection accuracy (%FS)	1
Options available :	
Autosampler	+
Vapour generator	+
Data station	+
Output :	
mV/V	
Digital display	+
RS232C	
Power supply {2}	1 ; 2
Dimensions (cm) :	
Width	15
Depth	28
Height	30
Weight (kg)	10
Operating environment :	
Maximum temp.(°C)	
Minimum temp.(°C)	
Humidity (%RH)	
Price range (U.S.$)§	
INSTRUMENT	1

INSTRUMENT :

1.PSA Merlin Fluorescence Detector :P.S.Analytical Ltd.

{1}Detection method :F=Cold vapour fluorescence

{2}Power supply :1=110/120VAC 50-60Hz

2=220/240VAC 50-60Hz

§ See Introduction

PLANT/SOIL/WATER ANALYSIS	
COMPANY	1
Substances analysed {1} :	S;W;V;Sa;F;G
Laboratory set-ups available {2} :	C;P;M;A
Reagents & equipment available for the following tests :	
pH	+
Electrical conductivity	+
Redox	+
Available nitrogen (N-NO3,NH4)	+
Available phosphorus (P-PO4)	+
Available potassium (K)	+
Available magnesium (Mg)	+
Available calcium (Ca)	+
Total N,P,Mg,Ca	+
Uric N,urea,ammonium N	+
Aluminium (Al)	+
Boron (B)	+
Cobalt (Co)	+
Copper (Cu)	+
Iron (Fe)	+
Manganese (Mn)	+
Molybdenum (Mo)	+
Zinc (Zn)	+
Total Cu,Fe,Mn,Zn	+
Chlorides (Cl)	+
Sulphates (S-SO4)	+
Organic matter	+
Humus	+
Particle size distribution	+
Free carbonates	+
Total carbonates	+
Cation exchange capacity (CEC)	+
Sodium absorption ratio (SAR)	+
Lime requirement	+
Sodium requirement	+
Chemical oxygen demand (COD)	+
Crude protein	+
Acid detergent fibre	+
Acid detergent insoluble protein	+
Non protein nitrogen	+
Sodium chloride	+
Moisture determination	+

SUPPLIER	1

SUPPLIER :

1.Agri-Diagnostics System :Marton Products Ltd.

{1}Substances analysed :S=Soil ;W=Water ;V=Vegetable tissues ;
Sa=Sap ;F=Fertilizers ;G=Growing media

{2}Laboratory set-ups available :C=Convertible (for field/lab. use) ;P=Portable ;
M=Multi-range permanent lab. installations for analysis of most agricultural material ;
A=Automated computer controlled unit

PLASMA SPECTROMETERS			
INSTRUMENT	1	2	3
Optics :			
Focal length (m)	1	1	0.7
Wavelength range (nm):			
First order	345-780	300-800	180-800
Second order	170-385	160-300	
Entrance slit width (μm)			60
Grating (grooves/mm)	1440	1800	79
No. exit slits	260	1	
Detectors {1}	Pm	Pm	
Simultaneous	+		
Plasma source :			
RF generator (MHz)	40	40	40.68
Automatic impedance matching	+	+	
Internal recirculation	+	+	
Software :			
Controller {2}	B		Ph
Data station {3}	I	I	I
Software {4}	P		
Nebuliser :			
Detection limits[with UDX](μg/l)	As=3; Pb=2 Se=3; Mg=0:02	As=3; Pb=2 Se=3; Mg=0.02	
Aerosol generation {5}			
Options/accessories {6}	M;A;S;N;R;C;H	N;R;C;H	C;A
Power supply {7}	1;2	1;2	2
Dimensions (cm) :			
Length	119	104	
Width	152	157	
Height	142	137	
Weight (kg)	750	725	
Operating environment :			
Maximum temp. (°C)	18	18	
Minimum temp. (°C)	29	29	
Humidity (%RH)	60	60	
Price range (U.S.$)§			
INSTRUMENT	1	2	3

INSTRUMENTS :

1.ICP 2000 Simultaneous Plasma Emission Spectrometer :Baird Europe B.V.
2.IPC 2070 Sequential Plasma Emission Spectrometer :Baird Europe B.V.
3.7000 Series ICP Spectrometer :Unicam Ltd.

{1}Detectors :Pm=Photomultiplier tubes
{2}Controller :B=Baird MC-20 ;Ph=Unicam Analytical Data Station
{3}Data station :I=IBM System 2
{4}Software :P=Plasmacomp V inc. multiple order curve fitting;line switching;dilution corre
{5}Aerosol generation :Pi=Piezoelectric
{6}Options/Accessories :M=Monochromator ;A=Autosampler ;S=IDS injection & dilution
N=UDX ultrasonic nebuliser ;R=Random access sample chamber
C=Ceramic injector for HF acid solutions ;H=Hydride analysis
{7}1=110/120VAC 50-60Hz ;2=220/240VAC 50-60Hz
§ See Introduction

U.V./VIS SPECTROPHOTOMETERS					
INSTRUMENT	1	2	3	4	5
Detector {1}	P	P	P	Ph	Ph
Maximum wavelength (nm)	1100	1100	1100	900	900
Minimum wavelength (nm)	195	195	195	200	200
Wavelength accuracy (nm)	1	1		0.5	1
Bandwidth :					
Fixed (nm)	8	8	2	0.5 - 8	0.5 - 8
Variable			+	+	+
Cell path lengths (mm) :					
Maximum	50			10	
Minimum	0.1				
Scan speed (nm/min)			25 -2000		
Accessories available {2}	F	A,C,Pr			
Output :					
mV/V	+	+	+		+
RS232C	+	+	+		
LCD display	+	+		+	+
Recorder interface	(+)	(+)	+		+
VDU			+		
Dimensions (cm) :					
Width		59			68
Depth		37.4			34
Height		27.8			30
Weight (kg)		19.3		20	40
Power supply {3}	1;2	1;2	1;2	2	2
Operating environment :					
Maximum temp.(°C)					
Minimum temp.(°C)					
Humidity (%RH)					
Price range (U.S.$)§					
INSTRUMENT	1	2	3	4	5

INSTRUMENTS:

1.8625 Series UV/VIS Spectrophotometer :Unicam Ltd.
2.8620 Series UV/VIS Spectrophotometer :Unicam Ltd.
3.8700 Series UV/VIS Spectrophotometer :Unicam Ltd.
4.Model CL-54D 'Ultraspec' UV-VIS Spectrophotometer :Elico Private Ltd.
5.Model CL-58 Autoscan Microprocessor based UV-VIS Spectrophotometer :Elico Private Ltd.

{1}Detector :P=Photodiode ;Ph=Photomultiplier tube
{2}Accessories available :F=Flow action sample throughput ;A=Automatic sample changer ; C=Cell temperature control ;Pr=Printer
{3}Power supply :1=110/120VAC 50-60Hz ;2=220/240VAC 50-60Hz
§See Introduction

FIELD DATA LOGGERS

INSTRUMENT	1	2	3	4	5	6	7	8
Analogue inputs :								
Maximum no.	60	4+	16+	196+	12+	10	12	4
Options {1}	Tr,Te, Re,V	V,C,O	V,Te,Tr, Re,O	V,Te,Tr, Re,O	V,Te,Tr, Re,O	V,Te,C, Tr,O	V,Te,O, Tr,Re	V,O
Precision (%FSD)	0.07-0.2		0.1*	0.04*	0.2*	0.2-0.3	0.5	0.02
Scan interval :								
Maximum (h)[I=Infinite]	24	12	I	I	I	24	24	672
Minimum (s)	1	30	0.01	0.01	0.016	60	60	1
Interface bit size	n.a		14	16	13	14	12	12
Channel skip	n.a		+	+	+		+	+
CMOS processor	+	+	+	+	+	+	+	+
Storage {2} :	Ra	E	Ra/Ca	Ra/Ca	Ra/Ca/E		Ra (E)	E
Memory (K bytes)	256		40	88	64	32	32(128)	64
RS232C interface	+		+	+	+	+	+	+
Min. baud rate	150		300	300	300	300	300	300
Max. baud rate	9600		76800	76800	76800	4800	9600	9600
Minimum power consumption(mA)	0.05	0.4	1	3.5	0.5		0.2	0.2
Power supply voltage (VDC)	7 - 15	8 - 15	9 - 15	9 - 15	9 - 16	12	12	
Battery type {3}	Al/L	Al/R/L	Al/R/S	R/S	Al/R/S	Al	Al/R/S/L	Al
Dimensions (cm) :								
Width	28	23	14.5	43	19.8	11.4	16	12.4
Depth	22	15	8.4	15.2	3.8	6.9	9	8.1
Height	14	12	20.8	30.5	8.9	21	24	21.8
Weight (kg)	2.7	1.5	2.8	18.2	0.9	1.6	1.7	1.6
Weatherproof case	+	+	(+)	+	(+)	+	+	+
Max. unattended operating period (days)	365+	60	300+	30+	600+	300	180	180
Availability of processing	Ib		Ib,Ep,	Ib,Ep,	Ib,Ep,	Ib		Ib/A/Ep

Maximum temp. (°C)	60	50	85	60	85	55	50	50
Minimum temp. (°C)	-20	-20	-55	-40	-55	-25	-30	-30
Humidity (%RH)	100	100				100	100	100
Price range (U.S.$)§	E	C	E	G	D	C/D	E/F	D/E
INSTRUMENT	1	2	3	4	5	6	7	8

INSTRUMENTS :

1.Delta-T Logger :Delta-T Devices Ltd.
2.SDL410 Integrating Printer Logger :Skye Instruments Ltd.
3.21X Micrologger :Campbell Scientific Ltd.
4.CR7 Measurement & Control System :Campbell Scientific Ltd.
5.CR10 Measurement & Control System :Campbell Scientific Ltd.
6.LI-1000 Datalogger :LI-COR,inc.
7.Easy Logger :Omnidata International,Inc.
8.Datapod II Data Logger :Omnidata International,Inc.

{1}Analogue input options :V=Voltage ;Te=Thermocouple ;Re=Resistance ;C=Current ;Tr=Thermistor ;O=Others

*Guaranteed measurement accuracy over -25/+50°C temperature range.

{2}Storage :Ra=RAM ;Ca=Cassette tape ;E=EPROM

{3}Battery type :Al=Alkaline ;R=Rechargeable ;S=Solar cell ;L=Lithium

{4}Availability of processing software :Ib=IBM ;Ep=EpsonHX20 ;HH=Husky Hunter ;A=Apple

§ See Introduction

PORTABLE CHART RECORDERS				
INSTRUMENT	1	2	3	4
No. channels	1	2	1	3
Chart width (cm)	12	12	6	12
Max. chart length (m)	30	15	19	15
Writing system {1}	F	F	S	F/C/E
Accuracy (%FS)	1	1	2	0.25
99% response time (s)	0.5	0.5	2	
No. ranges	5	4		
Maximum range (V)	100	50	100	
Minimum range (mV)	1	10	100	
Max. unattended period (days)	250	383		150
Maximum chart speed (cm/min)	10	16	0.5	12
Minimum chart speed (cm/h)	0.5	0.5	0.6	2.5
No. speeds	10	12	I#	10
Dimensions (cm) :				
Width	32	28	9.2	
Depth	31	23	11	
Height	18	13	14.3	
Weight (kg)	8	6	2	
Power supply {2}	1;2;D;R	1;2;D;R	2(1;D)	1;2;D
Battery life (h)*	80	5 - 8		
Casing {3}	M	M	M	
Weatherproof	+			
Operating environment :				
Maximum temp. (°C)	60	40	40	50
Minimum temp. (°C)	0	5	0	0
Humidity (%RH)		80	80	90
Price range (U.S.$)§				
INSTRUMENT	1	2	3	4

INSTRUMENT :

1.No.360WA100 Portable,Single Channel,Watertight POSIWAR Recorder :Kahlsico Int.Corp.
2.No.6723 2-pen Portable AC/DC Recorder :Soltec Distribution
3.Rustrak Miniature Strip Chart Recorder :Gulton Europe Ltd.
4.No.301 Recorder :Chessel Ltd.

{1}Writing system :F=Fibre-tip pen ;S=Pressure sensitive chart ;C=Capillary ink ;E=Electric
I=Interchangeable
{2}Power supply :1=110/120VAC 50-60Hz ;2=220/240VAC 50-60Hz ;D=DC ;
R=Rechargeable batteries
*Dependant on chart speed
{3}Casing :M=Metal
§ See Introduction

TRANSIENT RECORDER	
INSTRUMENT	1
Size of memory (K bytes)	2
Accuracy (bits)	10
Record times :	
Maximum (min)	1
Minimum (msec)	60
Replay times :	
Maximum (min)	10
Minimum (sec)	0.6
Output :	
mV/V	+
RS232C	
Power supply {1}	1 ; 2
Dimensions (cm) :	
Width	
Depth	
Height	
Weight (kg)	2.25
Operating environment :	
Maximum temp. (°C)	50
Minimum temp. (°C)	5
Humidity (%RH)	80
Price range (U.S.$)§	D
INSTRUMENT	1

INSTRUMENT :

1.TRI Transient Recorder :Hansatech Instruments Ltd.

{1}Power supply :1=110/120VAC 50-60Hz
2=220/240VAC 50-60Hz

§ See Introduction

Manufacturers' Addresses

ALCYON DATA S.A., 1 Chemin de la Roche, Renens, P.O. Box 6, CH-1000, Lausanne 16, SWITZERLAND.
ANALYTICAL DEVELOPMENT CO. LTD., Pindar Road, Hoddesdon, Herts. EN11 0AQ, UK. Telex 266952 ADC G Fax +44 (992) 444567
ASSEMBLED ELECTRICS, 66 Smith Street, Yagoona, NSW 2199, AUSTRALIA. Fax +61 (2) 645-4597
BAIRD EUROPE B.V., Producktieweg 30, 2382 Zoeterwoude, NETHERLANDS. Telex 35498 Fax +31 (71) 414899
BEIJING ANALYTICAL INSTRUMENT FACTORY, Wen Quan, Hai Dian District, Beijing, P.R. CHINA.
BELFORT INSTRUMENT CO., 727 South Wolfe Street, Baltimore, MD 21231-3513, USA.
BRONKHORST HIGH TECH B.V., Nijverheidsstraat 1A, 7261 AK Ruurlo, NETHERLANDS. Telex 49066 HITEC NL Fax +31 (5735) 3297
BROOKS INSTRUMENT DIVISION, EMERSON ELECTRIC UK LTD., Brooksmeter House, Stuart Road, Bredbury, Stockport SK6 2SR, UK.
CAMPBELL SCIENTIFIC LTD., 14-20 Field Street, Shepshed, Leics. LE12 9AL, UK. Telex 94016393 CAMP G Fax +44 (509) 601091
CASELLA (LONDON) LTD., Regent House, Britannia Walk, London N1 7ND, UK.
CHANNEL ELECTRONICS (SUSSEX) LTD., Cradle Hill Industrial Estate, P.O. Box 58, Seaford, Sussex BN25 3JE, UK.
CHESSEL LTD., Southdown View Road, Worthing, West Sussex BN14 8NL, UK.
CIBA CORNING ANALYTICAL LTD., Halstead, Essex CO9 2DX, UK. Telex 98204 Fax +44 (787) 475088
C.I.D. INC., P.O. Box 9008, Moscow, ID 83843, USA.
COMOY COMPONENTS LTD. – PODDYMETER, 27-19 Brunel Road, Acton, London W3 7UW, UK.
CONTROLLED ENVIRONMENTS LTD., 1461 St. James St., Winnipeg, Manitoba RBH 0W9, CANADA. Telex 757777 Fax +1 (204) 786-7736
DASIBI ENVIRONMENTAL CORP., 515 W Colorado Street, Glendale, CA 91204, U.S.A. Telex 691629 Fax +1 (818) 247-7614
DELTA-T DEVICES CO. LTD., 128 Low Road, Burwell, Cambridge CB5 0EJ, UK. Telex 817670 ASABSE G (Attn. Delta-T Devices) Fax +44 (638) 743155
DIDCOT INSTRUMENT CO. LTD., Station Road, Abingdon, Oxon. OX14 3LD, UK. Telex 83343 ABTLX
E.G. & G. ENVIRONMENTAL EQUIPMENT, 151 Bear Hill Road, Waltham, MA 02154, USA.
EIJKELKAMP AGRISEARCH EQUIPMENT, P.O. Box 4, 6987 ZG Giesbeek, NETHERLANDS. Telex 35416 EYKEL NL Fax + 31 (83) 362167
ELICO PVT. LTD., 309 Model House, 6-3-456/A/1 Punjagutta, Hyderabad 500 482, INDIA. Telex 4256714 PHHD IN
ENVIRODATA, P.O. Box 395, Warwick, QLD 4370, AUSTRALIA. Fax +61 (76) 612485
EVEREST INTERSCIENCE, P.O. Box 345, Tustin, CA 92634-3640, USA. Telex 295368 Fax +1 (714) 992-0178
FISCHER & PORTER GmbH, 1 Gibraltar Plaza Building, Horsham, PA 19044, USA.
GALLENKAMP INDUSTRIAL, Belton Road West, Loughborough, Leics. LE11 0TR, UK.
GENERAL EASTERN INSTRUMENTS CORP., 50 Hunt Street, Watertown, MA 02172, USA.
GULTON EUROPE LTD., The Hyde, Brighton BN2 4JU, UK.
HAENNI & CIE. AG., CH-3303, Jegenstorf, SWITZERLAND.
HANSATECH INSTRUMENTS LTD., Narborough Road, Pentney, King's Lynn, Norfolk PE32 IJL, UK. Fax +44 (760) 337303
HEREAUS-VVTSCH GmbH, Beethovenstrasse 34, W-7460 Balingen-Frommern, GERMANY. Telex 763629 HVB; Fax +49 (7433) 303112
HNU SYSTEMS LTD., 254 Europa Boulevard, Gemini Business Park, Warrington WA5 5TN, UK. Fax +44 (925) 445940
HORIBA INSTRUMENTS LTD., 1 Harrowden Road, Brackmills, Northampton NN4 0EB, UK. Telex 311869
INTERNATIONAL LIGHT, INC., Dexter Industrial Green, Newburyport, MA 01950-4092, USA. Telex 947135 Fax +1 (508) 462-0759
JENWAY LTD., Gransmore Green, Felsted, Dunmow, Essex CM6 3LB, UK. Telex 817766 Fax +44 (371) 821083
KAHLSICO INTERNATIONAL CORP., P.O. Box 947, El Cajon, CA 92022-0947, USA.
KANE-MAY LTD., Burrowfield, Welwyn Garden City, Herts. AL7 4TU, UK.
KDG FLOWMETERS, Rotameter Works, 330 Purley Way, Croydon, Surrey CR9 4PG, UK.
KENT INDUSTRIAL MEASUREMENTS LTD., Hanworth Lane, Chertsey, Surrey KT16 9LF, UK.
KROHNE MESSTECHNIK GmbH & CO. KG, Ludwig-Krohne-Strasse, Postfach 100970, D-4100 Duisburg 1, GERMANY.
LAND INFRARED LTD., Dronfield, Sheffield S18 6DJ, UK. Telex 54457
WILH. LAMBRECHT GmbH, Friedlander Weg 65/67 P.O. Box 76, D-3400 Gottingen, GERMANY.
LEE-INTEGER LTD., Desborough, Kettering, Northants NN14 2QW, UK.

LI-COR, INC., Box 4425, Lincoln, NB 68504-0425, USA. Telex 910-621-8116 Fax +1 (402) 467-2819

MACAM PHOTOMETRICS LTD., 10 Kelvin Square, Livingstone EH54 5DG, UK. Fax +44 (506) 38543

MARTON PRODUCTS LTD., Assets House, 17 Elverton Street, London SW1, UK. Telex 916832 ASSETS G Fax +44 (71) 876-8790

MBW ELEKTRONIK AG., Seminarstrasse 55, CH-5430, Wettingen, SWITZERLAND.

MKS INSTRUMENTS, INC., 34 Third Avenue, Burlington, MA 01803, USA.

NOVASINA AG., Thurgauerstrasse 50, CH-8050 Zurich, SWITZERLAND.

OMNIDATA INTERNATIONAL, INC., P.O. Box 3489, Logan, UT 84321, USA. Telex 3725960 Fax +1 (801) 753-6756

OPTRONIC LABORATORIES, INC., 4470 35th Street, Orlando, FL 32811, U.S.A. Telex 910-240-5263 Fax +1 (407) 648-5412

PLATON FLOWBITS LTD., Wella Road, Basingstoke, Hants. RG22 4AQ, UK.

P.M.S. INSTRUMENT CO., 2750 N.W. Royal Oaks Drive, Corvallis, OR 97330, USA. Telex WU EASYLINK 402320 PMS INST UD Fax +1 (503) 752-7929

P.P. SYSTEMS, 24/26 Brook Street, Stotford, Hitchin, Herts. SG5 4LA, UK. Telex 934999 TXLINK G Fax +44 (462) 731807

PROSSER SCIENTIFIC INSTRUMENTS LTD., Lady Lane Estate, Hadleigh, Suffolk, UK. Telex 987703 CHACOM

PROTIMETER plc, Meter House, Fieldhouse Lane, Marlow, Bucks. SL7 1LX, UK. Telex 849305 PMETER G Fax +44 (6284) 74312

P.S. ANALYTICAL LTD., Arthur House, B4, Chaucer Business Park, Watery Lane, Kemsing, Sevenoaks, Kent TN15 6QY, UK. Telex 957645 PSA G Fax +44 (732) 61340

QUALIMETRICS, INC., 1165 National Drive, Sacremento, CA 95834, USA. Telex 377310

ROTA GmbH, Postfach 20, D-7867 Wehr/Baden 2, GERMANY.

SIERRA INSTRUMENTS, INC., P.O. Box 909. Sierra Building, Carmel Valley, CA 93924, USA.

SKYE INSTRUMENTS LTD., Unit 5, Ddole Industrial Estate, Llandrindod Wells, Powys LD1 6DF, UK. Telex 94070926 SKYE G Fax +44 (597) 824812

SOLTEC DISTRIBUTION, 11684 Pendleton Street, Sun Valley, CA 91352, USA.

TELEDYNE HASTINGS-RAYDIST, P.O. Box 1275, Hampton, VA 23661, USA. Telex 710-882-0085 TDYHAST HAMP Fax +1 (804) 723-3925

ADOLF THIES GmbH & CO. KG GOETTINGEN, P.O. Box 3536 & 3541, Hauptstrasse 76, D-3400 Goettingen, GERMANY. Telex 96722

UNICAM LTD., York Street, Cambridge CB1 2PX, UK. Telex 817331 UNICAM G Fax: +44 (223) 312764

UNIPROBE INSTRUMENTS LTD., Clive Road, Cardiff CF5 1HG, UK. Telex 497492 CHACOM G (Uniprobe) Fax +44 (222) 222735

VAISALA OY, PL26 SF-00421, Helsinki 42, FINLAND.

VECTOR INSTRUMENTS, 115 Marsh Road, Rhyl LL18 2AB, UK. Telex 669755 OFFICE G (Attn. Vector) Fax +44 (745) 344206

WALLACE & TIERNAN LTD., Priory Works, Tonbridge, Kent TN11 0QL, UK.

HEINZ WALZ GmbH, Eichenring 6, D-8521 Effettrich, GERMANY. Telex 629783 WALZ D Fax +49 (91) 335395.

WESCOR, INC., 459 South Main Street, Logan, UT 84321, USA. Telex 4930393 WESC UI

WISSENSCHAFTLICH-TECHNISCHE WERK-STATTEN GmbH (W.T.W.), Dozent Dr. Habil. K. Slevogt, D-8120 Weilheim, GERMANY.

YELLOW SPRINGS INSTRUMENT CO., INC., Industrial Division, Yellow Springs, OH 45387, USA.

R.M. YOUNG CO., 2801 Aero-Park Drive, Traverse City, MI 49684, USA Telex 810-291-3366 Fax +1 (616) 946-4772

Appendix B
Experimental design and presentation of results

J.M.O. Scurlock and S.P. Long

Whether performing a laboratory or a field experiment, accurate observations are required in order to obtain useful information, which can then be analysed easily and later presented clearly. The following simple schedule may be of help in carrying out experimental work:

1. Design.
2. Perform.
3. Observe.
4. Record.
5. Conclude.

The importance of these experimental elements applies to every subject described in this book. The following suggestions concern particularly the UNEP series of Training Courses on Photosynthesis and Productivity, but they also form a series of general recommendations for research work in this area:

1. Design
   (a) Define questions to be answered.
   (b) Choose adequate healthy plant material, and identify it exactly. Do not harvest plant material or attach equipment until you have designed the experiment thoroughly.
   (c) Think through each experimental step, and work out what equipment and chemicals are required before you start.
   (d) If using plant material over a period of time, be systematic; identify all inputs and note any disease, pests or treatments.
   (e) Have adequate controls.
   (f) Be simple in approach; answer one point at a time.
2. Perform
   (a) Be accurate.
   (b) Avoid contamination.
   (c) Avoid unintentional stresses or artefacts.
3. Observe
   (a) Be critical.
   (b) Be honest.
   (c) Be accurate.
   (d) Be objective.
4. Record
   (a) Be accurate.
   (b) Be systematic.
   (c) Be thorough.
5. Conclude
   (a) Be objective. Is the conclusion justified or is it biased towards a preconceived idea?
   (b) Summarise conclusions precisely and clearly.

It is essential that a clear and concise record is kept of all experimental work, however

Photosynthesis and Production in a Changing Environment: a field and laboratory manual
Edited by D.O. Hall, J.M.O. Scurlock, H.R. Bolhàr-Nordenkampf, R.C. Leegood and S.P. Long
Published in 1993 by Chapman & Hall, London.
ISBN 0 412 42900 4 (HB) and 0 412 42910 1 (PB).

trivial it may seem at the time. Unexpected or interesting results may result in the need to repeat an experiment exactly, so a laboratory or field notebook may turn out to be a very useful document.

Use both sides of the pages in the notebook, or keep two notebooks (preferably about 21 × 30 cm, A4 size, which is convenient for photocopying any important pages). It is suggested that the left-hand side of the open book is used for recording observations as they are made, calculating how much of a given chemical to weigh out when making solutions, etc., and the right-hand side for writing up all experiments *whilst the work is under way*. In order to encourage good practice during the UNEP Training Courses, experimental notebooks are assessed by the lecturers in order to provide feedback to participants.

Experiments should be set out under the following headings:

1. Title – together with a date and experiment reference number, which should be repeated at the top of every page.
2. Objectives – a description of the aims of the experiment;
3. Plant material
4. Methods
5. Results – given numerically, with a note of any unusual delays or occurrences which might have contributed to spurious results;
6. Calculations – derived from the primary data (e.g. leaf area, rates of reaction, concentration of chlorophyll);
7. Conclusions and Summary.

In the conclusions it should be noted whether these have answered the questions outlined in the aims of the experiment. Suggestions for further or better experiments should also be given here as well as any references.

In recording results care should be taken in respect to the following:

(a) Replication. Most measurements should be taken for a number of samples. This is essential if the results are to be statistically significant. Replicate measurements, whether taken in the field or laboratory, must be made in such a way as to avoid systematic or subjective bias. This may be achieved through randomisation in experimental or sampling design. Appropriate designs are highly dependent on the type of experiment: it cannot be over-emphasised that the choice of appropriate design ultimately determines the value of the results and the validity of the conclusions.

(b) Significance. Results should not be recorded to a greater accuracy than the variations between replicates justify; e.g. for two weighings of 11.349 g and 12.015 g an average of 11.7 g should be recorded rather than 11.682. Three significant figures are sufficient for most measurements and calculations, despite the ability of computers and calculators to give results to 8 digits!

(c) Statistics. It is not sufficient to report the difference between a treatment and a control: it is necessary to show that the difference is statistically significant, rather than the result of random variation. Statistical tests calculate the probability of such a difference being due to chance. Normally, a probability of 0.05 (1 in 20) is taken as the threshold; a difference with a probability of less than 0.05 is said to be significant. The appropriate statistical test depends on the nature of the experiment but there are many suitable books available [3–8], as well as guides on the use of computer programs [9,10]. Note that *parametric* statistical tests (e.g. Student's *t*-test, Pearson's correlation, least squares regression) are only suitable where the frequency distribution of the measurements corresponds to the Normal Distribution. Otherwise, *non-parametric* tests should be applied; these make no assumptions about the frequency distribution.

Depending upon the aims of the experiment, measurements of rates of CO_2 assimilation, for example, should be combined with determinations of leaf area, fresh weight, dry weight and chlorophyll content, to give rates of photosynthesis in terms of leaf area, weight or chlorophyll content for a given light intensity. Again, in biochemical experiments, results should be treated in terms of weight, chlorophyll or protein content. For enzyme experiments results should be expressed in terms of the amount of substrate converted in a given time by a given amount of protein.

REFERENCES

1. Fisher, R.A. (1966) *The Design of Experiments*, 8th edn. Oliver and Boyd, Edinburgh.
2. Leaver, R.H. and T.R. Thomas (1974) *Analysis and Presentation of Experimental Results*. Macmillan, London.
3. Snedecor, G.W. and W.G. Cochran (1989) *Statistical Methods*, 8th edn. Iowa State University Press, Ames, Iowa, USA.
4. Sokal, R.R. and F.J. Rohlf (1981) *Biometry: the principles and practice of statistics in biological research*, 2nd edition. W.H. Freeman, San Francisco and Oxford.
5. Gilbert, N. (1989) *Biometrical Interpretation*, 2nd edn. Oxford University Press.
6. Mead, R. and R.N. Curnow (1983) *Statistical Methods in Agriculture and Experimental Biology*. Chapman and Hall, London.
7. Wardlaw, A.C. (1985) *Practical Statistics for Experimental Biologists*. John Wiley, Chichester and New York
8. Webb, N. and R. Blackmore (1985) *Statistics for Biologists: a study guide*. Cambridge University Press.
9. Lee, J.D. and T.D. Lee (1982) *Statistics and Numerical Methods in BASIC for Biologists*. Van Nostrand Reinhold, New York.
10. Ireland, C.R. and S.P. Long, eds. (1985) *Microcomputers in Biology: a practical approach*. IRL Press, Oxford.

Appendix C
Biomass production and data

D.O. Hall and J.M.O. Scurlock

In many tropical and subtropical countries it is possible to use plant material as a source of fuel, fodder and fibre at the same time as providing sufficient food. Types of processes which might be used are summarized in some of the tables below, where it can also be seen that the global annual products of photosynthesis have an energy content of about 8–10 times the world's present annual energy use. Currently, biomass energy accounts for 14% of world primary energy use, comprising 55 EJ per annum (1 EJ = 10^{18} J) or the equivalent of 25 million barrels of oil per day [24]. Furthermore, world standing biomass is comparable with proven reserves of fossil fuel (Table C.1). The total requirement for biomass energy is unlikely to decrease in the near to medium term in many countries where biomass is the fuel of development. Increased use of biomass with modern technology may also provide the least expensive pathway towards controlling the amount of CO_2 in the atmosphere. The majority of terrestrial primary production occurs in the warmer regions of the world; it is also in these regions that species with the highest rates of production are found.

It is important to consider all the inputs and outputs when producing biomass whether it is used for food, fuel, fodder, fibre, etc. Thus the aim is to have sustainable production over the long term which takes into account all the external costs, e.g. energy, fertilisers, pesticides, which should be included when accounting for the total balances of the system.

In order to achieve the full potential of biomass for food as well as biomass energy and other uses, considerable scientific and technological inputs are being applied. The aims of any system designed for sustained use of plant material are:

(a) High yields;
(b) Minimum inputs;
(c) Utilisation of most plant material;
(d) Use of process wastes;
(e) Optimum use of land, water, fertilizer, etc.;
(f) Selection of plants for non-food as well as for food biomass;
(g) Environmental sustainability.

In general, much agriculture at subsistence level is characterized by:

(a) Low crop yields;
(b) Absence of cash inputs;
(c) Poor credit facilities for investment;
(e) Decreasing soil nutrient content;
(e) Shifting cultivation, often with destruction of vegetation, followed by degradation of the soil;
(f) Production increases which result from an expansion of the cultivated area with

Photosynthesis and Production in a Changing Environment: a field and laboratory manual
Edited by D.O. Hall, J.M.O. Scurlock, H.R. Bolhàr-Nordenkampf, R.C. Leegood and S.P. Long
Published in 1993 by Chapman & Hall, London.
ISBN 0 412 42900 4 (HB) and 0 412 42910 1 (PB).

yields per hectare remaining static or falling;
(g) Inadequate infrastructure and research and development (R & D).

In contrast, intensified agriculture is characterised by many inputs such as:

(a) Inorganic fertilizers;
(b) (Chemical) pest, disease and weed control;
(c) Mechanical cultivation and harvesting;
(d) Storage and process facilities;
(e) Access to credit for investment;
(f) Extensive infrastructure and R & D.

If these inputs are available and put into subsistence-level farming, an exponential increase in production can be obtained. However, this is often correlated with a decrease in energy input/output ratios (i.e. a higher energy requirement) and mechanization (i.e. a lower use of manpower). The socio-economic effects of the latter should always be taken into account, whilst the importance of energy inputs in their various forms cannot be too strongly emphasised, particularly when considering fuel crops. Although the energy available from a biomass crop may be in a more convenient form than the energy used up in its production (e.g. concentrated liquid fuels such as ethanol produced with bulky, diffuse solid fuel inputs), the overall energy balance deserves attention along with the financial and socio-economic analyses of the overall process.

An alternative approach to improvement of subsistence-type agriculture is to concentrate on scientific aspects of the system, such as use of biological nitrogen fixation and recycling of organic material to increase soil fertility and give higher yields. Other factors such as disease and pests, extremes of temperature and deficiencies of water, nutrients or light can limit yields, often through their effect directly on photosynthesis or on the plant's ability to carry out photosynthesis and store its products. Once again, a careful study of these limiting factors can lead to inexpensive improvements in yield.

Table C.1. Fossil fuel reserves and resources, biomass production and CO_2 balances [13,16,25]

	Billion tonnes oil equivalent (Gtoe)	Energy content (EJ)
PROVEN RESERVES [1]		
Coal	572	25,168
Oil	137	6,028
Gas	109	4,796
	818	35,992
ESTIMATED RESOURCES [12]		
Coal	5,409	238,000
Oil	227	10,000
Gas	227	10,000
Uranium	30	1,324
Tar sands + oil shale	36	1,600
	5,930	260,924
CARBON FLOWS		
Fossil fuel use (1860–1988) [4]	206 Gt C =	9,270 EJ
World annual energy use (1987)		399 EJ
	(5.9 Gt C from fossil fuels)	
Annual photosynthesis or gross primary production (terrestrial)	120 Gt C	
Net primary production (terrestrial + aquatic)	106 Gt C =	4,664 EJ
Stored in biomass		
total (80% in trees)	560 Gt C =	24,640 EJ
aquatic	3 Gt C	
CO_2 in atmosphere (1986)	730 Gt C	
CO_2 in ocean surface layers (75 m)	725 Gt C	
Soil carbon content	1,515 Gt C	
Deep ocean carbon	38,000 Gt C	

These data demonstrate that (i) the world's annual use of energy is only one-tenth the annual photosynthetic energy storage; (ii) stored biomass on the Earth's surface is presently equivalent to two-thirds the proven fossil fuel reserves; (iii) the total stored as fossil carbon represents only about 60 years of net photosynthesis; (iv) plants, the atmosphere and the ocean surface layers each contain approximately equal amounts of carbon, whilst soils contain about twice this amount.

Notes: Fossil fuel carbon emissions are calculated from UN energy statistical sources [4], assuming an energy yield of 45 GJ per tonne of carbon consumed for historical fossil fuel emissions. The average figure for current fossil fuel emissions is nearer 50 GJ/t C. The energy equivalent of biomass carbon is 44 GJ/t C, assuming a carbon content of 45% and an energy content of 20 GJ/t for oven-dry biomass. 1 EJ = 10^{18} joules. t C = tonnes carbon. 1 toe (tonne oil equivalent) = 44 GJ.

Table C.2. Global distribution of energy use, population and food intake (1987) [24]

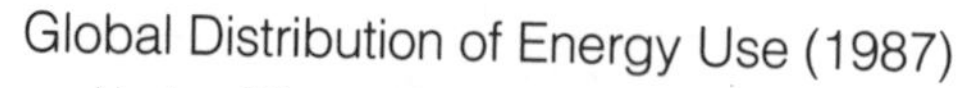

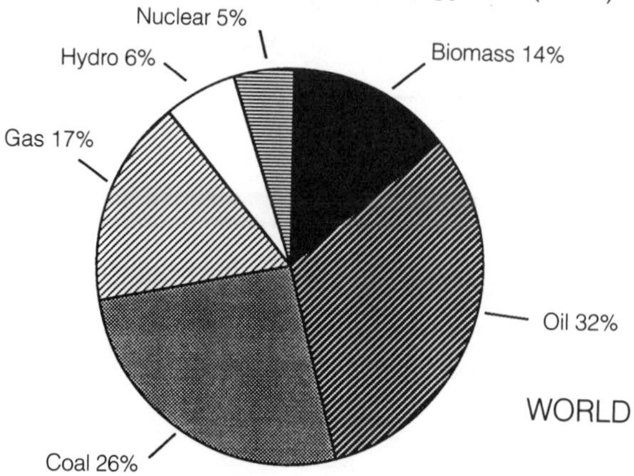

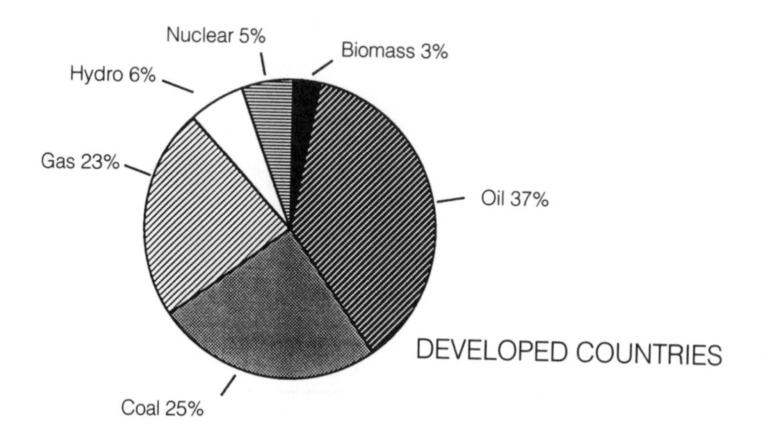

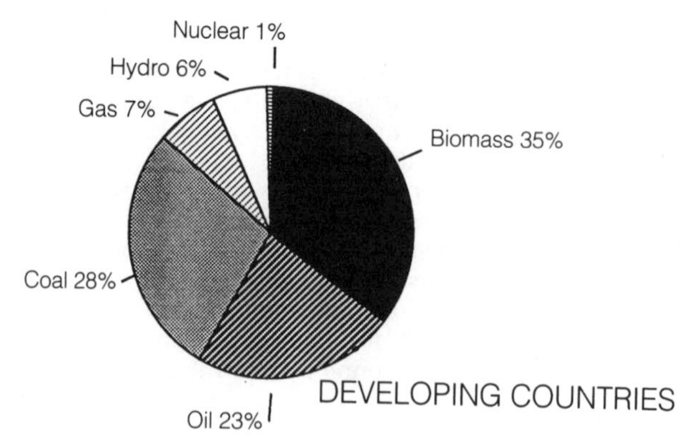

	Developed Countries			Developing Countries			World Total		
Population	1.2 billion (24%)			3.8 billion (76%)			5 billion (100%)		
Total energy use	262 EJ (66%)			137 EJ (34%)			399 EJ (100%)		
Average energy use per capita	218 GJ (4.96 toe)			36.0 GJ (0.82 toe)			79.8 GJ (1.81 toe)		
Energy use by source	EJ	Mtoe	%	EJ	Mtoe	%	EJ	Mtoe	%
Oil	97	2211	37	32	730	23	129	2941	32
Coal	66	1502	25	39	884	28	105	2386	26
Gas	59	1333	23	10	222	7	69	1555	17
Biomass*	7	169	3	48	1088	35	55	1257	14
Hydro	16	355	6	7	169	6	23	524	6
Nuclear	17	377	6	1	27	1	18	404	5
TOTAL	262	5947	100	137	3120	100	399	9067	100
Food intake per capita per day	3418 kcal (14.3 MJ)			2455 kcal (10.3 MJ)			2692 kcal (11.3 MJ)		
Per cent UN recommended daily basic diet† (2350 kcal; 9.83 MJ)	145%			104%			115%		

1.0 EJ = 10^{18} J (approximately equal to 1 Quad [USA], i.e. 10^{12} Btu)
toe = tonnes oil equivalent; 1 Mtoe (million tonnes oil equivalent) = 44×10^{6} GJ (44×10^{15} J)
*biomass includes woodfuel, charcoal, agricultural wastes, dung, bio-ethanol and other forms of biofuels.
†note that these average values conceal much wider regional and national differences

Table C.3. WORLD ECOSYSTEMS [15, and references therein]. Note: many authorities still fail to agree upon the major categories of world ecosystems and their respective areas. For example, the area of "tropical forests" cited here is larger than recent estimates of about 8–10 × 10^6 km^2 for "tropical moist forests". Likewise the distinction between "tropical woodlands", "tropical savannas" and "tropical grasslands" is not always clear. Figures in parentheses () indicate equivalent NPP in t ha^{-1} yr^{-1} or biomass in t ha^{-1}.

	Total terrestrial*	Tropical forests	Temperate forests	Boreal forests	Tropical woodlands	Temperate woodlands	Tropical grasslands	Temperate grasslands	Agriculture	Marine
NPP (Gt yr^{-1})										
Ajtay (1979)	133 (9.0)	30	8	7	13	5	26	4	15	–
Olson (1983)	134 (8.9)	21	16	7	19	5	16	5	27	–
Whittaker and Likens (1975)	118 (7.8)	49	15	10	_**	6	14***	5	9	55
Plant Biomass (Gt)										
Ajtay	1244 (33)	542	174	205	88	54	58	20	10	–
Olson	1242 (34)	364	281	318	177	49	50	26	48	7
Whittaker and Likens	1837 (40)	1025	385	240	_**	50	60	14	14	4
Area (10^6 km^2)										
Ajtay	149	15	6	9	10	5	12	13	16	–
Olson	151	13	12	12	17	5	17	17	16	360
Whittaker and Likens	149	25	12	12	_**	9	15	9	14	361

* Also included, but not separately listed, are tundra, deserts, wetlands, polar and extreme desert areas
** Included in tropical forests
*** Categorized as savanna

Comparison of Estimates	NPP (Gt)			Plant Biomass (Gt)			Area ($10^6\,km^2$)		
	Ajtay (1979)	Olson (1983)	Whittaker and Likens (1975)	Atjay (1979)	Olson (1983)	Whittaker and Likens (1975)	Atjay (1979)	Olson (1983)	Whittaker and Likens (1975)
Total Forests	45	44	74	921	963	1650	30	37	49
Total Woodlands	18 } 63	24 } 68	6** } 80	142	226	50	15 } 45	22 } 59	9** } 58
Total Grasslands	30	21	19	78	76	76	25	34	24
	93	89	99	1141	1265	1774	70	93	82

Estimation of World Annual Increment of Wood

Earl (1970 data)

Increment	$17.8 \times 10^9\,m^3$ on 3,800 Mha "global forests"
	= approx. 13×10^9 t (13 Gt) dry matter
	= approx. 260×10^{18} J (260 EJ) energy
Consumption	$2.4 \times 10^9\,m^3$ = 1.7 Gt = 34 EJ
Unused	$15.4 \times 10^9\,m^3$ = 11.3 Gt = 226 EJ

Openshaw (1990 estimates)

Increment	$12.5 \times 10^9\,m^3$ on 4,100 Mha "closed forests and open woodlands"
	= approx. 9 Gt dry matter
	= approx. 180 EJ energy
Consumption	$3.4 \times 10^9\,m^3$ = 2.5 Gt = 50 EJ
Unused	$9.1 \times 10^9\,m^3$ = 6.5 Gt = 130 EJ

Table C.4. Some high short-term dry weight yields of crops and their short-term photosynthetic efficiencies [27].

Crop	Country	$g\,m^{-2}\,day^{-1}$	Photosynthetic efficiency (% of total radiation)
Temperate			
Tall fescue	UK	43	3.5
Rye-grass	UK	28	2.5
Cocksfoot	UK	40	3.3
Sugar beet	UK	31	4.3
Kale	UK	21	2.2
Barley	UK	23	1.8
Maize	UK	24	3.4
Wheat	Netherlands	18	1.7
Peas	Netherlands	20	1.9
Red clover	New Zealand	23	1.9
Maize	New Zealand	29	2.7
Maize	USA, Kentucky	40	3.4
Subtropical			
Alfalfa	USA, California	23	1.4
Potato	USA, California	37	2.3
Pine	Australia	41	2.7
Cotton	USA, Georgia	27	2.1
Rice	S. Australia	23	1.4
Sugar cane	USA, Texas	31	2.8
Sudan grass	USA, California	51	3.0
Maize	USA, California	52	2.9
Algae	USA, California	24	1.5
Tropical			
Cassava	Malaysia	18	2.0
Rice	Tanzania	17	1.7
Rice	Philippines	27	2.9
Palm oil	Malaysia (whole year)	11	1.4
Napier grass	El Salvador	39	4.2
Bullrush millet	Australia, N.T.	54	4.3
Sugar cane	Hawaii	37	3.8
Maize	Thailand	31	2.7

Other yields: Loomis and Gerakis (see Cooper [5]) discuss figures for (a) sunflower–growth rates of 79–104 $g\,m^{-2}\,day^{-1}$ have been reported with a 3-week mean rate of 63.8 $g\,m^{-2}\,day^{-1}$ giving a photosynthetic efficiency of 7.5%. (b) carrot–growth rates of 146 $g\,m^{-2}\,day^{-1}$ and a dry matter yield of 54.5 $t\,ha^{-1}$ after 160 days were reported.
Note: yields in $g\,m^{-2}\,day^{-1}$ can be converted to $t\,ha^{-1}\,y^{-1}$ by multiplying by 3.65.

Table C.5(a). Average-to-good annual yields of dry-matter production [23,27].

	$t\,ha^{-1}\,yr^{-1}$	$g\,m^{-2}\,day^{-1}$	Photosynthetic efficiency (% of total radiation)
Tropical			
Amazon floodplain grass	99	–	3.5
Napier grass	88	24	1.6
Sugar cane	66	18	1.2
Reed swamp	59	16	1.1
Annual crops	30	–	–
Perennial crops	75–80	–	–
Rainforest	35–50	–	–
Temperate (Europe)			
Perennial crops	29	8	1.9
Annual crops	22	6	0.8
Grassland	22	6	0.8
Evergreen forest	22	6	0.8
Deciduous forest	15	4	0.6
Savanna	11	3	–
Desert	1	0.3	0.02

(b) Harvestable dry matter production ($t\,ha^{-1}\,yr^{-1}$)

(i) Average to good annual yields [17].

	Typical	High
Sugar cane	35	90
Maize	10	40
Wheat	5	20
Rice	4	16
Sugar beet	8	18
Temperate grass	7	25
Tropical grass	15	50
Cassava	8	35

(ii) Maximum photosynthetic productivity (USA)* [2].

Napier grass	139
Sugar cane	138
Sorghum	186
Sugar beet	113
Alfalfa	84

* All-year growth assumed (often not the case).

Theoretical maximum	
USA average	224
USA southwest	263

Table C.6(a). Photosynthetic efficiency and energy losses [27]

	Available light energy
At sea level	100%
50% loss as a result of 400–700 nm light being the photosynthetically usable wavelengths	50%
20% loss due to reflection, absorption and transmission by leaves	40%
77% loss representing quantum efficiency requirements for CO_2 fixation in 680 nm light (assuming 10 quanta/CO_2)* and that the energy content of 575 nm red light is the radiation peak of visible light	9.2%
40% loss due to respiration	5.5%
	Overall PS efficiency

*If the minimum quantum requirement is 8 quanta/CO_2, then this loss factor becomes 72% (instead of 77%) giving a final photosynthetic efficiency of 6.7% (instead of 5.5%).

(b) Losses of solar energy, traced from the solar constant in space to the energy fixed in a European woodland (50°N) producing 11 t $ha^{-1} y^{-1}$ of dry wood [6].

Solar constant in space, 117.5 MJ $m^{-2} d^{-1}$ = 26 000 t $ha^{-1} y^{-1}$ of wood

1. Reduced by earth's geometry (0.20) and atmosphere (0.40)

Incoming radiation at earth's surface, 9.4 MJ $m^{-2} d^{-1}$ = 2080 t $ha^{-1} y^{-1}$

2. Reduced by proportion of PAR (0.5) and leaf reflectance and transmittance (0.85)

Radiation available for photosynthesis, 4.0 MJ $m^{-2} d^{-1}$ = 884 t $ha^{-1} y^{-1}$

3. Reduced by quantum efficiency of photosynthesis (0.215)

'Theoretical' carbohydrate production, 0.86 MJ $m^{-2} d^{-1}$ = 190 t $ha^{-1} y^{-1}$

4. Reduced by the rate of diffusion of CO_2 to chloroplasts (0.4)

'Potential' carbohydrate production, 0.34 MJ $m^{-2} d^{-1}$ = 76 t $ha^{-1} y^{-1}$

5. Reduced because only a proportion of light is intercepted (at favourable temperatures (0.40))

'Actual' carbohydrate production, 0.137 MJ $m^{-2} d^{-1}$ = 30 t $ha^{-1} y^{-1}$

6. Reduced by respiration (0.6) and the proportion partitioned to wood (0.6)

Wood production, 0.0495 MJ $m^{-2} d^{-1}$ = 11 t $ha^{-1} y^{-1}$

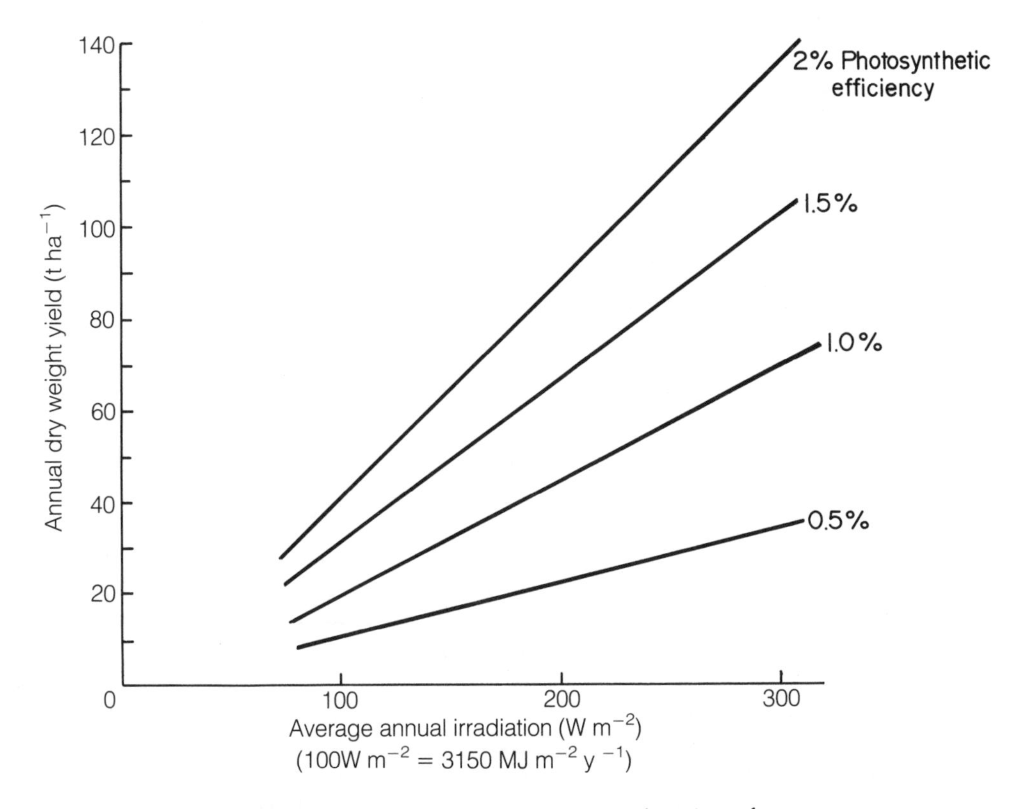

(c) Expected annual plant yields as a function of annual solar irradiation at various photosynthetic efficiencies [27].

Table C.7. Available land and current land use on a global basis [8]

	Land area (M ha)	Use (%)				
		Permanent	Arable	Pasture	Forests	Irrigation
World	13,442	0.8	10.0	24.0	30.0	1.7
Developed countries	5,614	0.4	11.6	22.3	33.3	1.1
Developing countries	7,768	1.0	9.3	25.2	28.1	2.2
USA	937	0.2	20.0	26.0	28.3	1.9
Canada	922	0.01	4.6	3.3	36.0	0.1
Europe	486	2.9	25.9	17.1	32.4	3.5
USSR (former)	2,240	0.2	10.2	16.6	42.2	0.9
Asia	2,757	1.1	15.3	25.0	19.0	5.3
S. America	1,781	1.5	6.5	27.0	50.3	0.5
Africa	2,964	0.7	5.8	27.0	26.5	0.4
Oceania	842	0.1	5.6	52.1	18.5	0.3

Table C.8. World land use and productivity classes – trends to year 2000 [5]

	Land use 1975	Land losses	Land reclamation	Land use 2000	Net change 1975/2000
Cropland					
High	400	100	45	345	−55
Medium	500	80	325	745	+245
Low	600	40	150	710	+110
	1500	220	520	1800	+300
Grassland					
High	200	30	0	170	−30
Medium	300	20	30	320	+20
Low	500	90	40	510	+10
Zero	2000	0	0	2000	0
	3000	140	70	3000	0
Forest					
High	100	25	45	30	−70
Medium	300	90	180	100	−200
Low	400	75	205	230	−170
Zero	3300	0	0	3140	−160
	4100	190	590	3500	−1600
Non-agricultural	400	200	0	600	+200
Other land	4400	100	0	4500	+100

See Buringh [5] for details of types of land losses and reclamation.

Table C.9. Areas of closed forest and size of growing stock [8].

	Closed forest (million ha)	Growing stock of closed forests		
		Total (million m^3)	$m^3\,ha^{-1}$	m^3 per capita
USA	220	20 200	92.1	95.3
Canada	250	17 800	71.2	809.1
Europe	144	14 900	103.5	29.1
USSR	770	81 800	106.2	324.6
Asia and Oceania	448	38 700	72.9	18.7
S. America	631	92 000	145.8	362.2
Africa	188	35 200	187.2	90.0

Table C.10. World crop production (1,000 tonnes) as harvested weight [10].

	1979	1989
Total Cereals	1,553,967	1,880,994
Wheat	428,373	541,765
Rice Paddy	377,363	517,565
Maize	420,430	470,646
Barley	158,075	167,897
Root Crops	588,064	598,978
Potatoes	299,133	277,131
Total Pulses	41,320	55,401
Vegetables	349,639	437,739
Fruits	293,730	344,820
Grapes	69,446	58,694
Citrus Fruits	59,119	74,920
Bananas	35,421	44,965
Apples	36,650	41,709
Tree Nuts	3,511	4,396
Oil Crops	51,369	71,822
Sugar (Centrifugal Raw)	88,361	105,050
Cocoa Beans	1,678	2,445
Coffee Green	4,982	6,078
Tea	1,820	2,454
Fibre Crops	19,951	23,044
Cotton Lint	13,937	17,056
Jute & Jute-Like Fibres	3,700	3,412
Tobacco	5,418	7,132
Natural Rubber	3,838	4,865
Total Meat	130,620	170,857
Total Milk	456,143	529,305
Hen Eggs	26,529	35,758
Wool, Greasy	2,717	3,192

Table C.11. Total production (10^6 tonnes) and yields ($t\,ha^{-1}$) of the major sugar and starch crops [10]. 1989 data.

	(a) Sugar				(b) Starch grains								(c) Starch roots and gubers							
	Cane		Beet		Cereals		Maize		Rice (Paddy)		Wheat		Roots and tubers		Potatoes		Sweet potatoes		Cassava	
	Total	Yield	Total	Yield	Total	Yield	Total	Yield	Total	Yield	Total	Yield	Total	Yield	Total	Yield	Total	Yield	Total	Yield
World	1,031	62	312	36	1,881	2.7	471	3.6	518	3.5	542	2.4	599	11.9	277	15.3	125	10.5	159	10.2
Developed countries	75	75	280	38	875	3.1	278	6.0	25	5.8	317	2.5	201	17	198	17	2	16.9	–	–
Developing countries	957	62	31	25	1,006	2.4	192	2.3	492	4.3	225	2.2	398	10.3	79	12.2	123	10.4	159	10.2
USA	24	78	23	44	284	4.5	191	7.3	7	6.4	55	2.2	17	31.3	17	32.4	0.5	14.7	–	–
Canada	–	–	1	38	48	2.2	6	6.4	–	–	25	1.8	3	24.9	3	24.9	–	–	–	–
Europe	0.3	64	156	45	289	4.2	53	4.8	2.1	5.1	127	4.6	100	20.9	100	21	–	–	–	–
USSR	–	–	97	29	20	1.9	15	3.7	2.6	3.9	92	1.9	72	12	72	12	–	–	–	–
Asia	421	60	28	24	835	2.7	115	3.0	474	3.6	192	2.3	235	12.2	63	13.1	115	11.3	55	13.4
S. America	332	62	3	53	79	2.1	37	2.1	17	2.5	19	1.9	44	12.1	11	12.6	1.5	10.2	30	12.5
Africa	72	60	4	44	94	1.2	38	1.8	11	2.0	12	1.4	120	7.9	8	9.3	6	5.1	73	8.2
Oceania	31	75	–	–	23	1.7	0.4	5.1	0.8	7.0	14	1.6	3	11.1	1	27.7	0.6	4.9	0.2	10.6

Table C.12. Sources of biomass for conversion to fuels [8].

Wastes and Residues	Land Crops	Aquatic Plants
Wastes	*Ligno-cellulose*	*Wetland plants*
Manures	Trees	Spartina
Industrial and commercial wastes	Eucalyptus	Cattails
Domestic rubbish	Poplar	Miscanthus
Food wastes	Willow	
Sewage	Pines	
	Acacias	
	Albizia	
Residues		
Wood industry residues	*Starch crops*	*Water weeds*
Bagasse	Maize	Water hyacinth
Cane tops		
Straw	*Sugar crops*	*Algae*
Stalks	Cane	Unicellular
Husks	Beet	Multicellular
Molasses	Sorghum	

Table C.13. Conversion techniques for producing fuels from biomass.

Process	Products	State of art
(a) Thermal		
Combustion	CO_2, water, ash, heat, steam, electricity	Well established
Pyrolysis	Char, oil, gas (CO, C_2H_4, H_2)	Small-scale established; Medium and large-scale under development and demonstration
Gasification	Mainly CO and H_2	
Gasification + catalytic synthesis	Methanol, hydrocarbons	
(b) Biological		
Yeast fermentation	Ethanol + CO_2	Established
Anaerobic digestion	Methane + CO_2	Established

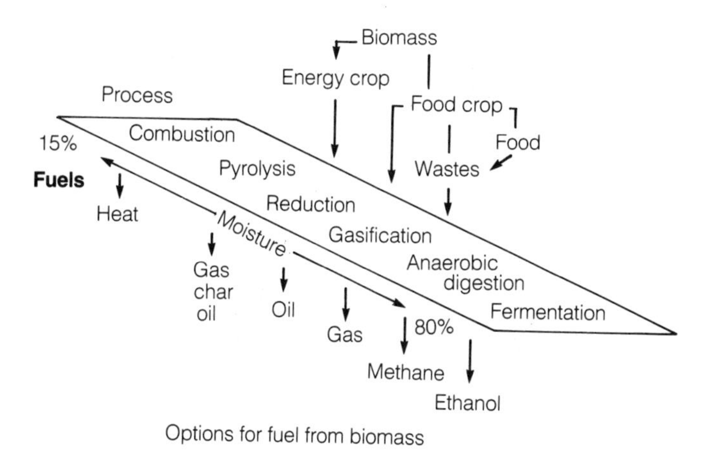

Options for fuel from biomass

Table C.14. Biomass use per capita, biomass and commercial energy use and percentage of biomass energy in selected developing countries [14].

Country	Biomass use per capita		Biomass use (10^6)		Commercial use (10^6)	Biomass as % of total energy
	twe	toe	twe	toe	toe	
Latin America						
Antigua	–	–	–	–	0.09	–
Argentina	0.18	0.06	5.75	1.99	41.55	5
Belize	–	–	–	–	0.06	–
Brazil	0.80	0.28	106.92	38.18	75.66	33
Costa Rica	0.79	0.28	2.01	0.74	0.97	43
Dominican Rep.	0.32	0.11	2.00	0.72	1.97	27
Guatemala	0.87	0.31	7.01	2.50	0.97	72
Haiti	0.66	0.23	3.43	1.22	0.21	85
Honduras	0.85	0.30	3.65	1.30	0.61	68
Jamaica	0.26	0.09	0.63	0.22	1.76	11
Mexico	0.34	0.12	27.00	9.62	98.33	9
Nicaragua	0.93	0.34	3.13	1.11	0.71	61
Panama	0.54	0.19	1.15	0.41	0.92	31
Guyana	1.44	0.51	1.14	0.41	0.33	55
St. Lucia	0.19	0.07	0.02	0.008	0.04	16
Uruguay	0.51	0.18	1.55	0.55	1.40	28
Africa						
Botswana	1.72	0.61	1.72	0.61	0.44	58
Burundi	0.76	0.27	3.61	1.29	0.07	95
Egypt	0.52	0.18	25.30	9.04	23.60	28
Gambia	0.80	0.28	0.60	0.21	0.07	75
Ghana	0.46	0.16	6.27	2.24	1.30	63
Kenya	1.32	0.47	26.91	9.61	1.57	86
Mauritius	0.96	0.34	0.96	0.34	0.40	46
Morocco	0.10	0.03	2.22	0.80	5.50	13
Mozambique	1.06	0.38	14.00	5.00	0.33	94
Nigeria	1.55	0.55	148.31	52.97	17.80	82
Rwanda	1.60	0.57	11.45	4.09	0.14	97
Seychelles	0.12	0.04	0.008	0.002	0.32	9
Somalia	1.03	0.37	4.76	1.70	0.28	86
Sudan	2.61	0.93	56.20	20.07	1.02	95
Tanzania	2.84	1.01	61.70	22.03	0.64	97
Tunisia	0.50	0.18	3.52	1.26	3.45	27
Zambia	0.94	0.33	6.25	2.23	1.30	63
Zimbabwe	1.15	0.41	9.56	3.41	4.50	43

Table C.14. *Continued*

Country	Biomass use per capita twe	Biomass use per capita toe	Biomass use (10^6) twe	Biomass use (10^6) toe	Commercial use (10^6) toe	Biomass as % of total energy
Asia						
India	0.75	0.27	569.52	203.40	154.00	57
Indonesia	1.08	0.39	177.00	63.21	32.90	66
Malaysia	2.84	1.02	44.20	15.79	14.50	52
Pakistan	0.86	0.31	83.06	29.67	19.40	60
Philippines	1.05	0.38	57.02	20.37	10.70	66
Sri Lanka	1.12	0.40	11.98	4.28	1.45	75
Thailand	1.61	0.58	13.74	4.90	18.33	21
Oceania						
Fiji	1.25	0.45	0.88	0.31	0.19	62
Vanuatu	–	–	–	–	0.02	78
W. Samoa	–	–	–	–	0.04	60
Other selected countries						
Bangladesh	1.02	0.37	101.50	36.25	4.80	88
Bolivia	1.12	0.40	7.18	2.57	1.45	64
China	0.59	0.21	619.14	221.12	559.00	28
Colombia	1.22	0.44	34.90	12.47	17.07	42
Ethiopia	0.80	0.29	34.12	12.86	0.85	94
Nepal	0.71	0.61	29.33	10.44	0.28	97
Peru	1.00	0.36	19.72	7.04	8.20	46
Senegal	0.44	0.18	2.87	1.03	0.66	61
Uganda	0.88	0.31	13.60	4.86	0.28	95
Zaire	0.79	0.28	24.10	8.61	1.45	86

twe = tonne wood equivalent (= 15 GJ air dry).
toe = tonne oil equivalent (= 42 GJ).
(Biomass data from various years and sources. Commercial fuel data from 1987. Population data from 1985).

Table C.15. Energy conversion factors

Units and electricity equivalents
1 gigajoule (GJ) = 10^9 joules
1 GJ = 278 kWh; 1 kWh = 3.6 MJ
1 GJ = 0.95×10^6 Btu
1 GJ = 0.239×10^6 kcal
1 calorie = 4.184 J
Biomass fuels
Figures for energy content are Lower Heating Value unless otherwise stated. Higher Heating Value (including condensation of combustion products) is approximately 5–15% greater, depending on fuel.
1 tonne (t) wood (air dry; 20% moisture) = 15 GJ
1 t wood (oven dry; <5% moisture) = 20 GJ
1 t agricultural residues = 13 GJ (usual range 10–17 GJ, depending on moisture content)
1 t charcoal (derived from 6–12 t wood) = 30 GJ
1 t wood = 1.4 m^3 wood (solid, not stacked)
1 m^3 wood (solid) = 0.73 t = 11 GJ ≈ 0.4 t coal
1 t ethanol = 28 GJ
1 t methanol = 20 GJ
Fossil fuels
1 t oil = 42–45 GJ = 7.3 barrels ≈ 0.85 t carbon
1 barrel oil = 159 litres = 5.8–6.1 GJ
1 litre gasoline = 35 MJ ≈ 0.66 kg carbon
1 t coal = 25–28 GJ ≈ 0.75 t carbon
1 t coal = 1.9 t air-dry wood = 2.7 m^3 wood
1 m^3 natural gas (methane) = 35.6/38.8 MJ (Lower/Higher Heating Value) ≈ 0.49 kg carbon

Table C.16. Approximate conversion factors for various light sources (PAR [Photosynthetically Active Radiation] waveband; 400–700 nm). After McCree [20].

	Light source					
	Daylight	Metal halide	Sodium (High Pressure)	Mercury	White fluorescent	Incandescent
To convert	Multiply by					
$W\ m^{-2}$ (PAR) to $\mu mol\ m^{-2}\ s^{-1}$ (PAR)	4.6	4.6	5.0	4.7	4.6	5.0
k lux to $\mu mol\ m^{-2}\ s^{-1}$ (PAR)	18	14	14	14	12	20
k lux to $W\ m^{-2}$ (PAR)	4.0	3.1	2.8	3.0	2.7	4.0

Table C.17. Conversion factors for radiation data [3].

Type of measurement	Instantaneous measurements	Full sun plus sky	Integrated measurements	One day's integration
Quantum	1 μE $m^{-2} s^{-1}$ = 1.0 μmol $m^{-2} s^{-1}$ 6.022×10^{17} photons $m^{-2} s^{-1}$ 6.022×10^{17} quanta $m^{-2} s^{-1}$ 6.022×10^{17} quanta $cm^{-2} s^{-1}$	Photosynthetic photon flux 2000 μE $m^{-2} s^{-1}$	1 Em^{-2} = 1 mol m^{-2} 6.022×10^{23} photons m^{-2} 6.022×10^{23} quanta m^{-2}	Photosynthetic photon exposure 60 E m^{-2}
Radiometric	1 W m^{-2} = 1.433×10^{-3} cal $cm^{-2} min^{-1}$ 1.433×10^{-3} langley min^{-1} 0.100 mW cm^{-2} 100 μW cm^{-2} 1.0 J $s^{-2} m^{-2}$ 1000 erg $s^{-1} cm^{-2}$ 0.317 BUT $ft^{-2} h^{-1}$ 5.283×10^{-3} BTU $ft^{-2} min^{-1}$	Total irradiance 1000 W m^{-2} Photosynthetic irradiance 500 W m^{-2} Near infrared (745–815 nm) irradiance 60 W m^{-2}	1 Wh m^{-2} = 0.0860 cal cm^{-2} 0.0860 langley 0.1 mWh cm^{-2} 100 μWh cm^{-2} 3600 J M^{-2} 0.3600 J cm^{-2} 3.600×10^{6} erg cm^{-2} 0.317 BTU ft^{-2}	Total radiant exposure 8000 Wh [illegible] $^{-2}$ Photosynthetic radiant exposure 4000 Wh m^{-2} Near-infrared (745–815 nm) radiant exposure 500 Wh m^{-2}
Photometric	1 lux = 1 lm m^{-2} 0.0929 lm ft^{-2} 0.0929 footcandle 0.001 klux	Illuminance 100 klux	1 lux h = 0.0929 footcandle h 0.001 klux h	Luminant exposure 800 klux h

The values shown for 'full sun plus sky' and 'one day's integration' are typical measurements taken mid-summer in Lincoln, Nebraska, USA (41°N), using cosine-type sensors. These values can vary greatly depending on atmospheric conditions, location (including latitude and elevation), time of year, and time of day.

REFERENCES

1. BP (1991) *BP Statistical Review of World Energy 1990*. British Petroleum Company, Moor Lane, London EC2Y 9BU.
2. Bassham, J.A. (1977) Increasing crop production through more controlled photosynthesis. *Science* **197**, 630.
3. Biggs, W. (1986) Radiation measurement. In: *Advanced Agricultural Instrumentation: design and use*. Martinus Nijhoff Publ., Dordrecht. pp. 3–20.
4. Boden, T.A., P. Kanciruk and M.P. Farrell, eds. (1990) *Trends '90: a compendium of data on global change*. Oak Ridge National Laboratory, Tennessee 37831, USA.
5. Buringh, P. (1987) Bioproductivity and land potential. In: *Biomass: Regenerable Energy* (D.O. Hall and R.P. Overend, eds.). John Wiley and Sons Ltd., Chichester, UK. pp. 27–46.
6. Cannell, M.G.R. (1989) Physiological basis of wood production: a review. *Scandinavian J. Forest Research* **4**, 459–490.
7. Grassi, G., A. Collina and H. Zibetta, eds. (1992) *Biomass for Energy, Industry and Environment*. Proc. 6th. EC Conf. Biomass Energy Ind. Env., Athens. 1430 pp. Elsevier Applied Science, London.
8. Coombs, J. (1980) Renewable sources of energy (carbohydrates). *Outlook on Agriculture* **10**, 235.
9. Cooper, J.P. (ed.) (1975) *Photosynthesis and Productivity in Different Environments*. 715 pp. Cambridge University Press.
10. FAO (1990) Agrostat-PC microcomputer database. U.N. Food and Agriculture Organisation, Via delle Terme di Caracalla, Rome.
11. Hall, D.O., G. Barnard and P.A. Moss (1982) *Biomass for Energy in the Developing Countries*. Pergamon Press, Oxford 220 pp.
12. Hall, D.O., M. Slesser and others (1987) Assessment of renewable and non-renewable energy resources. In: *Resources and World Development* (D.J. McLaren and B.J. Skinner, eds.) John Wiley and Sons, Chichester, UK. pp. 484–506.
13. Hall, D.O. (1989) Carbon flows in the biosphere: present and future. *J. Geol. Soc. Lond.* **146**, 175–181.
14. Hall, D.O. and F. Rosillo-Calle (1991) Why biomass matters. *Network News* **5(4)**, 4–15. Biomass Users Network, Montes de Oca, Costa Rica.
15. Hall, D.O., F. Rosillo-Calle, R.H. Williams and J. Woods (1992) *Biomass for energy: future supply prospects*. In: *Renewables for Fuels and Electricity* (Johansson, T.B., H. Kelly, A.K.N. Reddy and R.H. Williams, eds.) Island Press, Washington (in press).
16. Hall, D.O. and J.M.O. Scurlock (1991) Tropical grasslands and their role in the global carbon cycle. In: *Modern Ecology: basic and applied aspects* (G. Esser and D. Overdieck, eds.) Elsevier, Amsterdam. pp. 659–678.
17. Hudson, J.C. (1975) Sugar cane: energy relationships with fossil fuel. *Sugar J.* **37**, 25.
18. Kitani, O. and C.W. Hall (1989) *Biomass Handbook*. Gordon and Breach Science Publishers, New York. 963 pp.
19. McClure, T.A. and E.S. Lipinsky (eds.) (1982) *CRC Handbook of Biosolar Resources*. Vol. 2: Resource materials. CRC Press, Boca Raton, Florida.
20. McCree, K.J. (1981) Photosynthetically active radiation. In: *Physiological Plant Ecology*. Vol. 12A, Encyclopedia of Plant Physiology (New Series) (Lange, O.L., Nobel, P.S., Osmond, C.B. and Zeigler, H., eds.) Springer-Verlag, Berlin. pp. 41–55.
21. Mitsui, A. and C.C. Black (eds.) (1982) *CRC Handbook of Biosolar Resources*. Vol. 1: Basic principles. CRC Press, Boca Raton, Florida.
22. O.T.A. (1980) *Energy from Biological Processes*. Office of Technology Assessment, United States Congress, Washington, D.C.
23. Piedade, M.T.F., W.J. Junk and S.P. Long (1991) The productivity of the C_4 grass *Echinochloa polystachya* on the Amazon floodplain. *Ecology* **72**, 1456–1463.
24. Scurlock, J.M.O. and D.O. Hall (1990) The contribution of biomass to global energy use in 1987. *Biomass* **21**, 75–81.
25. Scurlock, J.M.O. and D.O. Hall (1991) The carbon cycle. *New Scientist* (Inside Science suppl., 2/11/91) **51**, 1–4.
26. Smil, V. (1983) *Biomass Energies; Resources, Links, Constraints*. Plenum Press, New York.
27. SES (1976) *Solar Energy – A U.K. Assessment*. 373 pp. Solar Energy Society, Atkins Building South 128, King's College, London W8 7AH, UK.

RECOMMENDED JOURNALS

Biomass and Bioenergy (eds. J. Coombs, D.O. Hall, R.P. Overend and W. Smith). Pergamon Press, Oxford.

Bioresource Technology. Elsevier Applied Science, London.

Energy (ed. S.S. Penner). Pergamon Press, Oxford.

Appendix D
Units, symbols and abbreviations

J.M.O. Scurlock

Since research on photosynthesis and productivity in relation to a changing environment now takes place in many countries throughout the world, it is more important than ever to keep to internationally agreed standards for reporting and writing papers. This enables comparison between work carried out in different countries, as well as between the studies of different institutions within the same country. The following is a set of recommendations or guidelines for reporting of experimental work.

Units

Systeme Internationale (SI) units should be used in preference to all others, where reasonable and in line with modern convention. Strict adherence to SI means that dm^3 should be used instead of l (litre), and cm^3 instead of ml. However, the latter terms are still the convention in much biochemical work, so they have been retained in this book. Note that units have neither plurals nor full stops (periods), and that compound units should be given in 'index' form (e.g. $g\,m^{-3}$, not g/m^3). Use either per cent (not percent) or % consistently. Particlar attention should be paid to the preferred units for photon flux, CO_2 assimilation rate and transpiration, viz. $\mu mol\,m^{-2}\,s^{-1}$ (or $mmol\,m^{-2}\,s^{-1}$). This allows direct comparison of their relative rates.

Abbreviations

All abbreviated terms should be given in full at the first mention, with the abbreviation in parentheses; thereafter the abbreviation should be used.

Symbols

Table D.1 lists many of the symbols used consistently throughout this book. They have been chosen to avoid ambiguity where possible, in line with the most recent established convention. Note that rate functions such as fluxes of photons, water vapour, etc. are indicated by bold letters (e.g. **A**, **E**). However, other symbols may be found in the scientific literature, and care must be taken to avoid confusion. In particular, note that F_c is sometimes used instead of **A** for rate of CO_2 assimilation, and B instead of W for dry weight of biomass.

Table D.2 is a supplementary list of selected symbols encountered in modelling (Chapters 13 and 14). There is less agreement here on the use of each symbol, and some of these symbols duplicate those used in other areas of photosynthesis and bioproductivity studies. Additional symbols used in determination of leaf gas exchange are given in Tables 9.1 and E.1.

Photosynthesis and Production in a Changing Environment: a field and laboratory manual
Edited by D.O. Hall, J.M.O. Scurlock, H.R. Bolhàr-Nordenkampf, R.C. Leegood and S.P. Long
Published in 1993 by Chapman & Hall, London.
ISBN 0 412 42900 4 (HB) and 0 412 42910 1 (PB).

Table D.1. Symbols used throughout this book.

Symbol	Parameter	Units
A	CO_2 assimilation rate	(μmol m^{-1} s^{-1})
c	concentration of CO_2 in air	(μmol mol^{-1} or μl l^{-1})
C	crop growth rate	(g m^{-2} day^{-1})
C	heat flux by convection	(W m^{-2})
D	dry weight of dead vegetation	(g m^{-2})
D	leaf area duration	(d)
D	saturation vapour deficit of air	(kPa)
d	loss or disappearance of plant material	(g m^{-2} d^{-1})
E	unit leaf rate	(g m^{-2} d^{-1})
E	rate of transpiration or evapotranspiration	(μmol m^{-2} s^{-1})
E	extinction coefficient of solutions	(dimensionless)
ε	photosynthetic efficiency of solar energy conversion	
ε	emissivity	
e	partial pressure of water vapour	(kPa)
F	leaf area ratio	(m^2 g^{-1})
F	cumulative leaf area index	(dimensionless)
F_m	maximum level of chlorophyll fluorescence	
F_v	variable chlorophyll fluorescence	
g_b	boundary layer conductance	(s m^{-1} or μmol m^{-2} s^{-1})
g_l	leaf conductance	(s m^{-1} or μmol m^{-2} s^{-1})
g_s	stomatal conductance	(s m^{-1} or μmol m^{-2} s^{-1})
I	irradiance, radiant energy flux	(W m^{-2})
$\mathbf{I}_a$	solar radiation absorbed by the canopy	(W m^{-2})
$\mathbf{I}_p$	irradiance in the range 400–700 nm	(W m^{-2})
k	foliar absorption coefficient	(dimensionless)
L, LAI	leaf area index	(dimensionless)
λ	wavelength	(nm or μm)
$\boldsymbol{\lambda}$	latent heat of vaporization of water	
P_g	gross primary production	(g m^{-2} d^{-1})
P_n	net primary production	(g m^{-2} d^{-1})
Q	quantum flux of photons	(μmol m^{-2} s^{-1})
q_P	photochemical fluorescence quenching	
q_{NP}	non-photochemical fluorescence quenching	
q_E	energy-dependent fluorescence quenching	
R	sum of respiration losses	(g m^{-2} d^{-1})
R*	relative water content	(dimensionless)
R	relative growth rate	(g g^{-1} d^{-1})
r_s	stomatal resistance	(s m^{-1})
s	photosynthetic surface area of leaf	(m^2)
T	temperature	(°C *or* K)
T_l	leaf temperature	(°C)
T′	wet-bulb temperature	(°C)
W	dry weight of biomass (live vegetation)	(g m^{-2})
ψ	water potential	(MPa)

Table D.2. Selected symbols used in modelling.

Symbol	Parameter	Units
a	albedo (reflectance)	(dimensionless)
A	leaf absorptance to solar radiation	(dimensionless)
δ	declination	(° or radians)
D_j	Julian date	(d)
Λ	latitude	(°)
η_s	solar azimuth	(°)
η_l	leaf azimuth	(°)
α	leaf inclination from horizontal	(°)
θ	solar zenith angle	(°)
Θ	volume fraction of water in soil	(dimensionless)
ι	angle between surface and line normal to direction of solar radiation	(°)
P	atmospheric pressure	(kPa)
r_d	non-photorespiratory rate of respiration in daytime	(μmol $m^{-2} s^{-1}$)
R	re-radiation of infra-red (long-wave) radiation	(W m^{-2})
C	sensible heat exchange	(W m^{-2})
LE	latent heat exchange	(W m^{-2})
M	metabolic heat	(W m^{-2})
G	stored heat energy	(W m^{-2})
G_b	canopy boundary layer conductance	(mol $m^{-2} s^{-1}$)
τ_A	atmospheric transmissivity	(dimensionless)
v	wind speed	(m s^{-1})
w	leaf width	(m)
x	mean ratio of horizontal to vertical projection of leaves in the canopy	(dimensionless)
ϕ	quantum yield of photosynthesis	(μmol CO_2 μmol^{-1} PPF)
ϕ_c	canopy light use efficiency	(mol CO_2 mol^{-1} PPF)
Φ_s	solar (short-wave) radiation	(W m^{-2})
Φ_L	infra-red (long-wave) radiation	(W m^{-2})
χ	inclination index	(dimensionless)
Γ_*	CO_2 compensation point	(μbar)

FURTHER READING

Anderton, P., and P.H. Bigg (1972) *Changing to the Metric System: conversion factors, symbols and definitions.* National Physical Laboratory: H.M. Stationery Office, London.

Goldman, D.T. and R.J. Bell, eds. (1982) *S.I.: The International System of Units,* 4th edn. H.M. Stationery Office, London.

Incoll, L.D., S.P. Long and M.R. Ashmore (1977) S.I. units and publications in Plant Science. *Current Advances in Plant Science* **28**, 331– 343.

Monteith, J.L. and M.H. Unsworth (1990) *Principles of Environmental Physics*, 2nd edn. Edward Arnold, London. pp. viii-xii.

Pearcy, R.W., R.W. Ehleringer, H.A. Mooney, and P.W. Rundel (eds.) (1989) *Plant Physiological Ecology*. Chapman and Hall, London. pp. 427–428

Royal Society (1975) *Quantities, Units and Symbols.* Report of the Symbols Committee. The Royal Society of London, U.K.

Appendix E

'GASEX': a program to study the influence of data variations on calculated rates of photosynthesis and transpiration

W.F. Postl and H.R. Bolhàr-Nordenkampf

E.1 INTRODUCTION

Measurements of photosynthesis and transpiration in the field are standard non-destructive methods. Some of the instruments in use are portable and measure both CO_2 gas exchange and transpiration rates in parallel, automatically controlled by a microprocessor (e.g. the portable IRGA systems *LCA-2, ADC, UK* or *LI-6200, LI-COR, USA*). According to the calculations of von Caemmerer and Farquhar [1] it is possible to estimate leaf net photosynthesis (**A**), transpiration rate (**E**), stomatal or leaf conductance for CO_2 ($g_l[CO_2]$) and water vapour ($g_l[H_2O]$), and finally the internal sub-stomatal CO_2 concentration (C_i) in the sub-stomatal cavities of a leaf. All these calculations can be done automatically by using the systems mentioned above. Sometimes it is necessary to recalculate input data, e.g. to correct input data which show possible inaccuracies. Especially in teaching it is necessary to look for dependencies of input data on results (e.g. differences in air pressure or boundary layer conductance). The features of the program 'GASEX' are, on the one hand, to calculate or recalculate gas exchange rates and, on the other hand, to see the influence of input data variations on the results. When using the program all data can be seen on the computer screen. Optionally, data can be stored on a disk or printed out via a matrix printer. The program is written in Basic (QuickBasic Ver. 4.5) and compiled to the stand-alone file 'GASEX.EXE'.

Photosynthesis and Production in a Changing Environment: a field and laboratory manual
Edited by D.O. Hall, J.M.O. Scurlock, H.R. Bolhàr-Nordenkampf, R.C. Leegood and S.P. Long
Published in 1993 by Chapman & Hall, London.
ISBN 0 412 42900 4 (HB) and 0 412 42910 1 (PB).

E.2 CALCULATIONS OF GAS-EXCHANGE PARAMETERS

E.2.1 Calculation of 'molar flow'

When using an 'open system', it is a basic requirement to measure the flow through the chamber as accurately as possible because of the linear relationship between the difference in the CO_2 (or H_2O) concentrations at the inlet and outlet of the leaf chamber (δCO_2, δH_2O) and the photosynthetic (or transpiration) rates. The majority of flowmeters (variable area flowmeters or

Table E.1. List of symbols used by the 'GASEX' programme

Symbol	Description	Unit
A	rate of CO_2 gas exchange	$\mu mol\ m^{-2} s^{-1}$
$AnCO_2$	CO_2 concentration in air leaving the leaf chamber	$\mu l\ l^{-1}$
$AnCO_2$ c.	corrected $AnCO_2$ value	$\mu l\ l^{-1}$
C_c	corrected $RefCO_2$ or $AnCO_2$ value	$\mu l\ l^{-1}$
C_i	sub-stomatal CO_2 concentration	$\mu l\ l^{-1}$
cf	correction factor for volume flow	
delta	difference in CO_2 concentration	$\mu l\ l^{-1}$
δt	difference in temperature between air and leaf	°C
E	transpiration rate	$mol\ m^{-2} s^{-1}$
e	actual water vapour pressure of air	mbar
e_i	water vapour pressure of air entering the leaf chamber	mbar
e_l	saturated water vapour pressure at leaf temperature	mbar
E_{max}	IRGA response to infinite water vapour	
e_o	water vapour pressure of air in the leaf chamber	mbar
e_s	saturated water vapour pressure at air temperature	mbar
Flow	volume flow into leaf chamber	$ml\ min^{-1}$
g_b	boundary layer conductance to water vapour	$mol\ m^{-2} s^{-1}$
g_t	total conductance to CO_2 transfer	$mmol\ m^{-2} s^{-1}$
$g_l\ (CO_2)$	leaf (stomatal) conductance to CO_2 transfer	$mmol\ m^{-2} s^{-1}$
$g_l\ (H_2O)$	leaf (stomatal) conductance to water vapour	$mmol\ m^{-2} s^{-1}$
H	energy absorbed by the leaf	$W\ m^{-2}$
Leaf Temp	measured or calculated leaf temperature	°C
MolFlow	molar flow of air per unit leaf area	$mol\ m^{-2} s^{-1}$
$MolVol_{20}$	molar volume of a gas at 20°C	l
P	atmospheric pressure	mbar
Quanta (**Q**)	photon flux density incident on leaf chamber	$\mu mol\ m^{-2} s^{-1}$
$r_b\ (H_2O)$	boundary layer resistance for water vapour	$m^2 s\ mol^{-1}$
$RefCO_2$	CO_2 concentration in air entering the leaf chamber	$\mu l\ l^{-1}$
$RefCO_2$ c.	corrected $RefCO_2$ value	$\mu l\ l^{-1}$
RH%in	% relative humidity of air entering leaf chamber	%
RH%out	% relative humidity of air in leaf chamber	%
$r_l\ (CO_2)$	leaf (stomatal) resistance to CO_2 transfer	$m^2 s\ mol^{-1}$
$r_l\ (H_2O)$	leaf (stomatal) resistance to water vapour	$m^2 s\ mol^{-1}$
s	projected leaf area surface	cm^2
T	air temperature in the leaf chamber	°C
Tr	conversion factor for quanta to absorbed energy	
V_{20}	flow rate corrected to 20°C and 1 bar	$ml\ s^{-1}$

thermal mass flowmeters) measure flow as a volume per time (e.g. ml min^{-1}, ml s^{-1}). To calculate data according to the equations of von Caemmerer and Farquhar [1] it is necessary to calculate the 'molar flow' (MolFlow) of air through the leaf chamber. Since most of the mass flow controllers/meters (e.g. *Tylan, USA*) are calibrated at 0°C and 1.013 bar, a correction factor (cf) has to be used. In order to calculate the flow at 1 bar and 20°C (V_{20}), Equations E.1 and E.2 may be used:

$$\mathrm{cf} = \frac{293.15}{273.15} \times \frac{1013.25}{1000} = 1.087 \qquad \text{(E.1)}$$

$$V_{20} = \mathrm{Flow} \times \mathrm{cf} \quad (\mathrm{ml\ min^{-1}}) \qquad \text{(E.2)}$$

$$0°\mathrm{C} = 273.15\,\mathrm{K},\ 20°\mathrm{C} = 293.15\,\mathrm{K}$$

Some of the air supply units using mass flow controllers (e.g. air supply unit *ASUM(MF); ADC, U.K.*) do this correction electronically. The readout corresponds to V_{20}.

Flow rates gained from variable area flow meters usually calibrated at 20°C and 1.013 bar have to be corrected for ambient air

Table E.2. Saturation water vapour pressure (e_s, mbar) between 0° and 50°C

T(°C)	0.0	0.1	0.2	0.3	0.4	0.5	0.6	0.7	0.8	0.9
0	6.1	6.2	6.2	6.3	6.3	6.4	6.4	6.5	6.5	6.6
1	6.6	6.6	6.7	6.7	6.8	6.8	6.9	6.9	7.0	7.0
2	7.1	7.1	7.2	7.2	7.3	7.3	7.4	7.5	7.5	7.6
3	7.6	7.7	7.7	7.8	7.8	7.9	7.9	8.0	8.1	8.1
4	8.2	8.2	8.3	8.3	8.4	8.5	8.5	8.6	8.6	8.7
5	8.8	8.8	8.9	8.9	9.0	9.1	9.1	9.2	9.3	9.3
6	9.4	9.5	9.5	9.6	9.7	9.7	9.8	9.9	9.9	10.0
7	10.1	10.1	10.2	10.3	10.3	10.4	10.5	10.6	10.6	10.7
8	10.8	10.8	10.9	11.0	11.1	11.1	11.2	11.3	11.4	11.4
9	11.5	11.6	11.7	11.8	11.8	11.9	12.0	12.1	12.2	12.2
10	12.3	12.4	12.5	12.6	12.7	12.7	12.8	12.9	13.0	13.1
11	13.2	13.3	13.4	13.4	13.5	13.6	13.7	13.8	13.9	14.0
12	14.1	14.2	14.3	14.4	14.5	14.6	14.6	14.7	14.8	14.9
13	15.0	15.1	15.2	15.3	15.4	15.5	15.6	15.7	15.8	15.9
14	16.0	16.2	16.3	16.4	16.5	16.6	16.7	16.8	16.9	17.0
15	17.1	17.2	17.3	17.5	17.6	17.7	17.8	17.9	18.0	18.1
16	18.3	18.4	18.5	18.6	18.7	18.8	19.0	19.1	19.2	19.3
17	19.5	19.6	19.7	19.8	20.0	20.1	20.2	20.3	20.5	20.6
18	20.7	20.9	21.0	21.1	21.2	21.4	21.5	21.7	21.8	21.9
19	22.1	22.2	22.3	22.5	22.6	22.8	22.9	23.0	23.2	23.3
20	23.5	23.6	23.8	23.9	24.1	24.2	24.4	24.5	24.7	24.8
21	25.0	25.1	25.3	25.4	25.6	25.7	25.9	26.1	26.2	26.4
22	26.5	26.7	26.9	27.0	27.2	27.4	27.5	27.7	27.9	28.0
23	28.2	28.4	28.6	28.7	28.9	29.1	29.3	29.4	29.6	29.8
24	30.0	30.1	30.3	30.5	30.7	30.9	31.1	31.2	31.4	31.6
25	31.8	32.0	32.2	32.4	32.6	32.8	33.0	33.2	33.4	33.6
26	33.8	34.0	34.2	34.4	34.6	34.8	35.0	35.2	35.4	35.6
27	35.8	36.0	36.2	36.4	36.7	36.9	37.1	37.3	37.5	37.7
28	38.0	38.2	38.4	38.6	38.9	39.1	39.3	39.5	39.8	40.0
29	40.2	40.5	40.7	40.9	41.2	41.4	41.6	41.9	42.1	42.4
30	42.6	42.9	43.1	43.4	43.6	43.9	44.1	44.4	44.6	44.9
31	45.1	45.4	45.6	45.9	46.2	46.4	46.7	47.0	47.2	47.5
32	47.8	48.0	48.3	48.6	48.9	49.1	49.4	49.7	50.0	50.2
33	50.5	50.8	51.1	51.4	51.7	52.0	52.3	52.5	52.8	53.1
34	53.4	53.7	54.0	54.3	54.6	54.9	55.2	55.6	55.9	56.2
35	56.5	56.8	57.1	57.4	57.7	58.1	58.4	58.7	59.0	59.4
36	59.7	60.0	60.3	60.7	61.0	61.3	61.7	62.0	62.4	62.7
37	63.0	63.4	63.7	64.1	64.4	64.8	65.1	65.5	65.8	66.2
38	66.6	66.9	67.3	67.6	68.0	68.4	68.7	69.1	69.5	69.9
39	70.2	70.6	71.0	71.4	71.8	72.1	72.5	72.9	73.3	73.7
40	74.1	74.5	74.9	75.3	75.7	76.1	76.5	76.9	77.3	77.7
41	78.1	78.6	79.0	79.4	79.8	80.2	80.7	81.1	81.5	81.9
42	82.4	82.8	83.2	83.7	84.1	84.6	85.0	85.5	85.9	86.4
43	86.8	87.3	87.7	88.2	88.6	89.1	89.6	90.0	90.5	91.0
44	91.4	91.9	92.4	92.9	93.3	93.8	94.3	94.8	95.3	95.8
45	96.3	96.8	97.3	97.8	98.3	98.8	99.3	99.8	100.3	100.8
46	101.3	101.9	102.4	102.9	103.4	104.0	104.5	105.0	105.6	106.1
47	106.6	107.2	107.7	108.3	108.8	109.4	109.9	110.5	111.0	111.6
48	112.2	112.7	113.3	113.9	114.4	115.0	115.6	116.2	116.8	117.3
49	117.9	118.5	119.1	119.7	120.3	120.9	121.5	122.1	122.7	123.3
50	123.9									

Values were calculated using the formula:

$$e_s = 6.13753 \times \exp(T \times (18.564 - T/254.4))/(T + 255.57))$$

where T is the temperature in °C [5].

Table E.3. Useful conversion factors

Photosynthetic rates	$\mu mol\ CO_2\ m^{-2} s^{-1}$
$1\ mg\ CO_2\ m^{-2} s^{-1}$	22.7
$1\ ng\ CO_2\ cm^{-2} s^{-1}$	0.227
$1\ nmol\ CO_2\ cm^{-2} s^{-1}$	10
$1\ mg\ CO_2\ dm^{-2} h^{-1}$	0.631
$1\ kg\ CO_2\ hectare^{-1} h^{-1}$	0.631
$1\ mm^3\ CO_2\ cm^{-2} h^{-1}$	0.114
Transpiration rates	$mmol\ H_2O\ m^{-2} s^{-1}$
$1\ mg\ H_2O\ m^{-2} s^{-1}$	0.0555
$1\ mol\ H_2O\ m^{-2} h^{-1}$	0.278
$1\ \mu g\ H_2O\ cm^{-2} s^{-1}$	0.555
$1\ \mu g\ H_2O\ cm^{-2} min^{-1}$	0.00925
$1\ mg\ H_2O\ dm^{-2} min^{-1}$	0.0925
$1\ g\ H_2O\ dm^{-2} h^{-1}$	1.542
$1\ kg\ H_2O\ dm^{-2} h^{-1}$	15.42
Units of pressure	kPa
1 bar	100
10 mbar	1
1 atmosphere	101.325
1 mmHg (0°C)	0.13332
$1\ mmH_2O$ (4°C)	0.0098064

temperature (T) and pressure (P) (Equation E.3).

$$V_{20} = \text{Flow} \times \frac{293.15}{(273.15 + T)} \times \frac{P}{1000} \quad (\text{ml min}^{-1}) \tag{E.3}$$

Without such corrections the flow rate (V_{20}) used for further calculations deviates by up to 10% or more. The molar volume of any gas is 22.41 at 0°C and 1.013 bar. The correction for 20°C and 1 bar is given in Equation E.4:

$$\text{MolVol}_{20} = 22.4 \times \frac{293.15}{273.15} \times \frac{1013}{1000} = 24.3531 \text{ at } 20°\text{C} \tag{E.4}$$

Using the results from Equations E.3 and E.4, the molar flow of air across a leaf area of 1 m^2 during one second can be calculated according to Equation E.5:

$$\text{MolFlow} = \frac{V_{20}}{1000} \times \frac{1}{24.353} \times \frac{1}{60} \times \frac{10{,}000}{s} \quad (\text{mol m}^{-2} \text{s}^{-1}) \tag{E.5}$$

($1.0\ m^2 = 10{,}000\ cm^2$, 1.0 l, = 1000 ml, 1 minute = 60 s)

E.2.2 Calculation of the transpiration rate

The transpiration rate (**E**) is calculated using the molar flow and the difference between the input (e_i) and output (e_o) humidity of the leaf chamber corrected for the actual air pressure (Equation E.6):

$$\mathbf{E} = \frac{(e_o - e_i)}{(P - e_o)} \times \text{MolFlow} \quad (\text{mol m}^{-2} \text{s}^{-1}) \tag{E.6}$$

The air humidity is often measured by capacitive humidity sensors which display the relative humidity (RH%). The absolute input (e_i) and output humidity (e_o) in mbar can be calculated according to Equations E.7–E.9:

$$e_s = B_1 \times \exp\left[T \times \frac{\left[B_2 - \frac{T}{B_3}\right]}{(T + B_4)}\right] \quad (\text{mbar})$$

$$B_1 = 6.13753$$
$$B_2 = 18.564$$
$$B_3 = 254.4 \tag{E.7}$$
$$B_4 = 255.57$$

$$e_i = e_s \times \frac{\text{RH\%in}}{100} \quad (\text{mbar}) \tag{E.8}$$

$$e_o = e_s \times \frac{\text{RH\%out}}{100} \quad (\text{mbar}) \tag{E.9}$$

[$w_i = \frac{e_i}{P}$; $w_o = \frac{e_o}{P}$; $w_{s[T]} = \frac{e_l}{P}$: see Chapter 9, Equations 9.32–9.34]

Some instruments use an inlet humidity of zero, since the inlet air is first dehumidified. This gives a fast response and simplifies calculations. The CO_2 exchange is compensated by the O_2 gas exchange. The transpiration rate enhances the molar flow escaping from the leaf chamber. This change in flow rate is corrected by the term ($P - e_o$) in Equation E.6. The deviation without correction is approximately −4%. Transpiration rates vary from 0.8 to 5 $mmol\,m^{-2}\,s^{-1}$. A value of 10 $mmol\,m^{-2}\,s^{-1}$ would be a very high transpiration rate.

E.2.3 Calculation of net leaf photosynthesis

All well-known gas analysers (IRGAs) measure CO_2 concentrations by the degree of absorption in the infrared wavelength bands. To correct for the interference with absorption bands of water vapour, filters are used and correction factors (e.g. E_{max}) are introduced in calculations. Depending on the instrument, at a high water vapour background pressure of, say, 20 mbar, the displayed CO_2 concentration will be 1–2 $\mu l\,l^{-1}$ too high (E_{max} = 1 to 2). The reason for this is the broadening of the major CO_2 absorption band at 4.5 μm caused by water vapour. Therefore, it is recommended to dry the reference and analysis air streams (E_{max} = 0) before they enter the IRGA, using 'Drierite' (self-indicating calcium sulphate), for example. Some manuals offer the possibility of calculating a correction (e.g. *LCA2; ADC, UK*) as shown below:

$$C_c = C - E_{max} \times \left[1 + \frac{7.87 \times C}{10{,}000}\right]^{(+)} \times [1 - \exp(-0.07 \times e)] \quad (\mu l\,l^{-1}) \qquad (E.10)$$

where C and C_c are the measured and corrected CO_2 concentration in $\mu l\,l^{-1}$ and e is the measured water vapour pressure in the air stream in mbar. The corrected inlet and outlet CO_2 concentrations are named $RefCO_2$ c. and $AnCO_2$ c.

This correction includes a term ($^{(+)}$) which compensates for the CO_2 background concentration because of its interference with the water vapour effect. All these constants used are empirical values, and are applied in the case of the *LCA2* and *LCA3* models of portable gas analyser (*ADC, UK*).

When calculating net leaf photosynthesis (**A**) the diluting effect of the transpired water vapour has to be taken into consideration:

$$\mathbf{A} = \left[RefCO_2\ c. - AnCO_2\ c. \times \frac{(P - e_i)}{(P - e_o)}\right] \times MolFlow \quad (\mu mol\,m^{-2}\,s^{-1}) \qquad (E.11)$$

Without this correction, photosynthetic rates are overestimated, depending on transpiration rates.

Net leaf photosynthesis can reach rates of up to 30 $\mu mol\,m^{-2}\,s^{-1}$. Maximal rates of up to 70 $\mu mol\,m^{-2}\,s^{-1}$ are reported for some C_4 plants (e.g. maize).

E.2.4 Calculation of leaf temperature

It is assumed that the relative humidity in the intercellular system of a leaf is near 100% [2]. Therefore, the absolute humidity can be calculated in mbar if the actual leaf temperature is known. The simplest option is to equate leaf temperature to the air temperature. Alternatively, leaf surface temperature may be measured directly by means of a fine thermocouple (Type T, 0.1 mm diameter) attached to the lower surface of the leaf by surgical tape. If flat expanded leaves are clamped like a diaphragm in the leaf chamber (e.g. *Parkinson Leaf Chambers; ADC, UK*), leaf temperature may also be calculated according to the energy budget of the leaf (Equations E.12a–c). The first step is to calculate the energy absorbed by the leaf (H):

$$H = \mathbf{Q}\ (\text{incident}) \times Tr \quad (W\,m^{-2}) \qquad (E.12a)$$

'Tr' is the conversion factor for changing quanta incident on the chamber ($\mu mol\,m^{-2}\,s^{-1}$) into the energy absorbed by the leaf. The value of 'Tr' depends on the visible:infrared ratio of the incident light, on the transmission and absorption characteristics of the leaf, and on the windows of the chamber. A value frequently used for 'Tr' is 0.175 [3]. The difference in temperature (δt) between the leaf and the surrounding air is given by Equation E.12b:

$$\delta t = \frac{(H - (\alpha \times E))}{(0.93 \times M_a \times \frac{C_P}{r_b} + 4.639 + 0.0583 \times T)} \quad (^\circ C) \qquad (E.12b)$$

where
α = latent heat of vaporization of water in J mol^{-1} (= 45064.3 − T × 42.9)
M_a = molecular weight of air (28.97)
C_p = specific heat at constant pressure (1.012 J g^{-1} K^{-1})
r_b = boundary layer resistance (m^2 s mol^{-1})

Finally the leaf temperature can be calculated:

$$\text{LeafTemp} = T + \delta t \quad (°C) \tag{E.12c}$$

E.2.5 Calculation of total conductance to water vapour and to CO_2

The flow of CO_2 into the sub-stomatal cavities of a leaf is determined by the diffusion gradient, the boundary layer resistance, r_b, and the stomatal or leaf resistance, r_l. The resistances can be calculated from the water vapour flux. The boundary layer resistance is slightly influenced by the anatomical characteristics of the leaf epidermis (e.g. hairiness) and is strongly reduced by an enhanced turbulent air flow in the chamber. To measure $r_b(H_2O)$, in most cases a wet filter paper can be used to simulate a leaf without stomata [4]. By analogy with Ohm's law, the total resistance of the flow into the leaf can be calculated by using the transpiration rate and the gradient of water vapour pressure, when the relative humidity in the intercellular spaces is assumed to be 100%:

$$r_l(H_2O) + r_b(H_2O) = \frac{(e_l - e_o)}{\mathbf{E}} \; (m^2\,s\,mol^{-1}) \tag{E.13a}$$

where e_l is the saturated water vapour pressure in the leaf air spaces at the leaf temperature (use Equation E.7 to calculate e_l by using LeafTemp instead of T).

Using Equation E.6, the varying stomatal resistance can be calculated according to Equation E.13b:

$$r_l(H_2O) = \frac{\frac{(e_l - e_o)}{(e_o - e_i)} \times \frac{(P - e_o)}{P}}{\text{MolFlow}} - r_b(H_2O)\;(m^2\,s\,mol^{-1}) \tag{E.13b}$$

This resistance shows how many square metres of leaf area are needed in order to have a water vapour flow of 1 mol s^{-1}. Most authors use stomatal or leaf conductance, g_l, instead of resistance, r_l. The conductance (g_l) can be simply derived as the reciprocal value of r_l and shows the number of moles of water vapour transpired by a square metre of leaf area per second:

$$g_l(H_2O) = \frac{1}{r_l} \; (mol\,m^{-2}s^{-1}) \tag{E.14}$$

Due to the different shape of water and CO_2 molecules, the ratio of diffusivities has to be used to convert water vapour resistance to CO_2 resistance:

$$r_l(CO_2) = r_l(H_2O) \times 1.61 \quad (m^2\,s\,mol^{-1}) \tag{E.15a}$$

$$r_b(CO_2) = r_b(H_2O) \times 1.37 \quad (m^2\,s\,mol^{-1}) \tag{E.15b}$$

$$g_l(CO_2) = \frac{1}{r_l(CO_2)} \quad (mol\,m^{-2}s^{-1}) \tag{E.15c}$$

$$g_b(CO_2) = \frac{1}{r_b(CO_2)} \quad (mol\,m^{-2}s^{-1}) \tag{E.15d}$$

The total conductance to CO_2 ($g_t(CO_2)$) will be given by:

$$\frac{1}{g_t} = \frac{1}{g_l(CO_2)} + \frac{1}{g_b(CO_2)} \tag{E.15e}$$

$$g_t = \frac{g_l \times g_b}{g_l + g_b} \tag{E.15f}$$

The ratio of 1.37 for r_b is only applicable if r_b is rather small, which also means that the chamber is well stirred.

Mesophytic plants show $g_l(CO_2)$ values from 0.08 to 1.2 mol m^{-2} s^{-1}. Coniferous trees and C_4 plants sometimes have smaller conductances.

E.2.6 Calculation of internal CO_2 concentration

According to von Caemmerer and Farquhar [1], net leaf photosynthesis (**A**) correlates with the CO_2 concentration in the intercellular system (C_i). This assumption is true if the CO_2 of the air (C_a) and/or the leaf conductance (g_l) is changing. If photosynthetic rates are increasing with increasing light intensities, internal CO_2 concentration will fall or remain unchanged. C_i is calculated using the photosynthetic rate for the consumption of CO_2 and the CO_2 gradient together with the resistances for the influx:

$$C_i = \text{AnCO}_2\ \text{c.} - \frac{\mathbf{A}}{g_t} \quad (\mu l\, l^{-1}) \qquad \text{(E.16)}$$

The efflux of water vapour is considerably higher than the influx of CO_2. This leads to an interaction of CO_2 and water vapour molecules in the stomatal pore and therefore a further correction is necessary:

$$C_i = \frac{\left[\left(g_t - \frac{\mathbf{E}}{2}\right) \times \text{AnCO}_2\ \text{c.} - \mathbf{A}\right]}{\left[g_t + \frac{\mathbf{E}}{2}\right]} \quad (\mu l\, l^{-1}). \qquad \text{(E.17)}$$

E.3. DESCRIPTION OF THE PROGRAM

All gas-exchange values are calculated for an 'open system' (Chapter 9). Three modules can be chosen:

1. Module no. 1: The leaf temperature is measured using a thermocouple attached to the lower surface of the leaf.
2. Module no. 2: The leaf temperature of a flat expanded leaf is calculated from the 'energy budget'.
3. Module no. 3: The leaf temperature is assumed to be equal to the air temperature. This applies if the chamber is well stirred and the photon flux density is low. Sometimes the true leaf temperature is difficult to measure, e.g. for conifer needles. In this case, Module no. 3 can be chosen.

E.3.1 Running the program

Type 'GASEX' and ⟨RETURN⟩ at the DOS-prompt. The program displays a menu

Scheme E.1: Hardcopy of the screen

```
--------- Module no. 1 ---------------------------------------------

                   INPUT PARAMETERS:

   Flow:         500.0   ml/min          Pressure:      989.0   mBar
   Temp.:         25.0   .C              Quanta:       1000.0   umol/m2 s
   In_Hum         20.0   %               Ref_CO2:       330.0   ul/l
   Out_Hum:       40.0   %               An_CO2:        300.0   ul/l
   gb:             3.0   mol/m2 s        Leaftemp.:      23.0   C
   Surface:       10.0   cm2             Emax:            2.0

--------------------------------------------------------------------

--------------------------------------------------------------------

                 CALCULATED PARAMETERS:

   V20            8.33   ml/s            MolFlow           0.342 mol/m2 s
                                         E                 2.23  mmol/m2 s
Gl (H2O)       149.5    mmol/m2 s        Gl (CO2)         94.3   mmol/m2 s
RefCo2 c.      329.1    ul/l             AnCO2 c:        298.5   ul/l
delta           30.6    ul/l
A                9.8    mol/m2 s         Ci              184.3   ul/l

--------------------------------------------------------------------
1 Calc   2 Print   3 Disk            eXit              Press a key.....
--------------------------------------------------------------------
```

Scheme E.2: Copy of the print out and data structure in *'GASEX.DOC'*

```
-----------------------------------------------------------------------
 Comment: TestCalc                              Module 1  Program GASEX
-----------------------------------------------------------------------
       INPUT Parameters:                        OUTPUT Parameters:
       -----------------                        ------------------
Flow            500.0   ml/min          V20              8.33  ml/s
Temp             25.0   .C              MolFlow          0.342 umol/m2 s
InHum            20.0   %               RefCo2 c.      329.1   ul/l
OutHum           40.0   %               AnCO2 c:       298.5   ul/l
gb                3.0   mol/m2 s        delta           30.6   ul/l
Surface          10.0   cm2             ----------------------------------
Pressure        989     mBar            E                2.23  mmol/m2 s
Quanta (Q)     1000     umol/m2 s       Gl (H2O)       149.5   mmol/m2 s
RefCO2          330.0   ul/l            Gl (CO2)        94.3   mmol/m2 s
AnCO2           300.0   ul/l            ----------------------------------
Leaftemp.        23.0   .C              A                9.8   mol/m2 s
Emax              2                     Ci             184.3   ul/l
-----------------------------------------------------------------------
```

Scheme E.3: Data structure in the file *'GASEX.DAT'*

Comment (max. 10 characters), Flow, T, RH%in, RH%out, g_b, Surface, Pressure, Quanta (Q), $RefCO_2$, $AnCO_2$, LeafTemp, E_{max}, V_{20}, Massflow, $RefCO_2$ c., $AnCO_2$ c., delta, E, $g_l(H_2O)$, $g_t(CO_2)$, A, C_i.

TestCalc, 500.0, 25.0, 20.0, 40.0, 3.0, 10.0, 989.0, 1000.0, 330.0, 300.0, 23.0, 2.0, 8.33, 0.342, 329.1, 298.5, 30.6, 2.23, 149.5, 94.3, 9.8, 184.3.

(Scheme E.1) of which the upper half forms the 'input screen'. By selecting a highlighted letter for the input parameters, data can be entered. When all the input parameters are entered, calculation ⟨C⟩ can be started. To get a print out ⟨N⟩ is used, and to store the data on disk ⟨D⟩ is used. Data are stored on the disk in the file 'GASEX.DOC' (Scheme E.2). The data base has the same format as the printout. In the file 'GASEX.DAT' the data are stored in one line separated by commas (Scheme E.3). This facilitates data transfer to other programs, e.g. dBASE, LOTUS 1-2-3, or Quattro.

REFERENCES AND FURTHER READING

1. von Caemmerer, S. and Farquhar, G.D. (1981) Some relationships between the biochemistry of photosynthesis and the gas exchange of leaves. *Planta* **153**, 376–387.
2. Sharkey, T.D., Imai, K., Farquhar, G.D. and Cowan, I.R. (1982) A direct confirmation of the standard method of estimating intercellular partial pressure of CO_2. *Plant Physiol.* **69**, 657–659.
3. Parkinson, K.J, Day, W. and Leach, J.E. (1980) A portable system for measuring the photosynthesis and transpiration of graminaceous leaves. *J. Exp. Bot.* **31**, 1441–1453.
4. Parkinson, K.J. (1984) A simple method for determining the boundary layer resistance in leaf cuvettes. *Plant Cell and Environment* **8**, 223–226.
5. Buck, A.L. (1981) New equations for computing vapour pressure. *J. Applied Meteorology* **20**, 1527–1532.
6. Farquhar, G.D. and Sharkey, T.D. (1982) Stomatal conductance and photosynthesis. *Ann. Rev. Plant Physiol.* **33**, 317–345.

Appendix F

'TISWAT.BAS': a program to calculate simple water-relations parameters from pressure chamber measurements

A. Davie, K.A. Badcock and C.L. Beadle

F.1 INTRODUCTION

The pressure chamber is a simple, though destructive, technique for the rapid measurement of total water potential, ψ (see Chapter 8 for the theoretical definition of ψ). The chamber can be used with twigs, leaves, needles or even roots, and ψ provides the experimental scientist with a crude measure of the level of water stress in the plant. A more precise measure of the level of stress is obtained by measuring ψ just prior to dawn. Pre-dawn ψ is the maximum water potential achievable at a given level of available water in the soil, as it is measured after the completion of the night period when the evaporative demand is at a minimum and the opportunities for the plant to saturate its water reservoirs have been at a maximum for several hours.

The pressure chamber can be used to establish the pressure–volume relationship of the same organs (see also Chapter 8). The pressure–volume curve defines several parameters of the water relations of the plant, which can be used to help understand the strategies taken to combat or tolerate water stress. Constructing such a curve is a stepwise procedure which limits the number of samples that can be investigated to only a few per day.

The following account summarises the procedure to be followed in the use of this program. Users are strongly encouraged to refer to the literature [1,2,3] for a more comprehensive analysis of the water relations of plant and their investigation with the pressure chamber.

F.2 CALCULATION OF WATER-RELATIONS PARAMETERS

F.2.1 A simple approach

The pressure–volume curve requires a series of measurements which describe the relationship between the inverse of the balancing pressure, the pressure required to return water to the cut surface after excision, and some measure of the weight or volume of water which has been expressed or lost from the organ under investigation to reach that balance point. In the first investigations, the organ (usually a shoot or leaf) was left in the chamber. This had the disadvantage of

Photosynthesis and Production in a Changing Environment: a field and laboratory manual
Edited by D.O. Hall, J.M.O. Scurlock, H.R. Bolhàr-Nordenkampf, R.C. Leegood and S.P. Long
Published in 1993 by Chapman & Hall, London.
ISBN 0 412 42900 4 (HB) and 0 412 42910 1 (PB).

committing the chamber to just one pressure–volume analysis measurement [3,4]. A simpler approach involves the removal of the leaf from the chamber between each measurement, and this program follows such a procedure. Several leaves (up to 6–8) can be investigated in any one day. The pressure–volume curve is constructed during the measurement phase to allow the investigator to meet the minimum requirements for a satisfactory description of the water-relations parameters.

F.2.2 Rehydration

The leaves (or shoots) to be used in a day's series of measurements should be excised with the surface to be cut under water. The leaves should then be stood with the petiole standing in distilled (air-free) water and placed in a humid chamber under low light intensity. The period required for rehydration can be investigated for different plant material but should normally not exceed 4 hours. Excessive periods of rehydration may result in changes to the water-relations parameters under investigation. In some instances, 4 hours may not result in full turgidity (W_t) (i.e. relative water content $R^* = 1.0$), but this can be estimated by extrapolation (see below).

F.2.3 Measuring water potential and relative water content

The first resaturated leaf to be studied should be quickly dried between several layers of tissue. To minimise weight loss while the leaf is in the pressure chamber, the leaf can be wrapped in plastic film of known weight before weighing. For an accurate measurement of R^*, most leaves will require a balance, preferably a top-pan balance with a wind shield, measuring to 1 mg. The weight of the leaf is determined immediately before (W_1) and after (W_2) the measurement of its water potential (ψ). R^* is then calculated as:

$$R^* = (W_f - W_d)/(W_t - W_d) \qquad \text{(F.1)}$$

where $W_f = (W_1 + W_2)/2$. As W_d remains unknown until the measurements are completed, an approximate estimate is required based on a preliminary study of the W_t/W_d ratio.

Between measurements, the leaf is allowed to dry free of its plastic sheath on the laboratory bench. Further determinations of W_f and ψ are made as above at appropriate intervals (−0.2 to −0.3 MPa) until R^* has declined to around 0.55–0.65. The relationship between $1/\psi$ and R^* is plotted and after an initial rapid decline which is curvilinear in form, a linear phase will be observed (see Chapter 8, Fig. 8.5). Approximately 5–6 points should be collected during this phase so that its slope and intercepts can be determined by least squares procedures. Upon completion of the experiment the leaf should be dried at 70°C to constant weight (W_d).

The rate of drying will determine the time interval between readings and therefore the number of leaves which can be investigated simultaneously. Rates of pressurisation should be slow; $0.005\,\mathrm{MPa\,s^{-1}}$ is recommended. Rates of depressurisation should also be slow to avoid damaging the leaf.

F.2.4 Calculation of turgid weight

The weight of the leaf at full turgor, W_t can be determined by extrapolation. The program incorporates a facility for observing the change in W_f with ψ. The first four or five points should form a linear phase. These are selected for least-squares regression and the intercept on the fresh weight axis at zero turgor is W_t. This can now be substituted into Equation F.1.

F.2.5 Calculation of osmotic potential at full turgor and apoplastic water content

Once the correct value of W_d has been obtained and R^* values recalculated, the pressure–volume curve can be redrawn and the linear regression fitted to the appropriate points on the lower part of the curve. The intercept on the y-axis is the reciprocal of the osmotic potential at full turgor, π_0 (at $R^* = 1.0$). This is a fundamental parameter of the water relations of the leaf. Changes in π_0 during water stress have been used to determine the capacity of plants to adjust their osmotic characteristics and thereby maintain turgor.

The intercept on the x-axis separates the relative water contents of the symplast and the apoplast. Pressure–volume analysis assumes that there is no change in the apoplastic water (including bound water) in tissue during pressurisation. As the estimation of the relative apoplastic water content R^*_a involves considerable extrapolation, it is estimated with less accuracy than π_0.

F.2.6 Calculation of turgor pressure

The fitted relationship between $1/\pi$ and R^* defines the value of π at any value of ψ and R^*. Within the linear part of the pressure–volume curve, the turgor pressure (P) is equal to zero. Within the curvilinear part, P can be calculated as the difference between ψ and π. Thus:

$$P = \psi - \pi \qquad (F.2)$$

The position at which the pressure–volume curve becomes linear is referred to as the point of zero turgor, $P = 0$. Above zero turgor, as both ψ and π are negative, and π is more negative than ψ, P in Equation F.2 is positive. Below the point of zero turgor, $\psi = \pi$.

The relationships between P and R^* and P and ψ can be defined. These describe the capacity of, say, a given species (or given species under an experimental treatment) to maintain a certain turgor pressure at given values of tissue water content or potential.

F.2.7 Calculation of bulk modulus of elasticity

The elasticity of the cell wall determines the change in volume of the cell as it loses and gains water. Thus the turgor pressure P is also dependent on the elasticity of the cell wall.

The pressure chamber is used to measure this elasticity as a weight-averaged bulk modulus of elasticity, ε. This is defined as the change in tissue turgor pressure (ΔP) for a given fractional change in symplastic water content. As the apoplastic water content R^*_a remains constant and the density of symplastic water is unity, this fractional change can be approximated by ΔR^*. Following Robichaux [5], ε is given by:

$$\varepsilon = \frac{\Delta P}{\Delta R^*}(R^*_x - R^*_a)$$

where R^*_x is the mean relative water content. To calculate the relationship between ε and $\bar{P}$, the linear slope of the first four points of the P vs. R^* relationship is calculated, R^*_x being the mean R^* of these points. This is repeated for points 2–5, 3–6, etc.; *viz.* each successive set of four points. The ε vs. $\bar{P}$ relationship can then be plotted.

If P declines less rapidly with R^* in one species compared to another (or after a period of treatment), ε is probably lower.

A lower ε at the higher values of R will probably result in P declining less rapidly with R^*. This property will also result in lower values of R^* when P reaches zero turgor.

F.3 DESCRIPTION OF PROGRAM

This program has been written in Microsoft Basic for an IBM PC microcomputer or equivalent.

F.3.1 The 'Main Menu'

The seven items of the program together form the Main Menu, which lists the following headings:

1 ENTER DATA
2 RECALCULATE R*

3 P/V CURVE
4 COMPONENTS OF WATER RELATIONS
5 BULK TISSUE ELASTIC MODULUS
6 CORRECT MISTAKES
7 EXIT TISWAT

Items (1) to (3) are used to assist data entry, (2), (4) and (5) to analyse the results, (6) to correct mistakes made during data entry if and when necessary, and (7) to exit from program.

F.3.2 How to input the data

A new data file is created by selecting '0' at the '#' prompt, and entering the date, e.g. JUL01.91. Selecting '1' at the '#' prompt provides a test data file named JAN01.91. The display then requests the species number from the species set (up to 9) which has been built into the program (the investigator can change these to generic names if desired) and assigns that chosen to LEAF 1. You are now ready to enter the balancing pressure, time of measurement and the fresh weights of your initially saturated leaf or shoot just before and just after your measurement of balancing pressure (or water potential, ψ) After the first measurement only you will be asked to enter a dummy dry weight (W_d) so that relative water content, R^*, can be calculated from the mean fresh weight (W_f). At this stage the turgid weight (W_t) is assumed to be your first measurement of fresh weight. For each subsequent set of measurements, a summary of the previous set is given and a self-checking procedure ensures that incorrect values (e.g. fresh weights greater than the previous reading) cannot be entered. New leaves can be started at any time and after each data set is entered the display will first request selection of the leaf number before allowing further data entry.

The program allows the experimental scientist to check the progress of fresh weight loss (Fig. F.1) and the current appearance of the pressure–volume curve (Fig. F.2) by accessing items (2) and (3) on the Main Menu, respectively. Certain components of (2) and (3) can be ignored at this stage as they relate to analysis on completion of the experiment. There should be a minimum of five points in the straight line part of the pressure–volume curve before the experiment is terminated.

F.3.3 Analysis of results

To establish the actual relative water content, the correct values of turgid weight (W_t) and dry weight (W_d) must be entered. Go back to item (2) on the Main Menu. After the relationship between fresh weight and balancing pressure has been plotted, you will be asked to draw a linear regression through a chosen number of points (say 1–3 in Fig. F.1). The intercept on the x-axis is the actual turgid weight (W_t). Enter dry weight (W_d) following drying of leaf.

Move to item (3) and replot the pressure–volume curve (Fig. F.2) with the corrected values of R^*. The values of inverse pressure, as 1/Psi (for $1/\psi$) and R^* are then printed out.

A linear regression is then fitted to the straight line part of the curve. Enter the number of the first and the last point on this line. The fitting procedure requires a minimum of five points for the regression and the line which gives the best correlation coefficient is chosen.

Item 4 re-plots the pressure–volume curve (Fig. F.3), superimposes the linear regression and prints out the y-intercept (the inverse osmotic potential, $1/\pi_0$ at $R^* = 1.0$) and the x-intercept (the apoplastic water content, R^*_a). This is followed by a plot of turgor pressure, P vs R^* and ψ (as Psi); (Figs. F.4 and F.5, respectively). The intercepts and complete data set including fitted and calculated values of osmotic potential, π (as Pi) and turgor pressure, P, respectively, are then summarised.

Finally, the relationship between the weight-averaged bulk tissue modulus of elasticity, ε, and average turgor pressure, $\bar{P}$, is calculated for each successive set of four

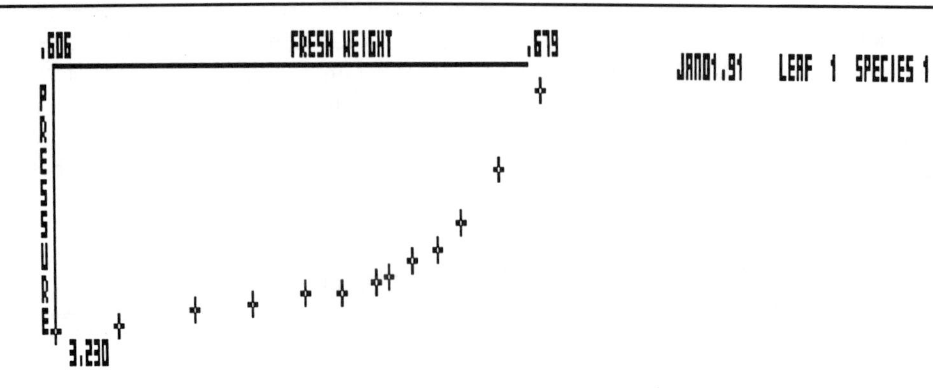

Fig. F.1.
Linear regression from point ? 1
to point ? 3
FW = 0.6823619 + (−7.207316E-03 * Psi)

Determination (R^2) = 0.9934916
Correlation = 0.9967404
Error of estimate = 6.856912E-04
New dry weight?

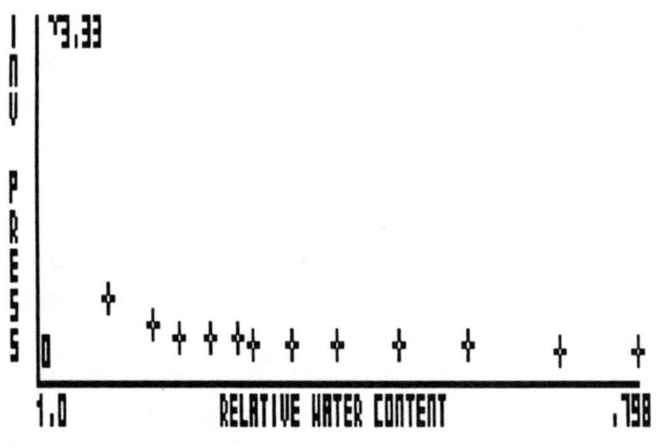

Fig. F.2.

#	1/Psi	R*
1	3.333	0.99469
2	0.769	0.97745
3	0.512	0.96286
4	0.444	0.95358
5	0.416	0.94297
6	0.387	0.93368
7	0.377	0.92838
8	0.363	0.91511
9	0.357	0.90053
10	0.344	0.87931
11	0.333	0.85676
12	0.317	0.82626
13	0.309	0.79973

Regressions from point ? 4
To point ? 13
Best correlation = 0.9963261
From 8 to 13
f(x) = −0.08105 + (0.48521* X)

points starting at full hydration and plotted (Fig. F.6).

F.4 ADDITIONAL NOTES AND INSTRUCTIONS

Requirements

IBM compatible PC with CGA, EGA or VGA colour display adapters (see note* below) and suitable floppy diskette drive
MS-DOS version 3.2 or later
Microsoft GW-BASIC

Loading and running 'TISWAT.BAS'

The program may be run from within GW-BASIC using the commands LOAD and RUN. However, it is simpler to run the 'TISWAT.BAS' program from the DOS

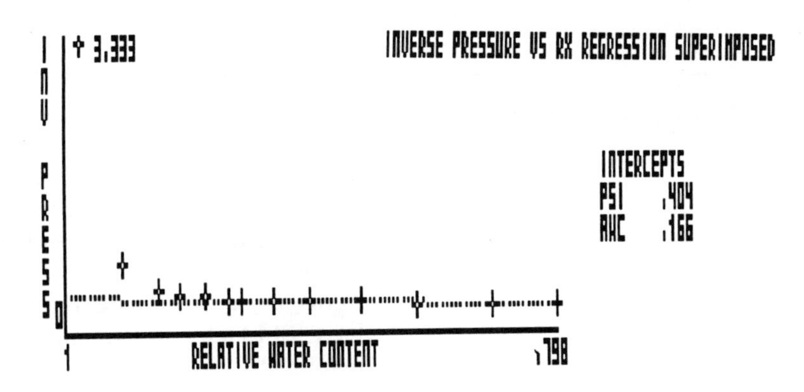

Fig. F.3.

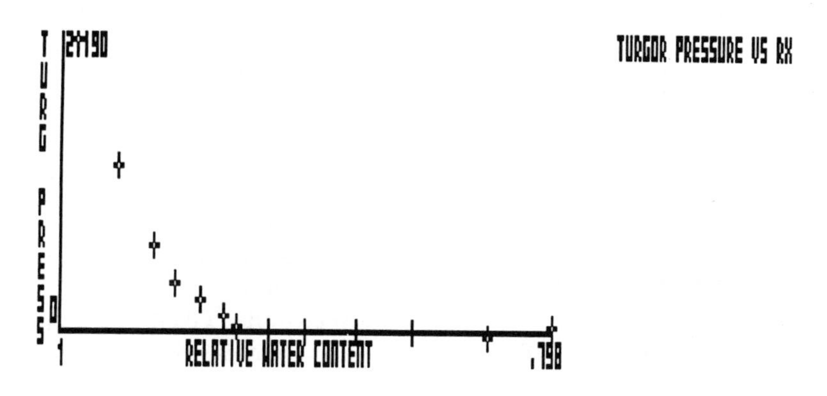

Fig. F.4.

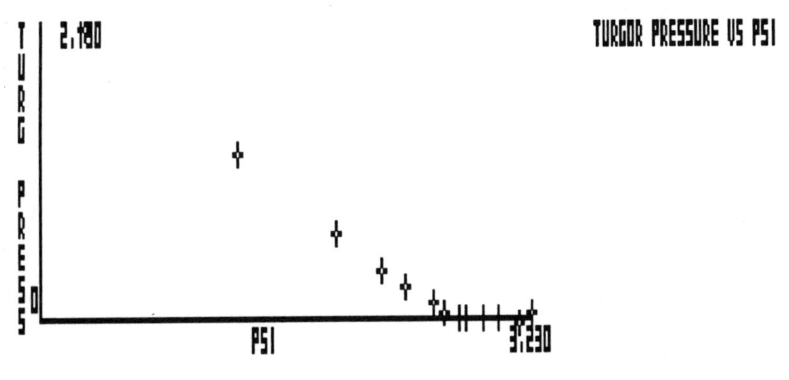

Fig. F.5.

Osmotic potential = −2.47425

Apoplasmic fraction = 0.16703

Psi	Pi	Tp	R*
−0.300	−2.491	2.190	0.994
−1.300	−2.544	1.243	0.997
−1.951	2.590	0.639	0.962
−2.250	−2.621	0.370	0.953
−2.400	−2.657	0.256	0.942
−2.580	−2.689	0.108	0.933
−2.650	−2.707	0.056	0.928
−2.750	−2.755	0.004	0.915
−2.800	−2.810	0.009	0.900
−2.900	−2.894	−0.007	0.879
−3.000	−2.989	−0.012	0.856
−3.150	−3.127	−0.024	0.826
−3.230	−3.258	0.027	0.799

Range	E	Mean P
1–4	35.636	1.110
2–5	22.674	0.627
3–6	14.229	0.343
4–7	9.605	0.197
5–8	6.879	0.106
6–9	2.234	0.045
7–10	0.956	0.016
8–11	0.208	−0.001
9–12	0.314	−0.009
10–13	−0.288	−0.004

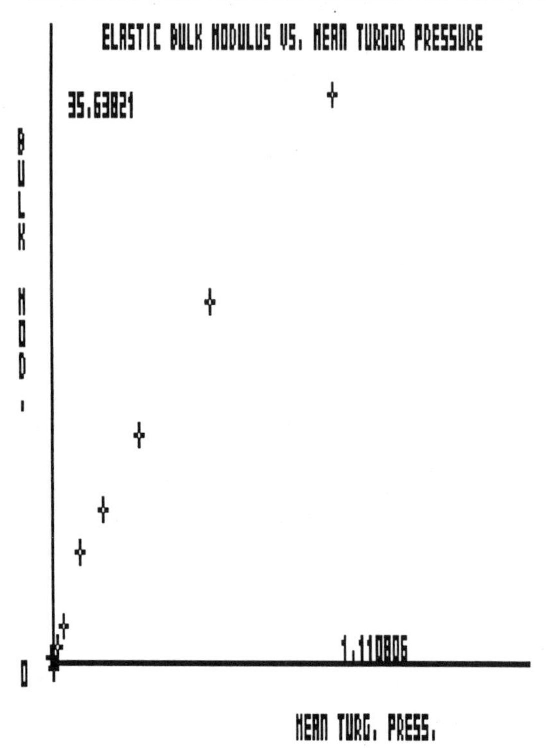

Fig. F.6.

prompt by typing GWBASIC TISWAT ⟨ENTER⟩.

Directing 'TISWAT.BAS' output to a printer

Output from the program can be sent to a printer using the 'print screen' key on the computer keyboard. The graphical image on the screen will be printed one line at a time. If you have difficulty obtaining sensible printed output, try using the MS-DOS program GRAPHICS.COM before loading and running 'TISWAT.BAS'. Your MS-DOS instruction manual should have more details.

Note*

If your computer has a Hercules monochrome display adapter, you may still be able to run 'TISWAT.BAS'. Software is available to run programs designed for the IBM colour graphics adapter on computers with the Hercules monochrome graphics adapter. A copy of the program HGCIBM.COM [Version 2.02; Copyright © Athene Digital] is provided with TISWAT.BAS. This program is distributed as 'Shareware' with all the usual rights reserved. Please refer to the text file HGCIBM.DOC for details regarding your responsibility to register for HGCIBM.COM.

If you wish to try out HGCIBM.COM on your computer, type HGCIBM/E ⟨ENTER⟩ to run in CGA Emulation mode. To 'uninstall' the CGA emulator, type HGCIBM/U ⟨ENTER⟩.

Trademarks mentioned

Hercules is the trademark of Hercules Computer Technology.

IBM is a trademark of International Business Machines.

Microsoft, GWBASIC and MS-DOS are trademarks of Microsoft Corporation.

REFERENCES

1. Ritchie, G.A. and T.M. Hinkley (1975) The pressure chamber as an instrument for ecological research. *Adv. Ecological Research* **9**, 165–254.
2. Tyree, M.T. and P.G. Jarvis (1982) Water in tissues and cells. In: *Encyclopedia of Plant Physiology* (New Series) Vol.12B (O.L. Lange *et al.*, eds.) Springer-Verlag, Berlin. pp. 35–37.
3. Turner, N.C. (1987) The use of the pressure chamber in studies of plant water status. In: *Proceedings of the International Conference on Measurement of Soil and Plant Water Status,* Vol.2. Utah State University, Logan. pp 13–24.
4. Hellkvist, J., G.P. Richards and P.G. Jarvis (1974) Vertical gradients of water potential and tissue water relations in Sitka spruce trees measured with the pressure chamber. *J. Applied Ecology* **11**, 637–688.
5. Robichaux, R.H. (1984) Variations in the tissue water relations of two sympatric Hawaiian *Dubautia* species and their hybrid combination. *Oecologia* **65**, 75–81.

Appendix G
'ENERGY', 'SOLAR' and 'TLEAF': Sun and leaf modelling programs

I.N. Forseth

The solar radiation, energy budget and single leaf photosynthetic models outlined in Chapter 13 have a long history. The ultimate origin of the solar radiation and leaf energy budget models lie in the pioneering work of Gates [6,7]. This type of approach was used by Ehleringer and Miller [2] for alpine plants, and since then has been extended to look at aspects of leaf energy balance in desert plants [3,8,9]. The energy budget approach of Gates [7] was modified with relationships presented in Campbell [1], and incorporated into a simulation model addressing the trade-offs involved in leaf orientation in two desert annuals by Forseth and Ehleringer [5]. This work also provided the general framework for the phenomenological photosynthetic model for single leaves presented in Chapter 13.

For teaching purposes, the model [5] was divided into three sub-models dealing with (i) solar radiation interception, (ii) leaf energy balance and (iii) photosynthetic response. This teaching version has been used for a number of years in ecology classes at the University of Maryland, and for the UNEP Training Courses held in India (1987) and Zimbabwe (1989). Tables 13.1–13.3 describe parameters and some results of the individual sub-models, while Figs. 13.1–13.3 present results from each of the sub-models on solar radiation, leaf energy balance and photosynthetic carbon, respectively.

Photosynthesis and Production in a Changing Environment: a field and laboratory manual
Edited by D.O. Hall, J.M.O. Scurlock, H.R. Bolhàr-Nordenkampf, R.C. Leegood and S.P. Long
Published in 1993 by Chapman & Hall, London.
ISBN 0 412 42900 4 (HB) and 0 412 42910 1 (PB).

REFERENCES AND FURTHER READING

1. Campbell, G.S. (1977) *An Introduction to Environmental Biophysics.* Springer-Verlag, New York.
2. Ehleringer, J.R. and P.C. Miller (1975) A simulation model of plant water relations and production in the alpine tundra, Colorado. *Oecologia (Berlin)* **19**, 177–183.
3. Ehleringer, J.R. and H.A. Mooney (1978) Leaf hairs: effects on physiological activity and adaptive value to desert shrub. *Oecologia (Berlin)* **37**, 183–200.
4. Ehleringer, J.R., H.A.Mooney and J.A. Berry (1979) Photosynthesis and microclimate of *Cammissonia claviformis*, a desert winter annual. *Ecology* **60**, 280–286.
5. Forseth, I.N. and J.R. Ehleringer (1983) Ecophysiology of two solar tracking desert winter annuals. IV. Effects of leaf orientation on calculated daily carbon gain and water use efficiency. *Oecologia (Berlin)* **58**, 10–18.
6. Gates, D.M. (1962) *Energy Exchange in the Biosphere.* Harper and Row, New York.
7. Gates, D.M. (1965) Energy, plants and ecology. *Ecology* **46**, 1–13.
8. Mooney, H.A., J.R. Elheringer and O.

Björkmann (1978) The energy balance of leaves of the evergreen desert shrub *Atriplex hymenelytra*. *Oecologia (Berlin)* **29**, 301–310.

9. Mooney, H.A. and J.R. Elheringer (1978) The carbon gain benefits of solar tracking in a desert annual. *Plant, Cell and Environment* **1**, 307–311.

Index